Lecture Notes in Computer Science 7812

Commenced Publication in 1973
Founding and Former Series Editors:
Gerhard Goos, Juris Hartmanis, and Jan van Leeuwen

Ariel M. Greenberg   William G. Kennedy
Nathan D. Bos (Eds.)

Social Computing, Behavioral-Cultural Modeling and Prediction

6th International Conference, SBP 2013
Washington, DC, USA, April 2-5, 2013
Proceedings

 Springer

Volume Editors

Ariel M. Greenberg
Nathan D. Bos
Johns Hopkins University
Applied Physics Laboratory, Research and Exploratory Development Department
11100 Johns Hopkins Road, Laurel, MD 20723, USA
E-mail: {ariel.greenberg, nathan.bos}@jhuapl.edu

William G. Kennedy
George Mason University
Center for Social Complexity, Department of Computational Social Science
Krasnow Institute for Advanced Study
4400 University Drive, Fairfax, VA 22030, USA
E-mail: wkennedy@gmu.edu

ISSN 0302-9743   e-ISSN 1611-3349
ISBN 978-3-642-37209-4   e-ISBN 978-3-642-37210-0
DOI 10.1007/978-3-642-37210-0
Springer Heidelberg Dordrecht London New York

Library of Congress Control Number: 2013933307

CR Subject Classification (1998): H.3, H.2, H.4, K.4, J.3, H.5

LNCS Sublibrary: SL 3 – Information Systems and Application, incl. Internet/Web and HCI

Typesetting: Camera-ready by author, data conversion by Scientific Publishing Services, Chennai, India

Printed on acid-free paper

Springer is part of Springer Science+Business Media (www.springer.com)

Preface

This proceedings volume contains the accepted papers and posters from the 2013 International Conference on Social Computing, Behavioral-Cultural Modeling, and Prediction. This was the sixth year of the SBP conference, and the third since it merged with the International Conference on Computational Cultural Dynamics.

In 2013 the SBP conference continued to grow. We received a strong set of 137 submissions, greatly exceeding the previous high of 88. SBP continues to be a selective single-track conference. Thirty-three submissions were accepted as full papers, for a 24% paper acceptance rate. We also accepted 27 posters, more than we had previously, for an overall 42% acceptance rate. All papers and posters presented at the conference are included in this proceedings volume, which are distributed to attendees and made available electronically as well as in print as part of Springer's *Lecture Notes in Computer Science* series.

This conference is strongly committed to multidisciplinarity, consistent with recent trends in computational social science and related fields. Authors were asked to indicate from a checklist which topics fitted the papers they were submitting. So many of the papers covered multiple categories that dividing this proceedings volume into topical sections presented a real challenge for the Program Committee. Of course, as a multidisciplinary conference, this is exactly the sort of problem we are glad to have. The topic areas that formed the core of past SBP conferences are all well represented: behavioral science, health sciences, military science and information science. There are also many papers that provide methodological innovation as well as new domain-specific findings.

There are a number of events that took place at the conference that are not well represented in the proceedings, but added greatly to the intellectual and collegial value of the experience for participants. The first day of the conference offered four free tutorials, from Marta C. Gonzalez (A Review of Human Mobility Models Based on Digital Traces of Human Activity), David Sallach (Categorial Analysis of Social Processes), Joey Harrison and Claudio Cioffi-Revilla (Building Agent-Based Models with the MASON Toolkit), and GeorgiyBobashev (Social Simulation: Introduction to Agent-Based Modeling). One of our excellent keynote speakers anchored each day's program: Myron Gutmann from the National Science Foundation and University of Michigan, Michele Gelfand from the University of Maryland, and BernardoHuberman from Hewlett-Packard Laboratories.

Nitin Agarwal and Wen Dong managed the second SBP Data Analysis Challenge Problem, designed around a rich and unique "reality mining" dataset generously provided and supported by the MIT Human Dynamics Laboratory. (These data continue to be available to the research community at reality-commons.media.mit.edu.)

Unique SBP traditions also continued: The "Cross-Fertilization Round Table" event created opportunities for interaction between technical specialists and domain experts, and a Q&A panel was held over lunch with program managers from a number of federal agencies. More information on all program activities is available on the conference website at SBP2013.org.

Activities such as SBP only succeed with assistance from many contributors. The conference itself was held during April 2-5 in downtown Washington, DC, at the University of California's DC Center on Rhode Island Avenue; we are grateful for their hospitality and many forms of logistical support. The Organizing Committee met early and often this year, lining up keynote speakers, working to publicize the conference, and making many decisions about programming, direction, and finances. Program Committee Co-chairs Ariel M. Greenberg and William G. Kennedy should be singled out for special recognition. They ably managed the submission, review, and proceedings production process, keeping it on an aggressive schedule without losing sight of the larger goals of promoting intellectual exploration and broad participation. Evaluating the large number of submissions could not have been accomplished without our volunteer reviewers, listed under the Technical Program Committee. Last but not least, we sincerely appreciate the support from the following federal agencies: Air Force Research Laboratory (AFRL), Army Research Office (ARO), National Institute of General Medical Sciences (NIGMS) at the National Institutes of Health (NIH), the National Science Foundation (NSF), and the Office of Naval Research (ONR). We also would like to thank Alfred Hofmann from Springer. We thank all for their kind help, dedication, and support in making SBP13 possible.

January 2013   Nathan Bos
                                                             Claudio Cioffi-Revilla

Organizing Committee

Conference Co-chairs

Nathan D. Bos   Johns Hopkins University / Applied Physics Laboratory, USA

Claudio Cioffi-Revilla   George Mason University, USA

Program Co-chairs

Ariel M. Greenberg   Johns Hopkins University / Applied Physics Laboratory, USA

William G. Kennedy   George Mason University, USA

Stephen Marcus   National Institutes of Health, USA

Steering Committee

John Salerno   Air Force Research Laboratory, USA

Huan Liu   Arizona State University, USA

Sun Ki Chai   University of Hawaii, USA

Patricia Mabry   National Institutes of Health, USA

Dana Nau   University of Maryland, USA

V.S. Subrahmanian   University of Maryland, USA

Advisory Committee

Fahmida N. Chowdhury   National Science Foundation, USA

Rebecca Goolsby   Office of Naval Research, USA

Joseph Lyons   Air Force Reasearch Laboratory, USA

John Lavery   Army Research Lab/Army Research Office, USA

Patricia Mabry   National Institutes of Health, USA

Tisha Wiley   National Institutes of Health, USA

Secretary

Chandler Johnson   Stanford University, USA

Treasurer

Guillermo Pinczuk   Johns Hopkins University / Applied Physics Laboratory, USA

Tutorial Chair

Shanchieh (Jay) Yang   Rochester Institute of Technology, USA

Challenge Problem Co-chairs

Nitin Agarwal   University of Arkansas, USA
Wen Dong   MIT Media Lab, USA

Workshop Co-chairs

Fahmida N. Chowdhury   National Science Foundation, USA
Tisha Wiley   National Institutes of Health, USA

Publicity Co-chairs

Donald Adjeroh   West Virginia University, USA
Patricia Mabry   National Institutes of Health, USA

Sponsorship Committee Chairs

Huan Liu   Arizona State University, USA

Student Arrangements Chairs

Patrick Roos   University of Maryland, USA
Wei Wei   Carnegie Mellon University, USA

Web Chair

Katherine Chuang   Drexel University, USA

Technical Program Committee

Mohammad Ali Abbasi   Wei Pan
Yaniv Altshuler   Shamanth Kumar
Asmeret Bier   Samarth Swarup
Huan Liu   Changzhou Wang
Wen Dong   Nicholas Weller
Michele Coscia   Matthew Gerber
Muhammad Ahmad   Rafael Diaz
Kiran Lakkaraju   Jiexun Li

Xiaofeng Wang
Jiangzhuo Chen
Shuyuan Mary Ho
Robert Mccormack
Nathalie Williams
Lei Tang
Nitin Agarwal
Lei Jiang
Paul Whitney
Bei Yu
Yuval Elovici
Inon Zuckerman
Jonathas Magalhães
Bonnie Riehl
Michael Fire
Sai Moturu
Peter Chew
Nasim Sabounchi
Donald Adjeroh
Vadim Kagan
Ruben Juarez
Keisuke Nakao
Christian Lebiere
Patrick Finley
Andrew Collins
Victor Asal
David Chin
Yu-Han Chang
Haiqin Wang
Terresa Jackson
Manas Hardas
Bethany Deeds
Jang Hyun Kim
S.S. Ravi
Myriam Abramson
Daniel Zeng
Robert Hubal
Edward Ip
Hasan Davulcu
Soyeon Han
Sujogya Banerjee
Wai-Tat Fu
Elyse Glina
John Salerno
Masahiro Kimura

Laurence T. Yang
Kouzou Ohara
Yanping Zhao
Zhijian Wang
Koji Eguchi
Alvin Chin
Juan Mancilla
Zeki Erdem
David L. Sallach
Mathew McCubbins
Byeong-Ho Kang
Rajiv Maheswaran
Fatih Özgül
Radoslaw Nielek
Hiroshi Motota
Joseph Lyons
Lashon Booker
Patrick O'Neil
Xintao Wu
Thomas Moore
Wen Pu
Kayo Fujimoto
Saulius Masteika
Gaurav Tuli
Alexander Outkin
Christopher Yang
William Ferng
Liang Gou
Alin Coman
Geoffrey Barbier
Kazumi Saito
Shanchieh Yang
Halimahtun Khalid
Tisha Wiley
Antonio Sanfilippo
Achla Marathe
Shibin Parameswaran
Kalin Agrawal
Jonathon Kopecky
Sedat Gokalp
Craig Vineyard
Anthony Ford
Hazhir Rahmandad
Rik Warren
Lei Yu

Amitava Das

Aleksander Wawer

Madhav Marathe

Armando Geller

Walter Hill

Stephen Verzi

Seyed Mussavi Rizi

Maciej Latek

Richard Fedors

Elizabeth Bowman

John Lavery

Laurie Fenstermacher

Chandler Armstrong

Jeffrey Ellen

Vincen Silenzio

Michael Lewis

Emrah Onal

Shusaku Tsumoto

Kaushik Sarkar

Bart Paulhamus

Michael Mitchell

Corey Lofdahl

Mi Zhang

Ma Regina Justina E. Estuar

Xiaofeng Wang

Xueqi Cheng

Alin Coman

Ariel M. Greenberg

Bill Kennedy

Nathan Bos

Table of Contents

Behavioral Science

Health Sciences

Information Science

Methodology

Military Science

The Evolution of Paternal Care

Mauricio Salgado

GSADI: Grup de Sociologia Analítica i Disseny Institucional, Departament de
Sociologia, Facultat de Ciències Poltiques i de Sociologia, Universitat Autònoma de
Barcelona, Edifici B, 08193 Bellaterra, Spain
muricio.salgado@uab.cat
http://gsadi.uab.cat/

Abstract. I describe an agent-based model to study the evolution of paternal care. The reported *n-person* Iterated Prisoner's Dilemma shows
that the relative differences in the reproductive effort between sexes can
explain the evolution of paternal care. When female reproductive costs
are higher than male reproductive costs, males cooperate with females
even when females do not reciprocate. Paternal care is thus an evolutionary achievement to compensate for the energy demands that reproduction involves for mothers. Paternal care, in turn, produces a sustained
population growth, since females can reproduce at higher rates.

Keywords: Agent-Based Modelling, Cooperative Breeding, Iterated
Prisoner's Dilemma, Parental Investment, Paternal Care.

1 Introduction

Paternal care, a suite of behaviours by a mature male (the genetic, putative or
social father of the immature young) which enhance the children's fitness, is a
rare phenomenon among mammals. Direct infant care by males occurs in fewer
than 5% of all mammals. In mammals, there is a differential *fitness trade-off*:
males tend to focus their reproductive behaviour on mating effort and away from
parental care, while most of the parental investment offspring require to survive
depends on females [1–3]. This difference is produced by the internal gestation
and obligatory postpartum suckling in mammals, which yields a reproductive
difference in the rate at which male and female can reproduce: during the long
period between gestation and children's maturity (usually after weaning) females
cannot reproduce, whilst males can. When this situation is combined with females' ability to care effectively for offspring, this reproductive difference results
in males focusing on mating effort, through male-male competition (or 'sexual
competition') and females focusing on parental investment.

When mothers cannot care effectively for offspring on their own, they confront
the challenge of producing viable offspring without having the ability to do so. This
is the case of human mothers, whose parental effort is much higher than males [4].
Female mothers produce the largest and slowest-maturing babies among primates
[5]. However, humans breed the fastest. Whereas interbirth intervals are estimated

A.M. Greenberg, W.G. Kennedy, and N.D. Bos (Eds.): SBP 2013, LNCS 7812, pp. 1–10, 2013.

at around 4 years in gorillas, 5.5 years in wild chimpanzees, and 8 years in orang-utans, in foraging societies the interbirth intervals average 3.5 years [6]. Thus, years before a mother's previous children were self-sufficient, she would give birth to another infant, and the care these dependent youngsters required would be far in excess of what a foraging mother by herself could regularly supply [7]. An evolutionary response to this challenge is for females to elicit cooperation from others, including males, in breeding activities. Trivers [8] analysed this scenario through his theory of *reciprocal altruism*: if individuals assist each other in turns, and the costs of cooperation are relatively low to donors while the benefits are high to recipients, reciprocal cooperation could evolve among related and even unrelated individuals. In *cooperative breeding* species, females and males help to raise offspring that are not necessarily their own [9]. Humans are the extreme example of this form of cooperation [10]. Men's investment in their offspring figures in many models of human evolution [11, 12]. But across cultures and within them, human males greatly vary in the manifestation of paternal care, ranging from complete absence or aloofness to great intimacy and direct care [13–15]. All in all, given their high energetic burden of reproduction, human mothers need the help of others, including real or possible fathers.

This paper reports the results of an agent-based model (ABM), which aims to simulate the emergence of paternal care. The model also allows us to explore the effects of this cooperative trait on population dynamics. My hypothesis is that differential reproductive costs between sexes set the selection pressure for paternal care. To test this hypothesys, the ABM is introduced, describing in detail its entities, properties and modelled mechanisms (Sec. 2). The model results and analyses are reported in the following section (Sec. 3). The paper finishes with some concluding remarks (Sec. 4).

2   The Simulation

The ABM allows us to study how differential reproductive effort between sexes influences the emergence of inter and intra-sex cooperation for breeding new agents. The objective of this simulation is to evaluate, on the one hand, the extent to which cooperative breeding and paternal care can endogenously emerge in an artificial society made of sexually different agents, as an evolutionary response to compensate for the high reproductive costs of mothers and, on the other, to study the effect of this cooperative behaviour on population dynamics. The emergence of paternal care, which allows females to increase their reproduction rates, should yield sustained population growth.

2.1   Agents' Attributes

Let there be a set of agents $A = \{a_1, \ldots, a_i, \ldots, a_n\}$ and $i = \{1, \ldots, n\}$. At the beginning of the simulation, the population size is N_t. Agents can be either males or females. They have complete life histories, measured by the parameter *Age*, so they are born, become older with each simulation step and live until the maximum agent age is reached, a model parameter which is set by *Max Age*. Agents have thereby more realistic features which replicate real agents' life histories, with a parameter called *Reproductive Age*, so at that age agents are considered 'adults', reaching sexual maturity; therefore they can reproduce. They are also endowed with memory, m, so they can remember their last interaction with each other agent.

Agents have *Energy Levels*. The only difference between sexes lies in their reproductive costs. *MRC* stands for *Male Reproductive Costs* and *FRC* for *Female Reproductive Costs* and these are simulation parameters. Females can produce offspring, provided they interact with one agent of opposite sex and both of them are fertile. An agent is fertile if he or she is an *adult* —when the agent has reached sexual maturity—and if he or she has enough energy to pay for its reproductive costs. To pay the reproductive costs, agents must gain energy-points by playing the *Iterated Prisoner's Dilemma* (IPD) with other agents. This means that the agents with the best strategies for playing the IPD will have the most offspring. The points so gained are equivalent to *fitness*. At every time step, agents are randomly paired to play the IPD (but some agents might not find a partner), so the mechanism it is implemented in this simulation is *direct reciprocity* [16].

2.2 Strategies and Reproduction

It cannot be assumed that a female agent will behave in the same way with another female as she would with a male. Four possible interactions could arise: 1) males vs. females; 2) females vs. females; 3) females vs. males; 4) males vs. males. Each agent i carries a strategy string Θ, made of four strategies, one for each of these possible situations, which are encoded as genetic information to be transferred to their offspring. Although a male agent, for instance, only requires strategies 1) and 4), his potential daughters will require information from strategies 2) and 3). By carrying all four strategies an agent contributes to the behaviour of its offspring regardless of their sex.

In this ABM, there is a set S of five strategies, which are defined as follows: 1) *tit-for-tat*: cooperate in the first encounter and then choose the same action the opponent chose; 2) *unforgiving*: cooperate unless an opponent defects once, then always defect in each interaction with it; 3) *random*: randomly cooperate or defect; 4) *defect*: always defect; and 5) *cooperate*: always cooperate.

When a new offspring is produced, its strategy string Θ_i will be made by crossing over her parents' strategy strings. To include errors in transmission, mutations occur at rate μ. Offspring are randomly assigned a sex, their age is set to zero and their memory to null.

2.3 Initialisation

The ABM allows an exploration of the effect of varying two parameters, MRC and FRC, on the evolution of cooperation and the consequent population dynamics. At $t = 0$, the population consists of 250 agents, made up of 50% females, the agents' memory is empty and *Age* and *Energy Levels* are randomly set, as are the strategies in each agent's string Θ_i. Each simulation step t corresponds to 0.02 years and agents die when they are 40 years old (so agents live for 2,000 time steps). Each agent has its strategies encoded in a strategy string of 4 bits. During reproduction the strategy strings of each parent are 'crossed over' and may mutate, with a chance $\mu = 0.006$. Agents reach sexual maturity—i.e., they can reproduce—when they are 15 years old (so, *Reproductive Age*= 15). Both *Reproductive Age* and *Max Age* were set taking into account the current wild chimpanzee reproductive schedule and lifespan, which are likely to be similar to early humans [17]. Two stop conditions were established: 1) a simulation stops once the population size reaches 2,500 agents (10 times the initial population size); and 2) it stops once the simulated time reaches 4,550 years (or 227,500 steps), which corresponds to a maximum of 114 generations. The ABM was built using NetLogo 4.1.3 [18], and the model itself is an extension of the N-person IPD model available in the NetLogo Library [19].

3   Results and Analyses

I explored the parameter space given by the combination of MRC and FRC, varying MRC from 200 to 2,000 by 200 energy units, and keeping FRC constant at 2,000. By so doing, the simulation addressed the effect of increasing or decreasing the ratio of reproductive effort between sexes on the evolution of cooperation and population dynamics. It was assumed in this set of experiments that FRC are high, given the energy burden that reproduction implies for human mothers. To have more robust results, each parameter combination was run 25 times. Two hundred and fifty such experiments were performed in order to study the evolution of cooperation and population dynamics over time.

Figure 1 reproduces the main results of the experiments. The analysis considered the ratio of MRC to FRC, which establishes a continuum between no differences and maximum difference in the reproductive costs of both sexes. Thus, $\frac{MRC}{FRC} = 0.1$ corresponds to the maximum difference, i.e., MRC= 200 and FRC= 2,000. When $\frac{MRC}{FRC} = 1.0$, MRC=FRC= 2,000. Plot 1(a) depicts the percentage of simulations that, in each parameter combination, reached the maximum number of agents ($N = 2,500$) and the percentage of simulations that became extinct ($N = 0$). Plot 1(b) depicts the average points or scores earned per player per game of the prisoner's dilemma played during the set of experiments at different values of the ratio MRC to FRC.

Plot 1(a) and Plot 1(b) in Figure 1 reveals that larger differences in the reproductive effort of males and females (differences in which males have lower reproductive costs than females) are associated with faster population growth rates. Low or non-existing differences in reproductive costs between sexes produced more cooperation among agents, regardless of the opponents' sex. Thus, when $\frac{MRC}{FRC} \geq 0.8$, average scores converge to 3 points for all four types of interactions. In these cases, few (if any) simulations described population growth and most of them became extinct, as can be seen in Plot 1(a). Conversely, high differences in the reproductive efforts of both sexes (i.e., $\frac{MRC}{FRC} \leq 0.4$) led to weak

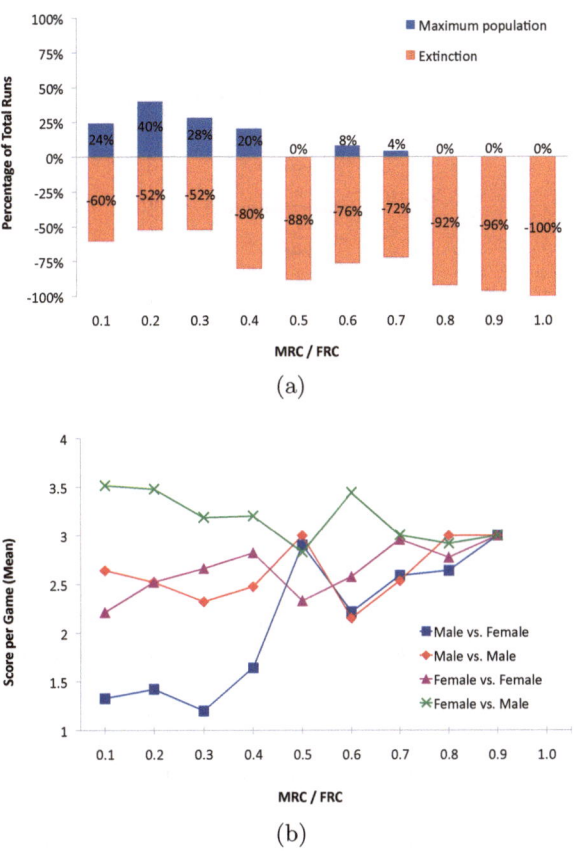

Fig. 1. Main simulation results. Plot 1(a) depicts the percentage of simulations that reached the maximum population and the percentage of those which went extinct. Plot 1(b) summarises the average score per player per game of the prisoner's dilemma. Scores are shown separately for different interaction types: male vs. male, female vs. female, male vs. female and female vs. male.

cooperation among same sex agents and produced a remarkable divergence in inter-sex interactions, as Plot 1(b) shows, with males unconditionally cooperating with females, and females mainly exploiting males by defection. At the maximum difference, i.e. $\frac{MRC}{FRC} = 0.1$, females gained, on average, 3.51 points from their games against males, whilst males obtained just over 1.33 points in return. Thus, when the reproductive effort of females is much higher than that of males, evolution will favour the emergence of *non-reciprocal altruism* [9] or pure altruism between sexes: males will tend to cooperate with females even though the latter do not reciprocate. This scenario yielded more simulations to reach the maximum number of agents and fewer populations became extinct.

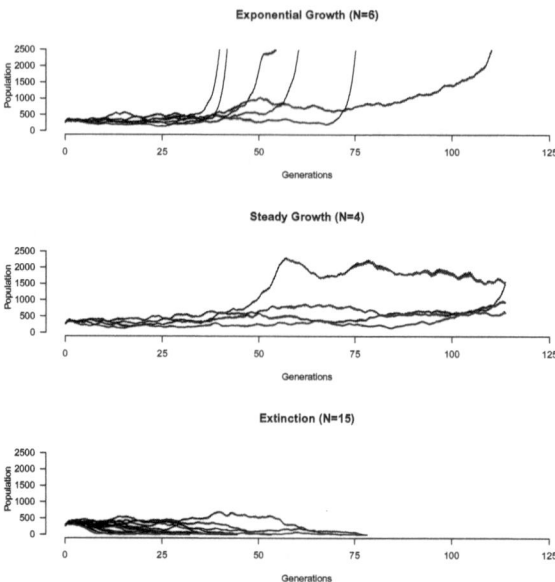

Fig. 2. Three patterns of population growth during the 25 simulations for $\frac{MRC}{FRC} = 0.1$. The patterns correspond to 'Exponential growth', 'Steady Growth' and 'Extinction'.

Therefore, once male non-reciprocal altruism emerges, the fitness of exploitative females is enhanced, so they can produce more offspring, leading to higher rates of population growth.

Since the emergence of males' non-reciprocal altruism is related to the differences in the reproductive costs between sexes, I shall focus on the results obtained at the maximum difference in the reproductive effort between sexes (i.e., when $\frac{MRC}{FRC} = 0.1$). Figure 2 shows three different types of population dynamics that were observed in that combination of parameters, namely: 'Exponential Growth', 'Steady Growth' and 'Extinction'. The simulations that experienced exponential growth (first Plot in Figure 2) show a *phase transition*. The usually long period of time in which the simulations had steady growth indicates that some behaviour at the micro-level is being *incubated*, and after some generations, the incubating micro-behaviour produces a 'tipping-point' so the population dynamics are dramatically altered, growing exponentially. The second plot in Figure 2 depicts the population dynamics of those simulations that neither reached the maximum population nor became extinct during the analysed period of time ('Steady Growth'). The last plot in Figure 2 depicts the extinct simulations. None of those simulations described a period of rapid growth, so the micro-behaviour leading to exponential growth never emerged. Here, no form of male non-reciprocal altruism was observed, as I discuss next.

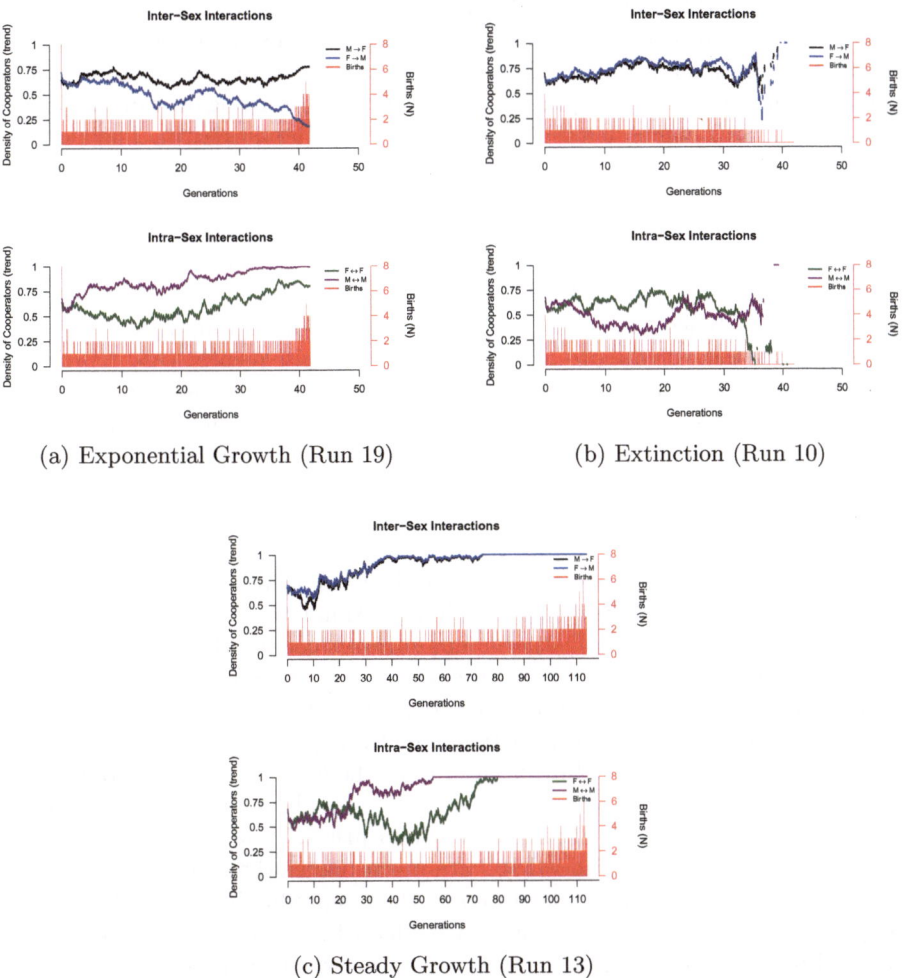

Fig. 3. Evolution of cooperation in inter- and intra-sex interactions in three different simulations. Vertical spikes (red) represent births and lines represent the moving average (with a window of ± 81) of the average cooperation for each type of interaction.

Figure 3 depicts the evolution of cooperation in inter- and intra-sex interactions followed during three different population dynamics. Each plot also shows the number of births during these simulations (vertical spikes). Subfigure 3(a) shows the evolution of cooperation between sexes in a simulation that described exponential growth (i.e., run 19). As can be seen, there is a clear divergence in the trajectory of cooperation between sexes (see first plot in Subfigure 3(a), entitled 'Inter-Sex Interactions'). Males tend to maintain relatively high levels of cooperation with females, and during the last 10 generations (before the simulation reached the maximum number of agents) male cooperation towards females increased from 64% to 78% on average. Instead, females decreased their cooperation towards males from the early stages of this simulation. During the last 10 generations, female cooperation towards males decreased from 42% to just 18% on average. Consequently, whilst females playing against males earned, on average, 3.13 points, males obtained just 1.4 points in return. Therefore, an increasing percentage of females exploited males—who did not decrease their cooperative levels. At the same time, intra-sex cooperation steadily increased over time for both sexes (see second plot in Subfigure 3(a)). Males cooperated with each other almost always. However, female cooperation, although increasing over time, reached a maximum of 85% and had a small drop at the end of the simulation to 80%. Since females were obtaining more points from males, they were able to cover their high reproductive costs faster; consequently, they reproduced faster (see vertical spikes). During the last 10 generations, the annual number of births in this simulation increased from 7 to 79, with annual average births equalling 19.4 during the last 10 generations, 29.56 during the last 5 generations, and 61.7 during the last generation.

Subfigure 3(b) depicts the evolution of inter- and intra-sex interactions during a simulation that became extinct (i.e., run 10). It is possible to see remarkable differences with the simulation that described exponential growth. In run 10, inter-sex cooperation converged early after the simulation began and it was similar most of the time. On the other hand, intra-sex cooperation was very erratic, although females tended to cooperate with each other more than males did among themselves (see second plot in Subfigure 3(b)). All in all, the cooperation structure here reinforces the idea that females require male non-reciprocal altruism to reproduce faster. In this case, since the levels of inter-sex cooperation were similarly high (70% on average), females were obtaining just 2.63 points on average when they played against males (and just 2.7 points on average during the last 10 generations). Since FRC are much higher than MRC, inter-sex cooperation does not help females to cover their reproductive effort. This fact is expressed in the low annual number of births, which during the last 10 generations dropped from just 3 births in year 1, 278—when the population total just 166 agents—to no further births from year 1, 638 onwards (with a maximum of 7 annual births by year 1, 281). The population became extinct by year 1, 677. This adverse evolutionary scenario was magnified by the fact that intra-sex cooperation, although still high, was not enough for females to offset the low points earned in inter-sex interactions.

Subfigure 3(c) shows the evolution of cooperation in inter- and intra-sex interactions during a simulation (run 13) that described a steady population growth. This simulation covered the maximum period of time. In this simulation, with a steady population growth, inter-sex cooperation was for most of the time at similar levels and, by generation 75 full cooperation had emerged. Just a few generations after that, full intra-sex cooperation emerged. From then onwards, all the agents were cooperating with each other, regardless of their opponent's sex, and consequently all the agents in the simulation earned 3 points when they played the prisoner's dilemma. Subfigure 3(c) also shows that full cooperation can sustain a steady population growth. Thus, during the last 25 generations, the annual number of births increased from 10 to 28 by year 4,550, when the simulation stopped because the time limit was reached.

4   Concluding Remarks

The results showed that the greater the differences in the reproductive costs between sexes are, the more likely a simulation will reach the maximum number of agents. This association is given by the kind of cooperation that emerges when the reproductive effort of males and females is considered. When male reproductive effort is less than female reproductive effort, males will tend to cooperate unconditionally with females ('non-reciprocal altruism'), and females will tend to exploit males' unconditional cooperation. In this exploitative scenario, females are able to cover their high reproductive costs and, consequently, reproduce faster, so the agent population grows exponentially. Conversely, when the reproductive efforts are equally high, inter and intra-sex cooperation will tend to emerge, although this reciprocal structure of cooperation does not always allow female agents to reproduce faster, so fewer simulations will reach the maximum number of agents—most of them will become extinct. The simulation predicts the evolution of different strategies to cope with the challenge of reproduction when female reproductive effort is high. Primates and humans are ideal to test this prediction due to their variability in cooperative relationships.

Acknowledgments. This work was supported by the National Plan for R&D through the CONSOLIDER-INGENIO 2012 Programme of the Spanish Ministry of Science and Innovation (MICINN, Grant No. CSD2010-00034-SimulPast).

References

1. Smith, J.: Parental investment: A prospective analysis. Animal Behaviour 25, 1–9 (1977)
2. Trivers, R.: Parental investment and sexual selection. In: Campbell, B. (ed.) Sexual Selection and the Descent of Man, London, Heinemann, pp. 136–179 (1972)
3. Clutton-Brock, T.H.: Review lecture: Mammalian mating systems. Proceedings of the Royal Society of London. B. Biological Sciences 236(1285), 339–372 (1989)
4. Aiello, L.C., Wells, J.C.K.: Energetics and the evolution of the genus Homo. Annual Review of Anthropology 31, 323–338 (2002)
5. DeSilva, J.M.: A shift toward birthing relatively large infants early in human evolution. Proceedings of the National Academy of Sciences 108(3), 1022-1027 (2011)
6. Key, C.A.: The evolution of human life history. World Archaeology 31(3), 329–350 (2000)
7. Hrdy, S.B.: Mothers and Others: The Evolutionary Origins of Mutual Understanding: The Origins of Understanding, 1st edn. Harvard University Press (2009)
8. Trivers, R.L.: The evolution of reciprocal altruism. The Quarterly Review of Biology 46(1), 35–57 (1971)

9. Bergmüller, R., Johnstone, R.A., Russell, A.F., Bshary, R.: Integrating cooperative breeding into theoretical concepts of cooperation. Behavioural Processes 76(2), 61–72 (2007)
10. Bogin, B.: Evolution of human growth. In: Muehlenbein, M.P. (ed.) Human Evolutionary Biology, pp. 379–395. Cambridge University Press (2010)
11. Anderson, K.G., Kaplan, H., Lancaster, J.: Paternal care by genetic fathers and stepfathers I: Reports from Albuquerque men. Evolution and Human Behavior 20(6), 405–431 (1999)
12. Fernndez-Duque, E., Valeggia, C.R., Mendoza, S.P.: The biology of paternal care in human and nonhuman primates. Annual Review of Anthropology 38, 115–130 (2009)
13. Geary, D.C.: Evolution and proximate expression of human paternal investment. Psychological Bulletin 126(1), 55–77 (2000)
14. Sear, R., Mace, R.: Who keeps children alive? A review of the effects of kin on child survival. Evolution and Human Behavior 29(1), 1–18 (2008)
15. Hewlett, B.S.: Intimate Fathers: The Nature and Context of Aka Pygmy Paternal Infant Care. University of Michigan Press (March 1993)
16. Nowak, M.A.: Five rules for the evolution of cooperation. Science 314(5805), 1560–1563 (2006)
17. Thompson, M.E., Jones, J.H., Pusey, A.E., Brewer-Marsden, S., Goodall, J., Marsden, D., Matsuzawa, T., Nishida, T., Reynolds, V., Sugiyama, Y., Wrangham, R.W.: Aging and fertility patterns in wild chimpanzees provide insights into the evolution of menopause. Current Biology 17(24), 2150–2156 (2007)
18. Wilensky, U.: NetLogo. Center for Connected Learning and Computer-Based Modeling, Northwestern University, Evanston, IL (2011), http://ccl.northwestern.edu/netlogo
19. Wilensky, U.: NetLogo PD N-Person iterated model (2011), http://ccl.northwestern.edu/netlogo/models/PDN-PersonIterated

Parent Training Resource Allocation Optimization Using an Agent-Based Model of Child Maltreatment

Nicholas Keller and Xiaolin Hu

Department of Computer Science, Georgia State University, Atlanta, GA 30303

Abstract. This paper uses our previously created agent-based model to study the optimal way of allocating parent training resources to address unmet child need, a measure of child maltreatment. We consider parent training resource allocation prioritization based upon: a family's own resources and the resources they have access to through their social network. Simulation results show that when targeting a community with a parent training intervention program, ignoring heterogeneity within a community is a mistake. This work demonstrates the utility of the developed agent-based model and suggests that child maltreatment research can benefit from a complexity science approach.

1 Introduction

Child Maltreatment (CM) likely affects over one million children a year in the United States [1]. Although each year more than $20 billion is spent on Child Protective Services (CPS) [1] there has been little research into the effectiveness of CPS interventions. To our knowledge our model is the first agent-based model of child maltreatment. This paper models CM rates by looking at the level of unmet child need. Most child maltreatment studies involve collecting data from communities over many months or years, which is expensive and slow. Additionally, the transient nature of those studied often means that they are difficult or impossible to contact for followup surveys. Child maltreatment research can benefit from computer modeling, because CM occurs in an inherently complex system. Computer models give us the ability to move beyond our mental models. Much of this complexity is due to the interactions of agents in a social graph.

In our previous work [4, 5], we developed an agent-based model of child maltreatment, where risk and protective factors of CM at different levels of the social ecology exert influences through several major pathways and feedback loops to determine the likelihood of child maltreatment. In the developed model, each agent represents a family unit in a community. At each step of the simulation parents have to decide whether or not they will meet their child's needs. The model takes into account caregivers' dynamic cognitive decision making process and the impact of their ability to both ask for help from their social network and their ability to provide help to other caregivers who request it. This paper utilizes the previously developed model to study CM intervention/prevention. Specifically, we focus on the problem of parent

A.M. Greenberg, W.G. Kennedy, and N.D. Bos (Eds.): SBP 2013, LNCS 7812, pp. 11–18, 2013.
© Springer-Verlag Berlin Heidelberg 2013

training resource allocation and use the developed agent-based model to help optimize parent training resource allocation to minimize child maltreatment.

Often child maltreatment is a result of poor parenting. Parents and children perceive child maltreatment, such as hitting, as discipline, which is why they often freely admit to it when asked [3]. Therefore, CM interventions often focus on improving parenting skill. Parent training programs such as the Triple P-Positive Parenting Program have been shown to be effective at reducing child maltreatment [6]. Parent training is a common CM intervention/prevention strategy [1].

Given the success of parent training programs we asked how can the limited funds devoted to parent training programs be more effectively invested to increase the return on investment? Programs like Triple P are expensive and time consuming ways to test parent training resource allocation strategies, although Triple P attempts to control costs by using a tiered service delivery system [7]. Simulations like ours cannot replace field trials, but they can help to guide them. Agent-based models as opposed to other modeling techniques have the advantage that they can model the impact of the complex social graph.

To study parent training resource allocation strategies, we identified several different strategies for allocating parent training resources and simulated their impacts on the overall rate of CM using the developed agent-based model. We modeled different types of communities and compared the outcomes of each strategy in these different communities. By showing that the model can produce potentially valuable conclusions we seek to justify future validation research.

2   Model of CM and Parent Training Intervention

Our previous paper [4, 5] has a detailed explanation of our model; therefore we will only give a cursory overview of the model. Fig. 1 is a flowchart representation of the model. Currently sibling interactions are not modeled; however, child need could be used to represent the needs of more than one child. Child need is an abstraction of the various types of child needs such as physical and emotional. Similarly, physical resources are an abstraction of different types of resources. A family's resources are modulated by the parenting skill of the caregivers. The interactions between multiple caregivers are not modeled. The community resources perceived and available to a family depends upon how many neighbors a family has, what their relationships to those neighbors are, and what resources those neighbors have and how willing they are to give them. In the simulations, families are divided into communities and groups within communities based upon either their families' personal resources or the resources available to them in the social graph. Groups within communities are selected based upon some criteria and are singled out to receive parent training benefits; however, families in a group are not necessarily more connected to each other than to other families in the community.

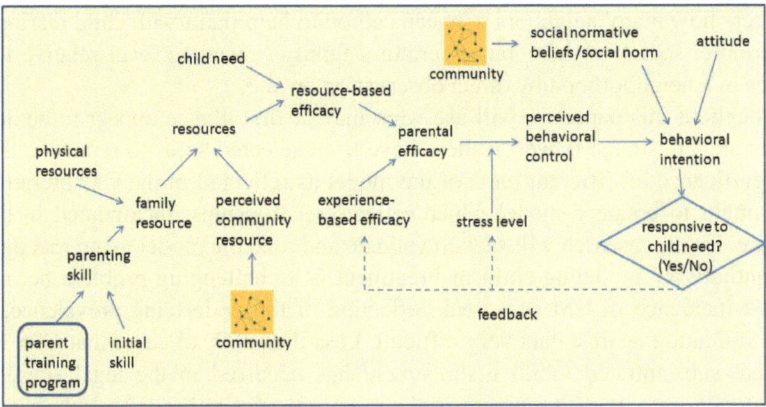

Fig. 1. Model of caregiver's decision making process with parent training program added

The variety of needs children have such as: physical, emotional, and psychological, are combined in our model into one category we call "child need". Similarly, the many resources that families can bring to bear are combined into one category called "physical resources." Perceived community resources represent resources available to families in the social graph, such as from neighbors. A child's need is said to be "not met" when the resources that are devoted to them are less than their need.

There are many forms of child maltreatment; however, our model abstracts these categories of maltreatment into one called "unmet need." It is not necessarily true that each time a child's need is "not met" that child maltreatment has occurred; rather the number of times a child's need is "not met" is a measure of the quality of child care. At some point a family will not meet their child's need often enough and it can be said that CM has occurred. Although this point is arbitrary within the context of our model, this does not reduce the value of the model, because the model's value lies in its ability to predict qualitative patterns that would otherwise remain hidden. These patterns can then be used to justify quantitative and expensive field studies.

The goal of our simulation is to demonstrate the value of our model and study the best way to allocate scarce parent training resources given communities' characteristics. Communities are defined by their average level of family resources. The average number of caregivers in a household represents that community's average family resources; however, we do not assign a certain number of caregivers to a family, instead we assign each family an abstraction of caregiver resources called "family resources". We look at communities along the spectrum between one parent households and two parent households. Although the model uses the number of caregivers as a representation of the amount of family resources it could just as easily have used some other measure of family resources, such as income.

Communities are a heterogeneous collection of interconnected families (agents). For the purpose of this simulation families will be differentiated by the number of resources they have and by the number of neighbors they have. More complex social graph information is available in the model, but this information will not be readily accessible to social workers in the field. It is cheap and quick for social workers to ask caregivers how many neighbors they can call on to help them with child rearing tasks. Furthermore, social workers can ascertain a family's resource level relative to other families in a neighborhood by direct observation on site.

Throughout this paper we will use what may at first glance appear to be arbitrary weights and constants; however, these have been selected so as to represent the relative significance of different parts of the model as reflected in the CM literature. We have sought to create a model which reflects relationships documented in the CM literature. Future research will seek to validate and tune the model using real data.

Quantitatively modeling child maltreatment is a challenging problem because the reported incidence of CM is a weak reflection of the underlying prevalence, which makes validation against data very difficult. Less than 30% of cases that CPS investigates are substantiated (child maltreatment has occurred in the legal sense)[2]. A substantiated case of maltreatment is a bureaucratic definition with little bearing on whether CM occurred or not [2]. This suggests that CPS cannot effectively detect CM, which in turn implies that detecting CM through other means (such as hospital reports) is also likely to have a high false negative rate.

Each parent has a baseline parenting skill, which can be increased through parent training. Parenting skill is an abstraction of various parenting skill metrics and varies between 0 and 100. The different types of parent training a caregiver can receive are abstracted into one number. When parent training is given to an agent it is summed with their old parenting skill to give their new parenting skill. In the experiments represented by Fig. 2, the total amount of parent training distributed was equal to 25 multiplied by the number of agents, so each agent increased their parenting skill by 25 if the resources were distributed equally. In Fig. 3 the amount of parent training was varied for the purposes of sensitivity analysis. Parenting skill impacts family resources as shown in equations 1 and 2.

$$parentingSkillCoefficient = 1 + 0.3 * \left(\left(\frac{parentingSkill + parentTraining}{50} \right) - 1 \right) \qquad (1)$$

$$familyResources = physicalResources * parentingSkillCoefficient \qquad (2)$$

A family's *physicalResources* and their *familyResources* are always between 0 and 100.

At the beginning of the simulation the parent training resources are doled out to families depending upon the strategy being simulated in the experiment. After parent training resources have been given to families, the simulation runs for 1000 iterations to study the impact of the parent training.

In this work, we consider five different strategies as described below.

The first strategy is to select families at random until parent training resources run out and give the selected families enough training to max out their parenting skill. This is akin to a first come first-served parent training program that provides parent training until the money has runs out.

The second strategy is to give each family an equal amount of training. This has the advantage that every family gets some parent training, but it is spread more thinly than in the first strategy. This would be like picking a neighborhood and offering each family parent training regardless of individual families' characteristics. It seems intuitive that this sort of strategy could be used to target high risk neighborhoods, but as will be shown in the results section ignoring community heterogeneity is a mistake.

The third strategy is to give maximum parent training to those families that have relatively low family resources on a first come first serve basis. This is similar to the first strategy in that it maxes out the parent training of those that get assistance, but it differs in how it selects those families. This strategy appears preferable to the first one, but might be harder to implement, so it is important to compare this to the first strategy to see if the advantages of this strategy (if they exist) are significant enough to outweigh their potentially higher administrative costs.

The fourth strategy is to equally distribute parent training among well connected families; three or more neighbors. The fifth strategy is like the fourth strategy except that it equally distributes parent training resources among the poorly connected families; two or fewer neighbors. If a parent cannot meet a child's need on their own they ask their neighbors for assistance. Those neighbors are more likely to give assistance if they have already served the needs of their own children and they have a significant surplus. Strategies four and five study the tradeoff between helping families with few social resources and helping well connected families, so that they will have more resources to give to their neighbors.

3   Simulation Results

For each community five parent training resource allocation strategies were studied along with one baseline experiment with no parent training, and for each of the strategies studied 100 trials were run and averaged. For each trial each agent made 1000 decisions about whether to provide for their child's need or not. There were a total of 50 families (agents).

In all trials the amount of parent training available was the same (25 * 50 (number of agents) = 1250). In all trials the initial parenting skill for each agent was randomly set to a number between 35 and 55 (45 $\pm$10).

The physical resources given to a family at the beginning of a simulation are defined below.

$$average\ physical\ family\ resources = 60 + (averageParentsPerHouse - 1) * 30 \qquad (3)$$

$$physical\ resources\ for\ family\ k = (average\ physical\ family\ resources - 8) + 16 * rand \qquad (4)$$

Where $rand$ is a function that returns a random number between 0 and 1.

Each family's average child need is a randomly selected value between 60 and 70. At each timestep of the algorithm each family calculates the child need for that timestep as follows:

$$child\ need\ for\ \text{child c at}\ timestep\ i = averageChildNeed + (rand - 0.5) * 10 \qquad (5)$$

Fig. 2 below shows the effectiveness of different parent training resource allocation strategies across communities with different levels of family resources as represented by the average number of caregivers per family. Fig. 2's Y-axis is a measure of the total number of times children have not had their need met over the course of the experiment. The most important part of the graph is the middle part because most real life communities with high rates of child maltreatment will have a caregiver to child ratio somewhere in there. Communities with an average near two parents per family are not interesting, because they will not have high rates of child maltreatment, so will be less likely to receive, in real life, the parent training intervention being studied.

It can be observed from Fig. 2 that generally as the average number of parents per household in the community increases, the incidence of unmet need decreases. This makes sense because the number of parents per household represents the physical resources in Fig. 1. We can also see that as expected, all five strategies generated positive impacts in the communities because they reduced the average number of incidence of unmet need compared to the case of no parent training. Nevertheless, the levels of impact of these strategies are different. For example, when the parents per household is 1.5, the strategy of equally distributing the resources results in the worst performance, while the strategy of allocating resources to the poorly connected families resulted in the best performance. The simulation results also show that the relative performance of these strategies varies as the average number of parents per household changes. For example, the performance of the equal distribution of resources goes from the worst to first as the average number of parents goes from 1.5 to 2.0. This is because when the average number of parents is 1.5, most families need a significant amount of extra resources before they can meet their child's needs. Equally distributing parent training resources in this case does not help much because the parent training each family receives is not enough to make a difference. However, when the average number of parents is 2.0, the equal distribution of resources gives the best results because it provides enough training for parents and covers every family in the community.

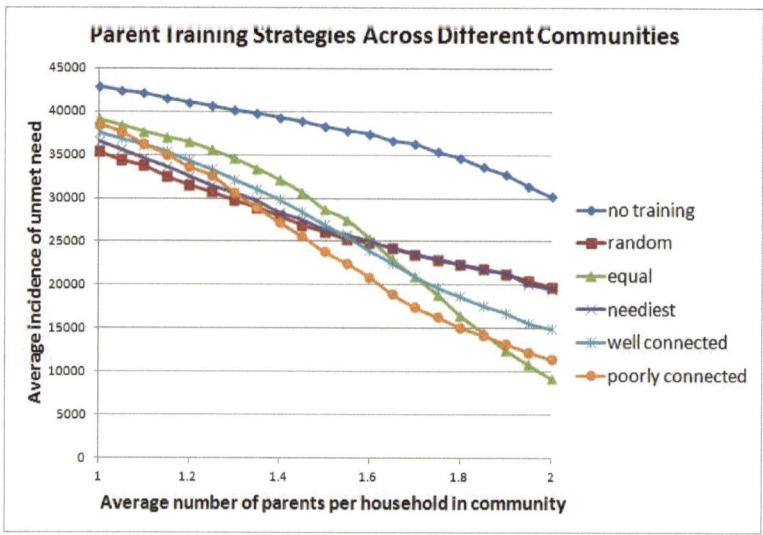

Fig. 2. Effectiveness of parent training strategies measured by total incidence of unmet need

We performed sensitivity analyses to confirm our findings the results of which are shown in Fig. 3 below. In Fig. 3 the amount of parent training was varied in a community with an average of 1.5 caregivers per family, but the parenting skill parameter was unchanged (for each agent it is set to a random value between 35 and 55). The X-axis is a measure of how much parent training each agent would receive if the parent training resources were distributed equally. For example, parent training = 40 means that the total training resource available is equal to 40 * 50 (number of agents) = 2,000.

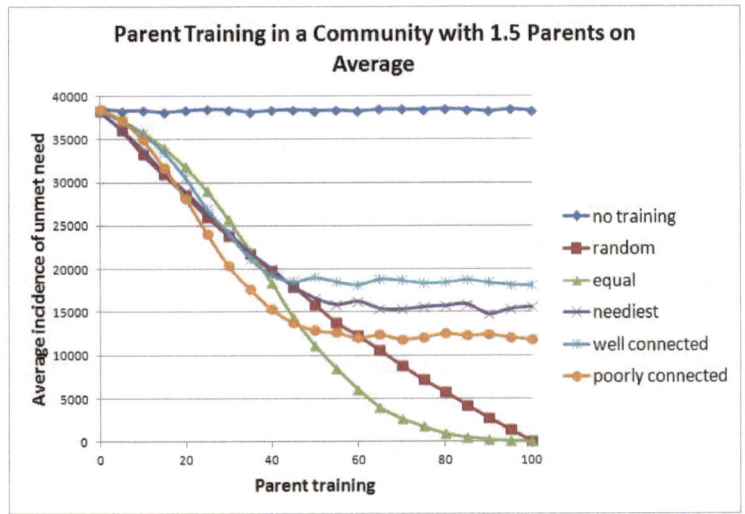

Fig. 3. Parent training sensitivity analysis; total incidence of unmet need

Fig. 3 supports our assertion that random and equal distribution strategies are inefficient. If the parent training provided is between 20 and 45, then giving resources to poorly connected families is the most efficient. Giving resources out on a first come first-served basis (random) has a slight advantage before this interval. Additionally before parent training = 20, equally distributing resources does worse than a first come first-served strategy (random), which is due to the parent training parents receive not being enough to make a difference. This suggests that it is sometimes better to help fewer people, but give them more help, than to help everyone insufficiently. Additionally, Fig. 3. shows that the equal distribution strategy is non-linear and has a tipping point around parent training =   35.

The right side of the graph in Fig. 3 looks "weird" because of edge effects, also called diminishing returns, that can be avoided in real life. As parent training increases unmet need decreases, which makes sense because increasing parenting skill allows a family to more effectively use their physical resources. The reason that the well connected, neediest, and poorly connected strategies level off at high levels of parent training is that they waste parent training resources. These strategies target a subset of the families in a community based upon either how many neighbors a family has (well connected and poorly connected strategies) or how many resources they have (neediest), so even if there are enough parent training resources to fulfill all families' needs some families have no opportunity to benefit from them because they are not in a group that is offered parent training. We included the right side of the graph in Fig. 3 for completeness, but it is unlikely that a real parent training program would ever turn families away when they realize that they set too stringent requirements on who they can help and as a result aren't using all of their money. The parameters of the experiments represented in Fig. 2 were tuned, so as to have no wasted resources.

4   Conclusions

The results suggest that child maltreatment research can benefit from a complexity science approach. The model suggests that equal and random parent training resource allocation strategies may not be optimal and that for many communities it makes sense to allocate resources to poorly connected families. This hypothesis warrants further research, but if it turns out to be true, then it suggests that parent training programs should not rely primarily on word of mouth for promotion, which targets the least socially isolated families. Furthermore, the results suggest that ignoring community heterogeneity as is the case with the equal and random parent training strategies is a mistake and that a small amount of information about a family's social resources and their own resources can be quite valuable. Our results are preliminary and may or may not be an accurate representation of the child maltreatment system; however, we have shown the potential value of computer simulations in the child maltreatment prevention domain. Future research will focus on validating and refining the model against real data.

References

1. The Future of Children. Preventing Child Maltreatment 19(2) (Fall 2009)
2. Hussey, J.M., Marshall, J.M., English, D.J., Knight, E.D., Lau, A.S., Dubowitz, H., Kotch, J.B.: Defining maltreatment according to substantiation: distinction without a difference? Child Abuse & Neglect 29(5), 479–492 (2005)
3. National Data Archive on Child Abuse and Neglect, Cornell University, National Survey of Child and Adolescent Well-Being (NSCAW): General Release Version, Appendix - Vol. III (August 2008)
4. Hu, X., Puddy, R.: Cognitive modeling for agent-based simulation of child maltreatment. In: Salerno, J., Yang, S.J., Nau, D., Chai, S.-K. (eds.) SBP 2011. LNCS, vol. 6589, pp. 138–146. Springer, Heidelberg (2011)
5. Hu, X., Puddy, R.W.: An agent-based model for studying child maltreatment and child maltreatment prevention. In: Chai, S.-K., Salerno, J.J., Mabry, P.L. (eds.) SBP 2010. LNCS, vol. 6007, pp. 189–198. Springer, Heidelberg (2010)
6. Prinz, R.J., Sanders, M.R., Shapiro, C.J., Whitaker, D.J., Lutzker, J.R.: Population-Based Prevention of Child Maltreatment: The U.S. Triple P System Population Trial. Prevention Science 10(1), 1–12 (2009)
7. Sanders, M.R., Markie-Dadds, C., Turner, K.M.T.: Theoretical, scientific and clinical foundations of the Triple P – Positive Parenting Program: A population approach to the promotion of parenting competence. Parenting Research and Practice Monograph 1, 1–21 (2003)

Influence and Power in Group Interactions

Tomek Strzalkowski[1], Samira Shaikh[1], Ting Liu[1], George Aaron Broadwell[1],
Jenny Stromer-Galley[1], Sarah Taylor[2], Veena Ravishankar[1], Umit Boz[1],
and Xiaoai Ren[1]

[1] State University of New York – University at Albany, NY 12222 USA
[2] Sarah M. Taylor Consulting, LLC
tomek@albany.edu

Abstract. In this article, we present a novel approach towards the detection and modeling of complex social phenomena in multiparty interactions, including leadership, influence, pursuit of power and group cohesion. We have developed a two-tier approach that relies on observable and computable linguistic features of conversational text to make predictions about sociolinguistic behaviors such as Topic Control and Disagreement, that speakers deploy in order to achieve and maintain certain positions and roles in a group. These sociolinguistic behaviors are then used to infer higher-level social phenomena such as Influence and Pursuit of Power, which is the focus of this paper. We show robust performance results by comparing our automatically computed results to participants' own perceptions and rankings. We use weights learned from correlations with training examples to optimize our models and to show performance significantly above baseline.

Keywords: computational sociolinguistics, online dialogues, social phenomena, linguistic behavior, influence, pursuit of power, multi-disciplinary artificial intelligence, social computing.

1   Introduction and Related Work

Our objective is to model high-level sociolinguistic phenomena such as Influence, Pursuit of Power, Leadership and Group Cohesion in discourse. This research aims to develop a computational approach that uses linguistic features of conversational text to detect and model sociolinguistic behaviors of conversation participants in small group discussions. Given a representative dialogue of multi-party conversation, our prototype system automatically classifies the participants by the degree to which they engage in such sociolinguistic behaviors as Topic Control, Disagreement, and several others discussed in this paper. These mid-level sociolinguistic behaviors are deployed by discourse participants in order to assert and maintain certain higher-level social roles such as leader, influencer, or pursuit of power, among others. Our approach to this problem combines robust computational linguistics methods and established empirical social science techniques. In this paper, we discuss robust detection of influence and pursuit of power in discourse.

A.M. Greenberg, W.G. Kennedy, and N.D. Bos (Eds.): SBP 2013, LNCS 7812, pp. 19–27, 2013.
© Springer-Verlag Berlin Heidelberg 2013

Following social science theory (e.g., [11], [4], [7]), we define an *influencer* as a group participant who has credibility in the group and who introduces ideas that others pick up on or support. Thus, an influencer exercises a degree of control over the topic and content of a conversation. This may be contrasted with the more explicit *pursuit of power* in the group, which is an attempt to seize control of the group's agenda or actions. Both behaviors often correlate with, and may be considered as dimensions of, group leadership ([4, 16]); however, they are manifested quite differently in interaction. Neither behavior necessarily implies leadership, and often may be seen as a challenge to the group leader. In order to fully understand the social dynamics of a group it is important to model these behaviors independently. This is the focus of our paper, and we are particularly interested in groups involved in online interactions.

Internet-enabled interaction is particularly interesting to study because in this reduced-cue environment, the only means of engaging in and conveying social behaviors is through written language. As such, online discussion relies on the more explicit linguistic devices to convey social and cultural nuances than is typical in face-to-face or telephonic conversations. Relevant recent research in this area includes Freedman et al. ([6]) who developed an approach to detect behaviors such as persuasion in online discussion threads, and Bracewell et al. ([2]) who categorize several types of social acts (e.g. agreement and disagreement) to detect pursuit of power in online groups. These approaches, however, depend on discovering specific linguistic markers that may indicate a type of behavior rather than looking for a more sustained demonstration of sociolinguistic behavior by each speaker over the course of entire discourse. Our research takes that latter approach, and the work presented here builds on Strzalkowski et al. ([15]) and Broadwell et al. ([3]), who also proposed the two-tiered approach to sociolinguistic modeling and have demonstrated that a subset of mid-level sociolinguistic behaviors may be accurately inferred by a combination of low-level language features. Our work successfully extends their approach to modeling of influence and pursuit of power in group interactions. The models discussed in this paper were developed and implemented based on online chat and threaded discussions in English and Mandarin.

2   Sociolinguistic Behaviors in Discourse

In the two-tier modeling approach, we use linguistic elements of discourse to first unravel sociolinguistic behaviors, and then, use the behavior models, in turn, to determine complex social roles, as shown in Figure 1.

It is important to note that, at both these levels, our analyses are solidly grounded in sociolinguistic theory. In this section, we briefly describe the mappings (which we call *measures*), between the sociolinguistic behaviors (mid-level in Fig 1) and the complex social roles. These measures operationalize the second tier in our system. We discuss the first tier only briefly in Section 4.

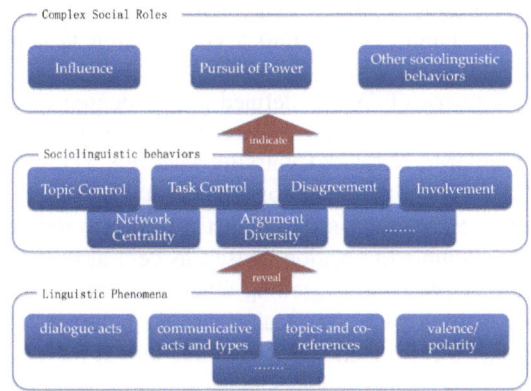

Fig. 1. Two-tier approach applied to model social roles in discourse

- **Topic Control Measure (TCM)** is defined as attempts of participants to impose a topic of conversation. In any conversation, whether it is focused on a particular issue or task or is just a social conversation, the participants continuously introduce multiple topics and subtopics. These are called *local topics*. Local topics, following the notion put forth by Givon ([8]), may be equated with any substantive noun phrases introduced into discourse that are subsequently mentioned again via repetition, synonym, or pronoun. Who introduces local topics into conversation and who continues to talk about them, and for how long are some of the indicators of topic control in dialogue.

- **Cumulative Disagreement Measure (CDM)** has a role to play with regard to influence and pursuit of power in that it is possible that a person in a small group engages in disagreements with others in order to control the topic by way of identifying or correcting what they see as a problem ([5], [13]). While each utterance where a participant disagrees with another is a vivid of expression of disagreement, we are interested in a sustained phenomenon where participants repeatedly disagree, thus revealing a social relationship between them.

- **Involvement Measure (INVX)** is defined as a degree of engagement or participation in the discussion of a group. A degree of involvement may be estimated by how much a speaker contributes to the discourse in terms of substantive content. Contributing substantive content to discourse includes introduction of new local topics, taking up the topics introduced by others, as well as taking sides on the topics being discussed. As previously defined, a local topic is a concept or a thing or an event referred to in a conversation, and is typically introduced with a noun phrase (a name, noun or pronoun).

- **Network Centrality Measure (NCM)** is the degree to which a participant is a "center hub" of the communication within the group. In other words, someone whom most others direct their comments to as well as whose topics are most widely cited by others.

- **(Measure of) Argument Diversity (MAD)** is displayed by the speakers who deploy a broader range of arguments in conversation. This behavior is signaled by the use of more varied vocabulary, including specialized terms and citations of authoritative sources, among others. A person who uses more varied vocabulary and introduces more unique words into a conversation is considered to have a higher degree of Argument Diversity.

- **Tension Focus Measure (TFM)** is defined as the degree to which a speaker is someone at whom others direct their disagreement, or with whose topics they disagree the most. Similar to Network Centrality Measure, Tension Focus reflects the actions of other members of the group towards the speaker.
- **Social Positioning (SPM)** is a sociolinguistic behavior defined as a degree to which the speaker attempts to position oneself as central in the group by committing to some future activity and by getting others to confirm or re-affirm what the speaker stated, as well as what the speaker already believes. This behavior is reflected in the speaker's conversational moves that aim at increasing their centrality in the group.
- **Involved Topic Control Measure (ITCM)** is defined with combined measure of Topic Control <u>and</u> Involvement. This measure identifies speakers who are both highly involved as well as control topics under discussion to a greater extent than other speakers in the group.

By computing the score of each *measure*, we obtain a full ranking of participants on each sociolinguistic behavior. The measures used to compute Pursuit of Power are ITCM, CDM, TFM and SPM. Measures used to compute Influence are TCM, CDM, NCM and MAD. We shall elaborate this further in Section 5.

3   Correlations between the Measures and Human Assessment

We computed the correlations among our proposed measures of Influence and Pursuit of Power. As described before, Topic Control Measure (TCM), Cumulative Disagreement Measure (CDM), Network Centrality Measure (NCM) and Measure of Argument Diversity (MAD) are calculated for Influence. Involved Topic Control Measure (ITCM), Cumulative Disagreement Measure (CDM), Tension Focus Measure (TFM), and Social Positioning Measure (SPM) are combined to predict Pursuit of Power.

Table 1 shows the correlations between all proposed Influence measures and proposed Pursuit of Power measures. We note that Cumulative Disagreement Measure correlations are lower than the other measures for influence, pointing to evidence of it being the discriminant variable. We have observed similar correlation patterns across the sessions we have looked at. Computing the correlation against human rankings elicited using survey questionnaire provides us with evidence that indeed the proposed behaviors are measuring the correct phenomena. The correlation between rankings produced by annotated data and ranking induced by participant ratings holds quite strongly across a significant proportion of data sets in our corpus with an average of over 0.80 Cronbach's alpha.

Using this evidence of high correlations among behaviors and their measures, as well as measures against human survey ratings, we can be confident about our approach in measuring and detecting Influence and Pursuit of Power. We demonstrate our evaluation and results in section 5.

Table 1. Correlation among measures of Influence and Pursuit of Power for a sample chat dialogue

Influence	NCM	MAD	TCM	CDM
NCM	1.0			
MAD	0.86	1.0		
TCM	0.98	0.86	1.0	
CDM	0.58	0.59	0.48	1.0

Pursuit of Power	ITCM	CDM	TFM	SPM
ITCM	1.0			
CDM	0.78	1.0		
TFM	0.67	0.91	1.0	
SPM	0.7	0.88	0.67	1.0

4   Linguistic Components on the First Tier

The first tier of our system comprises a series of basic linguistic components that support models for computing sociolinguistic behaviors in the mid-level of Fig. 1. These models are derived from a corpus of online dialogues that includes chat and threaded discussions. The chat portion of the corpus, the MPC chat corpus is described in ([14], [10]), consists of over 90 hours of online chat dialogues in English and Mandarin. The threaded discussions portion of the data consists of 70 asynchronous thread discussions collected from public Wikipedia discussions in English and Mandarin. A substantial subset of the corpus was annotated using trained annotators who are native speakers of the respective language. Annotators were trained extensively so that inter-annotator agreement level was sufficiently high (0.8 or higher Krippendorf's alpha). The annotated data was used to train the linguistic components and to calibrate mappings from linguistic components to sociolinguistic behaviors that constitute the first tier of our system. Space limitations do not allow us to discuss the first tier mappings here; instead, we briefly describe the key linguistic components involved: (1) communication links between participants, (2) dialogue acts, and (3) local topics in conversation.

Communication Links
It is important and very challenging to determine automatically who speaks to whom in multiparty discourse. In our annotation process, we ask annotators to classify each utterance in the chat by marking it as either a) addressed to someone or everyone; b) a response to someone else's specific prior utterance; or c) a continuation of one's own prior utterance. Using annotated data from this layer of annotation, we can train a communication link classification module, which uses context, inter-utterance similarity and proximity of utterances as some of the features in a Naïve Bayes classifier to automatically classify utterances in one of the above-mentioned three categories. The current performance of this module is 61% accuracy as measured against annotated ground truth data.

Dialogue Acts
We have developed a hierarchy of 15 dialogue acts in order to annotate the functional aspect of an utterance in discourse. The tag set adopted is based on DAMSL ([1]) and SWBD-DAMSL ([9]), but compressed to 15 tags tuned towards dialogue pragmatics and away from more surface characteristics of utterances. A detailed description of dialogue act tags and annotation procedure has been described in a separate publication. Some dialogue acts that are note-worthy are: Disagree-Reject, Offer-Commit,

Confirmation-Request, Assertion-Opinion. Annotated data from this process is used to train a cue-phrase based dialogue act classifier adapted from Webb and Ferguson's ([18]) approach, which currently performs at 64% accuracy. Our Cumulative Disagreement Measure (CDM) is calculated using the proportion of disagreement dialogue act utterances detected for each participant by this automatic module.

Local Topics
Local topics are defined as nouns or noun phrases introduced into discourse that are subsequently mentioned again via repetition, synonym, or pronoun. Annotators were asked to mark all nouns and noun phrases of import from the discussion. We use Stanford part-of-speech tagger ([17]) to automatically detect nouns from text. Princeton's Wordnet ([12]) is consulted to identify synonyms commonly used in co-references. Since POS taggers are typically trained on well-formed text, performance of POS tagging on chat text – where grammar may be disorganized, use of abbreviations and symbols etc. may be quite frequent – would affect the accuracy of POS tagging. Our automatic local topic detection module performance is at 70% in the current system prototype.

We note here that it is not the goal of this research to develop the best computational modules such as POS taggers or the most accurately performing dialogue act or communication link classifier. In spite of the shortcomings in modules that support our index calculations, we are able to achieve very robust performance in our intended task of modelling complex social roles. This is because we base our claims of sociolinguistic behavior on repeated counts of each linguistic phenomenon over the length of entire discourse. When computational modules such as local topic detection fail, such errors are systematic, and would be replicated consistently for each participant. If the count for each participant were not fully accurate, nevertheless, the *distribution* of counts for all participants would still hold, thus giving us the desired ranking or the degree of sociolinguistic behavior for each participant.

Having multiple indices for each behavior helps us account for the error introduced from our automatic modules. If the predictions on individual indices are not always consistent, we can still combine them into a single output by using different weighting schemes, albeit with lesser confidence. In order to validate our proposed indices and measures, we analyzed their correlation with each other, both from human annotated data as well as our automatic process, as we shall discuss next.

5   Evaluation and Results

We compute the scores for each participant for all proposed measures. Although we have a full ranking of participants, both from survey ratings, human assessment as well as system output, we are only interested in participants who have the highest Influence and Pursuit of Power. This means, the top-ranking participant on both rankings should match in order evaluate system performance. In cases where the top two individuals are quite close in the survey scores, we may consider top two participants.

In order to create a ground truth of assessments of sociolinguistic behavior, we needed certain information to be captured through questionnaires or survey following each data collection session. In the MPC corpus, at the conclusion of each chat session, participants were asked to fill out a survey consisting of a series of questions about their perceptions of and reactions to conversation that had freshly participated in.

The questions were focused on eliciting responses about sociolinguistic behavior. One such question pertaining to Influence is shown in Figure 2. There are similar questions regarding other behaviors we are interested in modeling and we refer the reader to the cited paper [14, 10] for a detailed discussion of these.

> *During the discussion, some of the people talking are more influential than others. For the conversation you just took part in, please rate each of the participants in terms of how influential they seemed to you?*
> *Scale: Very Influential --- Not Influential.*

Fig. 2. Post session survey question related to Influence in the MPC chat corpus

We calculate the Influence and Pursuit of power score for all participants by taking the mean of our measures and deriving an Influencer and Pursuit of Power score for each participant. We devised a weighting scheme that reflects the evidence found from our analysis of correlations against survey ratings. So, the weighting scheme for English dialogues is:

Influencer score $= (\alpha_{TCM} * TCM) + (\alpha_{CDM} * CDM) + (\alpha_{NCM} * NCM) + (\alpha_{MAD} * MAD)$

$$\text{Where } \alpha_{TCM} > \alpha_{NCM} > \alpha_{MAD} > \alpha_{CDM}$$

PoPScore $= \alpha_{ITCM} * ITCM + \alpha_{CDM} * CDM + \alpha_{TFM} * TFM + \alpha_{SPM} * SPM$

$$\text{Where: } \alpha_{SPM} < \alpha_{ITCM} = \alpha_{TFM} < \alpha_{CDM}$$

Similar combinations are derived for Chinese chat dialogues as well. For different data types and different languages, we have learnt different weighting schemes where the sociolinguistic behaviors may be combined differently to compute scores. In essence, the higher the correlation, the greater the weight given to the measure. As expected and shown in Table 2, different correlations hold across languages – hence possibly cultures, since the participants are native speakers. Where the scores are 0, it signifies that the behavior is found to not correlate well with the other measures that comprise the phenomena being modeled; hence we do not include these behaviors while taking linear combinations.

Table 2. Weighting schemes for combining social behaviors learnt from correlation analyses

Weights	TCM	ITCM	CDM	MAD	TFM	SPM	NCM
English Influence	0.75	0	0.15	0.4	0	0	0.5
Chinese Influence	0.75	0	0.1	0.75	0	0	0.4
English PoP	0	0.1	0.6	0	0.1	0.09	0
Chinese PoP	0	0.08	0.08	0	0.84	0	0

Table 3 shows performance accuracy of our automated system in detecting the top influencers and those who pursue power in group dialogues, computed across languages and compared to a random selection baseline. We could choose another baseline, such as selecting the participant with the most turns as an Influencer, etc.; however, we see similar performance for such baselines as the random one. The unweighted model represents the first approximation, untrained option of the system.

Table 3. Performance accuracy against random baseline, with and without weighting scheme

Performance	BaseLine	Without Weight	With Weight
English Influence	17.85%	71.4%	78.5%
Chinese Influence	12.5%	69%	90%
English PoP	10%	52.6%	84.2%
Chinese PoP	5%	60%	73.3%

6   Conclusion and Perspectives

We have shown a novel, robust method for modeling social phenomena in multi-party discourse. We have combined established social science theories with computational modeling to create a two-tier approach that can detect high-level sociolinguistic phenomena such as Influence and Pursuit of Power in language with a high degree of accuracy. In future work, we have planned for a larger scale evaluation, testing index stability, and resilience to errors in automated language processing, including topic detection, coreference resolution, and dialogue act classification. Current performance of the system is based on versions of these linguistic modules, which perform at about 70% accuracy, so these need to be improved as well.

The advantage of applying a two-tier approach is that we can add or remove mid-level sociolinguistic behaviors efficiently when applying our models to different data types and languages. This would be impractical in a straightforward machine-learning approach where one can add all features to a learning algorithm to decide how features may best be combined. A machine-learning approach modeled directly on linguistic features would not be easily transferable to other data types and could prove brittle. Some measures turn out to be more predictive in a given data genre, and when applied appropriately, perform well at predicting phenomena as rated and understood by human assessors. We note that there may be some variance as to how humans perceive the concept of Influence and Pursuit of Power and rate a participant based on their intuitive notion of the concept. The fact that we have multiple indicators in the form of measures helps us overcome the potential variance in this perception.

Acknowledgement. This research was funded by the Office of the Director of National Intelligence ODNI), Intelligence Advanced Research Projects Activity (IARPA), and through the U.S. Army Research Lab. All statements of fact, opinion or conclusions contained herein are those of the authors and should not be construed as representing the official views or policies of IARPA, the ODNI or the U.S. Government.

References

1. Allen, J., Core, M.: Draft of DAMSL: Dialog Act Markup in Several Layers (1997), http://www.cs.rochester.edu/research/cisd/resources/damsl/
2. Bracewell, D.B., Tomlinson, M., Wang, H.: A Motif Approach for Identifying Pursuit of Power in Social Discourse. In: Proceedings of the 6th IEEE International Conference on Semantic Computing (ICSC 2012) (2012)
3. Broadwell, G., Stromer-Galley, J., Strzalkowski, T., Shaikh, S., Taylor, S., Liu, T., Boz, U., Elia, A., Jiao, L., Webb, N.: Modeling SocioCultural Phenomena in Discourse. Journal of Natural Language Engineering (2012)
4. Bradford, L.P.: Group development. University Associates, La Jolla (1978)
5. Ellis, D.G., Fisher, B.A.: Small group decision making: Communication and the group process. McGraw-Hill, New York (1994)
6. Freedman, M., Baron, A., Punyakanok, V., Weischedel, R.: Language Use: What it can tell us? In: Proceedings of the Association of Computational Linguistics, Portland, Oregon (2011)
7. French, J.R.P., Raven, B.: The Bases of Social Power. In: Cartwright, D. (ed.) Institute for social research. Studies in Social Power, vol. 35, ch. 9, pp. 150–167 (1959)
8. Givon, T.: Topic continuity in discourse: A quantitative cross-language study. John Benjamins, Amsterdam (1983)
9. Jurafsky, D., Shriberg, E., Biasca, D.: Switchboard SWBD-DAMSL Shallow-Discourse-Function Annotation Coders Manual (1997), http://stripe.colorado.edu/~jurafsky/manual.august1.html
10. Liu, T., Shaikh, S., Strzalkowski, T., Broadwell, A., Stromer-Galley, J., Taylor, S., Boz, U., Ren, X., Jingsi, W.: Extending the MPC corpus to Chinese and Urdu-Multi-party Multilingual Chat Corpus for Modeling Social Phenomena in Language. In: Proceedings of the Eight International Conference on Language Resources and Evaluation (LREC 2012) (2012)
11. McClelland, D.C.: Human motivation, vol. 23. Cambridge University Press (1987)
12. Miller, G.A., Beckwith, R., Fellbaum, C.D., Gross, D., Miller, K.: WordNet: An online lexical database. Int. J. Lexicograph. 3(4), 235–244 (1990)
13. Sanders, R.E., Pomerantz, A., Stromer-Galley, J.: Some ways of taking issue with another participant during a group deliberation. Paper Presented at the Annual Meeting of the National Communication Association, San Francisco, CA (2010)
14. Shaikh, S., Strzalkowski, T., Taylor, S., Webb, N.: MPC: A Multi-Party Chat Corpus for Modeling Social Phenomena in Discourse. In: Proceedings of the 7th International Conference on Language Resources and Evaluation (LREC 2010), Valletta, Malta (2010)
15. Strzalkowski, T., Broadwell, G.A., Stromer-Galley, J., Shaikh, S., Taylor, S., Webb, N.: Modeling socio-cultural phenomena in discourse. In: Proceedings of the 23rd International Conference on Computational Linguistics, Beijing, China (2010)
16. Strzalkowski, T., Shaikh, S., Liu, T., Broadwell, G.A., Stromer-Galley, J., Taylor, S., Ravishankar, V., Boz, U., Ren, X.: Modeling Leadership and Influence in Multi–party Online Discourse. In: Proceedings of the 24th International Conference on Computational Linguistics, Mumbia, India (2012)
17. Toutanova, K., Klein, D., Manning, C., Singer, Y.: Feature-Rich Part-of-Speech Tagging with a Cyclic Dependency Network. In: Proceedings of HLT-NAACL 2003, pp. 252–259 (2003)
18. Webb, N., Ferguson, M.: Automatic Extraction of Cue Phrases for Cross-Corpus Dialogue Act Classification. In: The proceedings of the 23rd International Conference on Computational Linguistics (COLING-2010), Beijing, China (2010)

The Marketcast Method
for Aggregating Prediction Market Forecasts

Pavel Atanasov[1,*], Phillip Rescober[1], Eric Stone[1], Emile Servan-Schreiber[2],
Barbara Mellers[1], Philip Tetlock[1], and Lyle Ungar[1]

[1] University of Pennsylvania, 3720 Walnut Street, Philadelphia, PA 19104
[2] Lumenogic, 48 Rue du Cherche Midi, Paris 75006, France

Abstract. We describe a hybrid forecasting method called marketcast. Marketcasts are based on bid and ask orders from prediction markets, aggregated using techniques associated with survey methods, rather than market matching algorithms. We discuss the process of conversion from market orders to probability estimates, and simple aggregation methods. The performance of marketcasts is compared to a traditional prediction market and a traditional opinion poll. Overall, marketcasts perform approximately as well as prediction markets and opinion poll methods on most questions, and performance is stable across model specifications.

Keywords: Forecasting, Prediction Markets, Aggregation.

1 Introduction

Prediction markets, also known as ideas futures, have been shown to produce accurate forecasts for political and sports events [1]. Prediction markets serve two separable functions: elicitation and aggregation of individual judgments. We show that separating these functions is possible and practical. Namely, forecasts elicited through prediction markets can be aggregated using non-market mechanisms, producing what we call marketcasts. Marketcasts perform well even in their simplest forms. They can exploit information beyond the current price, for example using bids when no trades occur. As we demonstrate, marketcasts can also be incorporated into more sophisticated statistical algorithms including unequal weighting of forecasters, temporal smoothing and transformation, which have been shown to improve accuracy of forecasts elicited through opinion polls [2].

---

* This research was supported by a research contract to the University of Pennsylvania and the University of California-Berkeley from the Intelligence Advanced Research Projects Activity (IARPA) via the Department of Interior/ National Business Center contract number D11PC20061. The U.S. Government is authorized to reproduce and distribute reprints for Government purposes notwithstanding any copyright annotation thereon. Disclaimer: The views and conclusions expressed herein are those of the authors and should not be interpreted as necessarily representing the official policies or endorsements, either expressed or implied, of IARPA, DoI/NBC, or the U.S. Government. Please address any correspondence to apav@sas.upenn.edu.

A.M. Greenberg, W.G. Kennedy, and N.D. Bos (Eds.): SBP 2013, LNCS 7812, pp. 28–37, 2013.
© Springer-Verlag Berlin Heidelberg 2013

If the marketcast method demonstrates adequate performance, it could serve at least three valuable functions. First, the method could produce forecasts that are robust to market manipulation, a property that is especially beneficial in small, illiquid prediction markets where manipulation by a single individual could have long-lasting effects [3]. More generally, marketcasts could take advantage of all available data in the market rather than the latest matched orders. Examples of unused data include unmatched orders, heterogeneity of forecasting skills, expertise and risk preferences. To the extent such patterns persist over time, statistical methods could take advantage of them to produce better calibrated forecasts. Second, the marketcast method can provide a bridge between elicitation platforms and could be applied in organizations that use multiple platforms and need to aggregate individual-level forecasts across these. Third, the marketcast allows analysts to measure individual forecasting accuracy independently from individual earnings. This makes it possible to distinguish forecasters who gain advantage by placing accurate forecasts from those who simply exploit temporary market inefficiencies.

2   Prediction Markets vs. Survey Forecasts

Prediction markets offer one method for eliciting and aggregating crowd beliefs about uncertain events. An alternative method is to simply ask forecasters about their subjective probability of uncertain events, and average these values. Probably the most popular example of successful opinion pooling, albeit not of probability forecasts, was described by Galton [4], who showed that the median crowd estimate of an ox's weight was within 9 pounds (0.8%) of the correct answer. This method is known as opinion polling and the resulting values are referred to as survey forecasts.

2.1   Elicitation

The prediction market interface, in its various forms, has several useful features for eliciting probability forecasts. First, markets offer incentives, financial or otherwise, that encourage forecasters to learn new information about specific questions and communicate it by placing orders on the markets. Second, order size is a measure of how confident participants are in their beliefs, as measured by the size of the bet they place. Third, the prediction market interface provides information about crowd beliefs. Fourth, participation in prediction markets is a form of gambling and may lead to self-selection of participants who enjoy making such bets. Participants also face market selection, as consistent low-performers lose money and, unless they continue injecting funds, may lose liquidity and influence over market prices. In contrast, successful traders may gain influence if they choose to reinvest their winnings in future bets.

Opinion polls share some of the useful elicitation features of prediction markets. They may offer feedback about crowd beliefs (e.g. the mean of outstanding forecasts) and provide performance feedback using metrics such as Brier scores. Expertise self-ratings could help distinguish between the more and less knowledgeable forecasters [2]. Forecasts from prior low-performers could be removed or down-weighted in the analysis stage.

2.2   Aggregation

While elicitation methods influence who expresses beliefs and how these beliefs are expressed, aggregation methods deal with the problem of merging crowd beliefs into a single forecast. Prediction markets usually solve this problem by matching bid and ask orders to produce a market price. In the continuous double auction (CDA) used in this study, buyers place bid and ask orders, specifying desired price and volume. Other market-based mechanisms for scoring and aggregation of forecasts include pari-mutuel betting, dynamic pari-mutuel [5] and Robin Hanson's Market Scoring Rule [6]. A trade occurs if the bid price is higher than or equal to the ask price. Typically, the forecast is the last price at a certain time, although markets also provide related metrics such as typical (modal) and average (mean) price over the course of a day. In markets with few active participants, bid-ask spreads are often large and no trades occur for long periods. In a CDA market, a trade occurs if the bid price is higher than or equal to the ask price.

Market pricing is not the only way to aggregate beliefs among forecasts. An alternative method we propose and test in this study is to treat order prices as survey forecasts. Such marketcasts, as we call them, can potentially overcome many of the limitations of thin prediction markets. Table 1 shows the possibilities of eliciting and aggregating information from prediction markets and opinion polls.

In opinion polls, forecasters are asked the question: "What is the probability that event X will occur by date Y?" In addition to stating their probabilistic beliefs, forecasters often state their perceived expertise in the question. Forecasts could be updated until the day the question is resolved. Individual forecasts could then be aggregated using techniques of varying sophistication. The simplest method takes the mean or median of the most recent forecast by each participant.

Such forecasts could be imported to prediction markets using trading agent algorithms that translate probabilistic beliefs into prediction market orders. A widely used class of algorithms for this purpose is known as Zero-Intelligence-Plus (ZIP), and could be applied to Continuous Double Auction markets [7]. The method results in stable prices that approach "true" values in simulated prediction markets [8].

Table 1. Possible combinations between elicitation and aggregation methods

Elicitation	Aggregation	
	PM	Survey
PM	Core Prediction Market	Marketcast
Survey	Trading Agent	Survey Forecasts

Consider a binary prediction market in which a share pays $1 if an event occurs and $0 if it does not. If a participant submits a bid order at $0.60, a simple marketcast algorithm would impute the probability forecast of 60%. If two other traders submit orders at $0.70 and $0.74, the unweighted mean probability from these forecasts would be 68%, and the median, 70%. In contrast, prediction market orders are matched in the market and statistical processing is not necessary for aggregation.

3   Aggregation Parameters

Because aggregation of survey forecasts is performed after the fact, researchers face some important choices in the process of converting individual orders to probability estimates and aggregating these into a single forecast. We discuss the influence of seven aggregation parameters below.

Order Size. In its simplest form, marketcasts ignore considerable information about the orders and interactions among forecasters. For example, each order is weighted equally, independently of its size. Such simplification would be optimal only if large orders are just as informative as small orders. If the order quantity does provide useful information, larger orders should be given more weight in the aggregation phase. In a sensitivity analysis, we weight each order by the square root of the number of shares ordered. This weighting scheme is consistent with the intuition that large orders have more information value than small ones, but the value does not increase linearly with order size. A buy order of 100 shares at a given price, for example, was given ten times the weight of a one-share order at the same price.

An alternative interpretation of order quantity is that it represents the forecaster's view that buying or selling shares at the order price would bring a large profit margin. This intuition is shared by Wolfers and Zitzewitz [9] who model the desired number of shares in a given market as a function of difference between the forecaster's personal probability estimate and the market price at this time.

Bid vs. Ask Orders. The naïve marketcast method is insensitive to the distinction between bid and ask orders: all orders are taken at face value. It is possible that market participants act with a profit margin in mind. For example, a trader with a desired profit margin of $0.10 (10%) would submit a bid order at $0.60 if she believes that the probability of an event occurring is 70%, and pre-sell shares at $0.60 if she believes the event is 50% likely to occur. Sensitivity analyses with profit margins of 0%, 10% and 25% were performed to determine which of these most closely approximates the link between participant beliefs and outcomes. After adding or subtracting the assumed profit margins, the imputed probability values were forced to the [0.01, 0.99] probability range.

Order Matching. The naïve marketcast method ignores the distinction between matched and unfilled orders. In other words, each order is treated as a signal of belief, even if it is far from the consensus and is never matched. In practice, forecasters are discouraged from placing such orders because they limit the funds available for trading on other questions. A sensitivity analysis focuses on the sub-sample of matched orders, ignoring all orders that remain unmatched at the time of aggregation. A lower Brier score for this sub-sample would imply that unmatched orders provide more noise than signal in the aggregate.

Temporal Smoothing. Prediction markets and opinion polling aggregation methods deal with "stale information" in different ways. In prediction markets, orders are retained on the order book until they are canceled or executed. Unmatched orders do not affect the most recent price directly but may do so indirectly by influencing trader behavior. On the other hand, orders at prices close to consensus are quickly matched by new or existing orders, and are unlikely to stay on the order book very long. Survey forecasts lack this feature, so temporal smoothing is often used to limit the influence of old forecasts without ignoring them altogether. Exponential decay is a popular approach, in which forecasts are multiplied by a constant between zero and 1 for each day since they were refreshed. For example, if the exponential decay constant is set to 0.5, today's forecasts receive a weight of 1, yesterday's forecasts are given a weight of 0.5, two-days-old forecasts receive a weight of 0.25, and so forth. An alternative method is to retain only the most recent forecasts, while tossing out older ones. Our core method retains 15% of the most recent forecasts, a proportion that has proven efficient in opinion poll aggregation.

Central Tendency. We report the marketcast mean as our core measure of central tendency. Median, the measure advocated by Francis Galton, is influenced less by outliers and may perform better than the mean if forecasts far from the consensus are misinformed. Finally, we use the geometric mean of probability forecasts in the log-odds space. As Satopaa et al. [10] document, the logit aggregator is the maximum likelihood estimator of true probability and is computationally simple to implement.

Transformation. In its current use, transformation, also known as signal amplification, addresses the problem of miscalibration. For example, political prediction markets have been shown to exhibit long-shot bias: low probability events are overvalued, while high probability events are undervalued [11]. In practical terms, this means that aggregate predictions are less extreme than they should be. Extremizing forecasts improves accuracy in U.S. Presidential Elections prediction markets [12], as well as opinion polling-based forecasts of international events [13]. In the sensitivity analyses below, forecasts were extremized in the manner described by Baron et al., with the constant set to 2. For example, a 40% forecast will be transformed to 31%, while a 70% forecast will be transformed to 84%.

Expertise Weights. When placing market orders, participants in the prediction market are asked to provide their self-assessments expertise in the domain of the question they bid on. More specifically, they are asked to provide an estimate of their relative expertise compared to other forecasters on a five-point scale. If participants hold accurate beliefs about their relative competence, this information could be used to improve aggregate performance by placing higher weight on more competent forecasters and lower weight on their less knowledgeable counterparts.

4   Methods and Data

The study is conducted as part of a large ongoing forecasting tournament sponsored by the Intelligence Advanced Research Projects Activity (IARPA). Five teams, including ours, participate in the tournament. The main goal of this tournament is to develop innovative methods of assigning accurate probability estimates to events of national security interest. Each month, eight to ten new questions are added to the tournament, for a total of approximately 120 questions per year. While the teams are asked to suggest forecasting questions, an external party makes the final decision for inclusion in the tournament.

The current version focuses on forty-five binary (yes/no) questions that have resolved since the beginning of the 2012-2013 tournament year. Approximately seventy questions are expected to resolve by the end of March 2013, which may alter the current pattern of results.

Each question included in the tournament (e.g., "Will Victor Ponta resign or vacate the office of Prime Minister of Romania before 1 November 2012?") must satisfy the 10/90 rule: at the moment a question is posed, a hypothetical knowledgeable observers should not place probability estimates outside the range between 10% and 90%. In other words, questions with seemingly obvious answers are not included in the tournament.

Prediction market participants compete in a Continuous Double Auction market. Shares prices resolve to $0 if the event did not occur and $1 if the event occurred in the defined timeline. Dollar values represent play money so there no financial incentives for performance are provided. Participants, however, are given frequent feedback and face social incentives, including a leader-board for the top 20% participants in terms of total earnings. Financial incentives have been shown to exert minimal influence on prediction accuracy [14]. Forecasters are free to choose which questions to bid on, but are asked to submit at least one order on at least 30 questions over the course of the year, out of approximately 120 possible questions. Two markets are run in parallel for all questions. Forecasters are randomly assigned to one of two parallel prediction market conditions. In the first one, they receive basic training on prediction markets. In the second condition, participants receive an additional one hour of training on forecasting and probability reasoning. Mellers et al. (2012) show that such training improves performance [2].

The Brier scoring rule is used to assess forecast accuracy [15]. According to this strictly proper rule, the penalty is the squared difference between the forecast value and the outcome (0 and 1), summed over the two answer options (e.g. yes/no). The best score is 0, the worst score is 2, and with binary questions, a probability forecast of 50% always results in a Brier score of 0.5.

$$BS = \frac{1}{N} \sum_{i=1}^{N} BS_i$$

$$= \frac{1}{N} \sum_{i=1}^{N} \left(\frac{\sum_{k=1}^{D_i} \sum_{j=1}^{2} (f_{ijk} - x_{ij})^2}{D_i} \right)$$

where i refers to a question (out of N total questions), $j (= 1, 2)$ refers to an outcome, k refers to a specific day, D_i is the number of days that question i is open, and x_{ij} equals 1 if outcome j for question i occurs and 0 otherwise.

Note that in this variation of the Brier score, numeric values are exactly twice as large as the values in another commonly used version in which scores vary between 0 and 1 and at-chance performance is 0.25. Daily Brier scores are averaged over the period for which a question is open. Each question is equally weighted in the determination of the aggregate score.

We report Brier scores for three conditions. First, unweighted linear opinion poll (ULinOp) is used as a baseline condition in the tournament. The method takes a simple mean of the latest survey forecast for each participant for each question. Participants in this condition undergo no special training and receive no crowd feedback. Moreover, no temporal smoothing, weighting or transformation is applied to individual or aggregate forecasts. Second, the prediction market condition features both the PM interface and the CDA order-matching algorithm. Finally, marketcast uses values elicited through the prediction market but pooled using survey forecast aggregation methods. Sensitivity analyses demonstrate the impact of various marketcast specifications.

5 Results

In total, 524 participants submitted at least one order for at least one of the 45 binary questions they faced. On average, 132 individuals submitted at least one order on any given question, resulting in 357 orders over the course of a typical question. Questions were open for an average of 100 days, with approximately 5.94 unique orders submitted per day per question. The first day after a question opened attracted the most activity, and the number of orders usually stabilized after the first three to five days of trading.

Table 2 shows the mean Brier scores for the four conditions of interest: ULinOp, core prediction markets and various marketcast specifications. For ease of presentation, we start with a core marketcast condition and show the impact of varying settings, one change at a time. In the sensitivity analysis portion of the table, the core specification is repeated in the left-most column of every row. Standard deviations are shown in parentheses. We performed a series of paired t-tests to determine if the distributions of Brier Scores for marketcasts were significantly different from the core marketcast. All p-values are for the comparison between core marketcast specification and other methods. No Bonferroni adjustment for multiple comparisons was used, increasing the likelihood that some significant differences may have occurred by chance alone.

Overall, marketcast performance varied in a limited range. On the one hand, almost all marketcast methods yielded lower Brier scores than the ULinOp control condition. On the other hand, most marketcast specification produced slightly higher Brier scores than the core prediction market, while one specification resulted in a better score.

Sensitivity analyses, reported in the lower half of Table 2, revealed several notable patterns. First, ignoring order size yielded slightly lower Brier scores

Table 2. Mean Brier scores for 45 questions in the tournament

	Mean Brier Score		
ULinOp (Control)	0.368 (0.254)*		
Core Prediction Market	0.287 (0.374)		
Core Marketcast: Equal Weights, 10% Margin, All Orders, Most Recent 15%, Mean, Non-transformed	0.299 (0.368)		
	Marketcast Sensitivity Analyses		
1. Order Volume Weight	**Equal Weights** 0.299 (0.368)	**SqRt Weights** 0.301 (0.368)	
2. Profit margin (m)	**m=10%** 0.299 (0.368)	**m=0%** 0.302 (0.360)	**m=25%** 0.306 (0.359)
3. Order matching	**All Orders** 0.299 (0.368)	**Matched Orders** 0.296 (0.368) *	
4. Temporal smoothing	**Most Recent 15%** 0.299 (0.368)	**c=0.10** 0.308 (0.378)	**c=0.50** 0.300 (0.366) **c=0.85** 0.315 (0.347)
5. Measure of central tendency	**Mean** 0.299 (0.368)	**Median** 0.302 (0.386)	**Logit** 0.303 (0.410)
6. Transformation	**Non-Transformed** 0.299 (0.368)	**Transformed** 0.340 (0.508)*	
7. Expertise Weights	**Equal Weights** 0.299 (0.368)	**Expertise Weights** 0.295 (0.368)*	

* Denotes significant difference compared to core marketcast, using a two-tailed matched samples t test.

than weighting orders by the square root of order size, which implies that order size did not provide useful information. Second, larger profit margins improved forecast accuracy, a result consistent with the intuition that bid and ask orders of the same price reflect different beliefs. Third, marketcasts based on matched orders performed slightly better than those using all orders, which suggests that unmatched orders did not provide useful information.

Fourth, temporal smoothing parameters had a small impact on overall performance. A moderate level of exponential smoothing (c=0.50) yielded Brier scores on par with the core specification, which used the latest 15% of forecasts in place of exponential smoothing. Fifth, taking the mean marketcasts yielded slightly, but not significantly, lower Brier scores than either the median or the geometric mean of log odds (logit). However, the logit aggregator yielded lowest Brier score when the profit margin was set to zero. Finally, non-transformed marketcasts performed better than extremized ones, which suggested that marketcasts in this sample did not exhibit the long-shot bias.

In addition to the manual sensitivity analyses, we performed an optimization run using elastic net regularization (Zou & Hastie, 2005), in order to extract the optimal, Brier-score minimizing specification for aggregation parameters, while avoiding overfitting [16]. The optimal specification included only matched orders, used the logit aggregator with no transformation, made use of only the 15% most recent orders at a time, and gave higher weights to participants who provide higher self-rating of expertise and tend to submit more market orders. The mean Brier score for this combination was 0.277, slightly but not significantly better than the core prediction market, which yielded an average score of 0.287.

Figure 1 depicts performance of various methods by question in increasing order of Brier scores for the core marketcast specification. In other words, questions on the left side (1, 2, 3) were correctly forecasted by the marketcast, while those on the right side resolved in unexpected ways: Brier scores above 0.5 mean that forecasts were, on average, on the wrong side of 50%. Marketcast performance tracked core prediction very closely, and both methods are visibly more accurate than the ULinOp control condition for the majority of questions.

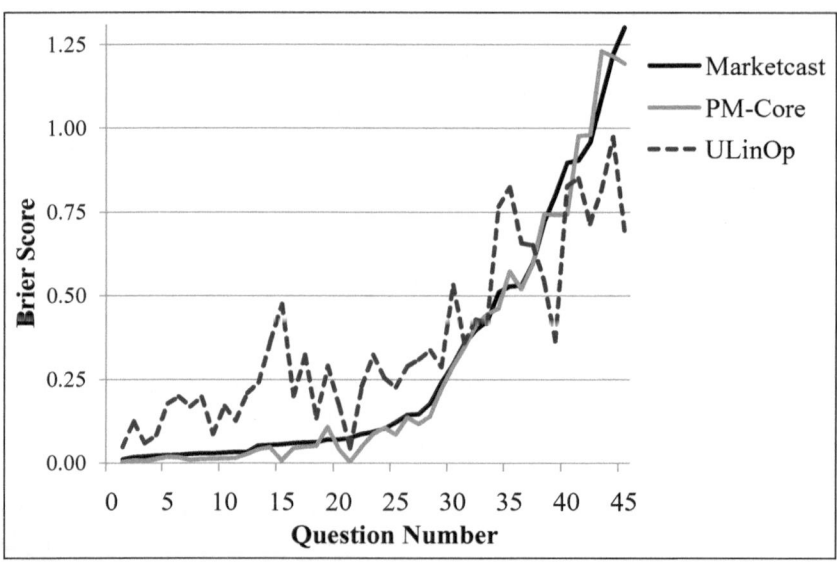

Fig. 1. Brier scores per IFP for ULinOp, core prediction market and marketcast

6   Conclusion

Marketcast analyses show that forecasts elicited by prediction markets perform well when used as inputs to non-PM aggregation algorithms. In other words, elicitation and aggregation elements are separable in principle and in practice. Some marketcast specifications slightly underperform traditional prediction markets, while the best specification produces forecasts that are 3% more accurate in terms of Brier score. Overall, the method produces stable and accurate forecasts conditions and specifications. Future research should examine the stability of the current results and illustrate novel applications of this promising method.

References

1. Wolfers, J., Zitzewitz, E.: Prediction Markets. J. Econ. Perspect. 18, 107–126 (2004)
2. Mellers, B.A., Ungar, L.H., Baron, J., Ramos, J., Gurcay, B., Fincher, K., Scott, S., Moore, D., Atanasov, P., Swift, S., Tetlock, P.E.: Improving Geopolitical Forecasting with Teamwork, Training and Algorithms (manuscript under review)
3. Christiansen, J.D.: Prediction Markets: Practical Experiments in Small Markets and Behaviours Observed. J. of Prediction Markets. 1, 17–41 (2007)
4. Galton, F.: Vox Populi. Nature 75, 450–451 (1907)
5. Pennock, D.M.: A Dynamic Pari-Mutuel Market for Hedging, Wagering, and Information Aggregation. In: EC 2004 Proceedings of the 5th ACM Conference on Electronic Commerce, pp. 170–179. ACM, New York (2004)
6. Hanson, R.: Combinatorial Information Market Design. Inform. Syst. Front. 5, 107–119 (2003)
7. Cliff, D., Bruten, J.: Zero Not Enough: On The Lower Limit of Agent Intelligence For Continuous Double Auction Markets. HP Laboratories Technical Report HPL (1997)
8. Othman, A.: Zero-intelligence agents in prediction markets. In: Proceedings of the 7th International Joint Conference on Autonomous Agents and Multiagent Systems, vol. 2, pp. 879–886. International Foundation for Autonomous Agents and Multiagent Systems (2008)
9. Wolfers, J., Zitzewitz, E.: Interpreting prediction market prices as probabilities (No. w12200). National Bureau of Economic Research (2006)
10. Satopaa, V.A., Baron, J., Foster, D.P., Mellers, B.A., Tetlock, P.E., Ungar, L.H.: Combining Multiple Probability Predictions Using a Simple Logit Model (Manuscript under review)
11. Page, L., Clemen, R.T.: Do Prediction Markets Produce Well Calibrated Probability Forecasts? Econ. J. (2012)
12. Rothschild, D.: Forecasting Elections: Comparing Prediction Markets, Polls, and Their Biases. Public Opinion Q. 73, 895–916 (2009)
13. Baron, J., Ungar, L.H., Mellers, B.A., Tetlock, P.E.: Two Reasons To Make Aggregated Probability Forecasts More Extreme (manuscript under review)
14. Servan-Schreiber, E., Wolfers, J., Pennock, D., Galebach, B.: Prediction Markets: Does Money Matter? Electronic Markets, 243–251 (2004)
15. Brier, G.W.: Verification of forecasts expressed in terms of probability. Monthly weather review 78, 1–3 (1950)
16. Zou, H., Hastie, T.: Regularization and variable selection via the elastic net. Journal of the Royal Statistical Society: Series B (Statistical Methodology) 67(2), 301–320 (2005)

Peer Nominations and Its Relation
to Interactions in a Computer Game

Juan F. Mancilla-Caceres, Eyal Amir, and Dorothy Espelage

University of Illinois at Urbana-Champaign, Urbana IL 61801, USA
{mancill1,eyal,espelage}@illinois.edu

Abstract. Peer nomination has been one of the main tools used by social scientists to study the structure of social networks. Traditionally, the
nominations are collected by asking participants to select a fixed number
of peers, which in turn are all considered for the analysis with the same
strength. In this paper, we explore several different ways of measuring
the popularity of peers by taking into consideration not only the nominations themselves but their order and total quantity in the context of a
computer social game. Using these different metrics, we explore the relationship between the nominations and the players' interactions through
text messages while playing the game. Although all five proposed metrics
can be used to find popular individuals among peers, they allow scientists
to measure different characteristics of the individuals as shown by the
correlations found between popularity scores and interaction variables.

Keywords: Group interaction and collaboration, Influence process and
recognition, Methodological innovation.

1 Introduction

Analyzing peer relationships in social networks has been a popular method for
explaining behaviors such as smoking [6], and bullying [1], among others. Traditionally, peer nominations are obtained by asking the participants to nominate
a fixed number of peers to which they are highly connected (e.g., friends, most
liked people, etc.). In this work, we explore a different way to collect peer nominations by taking advantage of current technology, and show the implications of
nominations in computer mediated communication (CMC). We explore the possible meanings in terms of popularity of each nomination by considering both the
total number of nominations made by each participant and the order in which
they are made, and explore how they are related to their general and pairwise
interactions.

Our work relies on a method to collect behavioral information from interactions between participants in a non-intrusive way. This is different from previous
approaches (see [4] and [5]) in which the actual peer interactions are hidden from
the researchers and reliance on self- or peer-reports is necessary. The method
increases the replicability of the experiments while allowing a more in-depth
analysis of the behavior of the participants, which is closer to reality.

A.M. Greenberg, W.G. Kennedy, and N.D. Bos (Eds.): SBP 2013, LNCS 7812, pp. 38–47, 2013.
© Springer-Verlag Berlin Heidelberg 2013

The nominations and interactions between participants are collected through a computer game used as a data collection tool [9]. The collected information is a network of nominations that can be considered an indirect observation of the actual social network of the participants, while the interactions consist of text messages that provide information about the relationships between participants.

Our results support a new way of obtaining peer nominations by exploring their implications when paired with CMC, and show that the order and total amount of nominations can be used to have a more detailed analysis of peer interactions and CMC. Another contribution of our work is that our results show, contrary to previous results [3], that aggressive individuals are not always rejected in peer nominations and that participants display both prosocial and aggressive behaviors, even toward peers that have been highly nominated. This suggests that there is value in interacting with aggressive individuals depending on the context (e.g., maybe they are stronger, better leaders, or more intelligent) and provides a justification for being aggressive while still being socially successful within the network. This is explained by Resource Control Theory [7] which suggests that aggressive individuals use both prosocial and coercive strategies in order to avoid the negative consequences of their aggressiveness while still being able to exploit some situations.

2   Related Work

Popularity in social networks (particularly in the context of schools) has been recently explored by Bramoulle and Rogers [2]. In their work, the authors explore the concept of popularity and its relationhip with diversity of friends using the degree of nominations received. In particular, they conclude that individuals with a higher degree (i.e., higher number of connections in the network) have a more diverse set of friends. In our work, we are not interested in exploring the diversity of the connections in the network, but in exploring the significance of the degree of nodes in order to account for different phenomena. One important aspect of our methodology that differs from previous work is that we store the order in which nominations happen and allow individuals to nominate as many people as they want.

In another study, Kerestes and Milanovic [8] studied the relationship between peer nominations and aggressive behavior. In their work, they relate peer rejection and acceptance (i.e., popularity) and find that it correlates with aggression in both males and females in grades 4th to 6th. One of the main differences with our study is that we do not focus explicitly on aggression but we study all kinds of interactions that are shown in the CMC, aggression just being one of them. To measure aggression, Kerestes and Milanovic used a scale that measures the behavior of individuals when angry and is obtained through a survey filled by the members of the network. This is different from our approach as we collect behavioral information from the actual interactions and not from reported values.

3 Method

3.1 Participants

Ninety-three students from six different 5th grade classrooms at two different Midwestern middle-schools took part in our study. Forty eight participants were male, their age ranging from 10 to 12 years ($\mu = 11$, $s = 0.62$). The specific choice of focusing on 5th graders was made with the advise of an Educational Psychologist with the purpose of obtaining engagement in the game while still showing complex social interactions.

3.2 Procedure

Participants completed eight different surveys designed to measure social behaviors including: willingness to intervene (when observing someone being aggressive against somebody else), unsafety at school, caring behaviors towards others, relational aggressiveness, how often they take part in name calling and social exclusion (i.e., verbal and social bullying), engagement in physical fights, how much they feel that they are victims of aggression in the school, and inclination towards dominating and influencing others. These surveys have been previously evaluated and used in the context of bullying detection[1].

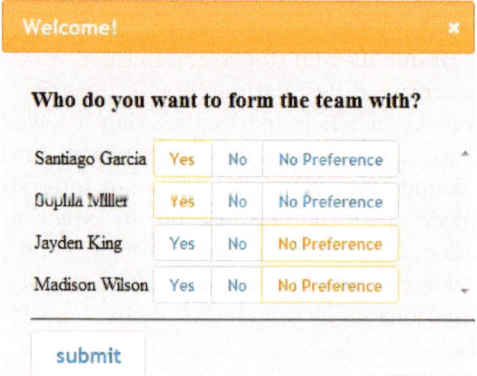

Fig. 1. Nomination phase of the game. Each player is allowed to select as many others as they want either positively, negatively or neutral. Their nominations are used as observations of popularity and possible friendship. Names on the image are fictitious.

After completing the surveys, participants took part in a computer social game designed as a data collection tool for players' interactions [9][10]. The game was played inside a computer lab and consisted in answering trivia questions while the players were allowed to communicate freely with each other through either a public or a private chat channel. Participants also interacted by trading in-game currency and either cooperated or competed for winning the game. All identities of the players were known throughout the game and all interactions were recorded for later analysis. The game started by showing the classroom roster and asking for nominations for team membership. Players selected with whom they would like to form a team by selecting *yes, no,* or *no preference* at the side of each name. Fig. 1 shows a screenshot of the interface.

After the nomination phase, participants played a trivia game that includes rules that promote collaboration and competition at different times. Players are allowed to communicate through a private chat channel (i.e., the messages would only be visible to the sender and the recipient) and through a public one (i.e., all members of the team would see the message). Most often, the private channel was used for personal communication about topics unrelated to the game, whereas the public channel was mostly used for game related topics (e.g., agreeing on an answer, discussing strategies, etc.).

In this study, we explored the relationship between peer nominations and text messages (and the results of the surveys collected before playing the game). We used an open ended coding framework and had each available text message classified by two independent raters who assigned one type and up to two descriptors (adjectives) to each message. The types and descriptors used are shown in Table 1. The inter-rater reliability was $\kappa = 0.61$ and it was calculated using the complete dataset with Cohen's kappa.

Table 1. Types and Descriptors used by the raters. Each message was assigned one type and up to two descriptors.

Types		Descriptors (Adjectives)		
1. Greeting	7. Accusation	1. Friendly	11. Cooperative	21. Rude
2. Question	8. Threat	2. Polite	12. Confirming	22. Frustrated
3. Request	9. Insult	3. Happy	13. Calmed	23. Defensive
4. Response	10. Statement	4. Helpful	14. Confident	24. Disagreeable
5. Offer	11. Emoticon	5. Playful	15. Emphatic	25. Bargaining
6. Command	12. Spam	6. Humorous	16. Informal	26. Argumentative
		7. Complimentary	17. Bored	27. Clarifying
		8. Encouraging	18. Concerned	28. Other
		9. Appreciative	19. Unconfident	
		10. Agreeable	20. Aggressive	

3.3 Popularity Metrics

We employed 5 different metrics of popularity based on the order and total amount of nominations, which provide different information and are defined as follows:

- *equal weights*: This method for handling nominations disregards the order and total number of nominations and assigns a weight of 1 to each nomination. Our intuition is that it represents how likable an individual is, because it indicates that the nominating player is OK to be in the same team as the nominee.

- *amount weights*: This metric normalizes the weight given to each nomination by the total number of nominations made by the nominating player, i.e., each nominated player receives a score of $1/n$ for each nomination, where n is the total number of nominations made by the nominating player. This is inspired by the intuition that if a player makes a lot of nominations, each nomination is less meaningful than if the player makes only one. This metric does not consider the order of nominations and we believe that it measures the strength of the friendship.
- *order weights*: This metric takes into account the order in which the nominations are made. It assigns a score of $1/x$ where x is the position of the nomination, so the first nominated person gets a score of 1, the second a score of $1/2$, the third $1/3$, etc. This metric emphasizes the preferred nominations.
- *combined weight*: For this metric we combine the previous two and assign a weight of $\frac{1}{nx}$ where n is the total number of nominations made by the nominating player and x is the position of the nomination. A limitation of this metric is that a player that was nominated first in a long list of nominations receives a lower score that one who was nominated first in a short list.
- *inverse log weights*: This metric also combines both the total amount of nominations and the order in which the nominations are made. A player receives a score of $\frac{1}{1+n\log x}$ where n is the total number of nominations and x the position of the nomination. This function assigns a weight of 1 to the first nomination and a very low weight to later nominations in long lists. A player with a high score with this metric is one that is highly preferred to be in the same team by the nominating player.

3.4   Statistical Analysis

The popularity metrics were used in two different analyses. In the first one, a global positive and negative score was computed for each participant by summing all the weights generated by all the nominations received using all five metrics. Positive and negative nominations were computed independently because it is our intuition that these are orthogonal variables (a participant may be highly nominated positively and highly nominated negatively at the same time). A correlational analysis was done with this score and all the scores obtained through the surveys and all the variables collected from the interactions of the participants during gameplay. The second analysis focused on the pairwise interactions of each player. For every pair of players that shared at least one nomination we performed a correlational analysis of the respective score and all the variables collected from the interactions during gameplay between the two participants. In the scenario where both players nominated each other, we analyzed both nominations independently, i.e., we explored what kind of interactions are seen when someone nominates another and when the other nominates the other one back.

4 Results

Figure 2 shows the distribution of positive and negative nominations. We observe a seemingly normal distribution of positive nominations, whereas the negative nominations show a skewed distribution, which indicates that most participants received at least few negative nominations and that few people received a large number of them. This supports our intuition that positive and negative nominations occur independently and therefore will be analyzed separately.

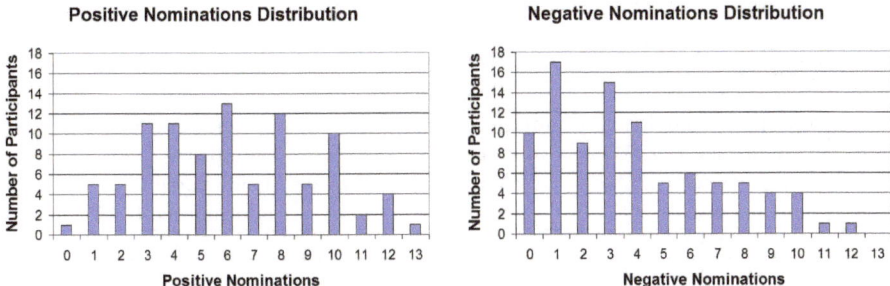

Fig. 2. Distribution of positive (left) and negative (right) nominations. Positive nominations show a normal distribution, whereas negative nomitations show a skewed distribution.

4.1 Global Analysis of Positive and Negative Nominations

The four strongest significant correlations found in the global analysis are shown in Table 2. In general, it was observed that positive nominations were highly correlated with prosocial attitudes (e.g., happy messages), particularly when using metrics that depend on the position of the nominations (*order*, *combined*, *log inverse*). Negative nominations were correlated with feelings of unsafety at the school and with victimization, which suggests that participants who received a lot of positive nominations do not feel unsafe at school, whereas those with few positive nominations tend to feel unsafe.

For positive nominations measured through the *equal weight* metric we observe positive correlations with the amount of private messages sent and received, and the amount of public messages sent. There was also a positive correlation with the score of the caring behavior survey suggesting that people that care about others tend to be nominated often (which supports our intuition that this metric measures likability of participants). This score was also positively correlated with receiving and sending messages with a prosocial tone, e.g., friendly, happy, and humorous messages. This metric was negatively correlated with the unsafety at school and relational aggression survey scores, showing again that *nice* people get a lot of nominations whereas aggressive people receive few.

Table 2. Most significant correlations between positive and negative nominations using 5 different metrics. The correlation coefficient is shown in parenthesis. The suffixes *.rec* and *.sent* stand for received and sent, respectively. * p-value < 0.05, ** $p-value < 0.01$.

	equal	amount	order
Positive nom.	private.rec (.49**)	private.sent (.39**)	private.rec (.23*)
	private.sent (.37**)	private.rec (.36**)	friendly.rec (.18**)
	caring behavior (.29**)	coins won (.29**)	encourage.rec (.13**)
	unsafe at school (-.25*)	unsafe at school (-.31**)	unsafe at school (-.27**)
	combined	log inverse	
	private.sent (.23*)	happy.rec (.14**)	
	happy.sent (.16**)	friendly.rec (.12**)	
	aggressive.sent (.15**)	encourage.rec (.12**)	
	unsafe at school (-.28**)	informal.rec (.10**)	
	equal	amount	order
Negative nom.	unsafe at school (.22*)	unsafe at school (.24*)	victim (.23*)
	happy.rec (-.22**)	victim (.21*)	happy.rec (-.11**)
	private.rec (-.25*)	coins won (-.22*)	confirm.rec (-.11**)
	coins won (-.26*)	private.rec (-.26*)	informal.rec (-.15**)
	combined	log inverse	
	victim (.23*)	victim (.22*)	
	confirm.rec (-.10**)		
	happy.rec (-.11**)		
	informal.rec (-.14**)		

When using the *amount weights* metric for positive nominations, we again observed a correlation with number of private messages sent and received, and with the number of public messages sent. We also observed a positive correlation with the number of coins obtained in the game, which suggests that when engaging in trade, participants with a high score in this metric tend to receive more coins than others. This supports our intuition that this metric measures how meaningful are the nominations, because people who were selected positively by themselves tend to have stronger ties with the players who nominated them and therefore receive more coins from them. With respect to the *order weights* metric, high positive scores are positively correlated with prosocial messages sent and received. This again suggests closeness between the players and the people that they nominated first.

The positive nominations measured by the *combined weights* are notably positively correlated with aggressive messages sent. This suggests that players who were only nominated first in short lists are either very close to the people who nominates them (and therefore can receive insults as jokes) or that such nominations are in reality not strong but very weak. This could happen in the scenario where the nominating player actually has no strong links in the classroom and simply selects a peer who "is not that bad".

When using the *inverse log weights* we only found a positive correlation with receiving some prosocial messages. The fact that this metric penalizes all nominations strongly (except the first one and the second one in short lists) means that we are only finding the kinds of interactions that are common with the first nomination that is supposed to be the most meaningful. Also, there is a correlation with the amount of informal messages sent. Informal messages use some type of slang or informal construction of sentences, e.g., *"hows it goin"*, in contrast to a more polite construction such as *"mary, how are you doing?"*. This again suggests that people with a high score in the *log inverse weights* metric are more strongly connected to the people that nominated them and therefore are closer and not required to be formal in their communication.

As mentioned above, the positive and negative nominations are independent of each other and tell us different things. In almost all metrics (with the exception of *equal weights*), having a high score in negative nominations is positively correlated with a high score in the victimization scale, which means that people that regard themselves as victims of the aggression of others do receive a high number of negative nominations. Also, negative nominations are negatively correlated with receiving prosocial messages (such as happy messages) and with receiving coins from others. This suggests that people with a lot of negative nominations are indeed targets of aggression or bullying victims.

4.2   Pairwise Analysis of Metrics and Interactions

We also explored the correlations of interacting pairs of players and the respective nomination score given by each other. Table 3 shows the results.

Most scores are correlated with prosocial behavior. The main difference lies in the order in which the variables are correlated to high popularity scores. With the exception of the *combined weight* metric, all scores that depend on the order of the nominatios (i.e., *order* and *log inverse*) emphasize friendliness, helpfulness and confidence; whereas the others (*equal* and *amount*) emphasize informality and questions sent. This suggests that, as expected, the order of the nominations are important when measuring friendliness, i.e., people who are selected first tend to receive more friendly messages and those who are selected late (regardless of

Table 3. Example of significant correlations between sent and received nominations in pairs of players using 5 different metrics. At the right of each variable the correlation coefficient is shown in parenthesis. * p-value < 0.05, ** p-value< 0.01.

equal	amount	order
question.send (.17**)	informal.rec (.15**)	friendly.send (.19**)
informal.send (.17**)	informal.rec (.13**)	informal.rec (.15**)
friendly.send (.17**)	happy.send (.13**)	confident.send (.15**)
happy.send (.15**)	question.send (.12**)	helpful.send (.14**)

combined	log inverse
appreciate.rec (.11*)	friendly.send (.17**)
	confident.send (.16**)
	humor.send (.14**)
	helpful.send (.14**)

the fact the nomination might be positive) are not really "as positive" as those who received an early nomination.

Because different classrooms may have different dynamics, the interactions among friends and popular individuals might be different. Thus, we repeated the previous analyses for each classroom separately. That is, instead of using the aggregated data of all 93 participants, we divided them by classroom (six classrooms in total). Although in general, prosocial behavior was correlated with positive nominations, in some cases we observed interesting differences such as having both prosocial and coercive (aggressiveness and frustrated) messages positively correlated with high scores in nominations. This phenomenon could be explained using Resource Control Theory, which suggests that proactive aggressors (e.g., bullies) use both prosocial and coercive strategies. These observations could possibly be instances of bullies applying such strategies through the chat channels.

5   Discussion

What we are observing in the network of nominations is not friendship, understood as a mutual dyadic relationship, but popularity, understood as the degree of acceptance by peers, related to willingness to play with the specific person. This popularity might be related to the expectation that the person will be a good addition to the team (i.e., the person is considered as someone with high problem-solving skills, but may or may not be a friend), to friendship (i.e., I don't care if the person is smart or not or if it will help me win the game, but I just want to play with him because I like him), to some feeling of aspiration (i.e., I will select the coolest guy in the classroom so I finally have a chance to interact with him and maybe earn his friendship) or simply to the fact that the person is the least bad of them all (i.e., I have no friends here but I have to choose someone so it might as well be him). This suggests that children might have a tendency to inappropriately include in their nominations others who have desirable characteristics and not only their friends. Our goal in this sense is to account for this phenomenon by comparing different metrics of popularity in our nomination network and relating it to the real interactions observed during gameplay.

Our results show that high scores in nominations are usually correlated with prosocial interactions, in terms of the tone and type of messages sent among participants. The popularity score is also correlated with psycometric scores obtained through surveys that measure how caring participants are, how victimized they feel, how unsafe they feel at school, among others. As expected, participants who score high in popularity also score high in caring behaviors; whereas those who score low (either by having few positive nominations or a lot of negative ones) tend to feel victimized and unsafe.

The metrics proposed in this paper are intended to measure different aspects of popularity, such as how likable is a participant, and how strong and meaningful are their relationships. Our results show different correlations of each metric with different types of interactions. Notably, all of them share positive correlations with prosocial messages. Our approach is applicable in cases where someone has to choose teams or report on friendships for particular tasks, which makes our method useful for peer nominations research in the social sciences and in organizational settings where the network is expected to remain unchanged during the relevant period of time.

As future work, we plan to include in the analysis the gender of the participants. Previous results [2] show that cross-gender interactions are related to popularity. Also, we plan to explore cultural differences in the way popularity and interactions occur in classrooms as suggested in [8]. We plan to explore these questions in the future by repeating our experiment in different countries.

References

1. Espelage, D., Green, H., Wasserman, S.: Statistical Analysis of Friendship Patterns and Bullying Behaviors Among Youth. New Directions for Child and Adolescent Development 118, 61–75 (2007)
2. Bramoulle, Y., Rogers, B.W.: Diversity and Popularity in Social Networks. CIR-PEE Working Paper No. 09-03 (2009) Available at SSRN: http://ssrn.com/abstract=1336634 or http://dx.doi.org/10.2139/ssrn.1336634
3. Cairns, R.B., de Cairns, B., Neckerman, H.J., Gest, S.D., Gariepy, J.L.: Social Networks and Aggressive Behavior: Peer Support or Peer Rejection? Developmental Psychology 24(6), 815–823 (1988)
4. Cillessen, A., Rose, A.: Understanding Popularity in the Peer System. American Psychological Society 14(2), 102–105 (2005)
5. Garandeau, C., Ahn, H., Rodkin, P.: The Social Status of Aggressive Students Across Contexts: The Role of Classroom Status Hierarchy, Academic Achievement, and Grade. Developmental Psychology 47(6), 1699–1710 (2011)
6. Ennet, S., Faris, R., Hipp, J., Foshee, V., Bauman, K., Andrea, H., Li, C.: Peer Smoking, Other Peer Attributes, and Adolescent Cigarette Smoking: A Social Network Analysis. Prevention Science 9(2), 88–98 (2008)
7. Hawley, P.H.: Prosocial and Coercive Configurations of Resource Control in Early Adolescence: A Case for the Well-Adapted Machiavellian. Merril-Palmer Quarterly 49(3), 279–309 (2003)
8. Kerestes, G., Milanovic, A.: Relations between different types of children's aggressive behavior and sociometric status among peers of the same and opposite gender. Scandinavian Journal of Psychology 47, 477–483 (2006)
9. Mancilla-Caceres, J.F., Pu, W., Amir, E., Espelage, D.: A computer-in-the-loop approach for detecting bullies in the classroom. In: Yang, S.J., Greenberg, A.M., Endsley, M. (eds.) SBP 2012. LNCS, vol. 7227, pp. 139–146. Springer, Heidelberg (2012)
10. Mancilla-Caceres, J.F., Pu, W., Amir, E., Espelage, D.: Identifying Bullies with a Computer Game. In: Proceedings of the 26th AAAI Conference on Artificial Intelligence, AAAI-2012 (2012)

Predicting Personality Using Novel Mobile Phone-Based Metrics

Yves-Alexandre de Montjoye[1,*], Jordi Quoidbach[2,*], Florent Robic[3,*],
and Alex (Sandy) Pentland[1]

[1] Massachusetts Institute of Technology - The Media Laboratory, Cambridge, MA
[2] Harvard University - Department of Psychology, Cambridge, MA
[3] Ecole Normale Supérieure de Lyon, Lyon, France

Abstract. The present study provides the first evidence that personality can be reliably predicted from standard mobile phone logs. Using a set of novel psychology-informed indicators that can be computed from data available to all carriers, we were able to predict users' personality with a mean accuracy across traits of 42% better than random, reaching up to 61% accuracy on a three-class problem. Given the fast growing number of mobile phone subscription and availability of phone logs to researchers, our new personality indicators open the door to exciting avenues for future research in social sciences. They potentially enable cost-effective, questionnaire-free investigation of personality-related questions at a scale never seen before.

Keywords: Personality prediction, Big Data, Big Five Personality prediction, Carrier's log, CDR.

1 Introduction

How much can one know about your personality just by looking at the way you use your phone? Determining the personality of a mobile phone user simply through standard carriers' log has became a topic of tremendous interest. Mobile cellular subscriptions have hit 6 billion throughout the world [1] and carriers have increasingly made available phone logs to researchers [2] as well as to their commercial partners [3]. If predicted correctly, mobile phones datasets could thus provide a valuable unobtrusive and cost-effective alternative to survey-based measures of personality. For example, marketing and phone companies might seek to access dispositional information about their customers to design customized offers and advertisements [4]. Appraising users dispositions through automatically collected data could also benefit the field of human-computer interface where personality has become an important factor [5]. Finally, finding ways to extract personality and, more broadly, psycho-social variables from country-scale datasets might lead to unprecedented discoveries in social sciences.

The idea of predicting people's personalities from their cellphone stems from recent advances in data collection, machine learning, and computational social

---

* These authors contributed equally to this work.

A.M. Greenberg, W.G. Kennedy, and N.D. Bos (Eds.): SBP 2013, LNCS 7812, pp. 48–55, 2013.

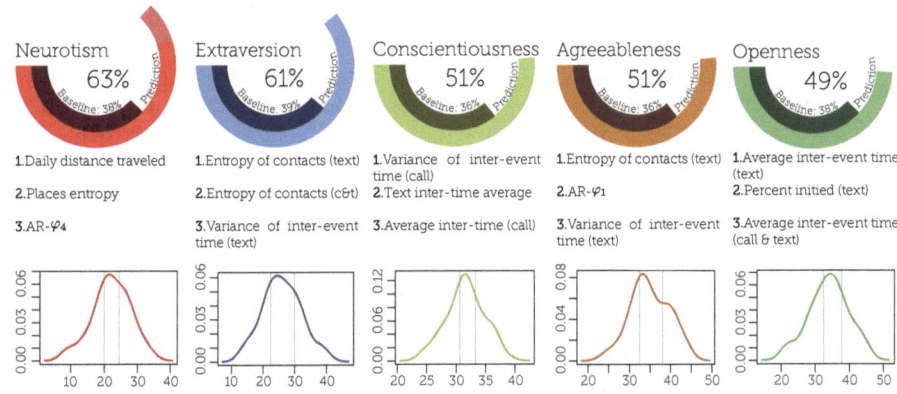

Fig. 1. (A) Accuracy of the prediction with respect to the baseline, **(B)** most useful features to predict personality traits, and **(C)** the distribution of personality traits across our dataset

science showing that it is possible to infer various psychological states and traits from the way people use everyday digital technologies. For example, some researchers have shown that pattern in the use of social media such as Facebook or Twitter can be used to predict users' personalities [6,7,8]. Others have used information about people's usage of various mobile phone applications (e.g., YouTube, Internet, Calendar, Games, etc.) or social network to draw inferences about phone owners' mood and personality traits [9,10,11,12,13]. Although these approaches are interesting, they either require to have access to extensive information about people's entire social network or people to install a specific tracking application on their phone. These constraints greatly undermine the use of such classification methods for large-scale investigations.

The goal of the present research is to show that users' personalities can be reliably inferred from basic information accessible from *all* mobile phones and to *all* service providers. Specifically, we introduce five sets of psychology-informed metrics–Basic phone use, Active user behaviors, Mobility, Regularity, and Diversity–that can be easily extracted from standard phone logs to predict how extroverted, agreeable, conscientious, open to experience, and emotionally stable a user is.

2   Results

Table 1 displays the different indicators and their respective contribution in predicting the big 5. Specifically, 36 out of our indicators were significantly related to personality and were all included in the final SVM classifier. As depicted in Figure 1, the model predicted whether phone users were low, average, or high in neuroticism, extraversion, conscientiousness, agreeableness, and openness with an accuracy of 54%, 61%, 51%, 51%, and 49%, respectively. The baselines being between 36 and 39%, we predict on average 42% better than random. For neuroticism, the predictive power of the model was further increased by including participants' gender as a predictor, increasing the accuracy to 63%. This finding is not surprising given that neuroticism is one of the traits that is most strongly associated with gender, with women having higher means levels than men in most countries world-wide [14].

An investigation of the most important feature to predict each trait revealed interesting associations. Indicators linked to users' mobility (i.e., distance traveled and entropy of places) were useful to predict Neuroticism. The entropy of participants' contacts helped predict both Extraversion and Agreeableness. These findings are inline with past research showing these traits both relate to different aspects of the diversity of one's social network: extraverts tend to seek more friends than introverts, agreeable individuals tend to be selected more as friends by other people [15]. Highly consistent with past research showing that conscientious individuals tend to like organization, precision, and punctuality [16], we found that the best predictor of Conscientiousness was the variance of the time between phone calls. Lastly, the strongest predictor of Openness was the average time between text interactions–a finding that remains be explained be future research.

3   Methodology

3.1   Participants and Procedure

The empirical sections of this work are based on a dataset collected from March 2010 to June 2011 in a major US research university [17]. Each participant was equipped with a Android smartphone running the open sensing framework *Funf* [18] While the framework is designed to collect a wide range of behavioral data from the user's phone, we voluntarily limit ourself to data available in standard carriers's logs such as phone calls, text messages sent and received, etc. These CDR (Call Data Record) have recently become widely use for computational social science research [2,19,20,21,22]. After removing participants who had less than 300 call or text per year and/or that failed to complete personality measures, our final sample was composed of 69 participants (51% male, Mean age = 30.4, S.D. = 6.1, 1 missing value).

3.2   Metrics

We developed a range of novel indicators allowing us to predict users' personality. To build our list of indicators, we examined theories and research in personality psychology and, more specifically, the literature five factor model of personality, the dominant paradigm in personality research [23]. The five-factor model is a hierarchical organization of personality traits in terms of five basic dimensions: Extraversion (i.e, the tendency to seek stimulation in the company of others, to be outgoing and energetic), Agreeableness (i.e, the tendency to be warm, compassionate, and cooperative), Conscientiousness (i.e., the tendency to show self-discipline, be organized, and aim for achievement), Neuroticism (i.e, the tendency to experience unpleasant emotions easily), and Openness (i.e, the tendency to be intellectually curious, creative, and open to feelings).

From this literature review, we generated novel indicators that can be easily computed from carriers logs and that we believed would meaningfully account for potential differences in personality (see Table 1). These indicators fall under 5 broad categories: Basic phone use (e.g., number of calls, number of texts), Active user behaviors (e.g., number of call initiated, time to answer a text), Location (radius of gyration, number of places from which calls have been made), Regularity, (e.g.,temporal calling routine, call and text inter-time), and Diversity (call entropy, number of interactions by number of contacts ratio). These indicators are detailed hereafter.

Entropy: Is a quantitative measure reflecting how many different categories there are in a given random variable, and simultaneously takes into account how evenly the basic units are distributed among those categories. For example, the entropy of one's contacts is the ratio between one's total number of contacts and the relative frequency at which one interacts with them. $H(a-c) = -\sum_c f_c \log f_c$ where c is a contact and f_c the frequency at which a communicates with c. The more one interacts equally often with a large number of contacts the higher the entropy will be. This work considers the entropy of calls, text, calls+text but also the entropy of places one visits.

Inter-event Time: Is the time elapsed between two events. This work then consider both the average and variance of the inter-event time of ones' call, text, call+text. call+text means that an interaction, a call or an text, happened between two users. Therefore, even though two users have the same inter-event time for both call and text, their mean inter-event times for call+text can be very different.

AR Coefficients: We can convert the list of all calls and texts made by a user into a time-series. We discretized time by steps of 6 hours. For example, the time-series X_t contain the number of calls made by a user between 6pm and 12am on Monday followed by the number of calls made by the same user between 12am and 6am on Tuesday and so on. We then train a *auto-regressive* model per user. This model takes the form $X_t = c + \sum_{i=1}^{p} \varphi_i X_{t-i} + \varepsilon_t$ where c is a constant and ε_t are noise terms. The coefficients φ_i can thus be interpreted as the extent to which knowing how many calls a person made in the previous 6 hours, the day before at the same time predicts the number of calls that person will make in the coming 6 hours. We only kept the coefficient that were statistically significant for at least 3 traits: $\varphi_{1,4,8,12,18,24}$. Note that while we see some patterns in the statistically significant coefficients, interpretation of such patterns requires caution given that (1) this analysis has been done post-hoc and (2) our relatively small sample size.

Response Rate and Latency (Text): We consider a text from a user (A) to be a response to a text received from another user (B) if it is sent within an hour after user A received the last text from user B. The response rate is the percentage of texts people respond to. The latency is the median time it takes people to answer a text. Note than by definition, latency will be less or equal to one hour.

Number of Places and Their Entropy: The dataset was collected using the open sensing framework *Funf* which prevent us from directly using cell phone towers. We instead empirically defined places by grouping together the GPS points of a user that are less than 50m apart and by defining their center of mass as the lat-long coordinate of the place. 50m made sense given the sampling resolution of our dataset. Finally, we only kept the places where a user spend more than 15 minutes in a row.

Radius of Gyration: This is the radius of the smallest circle that contains all the places a user have been to on a given day.

Distance Per Day: This is sum of the distance between the consecutive places a user has visited in a given day.

Home and Call Regularity: We look at regularity at which a user is coming back home (home regularity) or receiving/making a call (call regularity) using a neural coding inspired metric [24].

3.3   Personality

As part of a larger questionnaire, participant completed the Big Five Inventory (BFI-44 [25]), a 44-item self-report instrument scored one a 5-point Likert-type scale measuring the Big Five personality traits. The BFI-44 has been widely used in personality research and has been shown to have excellent psychometric properties [25]. As depicted in Figure 1, participants personality scores follow a normal distribution: Neuroticism (A = 0.3012, p = 0.5698), Openness (A − 0.2502, p = 0.7042), Extraversion (A = 0.2884, p = 0.6074), Conscientiousness (A = 0.4380, p = 0.2869), and Agreeableness (A = 0.4882, p = 0.2162).

3.4   Class Prediction

Because the relationship between personality traits and numerous behavioral and psychological factors can often be non-linear [26,27], we choose to use SVM over the more traditional GLM as the former automatically model non linear relationships. Consequently, following [28] we classified each user as low, average, or high on each on the five personality dimensions.

We then selected the most relevant features using a greedy method similar to [29]. At each iteration, features are ranked using the squared weight and the worst feature of the set is removed. We stop removing features when removing a subset of worst features of size less than 3 degrades the performance and report the 3 highest ranked features. We then classified using an SVM with a 10-fold cross validation.

4 Discussion

The present study provides the first evidence that personality can be predicted from standard carriers' mobile phone logs. Using a set of novel indicators that we developed based on personality research and that are available to virtually anyone, we were able to predict whether users were low, average or high on each of the big five from 29% to 56% better than random. These levels of accuracy were obtained while we purposefully adopted a restrictive approach only using phone logs.

To our knowledge, these predictions exceed all previous research linking psychological outcomes to mobile phone use. In particular, a previous study that used a combination of information from mobile phone logs *and* people's usage of mobile phone applications such as You-Tube, Internet, and Games predicted the personality of their owners with a mean accuracy of 15% [9]. In comparison, the mean accuracy in the present research is almost three times as high (i.e., 42%).

It is interesting to note that Extraversion and Neuroticism were the traits that were best predicted in our study. These two traits

Table 1. Metrics

Metrics	N	E	O	C	A
Regularity					
Average inter-event time (call)	•	•	•	•	•
Average inter-event time (text)			•	•	
Average inter-event time (c&t)			•	•	•
Variance of inter-event time (call)		•		•	
Variance of inter-event time (text)		•		•	•
Variance of inter-event time (c&t)	•	•		•	•
Home regularity		•			•
AR- φ_1				•	
AR- φ_4	•				•
AR- φ_8	•		•		
AR- φ_{12}				•	•
AR- φ_{24}			•	•	
Number of call regularity				•	•
Diversity					
Entropy of contacts (call)			•	•	•
Entropy of contacts (text)	•	•	•	•	•
Entropy of contacts (c&t)		•			
Contacts to interactions ratio (call)	•	•	•		•
Contacts to interactions ratio (text)	•	•		•	
Contacts to interactions ratio (c&t)	•	•	•		
Number of contacts (call)			•	•	
Number of contacts (text)		•		•	
Number of contacts (c&t)			•	•	
Spatial behavior					
Radius of gyration (daily)	•	•			
Distance traveled (daily)	•	•	•	•	•
Number of places	•	•	•	•	•
Entropy (places)	•	•	•	•	•
Active behavior					
Response rate (call)			•		
Response rate (text)	•		•	•	•
Response latency (text)			•		
Percent during the night (call)	•		•		•
Percent initiated (text)	•		•		
Percent initiated (call)	•	•	•	•	
Percent initiated (c&t)				•	
Basic Phone use					
Number of interactions (text)			•		
Number of interactions (call)	•	•	•		•
Number of interactions (c&t)	•	•		•	•

are the dimensions of personality that are the most directly associated with emotion. In particular, extraversion is a strong predictor of positive emotions and neuroticism is a strong predictor of negative emotion [30]. This raises the hypothesis that our indicators might be picking up on the emotional components associated with these two traits. It would be interesting to investigate whether our indicators can predict emotional variable such as happiness in future studies. In addition, contrasting cellphone-based vs. questionnaire-based measures of personality when predicting various psycho-social outcomes might lead to interesting asymmetries. In line with this idea, recent research in personality shows that ratings of one's personality that are made by oneself and ratings of one's personality that are made by others are both valid but different predictors of behavior. For example, self-ratings predict behaviors like arguing or remaining calmn, whereas other-ratings predict behaviors like humor and socializing [31].

Although more research is needed to validate our model and the robustness of our indicators for use on a large-scale and more diverse population, we believe that our findings open the door to exciting avenues of research in social sciences. Our personality indicators and the ability to predict personality using readily available mobile phone data may enable cost-effective, questionnaire-free investigation of personality-related questions at the scale of entire countries.

Acknowledgments. The authors would like to thanks Nadav Aharony, Wei Pan, Cody Sumter, and Bruno Lepri for sharing data.

References

1. CNET, 2011 ends with almost 6 billion mobile phone subscriptions, `http://news.cnet.com/8301-1023_3-57352095-93/2011-ends-with-almost-6-billion-mobile-phone-subscriptions/`
2. de Montjoye, Y.-A., Hidalgo, C., Verleysen, M., Blondel, V.: Unique in the Crowd: The privacy bounds of human mobility. Nature Sci. Rep. (2013)
3. CNN, Your phone company is selling your personal data, `http://money.cnn.com/2011/11/01/technology/verizon_att_sprint_tmobile_privacy/index.htm`
4. de Oliveira, R., et al.: Towards a psychographic user model from mobile phone usage. In: Proceedings of the 2011 Annual Conference Extended Abstracts on Human Factors in Computing Systems. ACM (2011)
5. Arteaga, S.M., Kudeki, M., Woodworth, A.: Combating obesity trends in teenagers through persuasive mobile technology. ACM SIGACCESS Accessibility and Computing 94, 17–25 (2009)
6. Back, M.D., et al.: Facebook profiles reflect actual personality, not self-idealization. Psychological Science 21(3), 372–374 (2010)
7. Counts, S., Stecher, K.: Self-presentation of personality during online profile creation. In: Proc. AAAI Conf. on Weblogs and Social Media (ICWSM) (2009)
8. Stecher, K., Counts, S.: Spontaneous inference of personality traits and effects on memory for online profiles. In: Proc. Int. AAAI Conference on Weblogs and Social Media (ICWSM) (2008)
9. Chittaranjan, G., Blom, J., Gatica-Perez, D.: Mining large-scale smartphone data for personality studies. In: Personal and Ubiquitous Computing (2012)

10. Do, T.M.T., Gatica-Perez, D.: By their apps you shall understand them: mining large-scale patterns of mobile phone usage. In: Proceedings of the 9th International Conference on Mobile and Ubiquitous Multimedia. ACM (2010)
11. Verkasalo, H., et al.: Analysis of users and non-users of smartphone applications. Telematics and Informatics 27(3), 242–255 (2010)
12. Staiano, J., et al.: Friends dont Lie–Inferring Personality Traits from Social Network Structure (2012)
13. Pianesi, F., et al.: Multimodal recognition of personality traits in social interactions. In: Proceedings of the 10th International Conference on Multimodal Interfaces. ACM (2008)
14. Lynn, R., Martin, T.: Gender differences in extraversion, neuroticism, and psychoticism in 37 nations. J. Soc. Psychol. 137(3), 369–373 (1997)
15. Selfhout, M., et al.: Emerging late adolescent friendship networks and Big Five personality traits: A social network approach. J. Pers. 78(2), 509–538 (2010)
16. MacCann, C., Duckworth, A.L., Roberts, R.D.: Empirical identification of the major facets of conscientiousness. Learning and Individual Differences 19(4), 451–458 (2009)
17. MIT Human Dynamics Lab, Reality Commons,
    `http://realitycommons.media.mit.edu/`
18. Aharony, N., et al.: Social fMRI: Investigating and shaping social mechanisms in the real world. In: Pervasive and Mobile Computing (2011)
19. Onnela, J.P., et al.: Structure and tie strengths in mobile communication networks. Proc. Natl. Acad. Sci. U S A 104, 7332–7336 (2007)
20. Meloni, S., et al.: Modeling human mobility responses to the large-scale spreading of infectious diseases. Nature Scientific Reports 1 (2011)
21. Balcan, D., et al.: Multiscale mobility networks and the spatial spreading of infectious diseases. Proc. Natl. Acad. Sci. USA 106, 21484–21489 (2009)
22. Gonzalez, M., Hidalgo, C., Barabasi, A.: Understanding individual human mobility patterns. Nature 453, 779–782 (2008)
23. McCrae, R.R., John, O.P.: An introduction to the fivefactor model and its applications. Journal of personality 60(2), 175–215 (1992)
24. Williams, M.J., Whitaker, R.M., Allen, S.M.: Measuring Individual Regularity in Human Visiting Patterns. In: ASE International Conference on Social Computing (2012)
25. John, O.P., Srivastava, S.: The Big Five trait taxonomy: History, measurement, and theoretical perspectives. In: Handbook of personality: Theory and Research 2, pp. 102–138 (1999)
26. Benson, M.J., Campbell, J.P.: To be, or not to be, linear: An expanded representation of personality and its relationship to leadership performance. Int. J. Select. Asses. 15(2), 232–249 (2007)
27. Cucina, J.M., Vasilopoulos, N.L.: Nonlinear personality performance relationships and the spurious moderating effects of traitedness. J. Pers. 73(1), 227–260 (2004)
28. MacCallum, R.C., et al.: On the practice of dichotomization of quantitative variables. Psychol. methods 7(1), 19 (2002)
29. Guyon, I., Weston, J., Barnhill, S., Vapnik, V.: Gene selection for cancer classification using support vector machines. Mach. Learn. 46, 389–422 (2002)
30. Gomez, A., Gomez, R.: Personality traits of the behavioural approach and inhibition systems: Associations with processing of emotional stimuli. Pers. Indiv. Differ. 32(8), 1299–1316 (2002)
31. Vazire, S.: Who knows what about a person? The self-other knowledge asymmetry (SOKA) model. J. Pers. Soc. Psychol. 98(2), 281 (2010)

Moral Values from Simple Game Play

Eunkyung Kim, Ravi Iyer, Jesse Graham, Yu-Han Chang,
and Rajiv Maheswaran

University of Southern California
Los Angeles, CA 90089
{eunkyung,raviiyer,jesse.graham,yuhan.chang,maheswar}@usc.edu

Abstract. We investigate whether a small digital trace, gathered from simple repeated matrix game play data, can reveal fundamental aspects of a person's sacred values or moral identity. We find correlations that are often counterintuitive on the surface, but are coherent upon deeper analysis. This ability to reveal information about a person's moral identity could be useful in a wide variety of settings.

1   Introduction

It is said that we leave behind a highly revealing digital trail from our myriad online behaviors. We investigate whether a small digital trace, gathered from simple repeated matrix game play data, can reveal fundamental aspects of a person's sacred values or moral identity. In particular, we conduct two studies: in the first, subjects use an online interface to play the "Social Ultimatum Game," a multi-player extension of the well-known ultimatum game [5]; in the second, subjects use a very similar interface to play a different multi-player sequential game. Further details of the games are given in Section 2. In both studies, we find small but significant effects between moral values, such as overall moral identity, Authority, and Fairness, and aspects of the game play, such as the choice of actions and the choice of who to play with in the multi-party game.

Offhand, one might conjure up various stereotypes about values and presumed game play. For example, one might expect that conservatives might be more likely to punish others for unfair actions when a society has already established a norm of fairness, since they typically believe in respect for authority and upholding traditions. However, we show that in fact, liberals' are much more likely to punish, due to their higher degree of desire for fairness. We show that some of these correlations hold across the same studies, under different populations and different game structures. While the effect sizes are not large, they are significant and show promise for further investigation of the degree to which we can extract fundamental aspects of a person's sacred values and personality through observation of simple game behaviors.

2   Background

Related Work. There is a substantial body of work relating observable behaviors to a person's innate personality [12,17,16]. When placed in exactly the same

A.M. Greenberg, W.G. Kennedy, and N.D. Bos (Eds.): SBP 2013, LNCS 7812, pp. 56–64, 2013.

setting, people usually exhibit different behaviors, often correlated with their personality traits. For example, some personality traits can be inferred from inspecting an employee's cubicle [7]. A person's personality not only affects behaviors in the real world but also in the virtual world [13]. A person's concerns, such as fairness and empathy, have also been used to explain behavior in classical game-theoretic matrix games such as Prisoner's Dilemma [18,2] and the Dictator game [3]. However, these investigations did not measure innate personality traits or moral values and relate those values to the observed game behavior; instead, they typically framed the situation [6] so that the subject would experience empathy, etc. In this paper, we specifically focus on simple repeated games, and examine the relationship between game play and moral values measured using a reliable instrument.

Moral Foundations. Moral values seem to fall within five general categories, or moral foundations [8]. Moral foundations are intuitive sensitivities to particular morally-relevant information. Table 1 shows the five moral foundations.

Table 1. Moral foundations [10]

Moral Foundations	Description
Harm/Care	A concern for caring for and protecting others
Fairness/Reciprocity	A concern for justice and fairness
In-group/Loyalty	A concern with issues of loyalty and self-sacrifice for ones in-group
Authority/Respect	A concern with issues associated with showing respect and obedience to authority
Purity/Sanctity	A concern for purity and sanctity

Based on past research that shows that empathy [15] and people's beliefs about fairness [4] relate to cooperation, we expected that both harm/care and fairness/reciprocity concerns would be significant in a repeated-trials ultimatum game task. We did not expect any relationship to exist due to any of the other three foundations. Past research suggests that political liberals place more emphasis on the harm/care and justice/reciprocity foundations relative to the other three foundations, whereas political conservatives place a relatively equal amount of emphasis on all five foundations [8]. Therefore, the different emphasis people place on these foundations can be used as a proxy for peoples political beliefs. We expected that people who place more emphasis on the harm/care and justice/reciprocity foundations ("liberals") would cooperate more than people who place an equal amount of emphasis on all five foundations ("conservatives"). The 32-item Moral Foundations Questionnaire (MFQ) [9] measures the degree to which people value each of five foundations. Research suggests that the MFQ is highly reliable and valid [9].

The Social Ultimatum Game. This multi-player extension of the classical Ultimatum game was used in the first study. The Social Ultimatum Game models the fact that people operate in societies of multiple agents and repeated pair-wise interactions. These interactions can be thought of as abstract economic transactions that result in surpluses that must be split between the two parties. In each round, we allow each player to propose one transaction with a partner of their choosing, where all transactions result in a $10 surplus that can be split. The proposer must propose a split of the $10. If the other party accepts the proposed split, then the transaction occurs, and the $10 surplus is split as proposed. If the other party rejects the proposed split, then the transaction does not occur, and neither party receives any money. Thus, in each round, a player can make one proposal, and a player can receive between zero and four proposals, and choose to accept or reject each proposal independently. Fig. 1 shows the web interface for the Social Ultimatum Game.

Fig. 1. The Social Ultimatum Game interface. The screen shown is where the player chooses the offer recipient and the offer value.

Trading Game. In the second study, we created another sequential game with a different payoff structure, but which also included notions of fair actions and punishment actions (Fig. 2). In each pair-wise interaction, there is a leader and a follower. The leader can choose either to Trade with or Steal from the follower. In response to a Trade, the follower can choose to complete a Fair or Unfair trade. In response to Stealing, the follower can choose to Punish or Forgive. The rewards intuitively follow the action labels. A leader can ensure a minimum payoff of $10 by choosing to Trade. Or, a leader can take a risk and choose to Steal, hoping that the follower will choose to Forgive, which is in the follower's own self-interest. From a purely economically-rational perspective, this is the equilibrium strategy. The risk is that the leader will receive a negative payoff if the follower decides to act against their own self-interest and Punish instead. As before, we designed a multi-player multi-round game, where in each round, each player can choose a partner and play Trade or Steal with that partner. Thus, in each round, it is possible for a player to receive no Trades or Steals, or up to four Trades or Steals, from the other players. Fig. 2 shows the web interface of the Trading game.

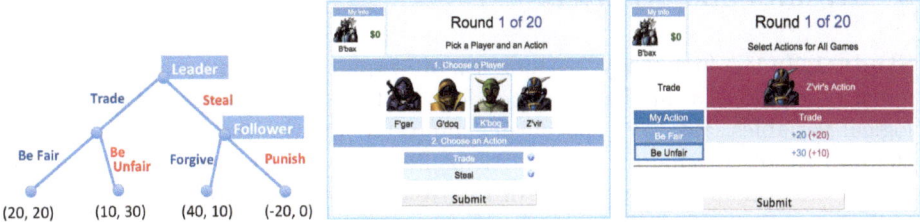

Fig. 2. (Left) Trading Game in extensive form; (Center) User interface for the Trading Game where the leader chooses who to play with, and an action to play; and (Right) Interface where the follower chooses an action

3   Experiments

We conducted the studies using Amazon Mechanical Turk. In both studies, all participants were invited to finish a compliance test first. This ensures that the experiment subjects understood our game rules and would provide useful data. In the sample game, we showed robot avatars for the agents and gave them screen names such as "Bot-1". After finishing the compliance test, we gave the user a short survey to get background information including gender, occupation, age, education and nationality. Then each participant was given the Moral Foundations Questionnaire. We looked at the timing of question answering and inserted questions with clear correct answers to ensure that the participants were filling out the questionnaire in good faith. In the first study, they would then play a series of four Social Ultimatum games, each time in a different simulated society of four other players. In the second study, they would play a sequence of two Trading games, where we again varied the behavior of the society over time.

In the first study, we focus on two of the societies encountered by the participants. In the "nice" T4T society, all the other agents played tit-for-tat, accepting all offers, and reciprocating whenever possible. In the "harsh" AF7 society, agents used an Adaptive Fairness model which was fit to data produced in an earlier study by an unusual group of subjects who made generous offers, but would not accept offers less than $7 [14].

For the second study, in the first Trading game in the sequence, the other agents in the society would play nicely 80% of the time for the first 10 rounds (playing "Trade" and responding "Fair" or "Forgive"), and not nice 80% of the time for the second 10 rounds (playing "Steal" and responding "Unfair" or "Punish"). In the second Trading game, the other agents would flip this behavior, playing mostly not nice for the first 10 rounds, and nicely the second 10 rounds.

Participants were not told they would play with artificial agents. To simulate that the participants were playing other humans, the avatars and screen names in the actual games were of the same class of that given to the player. We added randomized delays in response time adjusted to match the timings of all-human game play. We paid US$0.50 for all participants and an additional US$0.01 for each $100 earned in the games.

4   Results and Analysis

Study 1. In the first study, we took a more exploratory approach, varying the game types along a number of dimensions and measuring a variety of psychological variables. We were specifically interested in two aspects of game play, the average offer made to other participants and whether individuals chose to spread their offers to many people or to give them to specific others. Our results indicated two significant findings.

One of the most general measures of moral judgment, the Moral Identity Scale [1] interacted with the type of game played, particularly the internal subscale, which measures how central morality is to an individuals self-concept. Specifically, in games where the other players engaged in relatively harsh tactics (AF7), moral identity scores were significantly associated with lower offers ($r = -.20, p < .05$). In contrast, in T4T games, high moral identity scores were associated with higher average offers. This positive relationship was not significant ($r = .11, p = .29$), but the interaction between moral identity scores and game conditions was significant ($F = 3.96, p < .05$). Fig. 3 shows the correlation between Moral Identity Internalization scores and average offers across game conditions. Since tit-for-tat is nominally a fair strategy, this could be taken as indication that individuals who see themselves as more morally motivated reward fair behavior and punish unfair behavior.

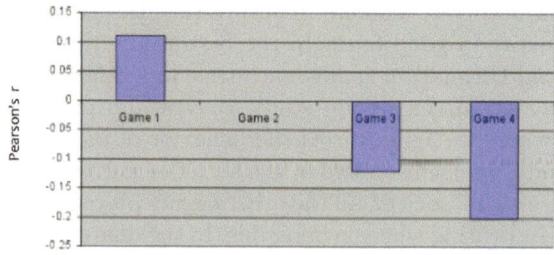

Fig. 3. Correlation of Moral Identity Scores with Average Offers across Games in Study 1: Game 1 is T4T, and Game 4 is AF7. Pearson's r values on Y axis tell us about the strength of the correlation and direction of the association. The closer the correlation value is to either +1 or -1, the stronger the correlation. A positive/negative correlation value means two variables have a positive/negative relationship. If the correlation is 0, there is no association between the two variables.

Many other variables exhibited similar patterns, with moral psychology variables relating to offers and entropy in a step-wise pattern, from "fair" to "unfair" environments though most relationships were not significant. Since our initial approach was exploratory, we did not vary the game environments specifically to create "fair" and "unfair" environments. As such, in Study 2, we sought to explicitly create more definitively "fair" vs. "unfair" game environments. As well, we noticed that variables tended to cluster in terms of the direction of the correlations, based on whether they were associated with a more liberal or conservative moral profile. For example, conservative values (see [8]) were differentially related to reciprocity and entropy between the T4T game and the AF7 game. As such, we sought to specifically examine whether these values (Ingroup Loyalty, Authority, and Purity) related to aspects of game player behavior differentially, depending on the fairness of the game environment.

We also examined several other behavior characterizations, which attempted to aggregate noisy actions over many rounds. A *window* $w(k, \tau)$ is the set of rounds $\{k, k+1, \ldots, k+\tau-1\}$. Here, we focus in windows involving all 20 rounds and the last 10 rounds. The features for these windows are (1) average offer amount, (2) total score, (3) offer value entropy, (4) offer recipient entropy and (5) reciprocity likelihood. Average offer amount looks at how much each agent offered to the chosen recipient. It is a standard metric for evaluating Ultimatum Game behavior [11]. The total score, i.e., the money made by the participant in the game, is also a standard metric in many economic games of this type [19].

The other metrics try to capture the variance in behavior over time. There may be variability in the offer amounts made by a single player over the course of a game. We introduce the notion of entropy dynamics to capture the changes in this variability over time. Offer amount entropy is a measure of the distribution of offer values over the the window considered. We normalize the standard information theoretic entropy so that the value is bounded above by 1. Similarly, offer recipient entropy is a measure of the distribution of who each player chooses as their offer recipient and is normalized.

Finally, we measure the degree to which players respond to offers by reciprocating with an offer in the next time period. A length-1 reciprocation is when a P_m chooses P_n in round k after P_n made an offer to P_m in round $k-1$, and P_m did *not* make an offer to P_n in round $k-2$. A length-2 reciprocation is when a P_m chooses P_n in round k after P_n made an offer to P_m in round $k-1$, P_m made P_n an offer in round $k-2$, and P_n did *not* make an offer to P_m in round $k-3$. A length-3 reciprocation is defined analagously. Reciprocation likelihood for a particular length is how likely a player engages in such an action given the chance.

The first result is the relationship between authority and recipient entropy. People who thought authority and respect were important tended to explore more in terms of choosing offer recipients. Table 2 shows the effect was stronger in AF7, i.e., the abnormal society, than in T4T, the society where all offers are accepted and reciprocity is high.

A second result was that people who valued in-group (loyalty and self-sacrifice to the group) also showed increased recipient entropy in AF7 (p-values of 0.1027, 0.1332). It seems reasonable that people who value authority and in-group, would find themselves searching more when facing with a society very different from their own. High values on authority and in-group also indicated a higher rate of exploration in terms of offer values with similar significance rates.

Table 2. P-values for the differences in authority value between the high/low recipient entropy clusters

		Clustered by recipient entropy	
		windows of 20	windows of 10
Society	T4T	0.1319	0.1144
	AF7	0.0040	0.0698

Interestingly, harm and fairness, that were initially hypothesized as being potentially key variables, did not seem to have a substantial effect on the measures considered. It is also interesting that for the traditional metrics, offer value and overall score, there were no big differentiators across morality dimensions, but the differences occurred in the temporal metrics.

The third and strongest result is that, again, authority and in-group values, are highly correlated with whether one is in the high and low reciprocation classes. When investigating reciprocation likelihoods, in the cases of length 1,2 and 3, higher authority and in-group values led to lower reciprocation rates. The reciprocation rates difference in the high and low groups were very large (see Table 3) and significant (p-values $\ll 0.001$). Table 4 shows p-values for the differences in authority value and in-group value between the reciprocation high/low clusters.

Table 3. Reciprocation rate difference

	Reciprocation likelihood		
	Length-1	Length-2	Length-3
High reciprocation cluster	0.4825	0.6752	0.6707
Low reciprocation cluster	0.2523	0.2064	0.2113

Table 4. P-values for the differences in authority and in-group values between the reciprocation high/low clusters

	Reciprocation likelihood		
	Length-1	Length-2	Length-3
Differences in authority value	0.0040	0.0015	0.0030
Differences in in-group value	0.0453	0.0337	0.1350

Study 2. Again, we evaluated the entropy measures described earlier, as well as some game-specific features such as: Trade Actions (%): Percentage of choosing Trade action as a proposer; Fair given Trade (%): Percentage of choosing Be Fair action for received Trade games; and Punish given Steal (%): Percentage of choosing Punish action for received Steal games. We calculated the above features for each phase: round 1-10, round 11-20, for each game.

The main result from these experiments is that "liberal" people are more likely to Punish when the leader Steals from them, in relative contrast to "conservative" people who are more likely to Forgive when the leader Steals from them. When the leader plays Steal, then the economically rational reaction by the follower is to Forgive, which provides a payoff of $10 vs. Punish, which provides $0 payoffs. However, liberal people tend to give up their own rewards in order to punish what they view as an unfair action by the leader. This can be interpreted as a tendency by liberal people to react more harshly to unfair game play. Fig. 4 shows this relationship between moral liberalness and the choice of Punish actions (p-value = 0.0218), in addition to showing the correlation between moral liberalness and fairness (p-value ≪ 0.0001).

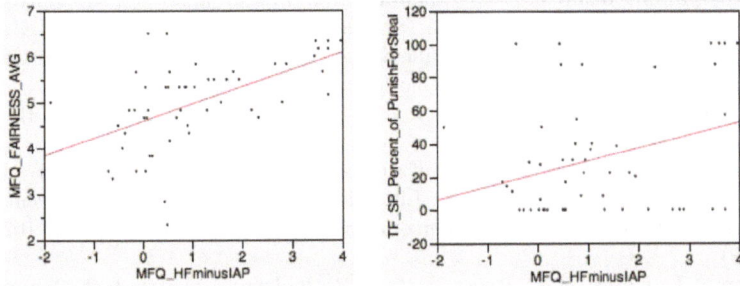

Fig. 4. Positive correlation between fairness and liberal tendency (left), punish action for steal game and liberal tendency (right)

Interestingly, the harm moral foundation, which was initially hypothesized as being a potentially key variable, did not seem to have a substantial effect on the measures considered. When we designed the game, we hypothesized the people with higher harm value on the risk/harm axis would be more likely to choose "Steal", "Be Unfair", "Punish". However, there were no significant effects between a particular subject's harm value and their in-game behaviors.

It is also interesting to note that there were two participants who never chose to Steal as a leader. Also, most of participants either always chose to Forgive, or always chose to Punish, when the leader chose to Steal from them. Finally, we found that there is a correlation between authority and offer recipient entropy, validating the relationship we discovered in Study 1 using the Social Ultimatum Game.

Conclusion. We showed that behavior in simple social games can indicate a player's underlying moral values. While the effect sizes are not large, they are significant and show promise for further investigation of the degree to which we can extract fundamental aspects of a person's sacred values and personality through passive observation of simple behaviors.

Acknowledgements. This material is based upon work supported in part by the AFOSR under Award No. FA9550-10-1-0569 and the ARO under MURI award No. W911NF-11-1-0332.

References

1. Aquino, K., Reed, A.: The self-importance of moral identity. Journal of personality and social psychology 83(6), 1423 (2002)
2. Batson, C., Moran, T.: Empathy-induced altruism in a prisoner's dilemma. European Journal of Social Psychology 29(7), 909–924 (1999)
3. Bolton, G., Katok, E., Zwick, R.: Dictator game giving: Rules of fairness versus acts of kindness. International Journal of Game Theory 27(2), 269–299 (1998)
4. Buchan, N., Croson, R., Johnson, E., Iacobucci, D.: When do fair beliefs influence bargaining behavior? experimental bargaining in japan and the united states. Journal of Consumer Research 31, 181–190 (2004)
5. Chang, Y.-H., Levinboim, T., Maheswaran, R.: The social ultimatum game. In: Decision Making with Imperfect Decision Makers (2011)
6. Glaeser, E.: Psychology and the market. Technical report, National Bureau of Economic Research (2004)
7. Gosling, S.: Snoop: What your stuff says about you. Basic Books (2009)
8. Graham, J., Haidt, J., Nosek, B.: Liberals and conservatives use different sets of moral foundations. Journal of Personality and Social Psychology 96, 1029–1046 (2009)
9. Graham, J., Nosek, B.A., Haidt, J., Iyer, R., Koleva, S., Ditto, P.H.: Mapping the moral domain. Journal of Personality and Social Psychology 101, 366–385 (2011)
10. Haidt, J., Graham, J.: When morality opposes justice: Conservatives have moral intuitions that liberals may not recognize. Social Justice Research 20, 98–116 (2007)
11. Henrich, J., Heine, S.J., Norenzayan, A.: The weirdest people in the world? Behavioral and Brain Sciences 33(2-3), 61–83 (2010)
12. Jaccard, J.: Predicting social behavior from personality traits. Journal of Research in Personality 7(4), 358–367 (1974)
13. Jackson, L., Zhao, Y., Witt, E., Fitzgerald, H., von Eye, A.: Gender, race and morality in the virtual world and its relationship to morality in the real world. Sex Roles 60(11), 859–869 (2009)
14. Kim, E., Chi, L., Ning, Y., Chang, Y.-H., Maheswaran, R.: Adaptive negotiating agents in dynamic games: Outperforming human behavior in diverse societies. In: Eleventh International Conference on Autonomous Agents and Multi-Agent Systems, AAMAS (2012)
15. Rumble, A., Van Lange, P., Parks, C.: The benefits of empathy: When empathy may sustain cooperation in social dilemmas. European Journal of Social Psychology 40, 856–866 (2010)
16. Snyder, M., Ickes, W.: Personality and social behavior. Handbook of social psychology 2, 883–947 (1985)
17. Staub, E.: Positive social behavior and morality: I. Social and personal influences. Academic Press (1978)
18. Tilley, J.: Prisoner's dilemma from a moral point of view. Theory and decision 41(2), 187–193 (1996)
19. Wellman, M.P., Greenwald, A., Stone, P.: Autonomous Bidding Agents: Strategies and Lessons from the Trading Agent Competition. MIT Press (2007)

An Agent-Based Model for Simultaneous Phone and SMS Traffic over Time

Kenneth Joseph[1,2], Wei Wei[1,2], and Kathleen M. Carley[1,2]

[1] iLab, Heinz College, Carnegie Mellon University, Pittsburgh, PA, USA
[2] Institute for Software Research, Carnegie Mellon University, Pittsburgh, PA, USA
{kjoseph,weiwei,kathleen.carley}@cs.cmu.edu

Abstract. The present work describes a utility-based, multi-agent, dynamic network model of phone call and SMS traffic in a population. The simulation is novel in its ability to generate interactions from both an asymmetric and a symmetric media simultaneously. Within the model, we develop and test a simple extension to the theory of media multiplexity, a well-known theory of how humans use the communication media available to them with different alters (friends). Model output qualitatively matches patterns in real data at the network-level and with respect to how humans use SMS and voice calls with different alters and thus shows general support for our theoretical claim.

Keywords: agent-based modeling, dynamic networks, interpersonal communication theory.

1 Introduction

Effects on how people use the repertoire of communication media (e.g. voice calling, emails and SMS) available to them to interact with others exist within a complex, nonlinear system across the individual, instantaneous, media, dyadic and network levels of analysis (see, e.g., [1]). The dynamic, multi-level nature of these effects has led to a variety of both competing and complementary theoretical and empirical claims of how humans use different communication media.

Precisely because of its ability to control effects at different levels of analysis, agent-based modeling presents a useful tool to develop and test theories of media use. In the present work, we describe an agent and utility-based dynamic network simulation model to analyze effects on media usage at three levels of analysis - the node, the relation, and the affordances of different media[1].

To show the practicality of such a model, we use it to test a new theoretical proposition that extends Haythornthwaite's argument for media multiplexity [2], a theory of human media usage. Haythornthwaite argues that strong social ties will interact on a variety of media, while weak ties will tend to stay on more "established" media within their social context- for example, email is an established

---

[1] Code for the model is available upon request to kjoseph@cs.cmu.edu

A.M. Greenberg, W.G. Kennedy, and N.D. Bos (Eds.): SBP 2013, LNCS 7812, pp. 65–74, 2013.

media within many organizations[2]. As a corollary to this argument, Haythornthwaite shows that an increase in tie strength promotes increased interaction on all media.

While an important contribution to the field, Haythornthwaite's work does not give clear guidance for analysis when no obvious social context or established media exist. This is often the case with datasets of call detail records (CDRs), which are typically anonymized sets of data from a large and diverse population. In the present work, we develop, implement and test a theory to extend media multiplexity to such situations. We hypothesize that where no obvious social context or establish media exist, people will increasingly rely on their preferred medium as tie strength decreases- correspondingly, they will be more open to using a variety of media when interacting with stronger ties. We title this theoretical claim the *weak-tie exaggerated choice theory* (WT-ECT).

We implement our theory in a model where agents develop a preference for one of two media, voice calling or SMS, based on what they have gained emotionally from communicating on these media in previous interactions. While each media has specific affordances in the model that present an obviously rational choice, agents are *boundedly rational* [3], meaning they make the most rational decision based on their view of the world, which may or may not be correct. Bounded rationality, combined with our implementation of WT-ECT, suggests a means by which people might prefer interacting on certain media even when it is not entirely rational to do so, and will rely on their (possibly irrational) preferences more heavily with weaker ties.

Our theory thus falls in line with recent thinking regarding the notion of the "domestication" of communication technology [4] and presents a possible means of connecting this theory with the concepts presented by Haythornthwaite. Additionally, it provides another perspective from which to interpret recent literature suggesting a strong effect of tie strength on the usage of SMS, in particular its use as a tool for communicating only with stronger ties [5,6].

Our main contributions in the present work are thus three-fold. First, we extend a previous model to simulate asymmetric interactions within an evolving social network. Second, our extensions allow us to simultaneously model interactions on two different media with distinct properties. Third, we propose, test, and find evidence for an extension of a well-known theory in the interpersonal communication literature, with ties to other relevant research on mobile communication.

2   Simulation Model

The model we present is grounded in the work of Du et al. [7], who develop the Pay-and-Call (PaC) Model, an agent and utility-based model of network creation over time. In the original PaC model, each agent wants to maximize their supply of "emotional capital", which they can do via interaction with others. Each agent

---

[2] When we refer to a social tie, we mean any connection between two people. The strength of a tie can be modeled in various ways - for example, by the number of interactions between them.

is defined by a uniformly distributed *friendliness* value that gives their aptness in social situations and a uniformly distributed *lifetime* value representing the likelihood that they are replaced on each turn of the model, as described below. For each interaction, agents receive a payoff, defined in Equation 1.

$$payoff = \sqrt{Fr_i * Fr_j} * \frac{1 - \alpha^{intLen+1}}{1 - \alpha} - CPM * intLen - initCost \qquad (1)$$

In Equation 1, Fr_i stands for the friendliness of agent i and $intLen$ is the length of the interaction (in minutes). The value α is a model parameter that represents the ability of a medium to convey the true benefit of social interaction to the agents. As time progresses, the effect of α increases, causing a limit in the amount of capital gained by agents on an interaction. Our model differs from the one presented in [7] by a factor of α - this is done in order to ensure that interactions lasting only one minute are differentiated by unique αs, as explained below. The benefit of an interaction is also mitigated by a linear cost defined by an intercept *initCost*, the cost of initializing an interaction, and CPM, the cost per minute of the interaction. In addition to the CPM mitigating the benefit from an interaction, it is also subtracted from the sender's capital each minute the interaction continues. An interaction ends when the interaction sender has no more capital to continue, or the payoff decreases from one minute to the next.

In our model, we differentiate the α values of voice calls and SMS- that is, we differentiate between their ability to relay emotional benefits as an interaction progresses. Early research in media choice suggested that the more "socially present" a medium is, the better it is able to convey emotion between sender and receiver [8]. Though these claims have since been questioned, it has also been shown that Americans interact more using voice calls than SMS when talking to core ties [9]. Thus, we would expect that agents in our model should be limited in the amount they can benefit emotionally from an SMS as compared to a voice call. From Equation 1, we see that by lowering the α value for SMS, we can model this theoretical concept- the precise values of α for each media are set after calibration, as discussed in the following section.

On each turn of the original PaC model, each agent, in succession, is tested to see whether or not they should be replaced based on their *lifetime*. If not replaced, an agent interacts with his alters in order of the payoff the agent received from each alter the last time the pair interacted. An agent will continue interacting with alters until his capital is exhausted or until the alter he will next interact with has a remembered payoff lower than the agent's expected payoff from talking to a new tie. This expected payoff is calculated as the average of the first payoff an agent received from his alters. At this point, the agent, i, will then interact with one new agent, j, that had the maximum payoff of all of i's alter's friends (i.e. friends of friends).

One vital difference between various communication media that cannot be modeled by the generator in [7] is their level of "synchronicity" [1] - phone calls are synchronous in that both agents must be present for a phone call to occur, while other media, such as email, SMS and IM, are asynchronous in that sends and replies need not occur concurrently. As shown in Algorithm 1, we modify

---

Algorithm 1. Time-scale Based PaC Model

```
1  foreach Minute, m do
2  |   foreach Agent, aᵢ in A (in random order) do
3  |   |   aᵢ.doTurn()
4  |   if (m modulo MinutesInDay) == 0 then
5  |   |   foreach  agent, aⱼ, in A do
6  |   |   |   if RandomZeroToOne()<agent.lifetime then
7  |   |   |   |   replace aⱼ with new agent
```

Algorithm 2. *Agent.doTurn*

```
1   if currInteraction != NULL then
2   |   Update currInteraction by one minute according to Equation 1 (only if sender)
3   |   if currInteraction is finished then
4   |   |   If asynchronous, only the sender obtains the payoff and then places the interaction
    |   |   on the receiver's queue. If synchronous, the recipient and sender both obtain the
    |   |   payoff
5   |   |   currInteraction = NULL
6   |   return
7   if capital<needed then
8   |   capital+ = storedCapital
9   |   storedCapital = 0
10  |   return
11  if Interaction Queue is Empty then
12  |   currInteraction = InteractionFactory.getNewInteraction()
13  |   Start currInteraction
14  |   return
15  nextInt = InteractionQueue.pop()
16  storedCapital+ = nextInt.getReceiverPayoff()
17  if nextInt.conversation is not over then
18  |   currInteraction = Interaction.replyTo(nextInt)
19  |   Start currInteraction
```

---

the original model to allow each agent one chance to act in each minute of the simulation. In making this straightforward modification, agents are now able to carry interactions across multiple turns, and may wait an indiscriminate amount of time before replying to an asynchronous interaction.

Algorithm 2 shows the process for each simulation turn for each agent. If the agent is currently on an interaction, the interaction is simply "updated" by one minute (the payoff equation is recomputed at the next minute). When an agent is on an asynchronous interaction, he is said to be constructing the message. This message construction, like in real life, occurs without the message receiver being made aware that it is occurring. In contrast, synchronous messages require two-way interaction - both the sender and the receiver must actively be on the interaction for it to commence. Consequently, when an agent begins a synchronous interaction, it is either accepted by the receiver if he is not currently on another interaction, or ignored otherwise. If the call is accepted, the receiver sets his current interaction to that call, and can have no other interactions until the call is over.

Ignored calls and completed asynchronous interactions are placed on the receiver's *interaction queue*. This interaction queue, akin to a "social to-do list", represents all interactions that the agent might respond to. The queue is sorted by a uniform probability associated with each interaction (as is done in [10]).

If an agent is not currently on an interaction, he will immediately attempt to begin a new one. In order to do so, however, he must have enough capital- if that is not the case, the agent will refresh his capital with the supply he has obtained from previous interactions. This delay between obtaining capital and being able to use it is done in order to ensure consistency with the original PaC model. Thus, agents act in "cycles", where a cycle is defined as the period between when an agent refills his capital from payoffs earned from previous interactions and when he does so again. In order to keep the distributional properties of the original PaC model with respect to the network, an agent can only interact with each alter once per cycle.

If the agent is not on a current interaction and has enough capital to begin a new one, he can do so in one of two ways. If the agent's interaction queue is not empty, he will obtain the first interaction off the queue, collect the payoff from the message, and determine whether or not to reply. The decision of whether or not to reply is based on the work of Wu et al. [11], who find that SMS conversation durations can be approximated by a power-law with respect to their temporal distributions. We therefore model the likelihood of agents responding to an interaction off of their queue as a power-law (with exponent 1) based on the number of times the interaction has been "bounced" back and forth between agents, using this as a proxy for conversation length. Thus, the likelihood of Agent b replying to an initial interaction from Agent a is 1. When a sends a reply message to b, b responds with probability .5, and so forth. If an agent obtains an interaction from his queue and chooses to reply, he will begin a reply using the same media on which the interaction was initially sent. Note that two agents may (and often do) end one conversation and begin a new one later in the simulation.

If the agent's interaction queue is empty, he will begin a new interaction. The agent will first select a new alter, using the mechanism described in the model from [7] (as described above). Once a partner has been selected, the agent will then determine whether or not to begin a phone call or text message with that alter. This decision is based on Equations 2 and 3, which model this decision as a function of both node (agent) and edge-level preferences. The agent, i, first calculates his average payoff from phone calls and SMS and uses these values to determine $p_i(Call)$, his base probability of making a phone call, as shown in Equation 2.

$$p_i(Call) = \frac{avgPayoffCall_i}{avgPayoffCall_i + avgPayoffSMS_i} \qquad (2)$$

According to the WT-ECT theory we propose, agent media preferences are exacerbated as the tie strength between two interacting agents decreases. We model tie strength here as the inverse of the "rank" of an alter in the agent's remembered payoffs. That is, the higher an alter's $tieRank$, the lower the strength of the tie. Each alter, j, of agent i has a dynamic $tieRank_{ij}$ which indicates j's place in i's list of remembered payoffs. For example, if i had alters j with last payoff .3 and k with last payoff .2, $tieRank_{ij}$ would be 1, and $tieRank_{ik}$ would be 2. The rank of a new tie is computed based on the agent's expected payoff from a new tie - thus, for example, if i's expected payoff from a new tie was .25, then a new alter, n would have a $tieRank_{in}$ of 2 (and $tieRank_{ik}$ would actually be 3).

$$p_{i,j}(Call) = \begin{cases} p_i(Call)^{\sqrt{tieRank_{ij}}} & p_i(Call) < .5 \\ p_i(Call)^{\frac{1}{\sqrt{tieRank_{ij}}}} & \text{otherwise} \end{cases} \tag{3}$$

Given this definition of tie strength, Equation 3 models the effect of tie strength on media preference to obtain a final likelihood of i calling (as opposed to text messaging) j, $p_{i,j}(Call)$. The equation, in accordance with our WT-ECT, increases an agent's likelihood to use their preferred media as tie strength decreases (or correspondingly, as $tieRank$ increases). If i's preference for (equivalently, likelihood to make a) phone call is less than .5, his preference for SMS moves towards 1 as $tieRank$ increases. Similarly, the likelihood of i making a phone call moves towards 1 if $p_i(Call) \geq .5$. For brevity, edge cases are omitted, however, they behave in the model as one would expect.

3   Results

In order to calibrate and test our model's relevance to real-world data, we utilize a dataset of approximately 110 million phone and SMS interactions from approximately 430,000 people spanning three months in early 2008 in an Asian nation. In the present work, we consider only moderately heavy users, which we define as those users having between 5 and 200 alters and having sent at least 30 text messages and 30 phone calls. Though this means we cannot extrapolate our findings to the entire population of study, we find that it is difficult to understand usage patterns for those people with less than 5 alters and 30 interactions per media, and that those having greater that 200 ties were relatively unlikely to be a single human. After pruning, we are left with approximately 65,000 users, which we split evenly into a training set for calibration and a testing set for evaluation.

Calibration was completed with a chief focus on α_{sms} and α_{phone}, as they are, in the present work, the only theoretically relevant parameters. However, we note that, with one exception, the model was reasonably robust to changes in the other parameters. The exception is sensitivity to moderately large changes in the CPM- though the reasons why are clear from Equation 1, modifications in future work are necessary to lower sensitivity to this parameter. During the calibration process, we also experimented with a variety of functional forms for Equation 3- we found that the square root function, our initial hypothesis, actually gave the best fit to our training data. Calibration resulted in parameter settings of $\alpha_{sms} = .6$ and $\alpha_{phone} = .8$. The CPM and $initCost$ parameters were kept at the values used by Du et al. [7]. We use a model with these settings and simulate interactions between 100,000 agents over 30 simulation "days" when contrasting results with the held-out (test) data.

Comparisons to the held-out data were made to understand the extent to which our simulation could generate a realistic social network and a realistic distribution of media preferences and the model's ability to capture evidence of the WT-ECT existent in the testing data. We here consider only one measure of

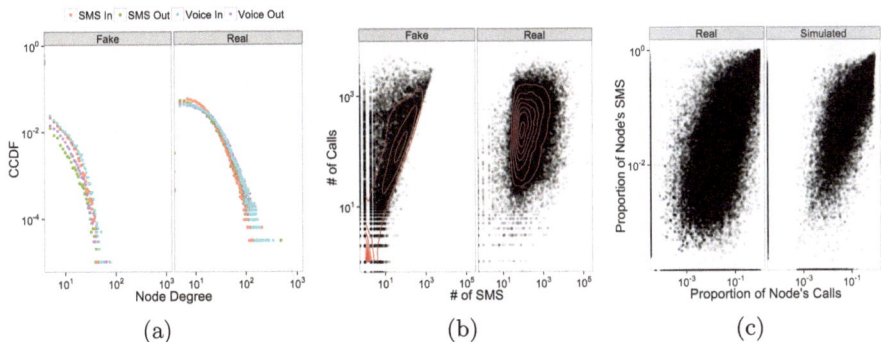

Fig. 1. a)The CCDF, in log-log scale, of the degree distributions of the real and simulated data; b) Number of phone calls (y-axis) versus SMS messages (x-axis) per person (agent). Each dot is a single node, and the red is a 2D density estimator; c) A random sample of all edges from 7,000 nodes in each dataset, where the y-axis is proportion of a node's total text messages sent that alter, and the x-axis is the proportion of a node's total phone calls made to that alter. In a),b) and c), the gray bar at the top indicates the dataset.

our model's ability to generate realistic social networks across both phone and SMS, and do so by plotting the degree distribution of the aggregate networks from the real and simulated data. As we see in Figure 1a, degree distributions are qualitatively quite similar. However, the graph belies one important difference that may exist between model output and the real data- due to our parsing mechanisms, it is unclear whether or not the simulated data correctly captures the head of the distribution or whether it generates too many low-degree nodes.

Regardless of the head of the distribution, our model captures the heavy-tailed nature of degree distributions well-known to exist in aggregate social networks from CDRs (e.g. [7]). We now consider aspects of our model (and the real-world data) that have received less attention in the literature. First, we consider how well the model represents media preferences at the node level. In order to do so, we consider the distribution of the number of phone calls agents (in the model) and people (in the real data) make in comparison to the number of text messages they send. As we see in Figure 1b, which plots a point for each node (agent or person) as well as a density estimator for the entire dataset, the vast majority of nodes preferred voice calling to SMS. As we know, a range of effects at various levels of analysis may have caused this in the real data, and though the point spread is quite similar, these two plots clearly show that our model does not capture significant effects. However, it is interesting to note that even when modeling node-level media choice as the resultant of only the technological affordances of the media and bounded rationality, the model suggests a qualitatively similar conclusion.

Our final comparison between the simulated and real data considers how well the two datasets conform to Haythornthwaite's theory of media multiplexity, and to the WT-ECT extension. The first claim of media multiplexity is that strong ties will tend to interact more on all media than weak ties - to this end, Figure 1c shows

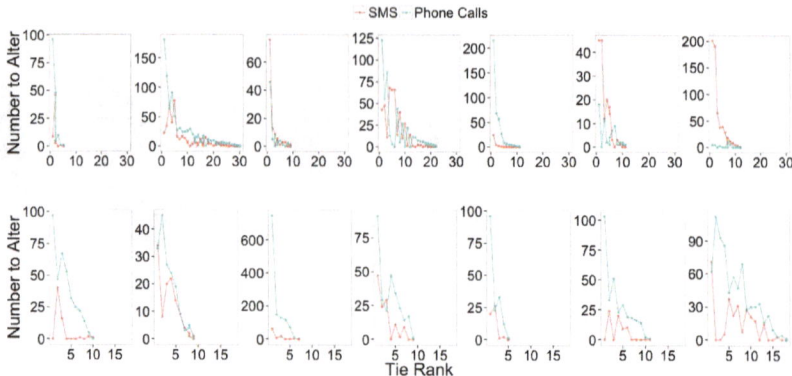

Fig. 2. Seven sample users from the real (top row) and simulated (bottom row) data, and their interactions via phone and SMS with alters of different tie ranks

that in both the real data and the simulated data, there exists an approximately linear correlation between the (log) proportion of a node's total SMS messages and the (log) proportion of their phone calls that go to a specific alter. Though the plot only show a small sub-sample of the data to avoid over-plotting, we find that the log proportion of a node's total calls to an alter explain 34.6% and 57.4% of the variance in the log proportion of text messages sent to that same alter in a simple linear model in the real and simulated data, respectively.

The fact that this simple linear model has such a clear predictive power lends significant support to Haythornthwaite's theory, and as is clear, our model definitively captures this aspect of interaction across multiple media. It is important to note, however, that the linear model only predicts 57.4% of the variance in our simulated data. Thus, our model contains a reasonable and desirable amount of stochasticity in agent interactions with different partners on different media. This variability can be seen in Figure 2, which shows sample users from the real (top) and simulated (bottom) datasets and the number of interactions they had via phone and SMS with alters of different "tie ranks" (discussed below).

The second claim of media multiplexity is that weaker ties will tend to stick to more established media. WT-ECT, however, suggests that with weaker ties, people will more heavily rely on the media they have a stronger preference for. If this were to be the case, then we would expect that with weaker ties, preferences would be more obvious- that is, we would see people use one media more heavily than they would under normal circumstances.

We use a three step process to obtain a set of statistics that can be used to understand the extent to which WT-ECT is supported. We first compute, for each edge $<i,j>$ in the network, the log odds of i calling (as opposed to texting) j as $log(\frac{NumCalls_{i,j}}{NumSMS_{i,j}})$ and subtract i's log odds of making a phone call to any of his alters. Thus, a positive value for an edge means that i used voice calling more heavily with j than he would on average, and a negative value the opposite. We then bin edges by their strength. To differentiate between ties of different strengths, we bin using the notion of tie "ranks", as above. Because there is no

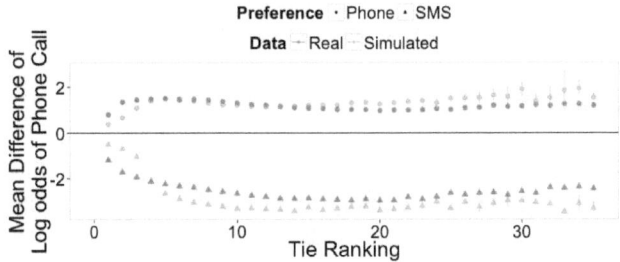

Fig. 3. The DLO statistic (y-axis) as tie rank increases (tie strength decreases) for the real and simulated data for agents (people) preferring both SMS and phone calling. All 95% confidence intervals shown are determined using the bootstrap method with 1,000 iterations.

notion of payoff in the real data, we compute rank here based on the number of interactions sent to each alter. Thus, the alter an agent communicated with most (on both SMS and phone combined) would have a tie rank of one, and so on (edges with the same strength are place into the same bin).

Finally, we compute the mean of this value (for convenience, the DLO, or difference of log odds, statistic) for all edges in each tie bin. As is clear, if the WT-ECT theory were to be supported, we would expect that as tie rank increases, the log-odds of an agent using their "preferred" media increases as well. Figure 3 plots our DLO statistic for the first 35 tie bins for the real and simulated data for agents preferring SMS (the triangles) and those preferring voice calling (the circles). We consider only the first 35 tie bins both because model estimates were highly varied outside this range and because we feel it is difficult to estimate differences in preference for the real data outside of this general range (the number of communications was, on average, around two).

The figure shows that our simulated data obtains pattern (and at some tie ranks, statistical) validity- this lends credence to our model of human communication, regardless of the underlying phenomenon. More importantly, though, the plot largely shows qualitative support for our WT-ECT- as tie rank increases (tie strength decreases), the likelihood of an agent using their preferred media increases to some extent before finally appearing to level out, as our square root function in Equation 3 predicts. In further analysis, we find that this "leveling out" extends to larger tie ranks, though as mentioned, such estimates are somewhat unreliable.

The one exception to support for the WT-ECT appears to be with agents who prefer phone calls and their ties to alters ranking from 10 to 20- in this range, the DLO statistic actually regresses back towards the node-level preference. Interestingly, there is also evidence of this regression in the simulated data, in which the WT-ECT is an explicit functional piece of the model. This observation can be accounted for by the fact that SMS is asymmetric, and thus will, by its nature, tend to have more interactions per conversation than phone calls. However, it also goes lengths to indicate the complexities and intricacies of human communication across media and, to this end, suggests the power of agent-based modeling in theory development.

4   Conclusion

The current work has several limitations- we provide mostly qualitative evidence of fits with real data and only a cursory model calibration. In addition, we model tie strength as a dynamic element- though it is true that the strength of human social ties with those around us are constantly fluctuating, our model may allow for too much fluctuation of tie strength in the described time span. Limitations aside, however, the model we develop shows promise, as it is the first we are aware of to attempt to model communication on different media simultaneously over time in an evolving network. Our results show that the WT-ETC deserves further attention, particularly because a derivative of the theory suggests that depending on the perceptions (preferences) of a person, (s)he may irrationally choose an affectively weak communication media (like SMS) when interacting with newer and/or weaker ties, leading to difficulties in forming strong social ties for emotional support.

References

1. Baym, N.K.: Personal Connections in the Digital Age. In: Polity (July 2010)
2. Haythornthwaite, C.: Strong, weak, and latent ties and the impact of new media. The Information Society 18(5), 385–401 (2002)
3. Simon, H.A.: Bounded rationality. The New Palgrave: Utility and Probability, 15–18 (1987)
4. Silverstone, R., Haddon, L.: Design and the domestication of ICTs: technical change and everyday life. In: Communicating by Design: The Politics of Information and Communication Technologies., pp. 44–74. Oxford University Press, Oxford (1996)
5. Ling, R., Deitel, T.F., Sundsy, P.R.: The socio-demographics of texting: An analysis of traffic data. New Media & Society 14(2), 281–298 (2012)
6. Xu, X.K., Wang, J.B., Wu, Y., Small, M.: Pairwise interaction pattern in the weighted communication network (March 2012), arXiv:1203.1105
7. Du, N., Faloutsos, C., Wang, B., Akoglu, L.: Large human communication networks: Patterns and a utility-driven generator. In: Proceedings of the 15th ACM SIGKDD International Conference on Knowledge Discovery and Data Mining, KDD 2009, pp. 269–278. ACM, New York (2009)
8. Short, J., Williams, E., Christie, B.: The Social Psychology of Telecommunications. John Wiley and Sons Ltd. (September 1976)
9. Hampton, K.N., Sessions, L.F., Her, E.J., Raine, L.: Social Isolation and New Technology. Pew Internet and American Life Project (November 2008) pewinternet.org/Reports/2009/18Social-Isolation-and-New-Technology.aspx (accessed January 10, 2011)
10. Barabasi, A.L.: The origin of bursts and heavy tails in human dynamics. Nature 435(7039), 207–211 (2005)
11. Wu, Y., Zhou, C., Xiao, J., Kurths, J., Schellnhuber, H.J.: Evidence for a bimodal distribution in human communication. Proceedings of the National Academy of Sciences (October 2010)

Reconstructing Online Behaviors by Effort Minimization

Armin Ashouri Rad and Hazhir Rahmandad

Grado Department of Industrial and Systems Engineering, Virginia Tech,
Northern Virginia Center, Falls Church, VA, 22043, USA
{Ashouri,Hazhir}@vt.edu

Abstract. Much social interaction is moving online, offering new opportunities
to analyze and understand fundamental patterns of human social behavior. One
of the challenges in using this data is lack of direct observations of users' online
activity in typical datasets. Building on the idea that people conserve their efforts in their online behavior, we develop a generic procedure for inferring user
online behavior from their observable interactions with online objects and apply
it to data from a social news website. We estimate which pages the users have
seen and what stories they have observed. We test the effectiveness of this method in increasing the accuracy of a regression model that attempts to predict
the number of votes a story is expected to receive, and show that the method
can significantly increase the precision of these regressions.

Keywords: Effort minimization, Social news website, Online behavior.

1   Introduction

An increasing share of our social interactions happen online in social networks, social
news websites, and other internet based media. In many of these settings people express themselves in their interactions with socially generated digital objects such as
postings, comments, stories, songs, and tweets. These interactions include different
forms and labels such as "like", "vote", "retweet", "share", and "Digg". They provide
a unique trail of individual choices and tastes in the online media, a valuable source of
data for researchers interested to better understand how people think and behave
online, change their tastes, and form and leave communities.

Yet a fundamental challenge in using this data efficiently is the fact that we usually
do not know what people have actually viewed online. The fact that a story did not get
a positive vote from user A could imply that user A did not like that story, or simply
that she did not see it. This problem is most acute for binary choices (e.g. "like" button in Facebook) where expression of a choice can only be seen as a positive signal
(compared to multi-category ratings that embed more information in a data point) [1].
The problem is exacerbated with the growing trend in customization and filtering of
websites that show the same item on multiple pages, or different items on the same
page for different users.

A few methods have been developed to address this challenge. These largely rely
on -statistical regularities in the user-story network to estimate the likelihood that lack

A.M. Greenberg, W.G. Kennedy, and N.D. Bos (Eds.): SBP 2013, LNCS 7812, pp. 75–82, 2013.
© Springer-Verlag Berlin Heidelberg 2013

of expression signifies lack of interest or lack of viewing. Counting the total number of friends who voted for a story [2], and counting the number of followers of a ret-weeted link [3] are two of these methods to estimate exposure of stories.

In this article we develop and test a new method for overcoming this challenge. Our method is rooted in the observation that people are lazy in their browsing activity and thus their behavior online could be approximated based on their interest in mini-mizing their effort. Therefore by defining the alternative pathways people could have taken, and minimizing their efforts in choosing those pathways while matching their observed expressions (e.g. votes), we can estimate and reconstruct their online activi-ty. We implement this idea on data from a social news website, and test its effective-ness in improving the predictive power of a regression model to estimate the number of votes a story receives on this website. We show sizeable improvements from the application of our method in the test case and discuss different applications which can benefit from this method.

2   Data and Methods

We use data from Balatarin, a social news web site where users can post stories (i.e. links to different news items and websites), read other users' stories, and vote or comment on stories posted. Balatarin is the largest social news website for the Per-sian-speaking community. It includes over 30,000 registered members, a million stories, thirty million votes and several million comments.

Balatarin promotes popular stories, those with votes more than a specified thre-shold, to its first page. Therefore a user visiting the website could choose between reading and voting on first page stories or the recently posted stories. If she chose to explore first page stories, she can sort them by date of promotion to first page or in the order of most voted stories in the last day, week, or month. If she picked recently posted stories, she can sort them by their posting time or descending number of votes. We call these options as different ordering pages. Each of these ordering pages are further broken down into multiple subpages each accommodating 25 stories. In sum-mary, a story cannot be seen in both first page and recently posted page at the same time, but in each of these pages it could be seen in more than one ordering page.

We have data on stories (Story ID, posting User's ID, Time of Posting, user identi-fied Story category (political, economics, sports, social, etc)) and votes (Story ID, Time of Vote, voting User's ID). We do not have data where a vote has been submit-ted (i.e. which page, ordering page, and subpage) or what stories have been viewed by the users. Estimating those missing characteristics is a major step towards better un-derstanding the online user behavior.

To this end we rebuild the history of Balatarin in a simulation environment. Using the data on posting and voting times as the inputs to this environment, we calculate the status of stories over time (i.e. the number of votes they have, and thus whether they could be on the recent or first pages at any point). We then solve an optimization problem with the purpose of finding ordering page and subpage that best fits overall user behavior in its totality. Specifically we assume that users minimize their browsing effort, i.e. they are less likely to jump between ordering page or different subpages than to vote for stories on the same ordering page and subpage. By minimizing user effort we find the most likely ordering page and subpage in which each vote could have been submitted. We then use this information to infer what stories the user should have seen and which stories were not observable to this user. For example if one has voted for the third and fifth story on a page, we assume she has seen the fourth story as well.

In order to validate our method we develop two Poisson regression models to estimate the number of votes a story gets during subsequent time intervals using the raw data available vs. using the inferred browsing trajectories from the method above. Using comparable sets of independent variables we compare the predictive power of the two regression methods and estimate the benefits of our approach for estimating browsing trajectories.

3   Results

We first used a simulated environment to rebuild Balatarin's history and find whether a story has been on the recent or first page at any time, and what positions it would have occupied on feasible ordering pages and subpages. The time for promotion to first page is deduced from the category-specific first page thresholds. These thresholds, not available directly in data, are also estimated from the discontinuity in voting rates for stories in recent vs. first page. Next, possible ordering pages and their subpages are generated, which takes into account the life time of stories on that ordering page (e.g. recent most voted ordering page includes only stories posted in the last 24 hours). Next, we use our effort minimization algorithm to determine in which ordering page and subpage each vote has been casted. Our algorithm finds the ordering page and subpage in which the user is most likely to have voted her next story minimizing her efforts from the current ordering page. The effort minimization algorithm penalizes changes in ordering page ($D_{Ordering}(i, i + 1)$: is one if a change in ordering is required), changes in subpages ($D_{Subpage}(i, i + 1)$: counts the number of subpages between subsequent votes), and the distant between two subsequent votes on an ordering page ($D_{Story}(i, i + 1)$: counts the stories between the two). Specifically we minimize the following penalty function for each user (u) examining different ordering pages in which she could vote for each story over all the stories she has voted for (N_u):

$$Cost(u) = \sum_{i=1}^{N_u} p1 * D_{Story}(i, i + 1) + p2 * D_{Subpage}(i, i + 1) + p3 * D_{Ordering}(i, i + 1) \quad (1)$$

After exploring the effect of different penalty parameters on the overall performance of the algorithm we set them at one unit for each story between two consecutive voted stories (p1), five units for changing the subpage (p2), and 200 for changing ordering page (p3). We solve the optimization problem for each user separately. We discuss the results for all stories on the first week of December 2009 but the algorithm is scalable and can be applied to any slice of the data.

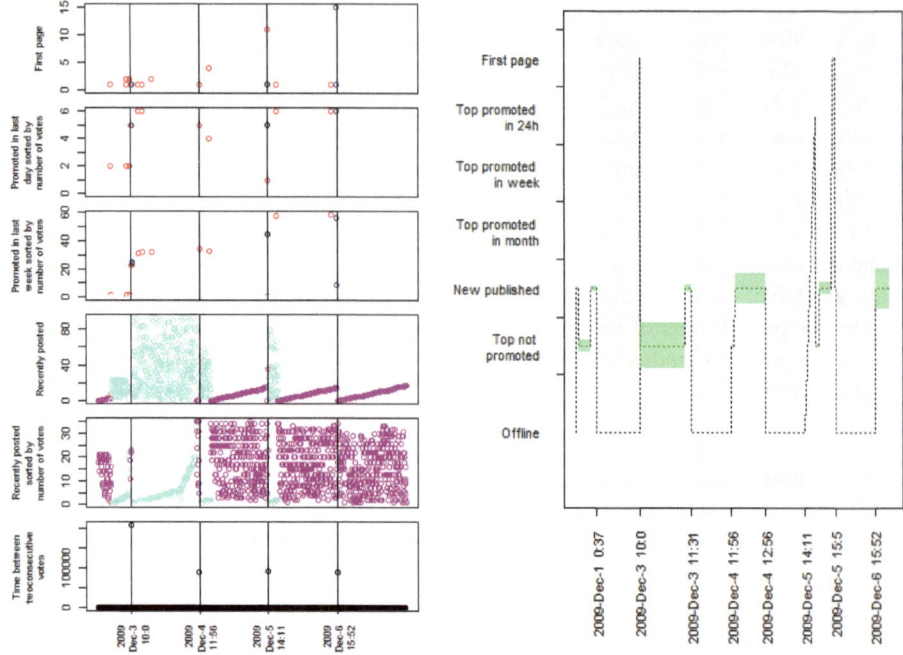

Fig. 1a. Feasible ordering (sub)pages   **Fig. 1b.** Inferred browsing history

Figure 1 represents feasible ordering pages and subpages (a) and inferred browsing history (b) for a single user. Panel a shows the five possible ordering pages and subpages (Y axes) for each story the user has voted on. Time between votes is also measured to distinguished different browsing sessions. Minimizing the effort we find the browsing trajectory for this user, specified with color codes in panel a and explicitly brought out in panel b. Here the area of each rectangle represents the time the user has spent on that ordering page with logarithmically scaled inter-session intervals. We assumed that the user is offline if the time between her two consecutive votes exceeds one hour.

Table 1. Percentage of votes casted on different ordering pages for all the users

Ordering page	Vote Percentage
First page (Promoted, Promotion time)	27%
Promoted, Most voted last day	6%
Promoted, Most voted last week	<1%
Promoted, Most voted last month	<1%
Recently posted, Posting time	42%
Recently posted, Votes	17%
Not categorized by the algorithm	8%

Table 1 provides an overview of the percentage of votes casted on different ordering pages for all the users, found through our algorithm. Most votes are cast on the recently posted page with chronological (default) sorting (42%), followed by first page (Promoted, chronologically sorted; 27%) and recently posted stories sorted by number of votes (17%). The algorithm is unable to make a precise categorization for 8% of the stories.

Once the user browsing history is recreated through the optimization method, we can specify, among other things, how long the user has spent on each of the ordering pages and subpages, which stories she has seen but not voted for, and what the neighboring stories for each voted stories are. In developing these metrics we assume that when a user votes for a story, the two stories above and below are also seen by the user. Also we assume that if a user votes for two stories in one subpage, all the stories in between the two are seen as well. If the user votes for two stories in subsequent subpages, we assume that the user has seen all the stories from the one voted to the end in the first subpage.

4   Validation

In order to assess the effectiveness of our method we devise a prediction test in which the same concept can be estimated directly using the Balatarin data, or indirectly using some of the variables inferred from the method above. Specifically, we develop parallel Poisson regression models to predict the number of votes each story gets in subsequent time intervals of 200 seconds each for the first hour after the story is posted. We use two parallel models with the direct data available from Balatarin and with the information estimated through our new method on what stories are seen by users.

Exposure parameters specify the base rate for the mechanism that generates the count item of interest (in this case voting) in Poisson regression. We can specify the exposure in two different ways. First we set Balatarin's activity (i.e. the total number of votes in the interval) as the exposure parameter around which Poisson regression estimates the changes in likelihood of votes. This model uses the best estimate for exposure available in the direct data in Balatarin. In the second model we quantify the exposure as the number of times the story is seen by users, calculated from our browsing reconstruction method. Keeping the other independent variables equal, we compare the accuracy of two regressions to assess the value of the new method for understanding the underlying browsing patterns.

Controlling for exposure (calculated in two different ways; see above) we include the following independent variables in our regressions: stories' categories, their promotion status, their current number of votes, and their place in each of ordering pages at the beginning of time interval as the independent variables. Equation 2 specifies the regression model(s) formally:

$$\ln\big(E(Number\ of\ votes\ received\ |X)\big) = \beta_0 + \beta_1 \times story\ promoted? + \beta_{2-7} \times$$
$$story\ subpage + \beta_{8-13} \times location\ on\ page + \beta_{14} \times\ number\ of\ votes + \beta_{15} \times$$
$$\log\ (number\ of\ votes) + \beta_{16-22} \times story\ category + \log\ (exposure) \qquad (2)$$

Table 2 provides regression coefficients for how the number of current votes may influence the prospects of getting more votes. These variables (and most of the control variables not reported due to space limits) are statistically significant. While the direction of effects are consistent across the two regressions, the magnitudes are different, providing different quantitative predictions for how the number of votes really influence a story's chances of getting more votes. We compare the predictive power of regression results using multiple metrics of accuracy. Table 3 reports these comparisons. Overall accuracy is the ratio of predicted most likely number of votes exactly matching the number of votes a story gets in the interval. Positive accuracy measures the same concept over all the stories with a positive number of votes (i.e. excludes the zero-vote stories). True positive indicator is the fraction of stories predicted to receive at least a vote, which do receive a vote. True negative (fraction correctly predicted not to receive any votes), false positive (fraction predicted to receive one or more vote but not receiving any), and false negative (fraction receiving vote despite a prediction to contrary) are also reported. Log likelihoods for the two models and Akaike Information Criterion [4] provide other metrics of accuracy. Another measure we use in our study is the F_1-measure [5]. This measure is frequently used to turn fit measures with two dimensions of precision and recall into a one-dimensional measure of accuracy, and is defined as:

$$F_1 = 2 \times \frac{\text{Precision} \times \text{Recall}}{\text{Precision} + \text{Recall}} \tag{3}$$

$$\text{Precision} = \frac{\text{True Positive}}{\text{True Positive} + \text{False Positive}} \tag{4}$$

$$\text{Recall} = \frac{\text{True Positive}}{\text{True Positive} + \text{False Negative}} \tag{5}$$

All metrics show improvements as a result of the use of our browsing behavior estimation technique. The improvements are most significant in log-likelihood and AIC measures. In fact a perfect model (i.e. correctly predicting the mean of the underlying Poisson process) would still get significant errors due to inherent randomness of the Poisson generating processes.

Table 2. Regression coefficients and standard errors for a subset of variables

Independent Variable	Overall Activity as Exposure	Estimated Viewings as Exposure
Intercept	159.4 (20.4)	113.4 (20.19)
promoted or not	-147.7 (20.3)	-105.9 (20.11)
number of votes	0.061 (0.002)	0.049 (0.002)
Ln(number of votes)	-0.144 (0.011)	-0.104 (0.011)

Table 3. Accuracy of regression models

Metric of Accuracy	Overall Activity as Exposure	Estimated Viewings as Exposure
Overall accuracy	62%	69%
Positive accuracy	33%	38%
True positive	65%	72%
True negative	76%	84%
False negative	17%	14%
False positive	24%	16%
Log likelihood	-113425	-90362.92
F-measure	0.760233918	0.827586207
AIC	226888	180764

5   Conclusion

With increasing spread and impact of online social interaction, new opportunities for using large datasets to analyze and understand topics such as social behavior, individual routines, group formation, and taste evolution are on the rise. One of the challenges in using this data is lack of direct observations of user online activity in typical datasets. In this paper we report on a generic procedure for inferring user online behavior from their voting patterns and apply it to data from a social news website. Building on the idea that people conserve their efforts in their online behavior, we develop an optimization approach to estimate user browsing, the pages they have visited, and the stories they have observed. We test the effectiveness of this method in increasing the accuracy of a regression model that attempts to predict the number of votes a story is expected to receive, and show that the method can increase the effectiveness of these regressions, providing additional evidence about the usefulness of the technique.

The specific insights from the regression results are worth noting. As predicted by previous research [6] an increase in the number of votes a story has enhances its chances of getting more votes. Yet this process is complicated by a non-linearity (first few votes are slightly less influential) and a step effect, where stories that are promoted are generally less likely to attract as many votes for each time they are observed. This effect may point to an interesting asymmetry in user behavior: people who visit the new stories are more likely interested in finding and promoting interesting stories, where as those visiting the promoted stories are more interested in reading the items without necessarily voting for those.

The method can inform many different applications. First, by indirectly estimating individual viewings of different pages and stories, we provide a way to distinguish between genuine no-vote and where data points are actually unavailable in binary categorical data (e.g. turning one/zero for getting/not getting a vote to one/zero/not available data). This significantly reduces the noise in binary data and offers better predictive power. Different regression analyses, data mining applications, and collaborative filtering studies can benefit from such data improvement. Moreover, by estimating individual online browsing behavior this method can offer new insights into how individuals build habits, explore a website, and change their browsing patterns over time. For example future research can look into exploration-exploitation tradeoffs in learning to browse social news sites.

The method can be broadly applied to data from different websites. Other social news websites such as Reddit (reddit.com) and Digg (digg.com) have very similar structures to Balatarin, so adaptation to that data is straight forward. Yet the method is more general in theory and can be applied to any data structure in which 1) The user input is measured as a binary choice. 2) The user can view the same content (on which s/he expresses her opinion) across a limited number of pages at any time. Therefore by customizing concepts used in Balatarin (page, ordering page, subpage) to a new application domain we can extend our method to other social websites such as Facebook and Twitter. For example, home page and each of "following pages" in Twitter or friends' walls in Facebook can be considered as an ordering page. Retweeting and liking in Facebook could be seen as voting in Balatarin. Therefore, our method for reconstructing online behaviors can be applied to many other online data sources.

In the current paper we used a greedy optimization algorithm and with some approximations reduced the computation time to scale linearly with the number of stories the user has voted for multiplied by the number of ordering pages. This kept the computational costs at a moderate level and made the algorithm scalable for large datasets. Future research can test alternative optimization methods to gain improved performance. Further research can also fine tune the penalty parameters and behavioral assumptions of the model based on prediction performance on a cross-validation dataset.

References

1. Leskovec, J., Huttenlocher, D., Kleinberg, J,: Predicting positive and negative links in online social networks. In: Proceedings of the 19th International Conference on World Wide Web. ACM (2010)
2. Steeg, G.V., Ghosh, R., Lerman, K.: What stops social epidemics? arXiv preprint arXiv:1102.1985 (2011)
3. Oken Hodas, N., Lerman, K.: How visibility and divided attention constrain social contagion. arXiv:1205.2736v2 (2012)
4. Sakamoto, Y., Ishiguro, M., Kitagawa, G.: Akaike information criterion statistics. D. Reidel, Dordrecht (1986)
5. Rijsbergen, C.J.: Information Retrieval, 2nd edn. Butterworths, London (1979)
6. Salganic, M.J., Dodds, P.S., Watts, D.J.: Experimental Study of Inequality and Unpredictability in an Artificial Cultural Market. Science 311, 854–856 (2006)

"Copping" in Heroin Markets: The Hidden Information Costs of Indirect Sales and Why They Matter

Lee Hoffer[1] and Shah Jamal Alam[2]

[1] Department of Anthropology, Case Western Reserve University, Cleveland, OH, USA
lee.hoffer@case.edu
[2] School of Geosciences, University of Edinburgh, Edinburgh, UK
sj.alam@ed.ac.uk

Abstract. Ethnographic research identifies brokering (a.k.a., "copping for others") as an important and popular way people who use heroin acquire the drug by making purchases for their peers. Brokering is when a customer buys drugs for a fellow customer using the buyer's money and is paid using drug the buyer purchases. This distributes heroin costs. Heroin dealers obviously manipulate price and/or drug purity to make profits and compete for buyers, but a hidden way they alter "price" is by adjusting the size of heroin packages they sell. Using an agent-based model, we simulate brokering and heroin package resizing to understand how these dynamics influence heroin consumption costs. High rates of dealer arrest are tested against these dynamics. Findings indicate the Quantity-Adjusted Price of heroin is greater than its retail price in all conditions, implying increased competition in heroin markets does not lower costs.

Keywords: Agent-based modeling, ethnography, heroin dealing, hidden costs.

1 Introduction

A fundamental assumption of U.S. drug policy is that the consumption costs of heroin are a linear function of the drugs retail price. Features of the market such as competition and law enforcement are, in turn, assumed to affect this price to increase or decrease costs. Daily heroin users are estimated to spend 60-72% of their monthly income on heroin consumption [1-3], spend more compared to cocaine users [3], and use cash as the number one commodity exchanged for heroin [3]. However, heroin users often report spending less on the drug than they report using [4] suggesting simplistic projections of annual costs are flawed, e.g., a $20 per-day drug habit costs $7300 annually or a gram per-day (at $120 per-gram) costs $43,800.

Documenting real-world cash spending on heroin is challenging because consumers employ a host of strategies to acquire the drug without cash or reduce expenditures [3-4]. People trade goods or sex for drugs; get money from friends and family; and pool resources with their peers to reduce their individual cash expenditures. Some customers acquire drugs on credit from their dealers, sell heroin or participate in other income generating crime to offset drug consumption costs.

A.M. Greenberg, W.G. Kennedy, and N.D. Bos (Eds.): SBP 2013, LNCS 7812, pp. 83–92, 2013.
© Springer-Verlag Berlin Heidelberg 2013

The agent-based model (ABM) we present estimates heroin consumption costs by characterizing two behaviors that are more consistent and basic to most, if not all, heroin markets: 1) brokering, a.k.a. "copping drugs for others" (by customers), and 2) drug package resizing (by dealers). These activities are selected because although documented independently, and quite common, the feedback between them and their influence on cost has been overlooked and never modeled. Brokering is how many heroin users describe acquiring heroin "for free."

2   Ethnographic Research on Heroin Markets

Unlike the immense literature on heroin (and opiate) addiction, a smaller literature in the social sciences describes how local illegal drug markets and drug dealers operate. Though the content of such studies is often unique, a challenge to theory development [6], the primary method used in the majority of this research is ethnography, which involves detailed longitudinal fieldwork with drug users and dealers to understand their decisions, as well as the social and political contexts of behaviors within these settings [7]. Since 1993, the anthropologist co-author of this paper (Hoffer) has conducted ethnographic research in Denver, Colorado, St. Louis, Missouri, and Cleveland, Ohio on illegal drug buying and selling activities among out-of-treatment drug users. As a result, the data represented in this ABM comes from different studies. Hoffer's fieldwork investigated an extensive range of operational activities associated with selling heroin. It addressed brokering as a process through which dealers accessed new customers and insulated their operations from the police, but also as an important way the market adapted to police interventions to dismantle it, a case simulated using ABM [5].

2.1   Brokering a.k.a. "Copping for Others"

A conventional face-to-face drug deal is a direct transaction between a heroin buyer and a seller. "Copping for others" is what heroin users refer to when one user purchases heroin for another user. Here, we call this *brokering*. Although brokering involves buying heroin from a dealer, it is an indirect transaction between customers: user *A* (the broker) takes user *B*'s (the buyer's) money, goes to a dealer, buys the drug, and returns to user *B*. Because the available cash is converted into heroin, user *B* then gives user *A* heroin as a "payment" for making the sale, and the transaction is complete. Buyers who use brokers pay more for heroin because they pay the broker in addition to purchasing the drug.

Brokering is important because it is a common strategy that heroin users employ to reduce their drug consumption costs [4, 8-9]. Brokers are not drug dealers. Although brokers may represent themselves to others as dealers, they do not invest in a quantity of drug to resell, they simply buy heroin for someone using that person's money. This also separates brokering from "juggling", a term describing a customer who buys heroin and repackages it into smaller quantities to resell [10-11].

Norms associated with brokering (i.e., copping) are clear: brokering is a service to a buyer not a seller; therefore, a buyer pays the broker. But despite payment expectations, because the buyer is not with the broker at the actual sale, brokers frequently hustle additional heroin "payments" during this transaction [8-9]. This makes brokering an attractive but risky drug acquisition strategy. In an extensive study on the economics of heroin use, copping combined with "touting and steering[1]" was estimated to save users $3,000 and occurred more frequently than direct drug sales [4]. Moreover, brokering also builds community between heroin users. For these transactions to work, which they do more often than not, users must trust one another. A buyer trusts the broker to make a purchase and return with drug; a broker trusts that if they do so the buyer will reward them [8, 12]. In this way, brokering is a "favor" and an economic service. Finally, because brokers often use heroin with buyers as part of the transaction, it increases HIV risks associated with injecting [13-14].

On one hand, buyers and dealers strive to make direct connections because they understand brokers manipulate the economics of transactions and it is not uncommon that a broker is cut out when buyers achieve this. On the other hand, brokers are highly motivated to maintain their position because it saves them money. The ways that brokering transmits market information is underappreciated in previous research. Through these dynamics, buyers learn about deals, products, and price, as well as identify access points for direct dealer relations.

Brokering redistributes wealth (heroin) from people with money and no dealers to people with dealers and no money. This commodified gatekeeping influences individual consumption costs, which is known. How brokering changes deal values in the marketplace (i.e., its aggregate influence on cost), through deal communications is unknown. Brokering is how buyers get information to make decisions. How much does sharing information about deals adjust and/or stabilize the size of heroin packages, i.e., dealer offers? And if we calculate the total cost of direct heroin sales, calculate the costs/savings of brokering, and adjust this to heroin package amounts sold, (i.e., true deal value) can we calculate a Quantity-Adjusted Price of heroin? And what affect does police intervention, i.e., arresting dealers, have in altering these dynamics?

2.2   Price Adjustment of Heroin in the Market

Heroin dealers also influence buyers' costs to consume heroin. Dealers raise or lower prices and/or "step-on" / "cut" the drug to stretch their supply, lowering potency to increase profits. Lowering potency requires customers to increase purchases, hence cost, but because they compete with other dealers, they risk losing sales or lowering the quality of drug. Heroin addicts frequently budget purchases and, like any consumer, desire stable prices. Our agent-based model simulates a less obvious strategy dealers employ in this situation; keeping prices constant but modifying the *amount* of drug sold in sales units. We label this *drug package resizing*.

---

[1] In touting and steering a seller (dealer) pays a customer to market their product, and although different than brokering, they were combined in this analysis.

Dealers sell different units of heroin, such as "bags", "pills", half-grams, grams, etc., standardizing prices by weight. However, in Hoffer's research [8], such sales units were rarely actually weighed. Instead, as the dealers received orders, they "eyeballed" the size of the unit to sell. Dealers were extremely accurate in estimating weights; on several occasions the researcher weighed "eyeballed" grams and found them to be within 1% of the true gram weight. But dealers also purposefully manipulated unit sizes to control customer behaviors, and compete for profits. If a customer acted in a way the dealer did not like, their next order would be "short" (i.e., less than the regular size). Alternatively, to reward customers, a dealer would make them a "fat" package (i.e., a larger than usual sale). Similarly, when sales were down, dealers made bigger units to attract/retain customers. When they had more sales, smaller units were made to extract more profit. Even though retail prices were constant, amounts were not.

Instances of dealers "shorting" customers appear in the ethnographic literature but detailed analyses of drug package resizing are rare. A notable exception is Lisa Maher's research on heroin trends in Australia in which she describes both drug package resizing and how dealers attract customers by up-sizing packages [15]. But evidence of this also comes in a more popular form: Dealers commonly reward good customers with extra bags of heroin or a better deal when they purchase larger heroin units.

Heroin dealers leverage the size of drug packages as a tool to take advantage of profit opportunities and control customer behavior, thus veiling price changes. To customers bigger units = better value/lower costs, and smaller units = worse value/ higher costs. Of course, product downsizing to lower cost and increase profits is not just for drug dealers. It is a common strategy manufacturers of legal goods use to hide price increases on commodities ranging from ice cream to toilet paper [16]; although it is unclear if these changes increase consumption, they do increase profits [17].

3   The Heroin Market Agent-Based Model[2]

Our model simulates interactions between people who buy and sell heroin to expand our existing understanding of local drug markets as complex systems [5] and connect micro-behavior and macro-market patterns associated with how much users spend to consistently consume heroin. We implement two types of agents: dealers (sellers) and customers (buyers). To reduce the parameter space, we exclude real world variables such as addiction, cash income, and variations in different drug units sold. Customer agents are automated consumers that: 1) never run out of money, 2) all purchase and consume the same unit of drug, one gram per transaction, and 3) always pay the same retail price, $120 per gram.

Customer agents schedule drug purchases based on a linear time scale set by the previous gram they consume. A gram is divided into equal units for this purpose; a 12-unit gram represents a full-sized gram. For customer agents, a 12-unit gram means an agent will make another purchase in eight hours; a 6-unit gram in six hours and so

---

[2] A technical description and source code is available at:
   http://code.google.com/p/drug-market

on[3]. This proxy's variation in the "size" of grams sold: more units = bigger grams (in real terms, each 1/12 increment equates roughly to an 8% change in gram size). It also allows agents to evaluate the *quality* of the product. Heroin is an "experience good," meaning consumers can only judge its quality *after* it is consumed [18]. A "good" deal is one in which the customer gets more heroin than a previous deal.

Customer agents have a (customer-to-dealer) *transaction* and a (customer-to-customer) *social* network. To purchase heroin they must have a direct link with a dealer or another customer (acting as a broker) who has a direct link with a dealer. At the start of the simulation, each customer agent is assigned some dealer agents. The agent selects the dealer from their *transaction* network offering the best value deal. If that dealer is unavailable, they go to the next best dealer, and so on. If a customer agent is not able to purchase from a dealer, they seek a deal from their *social* network.

In our model, seeking a deal from its *social* network, the customer agent uses a preferential ordering scheme based on the net good vs. bad deals shared by its social contacts over the past 30 days. The agent first asks the peer who has given them the best deal, the next best, and so on. A shared deal purchase is evaluated relative to all the customer agents' previous purchases. Social network ties dissolve when resulting deals are "not good" *n* number of times. Ties are assumed symmetric; dropping ties is mutual and we assume it will take time before they can re-establish their tie (by default, we assume a lag of 3 months).

Agents add a new dealer to their *transaction* network after purchasing a shared deal from another customer (a broker) *n* number of times (a model parameter). New *social* network ties are formed using a probability that increases based on the number of overlapping dealers two customer agents have in common. Customer agents who share deals (broker) receive a commission, fixed at $20 or variable depending upon model setup, from the agent they broker for and always offer the best deal from their *transaction* network. At the start, customer agents are linked in a *social* network constructed using one of the three network topologies Watts-Strogatz, Barabasi-Albert and Erdos-Renyi, see [19]. Parameters are set so approximately one new social network link is established per month. Customer agents dissolve ties with dealers in three ways: Customers drop dealer links if 1) they do not transact with the dealer for 7 days, 2) a dealer agent is "arrested" or fails (see below) or 3) if a dealer "franchises" their operation a customer link can be transferred to a new dealer, dissolving the old tie.

Dealer agents sell grams of heroin and resize grams in response to what is happening in their network of buyers. Dealers are supplied 12 grams at a time and are immediately resupplied when they sell out. Based on their sales activity, a dealer agent will 1) up-size their grams to attract customers and 2) downsize their grams if they have more customers and want to extract more profit. The simulation is configured so dealers change the size of their grams infrequently. If a dealer agent becomes too busy for too long they franchise their operations, randomly transferring half of their customers to a random customer, i.e., new dealer. New dealers sell gram units the same size as their benefactors or downsize to make more profit (a model parameter). Finally,

---

[3] Although somewhat arbitrary, this linear unit-to-time scale does reflect time frames heroin users report relative to their subjective experience injecting.

dealers drop out of the simulation if they: 1) do not have customers, 2) run a supply "surplus" for too long (a model parameter), or 3) get arrested (see below).

4 Simulation Results

All simulations are initially setup with 500 customer agents and 100 initial dealer agents and run for 1000 days. Here, we highlight the role of customer agents' social network in: 1) the diffusion of drug dealer information through brokering and 2) heroin cost adjustments in the market. We explore four specific settings with static and dynamic social network configurations together with low and high probabilities (0.1 and 0.75 respectively) of customer agent sharing deals. In all four settings, we assume an Erdos-Renyi configuration with an initial average degree of 12 for customer agent social network. The static network composition does not change whereas the dynamic network changes as the simulation proceeds, as described in the previous section. We also assume a high-risk situation with a baseline probability of a dealer agent being arrested (random removal) set high (0.01) but multiplied by the number of customer agents a dealer agent is linked to. This assumption simply reflects the more customers who know a dealer the more likely that dealer will be informed upon and consequently arrested. Although in real life dealers spend considerable effort avoiding arrest [8], this assumption reflects a fairly accurate baseline condition.

Fig. 1 shows the time-series chart for the number of dealer agents for respective simulations runs of the four configurations. As Fig. 1 shows, dealer agents survive with static networks and dynamic networks with low brokering. The market's collapse under the dynamic network configurations can be explained by the survival of dealer agents in the system, and how dealers' information is diffused among the customer agents through brokering of deals. In our model, dealer agents drop out through police arrests, or when they are unable to sell drugs to customer agents for some time (default: 7 days). On the other hand, new dealer agents enter into the system only when an existing dealer franchises. Under our high-risk situation, dealer agents get arrested at a much faster rate than they franchise. Also, in dynamic networks, when brokers finally share a dealer the chances are high that the dealer is already removed, earning the broker a 'bad' endorsement and more dropped ties. These intrinsic factors cause the system to fail. However, when brokering is low social ties are broken at a slower rate, suggesting this reluctance to share dealer information may not only be profitable to brokers (in terms of the taxes) but may also help the market.

Under a static network configuration, the number of dealers in the system stabilizes and reaches a dynamic equilibrium. This relates to the model parameter determining the 'dealing capacity' of a dealer agent, i.e., if a dealer agent is making a certain number of deals for a number of days it franchises its deals to a new dealer agent. Notice that here we are referring to the actual deals that a dealer agent is making per day and not the number of customer connections that it may have. A dealer agent may be known to many customer agents but may not get a deal at all. As Fig. 1 shows, even under a high risk of removal of dealer agents, the market sustains when the social network of the customer agent remains intact. Here brokerage plays a role in the

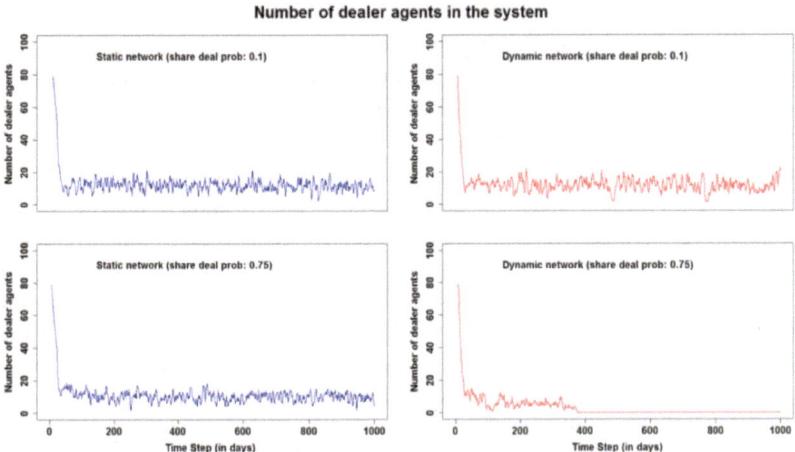

Fig. 1. Number of dealer agents for the four configurations based on the static and dynamic social network of customer agents and for low and high brokering

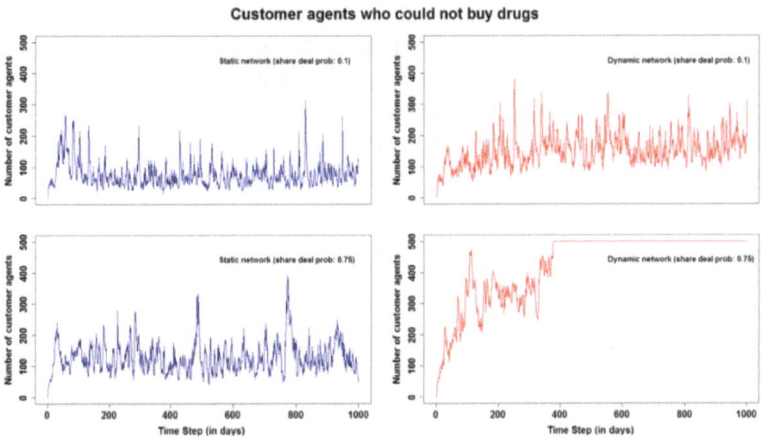

Fig. 2. Number of customers unable to purchase heroin for the four explored configurations

diffusion and redistribution of the links in the dealer-customer transaction network so that the removal rate of dealers through police arrests is balanced by the introduction of new dealers in the system. The system therefore continues to survive.

Fig. 2 shows time series of customer agents unable to buy drugs. In our model, customer agents learn about new drug dealers through their social ties. Customer agents have between 7-11 brokered deals before establishing a direct link to a dealer. Under our high-risk condition customers rely on brokering but we also find occasional spikes where a large number of customers are unable to buy drug. Counter-intuitively, high brokerage levels result in more episodic volatility and a greater number of customers failing to buy drugs. This happens because when sharing is high there is a high reliance on shared deals, information about dealers spreads faster, and dealer agents become more vulnerable to arrest, which increases their chances of being removed before a customer connects. Low sharing results in a slower diffusion of dealer information and relatively fewer customers unable to buy drug across the 5-year period.

In terms of *cost*, Fig. 3 notes the size of grams sold by units in the four conditions. In the fixed network time series, high brokering equals high competition with unit sizes occasionally touching both 9 and 14 units per gram. But despite increased volatility, the actual price paid by the customers remains comparable across all the four configurations. Here, a high rate of brokering results in higher volatility in the time series showing the number of customer agents who were not able to buy drugs (see Fig. 2) but increases dealers' competition in the market. For fixed network settings, high brokering (0.75) results in a higher variation in unit sizes, touching both 9 and 14 units per gram. Here dealer agents are competing for customers and adjusting prices frequently. For fixed networks with low sharing of deals (0.1), we see a narrower range of prices in the time series. Notice that in our model, dealer agents do not have information about the whereabouts and sales of other dealer agents in the market. Thus, dealer agents adjust their price based on the sales in the past and although they are unaware of other dealer agents' sales, they are compelled to adjust their price based on how many deals they were able to make on the previous day. Finally, the four histograms in Fig. 4 give the distribution of annual mean Quantity-Adjusted gram price paid in the four configurations. Effects of brokering can be seen in both the static and dynamic network configurations, with all distributions skewed above the $120 retail price.

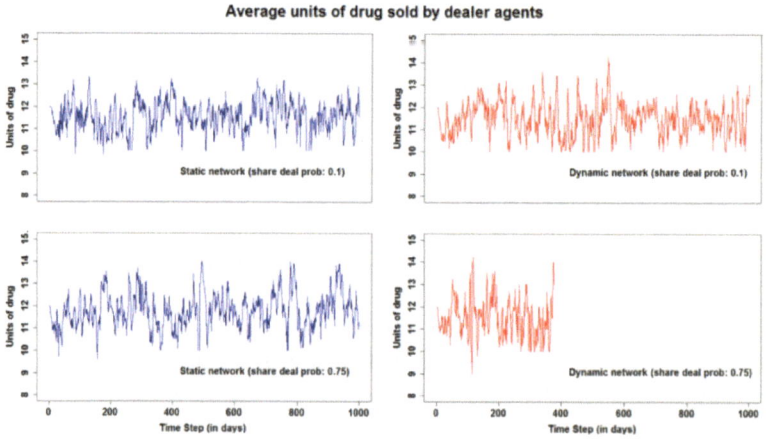

Fig. 3. Heroin unit size variation in grams being sold for the four explored configurations

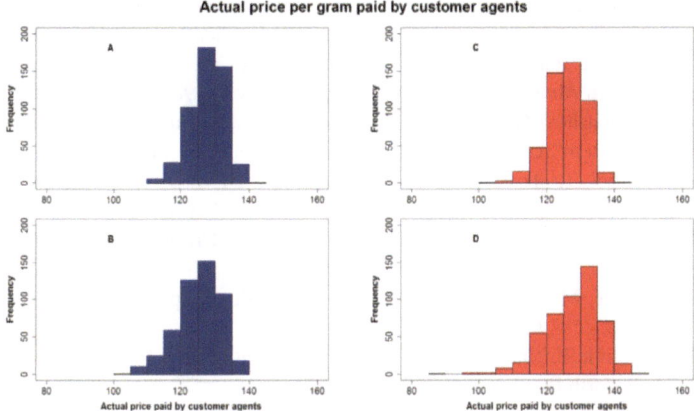

Fig. 4. Histogram of Quantity-Adjusted price per gram: static network and 0.1 share probability (A), static network and 0.75 share probability (B), dynamic network and 0.1 share probability (C) and dynamic network and 0.75 share probability (D)

5   Outlook

In real-world heroin markets, brokering is more popular than direct heroin sales and this collaborative effort (between an anthropologist who conducts ethnographic research with heroin users and dealers and a social simulation scientist) distinguishes why this is important. For consumers, brokering seems to overcome its advantage to influence better value deals in the heroin marketplace, i.e., increasing aggregate costs. However, in so doing it offers market stability under conditions of extraordinary (and unrealistic) outside pressure. In addition, brokering not only serves customer agents without dealers, it also affects competition in terms of price adjustments among dealers. Although detailed computational and social analyses remain to be presented, the findings are clear for policy: the dynamics of heroin price *do not* conform to orthodox economic models. Here logics supporting the war on drugs, i.e., arresting dealers as "supply reduction" are fundamentally defective. This paper emphasizes elevating the status of brokering as a social behavior transmitting market information and material, incorporating it into the epidemiology of drug addiction and in drug treatment / intervention efforts, in the resolve to develop effective drug policy.

Acknowledgements. LH was funded by a grant from the National Science Foundation, Cultural Anthropology Division (BCS 0951501).

References

1. Roddy, J., Steinmiller, C.L., Greenwald, M.K.: Heroin Purchasing is Income and Price Sensitive. Psychol. Addict. Behav. 25(2), 358–364 (2011)
2. Roddy, J., Greenwald, M.K.: An Economic Analysis of Income and Expenditures by Heroin-Using Research Volunteers. Substance Use and Misuse 44, 1503–1518 (2009)

3. Needle, H.R., Mills, A.R.: Drug Procurement Practices of the Out-of-Treatment Chronic Drug Abuser, National Institute on Drug Abuse, NIH Publication No. 94-3820 (1994)
4. Johnson, B.D., Goldstein, P.J., Preble, E., et al.: Taking Care of Business: The Economics of Crime by Heroin Abusers. Lexington Books, Lexington (1985)
5. Hoffer, L.D., Bobashev, G., Morris, R.J.: Researching a Local Heroin Market as a Complex Adaptive System. American J. of Community Psychology 44, 273–286 (2009)
6. Curtis, R., Wendel, T.: Toward the Development of a Typology of Illegal Drug Markets. In: Natarajan, M., Hough, M. (eds.) Illegal Drug Markets: From Research to Prevention Policy, Crime Prevention Studies, vol. 10, pp. 121–151. Criminal Justice Press, Monsey (2000)
7. Page, J.B., Singer, M.: Comprehending Drug Use: Ethnographic Research at the Social Margins. Rutgers Univ. Press, NJ (2010)
8. Hoffer, L.D.: Junkie Business: the Evolution and Operation of a Heroin Dealing Network. Thomson Wadsworth Publishing, Belmont (2006)
9. Goldstein, P.J.: Getting Over: Economic Alternatives to Predatory Crime Among Street Drug Users. In: Inciardi, J.A. (ed.) The Drug-Crime Connection, pp. 67–84. Sage (1981)
10. Preble, E., Casey, J.J.: Taking Care of Business: The Heroin User's Life on the Street. International J. of Addictions 4, 1–24 (1969)
11. Waldorf, D.: Careers in Dope. Prentice-Hall, Englewood Cliffs (1973)
12. Zule, W.A.: Risk and Reciprocity: HIV and the Injection Drug User. J. of Psychoactive Drugs 24(3), 243–249 (1992)
13. Page, J.B., Smith, P.C., Kane, N.: Shooting Galleries, their Proprietors, and Implications for Prevention of AIDS. Drugs and Society 5(1-2), 69–85 (1990)
14. Koester, S., Hoffer, L.D.: Indirect Sharing: Additional HIV Risks Associated with Drug Injection. AIDS and Public Policy J. 9(2), 100–105 (1994)
15. Maher, L.: Illicit Drug Reporting System (IDRS) Trial: Ethnographic Monitoring Component, National Drug and Alcohol Research Centre, University of New South Wales, Sydney, Technical Report No. 36 (1996)
16. Consumer Reports, February 18-21 (2011)
17. Megerdichian, A.: Product Downsizing and Hidden Price Changes in the Ready-to-Eat Cereal Market, dissertation chapter (2010)
18. Caulkins, J.P., Reuter, P.: What Price Data Tell Us about Drug Markets. J. of Drug Issues 28(3), 593–612 (1998)
19. Newman, M.E.J.: The structure and function of complex networks. SIAM Review 45, 167–256 (2004)

Cultural Polarization and the Role of Extremist Agents: A Simple Simulation Model

Shade T. Shutters

Center for Social Dynamics and Complexity and School of Sustainability,
Arizona State University, PO Box 875402, Tempe, AZ 85287-5402, USA
shade.shutters@asu.edu
www.public.asu.edu/~sshutte

Abstract. Cultural dynamics can be heavily influenced by extremists. To better understand this influence, temporal dynamics of an arbitrary cultural belief are simulated in a simple computational model. Extremist agents, holding an immutable and extreme belief, are used to examine the process of polarization – adoption of the extremist belief by the entire population. Two possible methods of counteracting polarization are examined, removal of the extremist agent and introducing a counter-extremist which holds an immutable belief at the opposite extreme. Eliminating the extremist agent is only effective at the onset of cultural transition, while introducing a counter-extremist is effective at any time and will lead to a dynamic intermediate belief. Finally, a parameter governing the society's willingness to adopt new beliefs is varied. As it decreases, extremist agents are unable polarize a society. Instead the population breaks permanently into two or more belief groups. The study closes with a possible pathway for extremists to nevertheless polarize a society not open to new beliefs.

Keywords: extremism, cultural transitions, consensus, networks, social simulation.

1   Introduction

Though conflict is an inextricable component of social organisms, humans are unique in that culture plays a central role in many conflicts [1]. To address conflict and to better understand social dilemmas more generally it is important to understand the dynamics of cultural beliefs and how those dynamics may be influenced. Previous studies have examined social diffusion of ideas [2], norms [3], innovations [4], social values [5], and diseases [6]. Others have focused on how the topology of the network governing a society influences the rate, dynamics, and efficacy of diffusion [7-11].

This study is largely a continuation of work presented in [11], which addresses cultural consensus and sources of perpetuated conflict. In the current paper, the effects of extremist agents are examined. Given a continuum of values representing an arbitrary cultural belief or norm, extremist agents hold a belief at one endpoint of the continuum and the belief cannot be changed. Because extremist agents may be sources of incitation to violence or other socially disruptive behavior it is important to

A.M. Greenberg, W.G. Kennedy, and N.D. Bos (Eds.): SBP 2013, LNCS 7812, pp. 93–101, 2013.

understand how they affect cultural dynamics and how they might respond to counter measures.

This study uses very simple and highly abstract social simulations to better understand how extremist agents affect the timing and ability of a population to become polarized. It further tests and contrasts two intuitive methods of preventing polarization: removing extremist agents from a population and introducing a counter-extremist.

2   Simulation Description

The base-case simulation initiates by embedding N agents in one of four social network structures. Each agent holds a single arbitrary belief that is assigned an initial random value, with uniform probability, on [0, 1]. The model then proceeds through a number of pairwise interactions until the population either converges to a single, universal belief or the simulation reaches the maximum allowable number of interactions.

During a single interaction, a member of the population is selected at random and paired randomly with one its immediate neighbors as defined by the network type. Let a_0 and b_0 represent the initial belief values of two interacting agents so that the initial difference between their beliefs is

$$T = |a_0 - b_0| \qquad (1)$$

The interacting agents influence each other's beliefs so that they are updated to a_1 and b_1 respectively. In this study, the new values are equal to each other and to the mean of their original beliefs

$$a_1 = b_1 = (a_0 + b_0) / 2 \qquad (2)$$

Given enough interactions, the population will converge to a single belief equal to the mean value of the initial population [11, 12].

A population level parameter D, determines whether the beliefs of two interacting agents are sufficiently similar for the agents to influence each other. This threshold represents the willingness of agents to adjust their beliefs towards others. One might also consider this parameter the degree to which a society is "open-minded" or dogmatic. During a pairwise interaction, the difference between agent beliefs T is compared to the threshold D. If the difference it too great, the interaction ends without any changes in beliefs

$$T \leq D: a_1 = b_1 = (a_0 + b_0) / 2$$

$$T > D: a_1 = a_0 \text{ and } b_1 = b_0 \qquad (3)$$

In the first series of simulations, $D = 1$ so that all agents change their beliefs when interacting with an agent that holds a different belief. In later treatments D is varied to understand how and when polarization is affected by a population's willingness to adopt new beliefs.

2.1 Extremist Agents

In some simulations a single agent is chosen randomly from the initialized population and converted to an extremist agent. Its belief value is set to 0 and it cannot be changed for the duration of the simulation. Given that an extremist agent is a participant in an interaction, let $a_0 = 0$ be the belief of the extremist agent and let b_0 be the initial belief of the other participant. The interaction results in $a_1 = a_0 = 0$ while $b_1 = b_0 / 2$. Thus the belief of the normal agent moves toward that held by the extremist agent, but the belief of the extremist agent remains unchanged.

In this case, given enough interactions, the belief of every agent in the population will converge to the belief of the extremist agent, a process referred to in this study as polarization.

2.2 Counter-Extremist Agents

In additional simulations a single extremist agent is again included in the initial population. After a number of interactions, which can be varied, an agent (other than the extremist agent) is selected randomly from the population and converted to what is referred to here as a counter-extremist agent. Its belief value is set to 1 and cannot be changed for the duration of the simulation. Thus its behavior during a pairwise interaction is identical to that of an extremist agent but causes beliefs to move towards the opposite extreme.

2.3 Network Structures

Four network topologies – complete, scale-free, small-world, and regular – are used to examine the role of social structure on polarization. All networks are unweighted and undirected. Scale-free networks are generated using a Barabási-Albert algorithm of preferential growth [13] with no nodal limit on links and in which each new node links to the existing network at two nodes. Small-world networks are generated using the Watts-Strogatz algorithm [14] in which each node in a ring substrate is linked to the two neighbors on either side and edges are randomly rewired with a probability $p = 0.05$. Regular networks are torroidal lattices in which each node has four adjacent neighbors – up, down, left, and right. For all networks other than complete mean degree $k = 4$, meaning differences in results among those networks are due to topological attributes other than mean degree or network density.

3 Results and Discussion: Time Until Polarization

Under all network topologies, the introduction of an extremist agent eventually led to polarization of the population (Table 1). However, the number of interactions required for polarization differed significantly among the four network types, both with population $N = 64$ (ANOVA, $F = 2,332$, $p < 0.001$) and $N = 400$ (ANOVA, $F = 516$,

$p < 0.001$). Compared to small-world networks, populations on complete, scale-free, and regular networks required relatively few interactions to become polarized. However, populations on small-world networks required up to 20 times more interactions to reach polarization than populations on a complete network.

In [11] we determined, for several network topologies, the number of interactions required for a population to converge to a single belief value. In comparison to these results, the addition of an extremist agent increased convergence time by as much as 160 times (Table 1). This suggests that polarization is a much slower phenomenon than forming a consensus at some intermediate belief value.

Table 1. Mean interactions (in thousands) until belief convergence, with and without extremist agents. Mean calculated from 100 runs.

Network type	N = 400		N = 64	
	With	Without	With	Without
Regular ($k = 4$)	3,975	176	77	5
Scale-free	3,330	37	87	5
Small-world	38,083	526	271	36
Complete	1,739	10	47	2

3.1   Societies Less Open to New Beliefs

Results shown in Table 1 were collected from simulations with $D = 1.0$. In additional simulations D was varied in increments of 0.1 to determine how a society's openness to new beliefs affects the ability of an extremist agent to polarize a population. Results (Fig. 1) show that at high values of D, an extremist agent will always result in a polarized society. At intermediate values of D (~ 0.4 to 0.6), the probability of polarization drops rapidly until when $D \leq 0.3$, polarization never takes place.

3.2   Further Considerations of Scale-Free Networks

Because scale-free networks are ubiquitous among physical systems as well as communication networks in social systems [15, 16], it is important to understand how cultural dynamics might be influenced by a scale-free topology. Thus, effects of scale-free networks on the ability of extremist agents to polarize a society are examined in more detail.

Results show that as the nodal degree k of the extremist agent increases the mean number of interactions required to achieve polarization decreases according to a power law (Fig. 2). This concurs with the intuitive notion that highly connected actors have a disproportionately strong influence on a society's cultural trajectory.

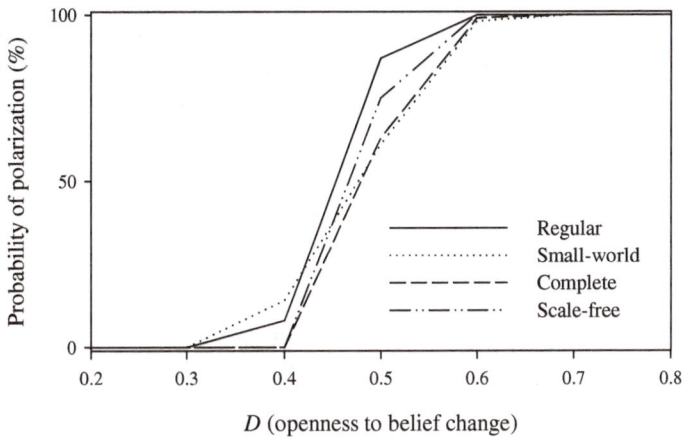

Fig. 1. Probability of polarization vs. D. Population $N = 64$. Probability calculated for 100 simulation runs per D value on each network type. As threshold D decreases below ~0.35, the probability of polarization drops to 0 on all networks.

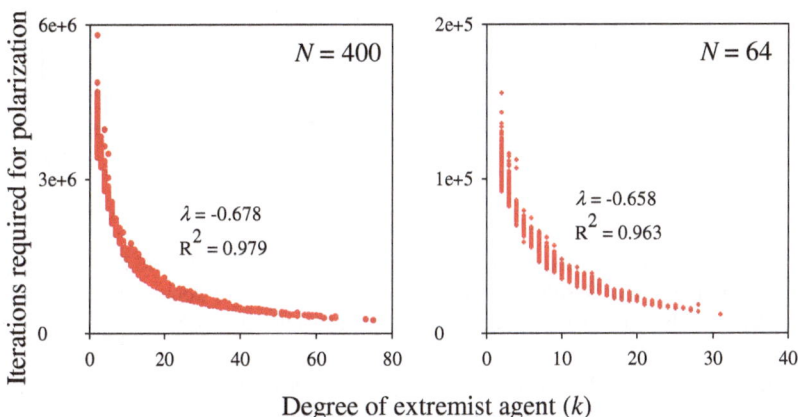

Fig. 2. Degree of extremist agent vs. time required for polarization on scale-free networks. Results are shown for 2,000 runs each at population sizes $N = 400$ and $N = 64$. Power law exponent λ and simple correlation coefficient R^2 are shown.

3.3   Counteracting Polarization

I now consider two intuitive methods for preventing polarization in scale-free networks. In both cases $D = 1$. The first method is removal of the extremist agent and the second method is introduction of a counter-extremist agent.

Removal of the extremist agent immediately halts further movement of the population's mean belief toward the extreme value. However, the future value to which the population will converge is fixed at time of removal and equal to the population's mean at that time. As shown in Fig. 3, this value drops rapidly after introduction of the extremist agent and so simply removing the extremist agent is only effective if done very early after it has begun to influence a society. Fig. 3 shows that when the extremist agent is removed after only 5,000 interactions, or about 6% of the interactions required for polarization, the population's mean belief has already fallen to 0.18, a value relatively close to the extremist belief.

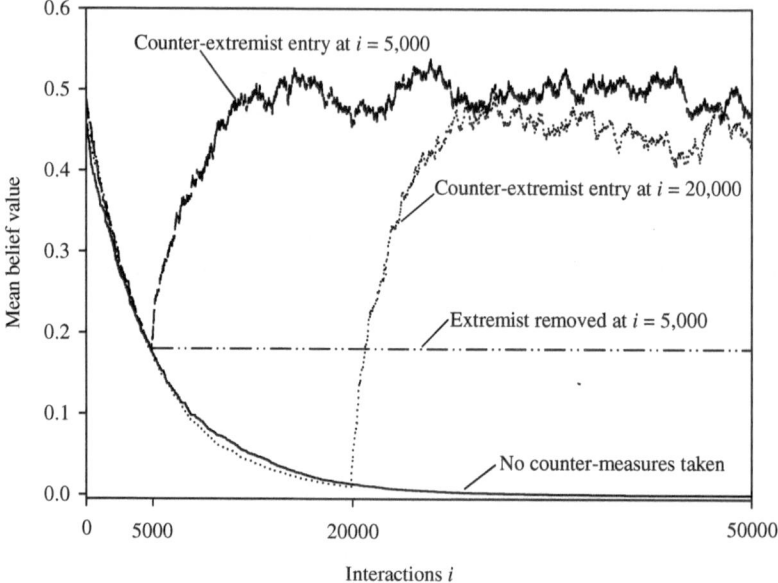

Fig. 3. Temporal dynamics of the population's mean belief. The first 50,000 interactions of four simulations are presented. Results are for a scale-free network with population size $N = 64$ and with the degree of both the extremist and counter-extremist agents $k = 5$.

On the other hand, the introduction of a counter-extremist agent will lead to a certain dynamic equilibrium belief value b_{eq} regardless of when the counter-extremist is introduced. In networks other than scale-free networks $b_{eq} \approx 0.50$. In scale-free networks b_{eq} is a function of the nodal degrees of the extremist k_e and counter-extremist k_c approximated by the belief-weighted relative probability of each being randomly chosen for an interaction

$$b_{eq} \approx (k_c+1) / [(k_e+1) + (k_c+1)] \qquad (4)$$

Within the constraints of this abstract model, the introduction of a counter-extremist agent offers a better counter-measure to polarization than simple removal of the extremist agent. This occurs when $D = 1$ and the effectiveness of introducing a counter-extremist would likely decrease over time in real world situations as the population slowly moves out of the counter-extremist's range of influence.

4   How Extremist Agents Can Overcome Low Willingness to Change

The concept of D is highly abstract and it is likely that any analog in real human societies is not only highly heterogeneous among individuals, but also relatively low. This presents the extremist agent with an obstacle to polarization. I close by discussing how an extremist agent may overcome this obstacle by introducing an additional agent type, the dogmatic agent. Like extremists and counter-extremists, dogmatic agents hold an immutable belief. However, the value of that belief lies at some intermediate point between extremes.

In the presence of such dogmatic agents, extremist agents have an opportunity to polarize a population even when D is low. The extremist must rely on dogmatic agents to convert the beliefs of those in the population that are far out of the extremists range of influence $(T \gg D)$. The extremist then must go through a stepwise process of eliminating dogmatic agents, starting first with those that differ most with respect to belief value. The extremist should also allow sufficient time between the elimination of dogmatic agents so that followers of recently-eliminated dogmatic agents will be fully drawn to the next closest dogmatic agent.

Consider a simple illustration. Let $D = 0.3$ for a certain population, meaning it is largely averse to change. As shown in Fig. 1, an extremist agent with belief 0 will be unable to polarize the society. Let dogmatic agents exist with beliefs at 0.75, 0.50, and 0.25. Given sufficient time, all agents originally holding beliefs on (0.75, 1] will be adjusted to 0.75 (or less). At that time, the extremist agent should take measures to eliminate the dogmatic agent with belief 0.75. This will lead to all agents holding a belief on (0.50, 0.75] to eventually hold beliefs of 0.50 or less, since that entire interval is within the range of influence of the next closest dogmatic agent (belief 0.50). At this point, the dogmatic agent with belief 0.50 is targeted for elimination and so on until the entire population is moved to the extremist belief.

Anecdotal evidence for such a strategy exists throughout history in cases where agents, once considered relatively extreme, come to be viewed as moderate in their beliefs and are eliminated. Well known examples include the 1922 assassination of Irish militant/politician Michael Collins by more extreme Irish nationalists and the 1917 Russian Revolution, in which dogmatic agents first overthrew the Tsarist regime but were then expelled some months later by the more extreme Bolsheviks. Contemporary examples might include the frequent execution of moderate Muslim clerics in Dagestan by more radicalized Islamic militants.

5   Future Directions

This study has used a simplistic binary method to determine the degree to which an agent may be influenced by another. An agent is either completely influenced or not at all. A more realistic assumption is that the ability of one agent to influence the beliefs of another decays as a function of the difference of their current beliefs.

Let i equal the degree to which agents may influence each other so that equation (2) becomes

$$a_0 \geq b_0; \quad a_1 = a_0 - iT/2 \quad \text{and} \quad b_1 = b_0 + iT/2$$

$$a_0 < b_0; \quad a_1 = a_0 + iT/2 \quad \text{and} \quad b_1 = b_0 - iT/2 \tag{5}$$

As with (2), this equation does not apply to extremist or dogmatic agents. In addition, implicitly $D = 1$ (though it may be explicitly set to values less than 1 to explore additional parameter space).

It is important then to understand how i decays as a function of the initial beliefs of two agents. A simple decay function might take the form

$$i = 1 - T^\lambda \tag{6}$$

Fig. 4 compares this decay function at three values of λ to the binary influence used above. However, evidence suggests that $\lambda = 0.2$ is the most realistic of the decay functions presented [17] and future empirical work should seek to refine this value. More importantly, the efficacy of extremists and extremist counter-measures discussed above should be reassessed under different decay models of influence.

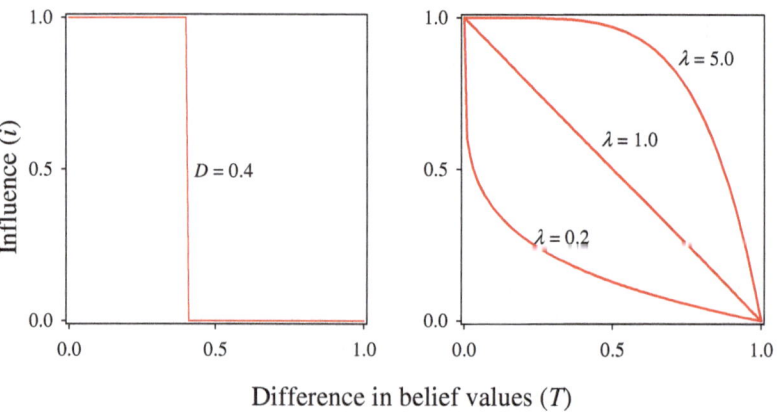

Fig. 4. Comparison of binary influence (left) and influence as a decaying function belief differences (right). In the binary model, $D = 0.4$ so that when $T > 0.4$, influence = 0. The decay model is presented for three values of λ. When $\lambda = 0.2$, an agent's beliefs can only be significantly influenced by another agent holding very similar beliefs.

6   Conclusion

This brief study has described a very simple and abstract model of the effects of extremist agents. It shows that, given enough time, extremists can polarize a population and that either removing the extremist agent or introducing counter-extremists can mitigate the extremist's polarizing effect. However, caution should be taken before drawing broad conclusions from such an abstract model. Future research should incrementally introduce more realistic parameters that are empirically grounded. In this manner, this and similar models will continue to make small but valuable steps toward a better understanding of cultural dynamics.

References

1. Bowles, S.: Warriors, Levelers, and the Role of Conflict in Human Social Evolution. Science 336, 876–879 (2012)
2. Buskens, V., Yamaguchi, K.: A new model for information diffusion in heterogeneous social networks. Sociol. Methodol. 29, 281–325 (1999)
3. Axelrod, R.: An Evolutionary Approach to Norms. Am. Polit. Sci. Rev. 80, 1095–1111 (1986)
4. Mahajan, V., Muller, E.: Innovation Diffusion and New Product Growth-Models in Marketing. J. Marketing 43, 55–68 (1979)
5. Bikhchandani, S., Hirshleifer, D., Welch, I.: A Theory of Fads, Fashion, Custom, and Cultural-Change as Informational Cascades. J. Polit. Econ. 100, 992–1026 (1992)
6. Santos, F.C., Rodrigues, J.F., Pacheco, J.M.: Epidemic spreading and cooperation dynamics on homogeneous small-world networks. Phys. Rev. E 72, 56128 (2005)
7. Cowan, R., Jonard, N.: Network structure and the diffusion of knowledge. J. Econ. Dyn. Control 28, 1557–1575 (2004)
8. Abrahamson, E., Rosenkopf, L.: Social network effects on the extent of innovation diffusion: A computer simulation. Organ. Sci. 8, 289–309 (1997)
9. Delre, S.A.: The Effects of Social Networks on Innovation Diffusion and Market Dynamics. Department of Economics and Business, vol. PhD, p. 148. University of Groningen, Groningen, Netherlands (2007)
10. Santos, F.C., Rodrigues, J.F., Pacheco, J.M.: Graph topology plays a determinant role in the evolution of cooperation. Proc. R. Soc. Lond. B. Biol. Sci. 273, 51–55 (2006)
11. Shutters, S.T., Cutts, B.B.: A Simulation Model of Cultural Consensus and Persistent Conflict. In: Subrahamanian, V.S., Kruglanski, A. (eds.) Proceedings of the Second International Conference on Computational Cultural Dynamics, pp. 71–78. AAAI Press, Menlo Park (2008)
12. Boyd, S., Ghosh, A., Prabhakar, B., Shah, D.: Randomized Gossip Algorithms. IEEE T. Inform. Theory 52, 2508–2530 (2006)
13. Barabási, A.L., Albert, R.: Emergence of scaling in random networks. Science 286, 509–512 (1999)
14. Watts, D.J., Strogatz, S.H.: Collective dynamics of 'small-world' networks. Nature 393, 440–442 (1998)
15. Barabási, A.L.: Scale-Free Networks: A Decade and Beyond. Science 325, 412–413 (2009)
16. Mendes, J.F., Dorogtsev, S.N., Povolotsky, A., Abreu, F.V., Oliveira, J.G. (eds.): Science of Complex Networks: From Biology to the Internet and WWW; CNET 2004. American Institute of Physics, Melville, New York (2005)
17. Cialdini, R.B., Trost, M.R.: Social influence: social norms, conformity, and compliance. In: Gilbert, D., Fiske, S., Lindzey, G. (eds.) The Handbook of Social Psychology, 4th edn., vol. 2, pp. 151–192. McGraw-Hill, New York (1998)

Using Imageability and Topic Chaining to Locate Metaphors in Linguistic Corpora

George Aaron Broadwell[1], Umit Boz[1], Ignacio Cases[1], Tomek Strzalkowski[1],
Laurie Feldman[1], Sarah Taylor[2], Samira Shaikh[1], Ting Liu[1], Kit Cho[1],
and Nick Webb[3]

[1] State University of New York – University at Albany, NY 12222 USA
[2] Sarah M. Taylor Consulting, LLC
[3] Union College, Schenectady, New York

Abstract. The reliable automated identification of metaphors still remains a challenge in metaphor research due to ambiguity between semantic and contextual interpretation of individual lexical items. In this article, we describe a novel approach to metaphor identification which is based on three intersecting methods: *imageability*, *topic chaining*, and *semantic clustering*. Our hypothesis is that metaphors are likely to use highly imageable words that do not generally have a topical or semantic association with the surrounding context. Our method is thus the following: (1) identify the highly imageable portions of a paragraph, using psycholinguistic measures of imageability, (2) exclude imageability peaks that are part of a topic chain, and (3) exclude imageability peaks that show a semantic relationship to the main topics. We are currently working towards fully automating this method for a number of languages.

Keywords: automated metaphor identification, imageability, topic chaining, semantic clustering, linguistic corpora, MRC psycholinguistic database, WordNet.

1 Introduction

Humans can reliably distinguish literal from metaphorical interpretations of words in discourse in a seemingly effortless way. For example, consider the word *minefield*. In some cases, it refers literally to a piece of ground in which mines have been placed. In others cases, it refers to some dangerous area or territory. A simple Google search for the word *minefield* will return over four million hits; millions of them will have literal interpretations and millions will have metaphorical interpretations. A human can usually distinguish them. But how could a computational system distinguish between the two readings to return only results relevant to mines? Or only results relevant to metaphors about dangerous situations?

Although the development of large linguistic corpora over the last two decades has greatly improved many aspects of the computational linguistic search processes, reliable automated identification of metaphors still remains a challenge in metaphor research. The pioneering work of Martin (1988) and Cameron (1999) in this field has

A.M. Greenberg, W.G. Kennedy, and N.D. Bos (Eds.): SBP 2013, LNCS 7812, pp. 102–110, 2013.

led to a number of techniques for metaphor identification, but such approaches also have their limitations. One is that such methods rely largely on the micro-level discourse analysis of the data with extensive interpretation of the semantic and contextual attributes of individual lexical items (Crisp et al., 2007; Steen et al., 2010).

In this paper, we describe a novel approach to metaphor identification which is based on three intersecting methods: imageability, topic chaining, and a semantic clustering analysis.

2   Metaphor Identification via Imageability

Our first method is based on the imageability of words, a property of words which is well-established in the psycholinguistic literature. A word is more imageable to the extent that it is possible to form a mental picture of its meaning. The imageability measure is based on rating data and values range from 100 to 700 (Paivio, Yuille, & Madigan, 1968; Gilhooly & Logie, 1980). Imageability has been shown to play an important role in memory, word recall, lexical decision tasks (Bleasdale, 1987; Nelson & Schreiber, 1992; Winnick and Kressel, 1965; Paivio and O'Neill, 1970; Reilly and Kean, 2007), and in experimental tasks of reading (Strain, Patterson, & Seidenberg, 1995). In our metaphor identification method, we used the original ratings determined via human subjects and available in the MRC psycholinguistic database (MRCPD) (Coltheart, 1981, Wilson 1988) and adapted the original 100 to 700 scale to a percentage scale. Our hypothesis is that metaphors are likely to use highly imageable words, and words that are generally more imageable than the surrounding context. Our method graphs the words in a paragraph according to their imageability and looks for peaks as potential metaphors.

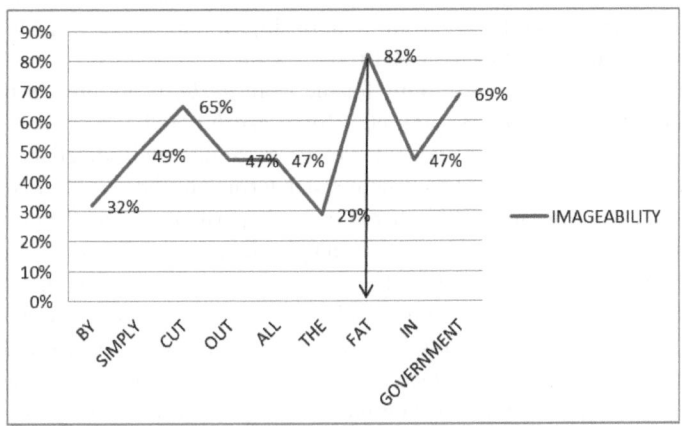

Fig. 1. Imageability values (%) of words in a sample metaphorical expression

For instance, Fig.1 shows the imageability ratings of the words in a portion of a paragraph (i.e., "…by simply cutting out all the *fat* in government") on government and bureaucracy [18]. In this example, the word *fat* has the highest imageability percentage, which, based on our approach, suggests a metaphorical usage in this particular context. As a first step in our metaphor identification method, imageability analysis allows us to identify such highly imageable words as potential metaphors in a given passage. However, considering that such words may be used literally, we further analyze the passage via topic chaining and semantic clustering to identify and eliminate the highly imageable words with literal meanings. Other metaphorical expressions that commonly employ imageable words can be listed as "navigating through the *maze* (79%) of bureaucracy", "being caught in a *web* (86%) of state and local government bureaucracy", "*wheels* (82%) of bureaucracy" etc.

2.1   MRCPD Expansion

Although the MRCPD contains data for over 150,000 words, a major limitation of the database is that not all words have ratings for all 26 variables. This finding is not particularly surprising because the MRCPD is composed from four separate sources of data, and not all sources collected values for all 26 variables. Of interest in the present study is the variable of *imageability* (i.e., how easily and quickly the word evokes a mental image) for which the MRCPD has ratings for only 9,240 (6%) of the total words in its database. We expanded the MRCPD database by adding imagery ratings for an additional 59,989 words (Cho et al., 2013). This was done by taking the words for which the MRCPD database has an imageability rating and using that word as an index to synsets as determined using the WordNet (Miller, 1995) database. New lexical items linked to those synsets were then added to the MRCPD data. The results were validated using regression analyses to compare how well imageability ratings and word frequency of words predict subjects' reaction time in identifying the word, as previous research reported a strong relationship among these variables (Baayen, Feldman, & Schreuder, 2006; Green & Brock, 2002).

For example, the imageability rating for the word *gerbil* is not present in the MRC database, but the word *rat* is present.   Since *gerbil* and *rat* share a WordNet hypernym, we developed a method in which (1) nouns and adjectives inherit the imageability scores of sister terms, where sister terms are those with a shared direct hypernym (e.g. where *rodent* is the shared direct hypernym of *gerbil* and *rat*) and (2) nominal and adjectival hyponyms inherit imageability scores from their hypernyms. By the logic of the second step, if we have imageability scores for *dog*, we can allow these scores to be inherited by words such as *terrier* and *poodle*. Because a single English word may appear in multiple WordNet synsets, we also face a question of which synset to select for inheritance purposes. The word *dog*, for example appears in the synset corresponding to the familiar meaning 'domestic dog; canis familiaris', but also in synsets where it serves as a synonym for 'unattractive girl or woman', 'morally reprehensible person', 'andiron', and 'sausage served in a bun', among others. WordNet generally structures lexical relations in such a way that the first synset corresponds to the most literal use of word, therefore we pursued a methodology in which we selected only the first synset for expansion.

A third step proved to be useful for verbs, in which we used verbal imageability scores to approximate the imageability of their hypernyms.   Thus from the imageability rating of the verb *to dog*, we approximate the imageability of hyponyms such as *chase,* and *hunt.*

By using a method of imageability rating expansion via first WordNet synset for nouns and adjective and hypernyms for verbs, we were able to construct an expanded imageability lexicon of 32,505 distinct words for English (or 59,989 if words which appear as multiple parts of speech are counted in each occurrence).

To validate the appropriateness of such an expansion, we used the previously-established negative relationship between lexical properties (e.g., lexical frequency, imageability) and reaction time (RT) in a lexical decision task where participants must judge whether letter strings are real words and indicate either "yes" or 'no" by a button press (Baayen, Feldman, and Schreuder 2006; Hargreaves and Pexman, 2012). Specifically, for words that appear infrequently in the literature or are less imageable, subjects require a longer period of time to correctly identify whether the string of letters is an English word. The English Lexicon Project (Balota, et al. 2007) (http://elexicon.wustl.edu/) provides RT data as well as lexical measures for over 40,000 words. We used simultaneous multiple regression methods to examine the how well word frequency and imagebility, individually and collectively, predict RT by selecting samples from the original and expanded concreteness database and comparing the predictive relation between RT and concreteness ratings. If the relationships of these variables are similar for both sources of words, i.e., the original and expanded lexicon, we would have confidence that our expansion method is indeed valid. For words in the original MRC database, the frequency and imageability variables are negatively related to RT, -.291, and -.624, respectively.[1] Conjunctively, these two variables accounted for 49.2% of the total variance of RT, which is statistically significant by conventional standards, i.e., $p < .05$, two-tailed.[2] These findings replicate those in the previous literature (e.g., Hargreaves and Pexman, 2012). Of current interest is whether words whose imageability rating is estimated would show a similar relationship. The results indicated the affirmative: frequency and imageability were both negatively related to RT, -.197 and -.747, respectively, and conjunctively they accounted for 58.4% of the variance of RT, which was statistically significant. These findings show that words in the original MRC database and words in the expansion via first WordNet synset (for nouns/adjectives) and hypernym (for verbs) show highly similar predictive values with respect to RT. This demonstrates the validity of our tool.

---

[1] These values represent Standardized β values which indicates the number of standard deviations that the outcome variable changes as a result of one standard deviation change in the predictor.

[2] The adjusted R^2 value ranges from 0-100%. The value is derived from squaring the correlation coefficient.

We take these results to be a strong indication that the method of imageability rating expansion via WordNet produces a lexicon in which words with inherited imageability ratings show very similar psycholinguistic properties to words where imageability rating were derived by experiments with human subjects. With the larger imageability lexicon, we are thus able to approach the problem of linguistic metaphor identification with a more robust tool which is able to plot imageability values for the great majority of English text. We should also note that the imageability analysis along with other two methods reported in the following sections is currently being applied to metaphor identification in Spanish. Comparable imageability results are available in the Spanish psycholinguistic literature at Davis and Perea (2005) and other sources. We have used an expansion technique parallel to that just described for English with the Spanish imageability results to make them usable for a larger portion of the Spanish lexicon.

3   Topic Chaining

We combine the search via imageability with a second method which seeks to establish the topic chains in a discourse. We follow the definition of topic chain in Broadwell et al. (2012) where a topic chain is a noun phrase along with all subsequent noun phrases which refer to the first noun phrase (by pronominal mention, repetition, or synonym). According to Broadwell et al., each noun phrase in a topic chain can be considered as a *local topic*, whereas a noun phrase outside of any topic chain (a word that is mentioned only once in data) is conceptualized as a non-local topic. Drawing on this definition, our hypothesis is that metaphorical words do not tend to have a topical association with the surrounding context, thus standing out from the topical structure of discourse like islands on the ocean.

For instance, we want to exclude highly imageable phrases (e.g. *the fat*, as in *cut the fat*) which occur in a paragraph on grilling meat, since the word *fat* is likely to be topically related to the previous mentions of *fat*, thus being used with its basic meaning. However the same phrase, *cut the fat*, should be counted as metaphorical in a paragraph on economics (e.g., there really was no *fat* left to *cut* from the budget), since our hypothesis is that neither imageable word is likely to appear outside a metaphorical context.

Table 1 illustrates our topic chaining method by indicating the majority of the topic chains extracted from a text on government and economy. In that particular text, the distribution of noun phrases by topicality was as follows: topic chains 68% (103 counts), non-local topics 32% (48 counts). Upon filtering out the topic chains, our imageability analysis indicates the non-local topics *armies*, *cancer*, and *fat* as having the highest imageability scores (83%, 81%, 82% respectively) in the passage, which are all in fact used metaphorically.

Table 1. Longest topic chains in a passage on government and economy

Topic Chains	Total Number of Local Topics
bureaucracy	22
government	19
tax	7
conservatives	6
city	5
waste	5
budgets	3
agencies	3
programs	3
man	3

One limitation of this method is that in certain cases, a metaphorical word may be repeated more than once, thus forming a topic chain with its subsequent mentions. In that case, our method would disregard such instances for further consideration as a metaphor. While this phenomenon may be highly infrequent, we plan to address this limitation in our future efforts.

4   Semantic Clustering

The third method is a semantic clustering technique, in which we identify and cluster words in the data that are semantically related to the main topics of the paragraph. As with topically related words, we exclude semantically related words from the set of likely metaphors. In order to implement this particular analysis, we are developing an automated technique which incorporates search processes from the Corpus of Contemporary American English (COCA) and Princeton WordNet.

As a sample analysis, let's consider the following passage that includes a metaphorical phrase:

Seven Morgan County transportation enhancement projects awarded over the past two years remain in the planning stages, *seemingly ensnared in a web of state and local government bureaucracy.*

In this example, *ensnared* and *web* are imageability peaks with their scores 73% and 86% respectively; *award* (img:75%) is also high in imageability, but is excluded as a potential metaphor because of its semantic association with *project* (img:65%). The words *transportation* (img:61%), *state* (img:73%), *local* (img:69%), and *government* (img:69%) are also filtered out as they are part of a topic chain in the greater context.

As previously discussed, topic chaining and semantic clustering methods act essentially as filters on the first method to exclude highly imageable lexical items where contextual clues point toward a literal reading. In this respect, we believe our method results in fewer false positives that the other methods of semantic annotation (e.g. Koller et al 2008) which do not incorporate a contextual component.

Our method is thus the following: (1) identify the highly imageable portions of a paragraph, using both existing psycholinguistic measures of imageability and an expanded lexical database of imageability, (2) exclude imageability peaks that are part of a topic chain, and (3) exclude imageability peaks that show a semantic relationship to the main topics.

5   Evaluation

In order to create a ground truth for the accurate retrieval of metaphorical examples, we collected human assessments both from expert judgments as well as crowd-sourcing interfaces like Amazon Mechanical Turk. We first presented participants with a brief set of instructions on how to identify a number of metaphors in a given context, which was immediately followed by the actual assessment task. For the actual task, participants were given a set of text fragments for which they were asked to judge the degree to which "metaphorical language is present or not". The validation task involved a set of other questions such as how imagable and common the expression was. Once we had enough human assessment collected, we could establish a ground truth against which system performance could be judged. Additionally, crowd-sourced data was validated using a grammar test score as well as ability to correctly classify known metaphorical and literal examples.

Once ground truth was established in this manner, we could adjust the performance of system in correctly classifying metaphorical vs. literal language. Our automated prototype system performs at 71% accuracy in detecting metaphors in English language data and 80% for Spanish language data. It should be noted that this result is obtained by applying the system to a series of approx. 200 examples (100 English and 100 Spanish) where only about 50% are determined to be metaphorical by our assessors. We anticipate that this initial performance can be substantially improved by optimizing the automated underlying processes, including dependency parsing, local topic co-reference tracking, topical clustering, and word sense disambiguation. We have not yet attempted these optimizations, focusing instead on feasibility of the overall method. Part of our future work is also to compare the contribution of each of the three methods of metaphor identification described here on the overall performance of our system.

6   Conclusion

Using automated tools to identify metaphorical language is a challenging task. However, we believe that our approach addresses this challenge by incorporating a multilayered analysis where lexical, semantic, and contextual properties in a given text are captured and formulated towards making a reliable distinction between literal and metaphorical readings. Additionally, unlike traditional machine learning approaches, our system is not reliant on large amounts of training data. The logic of our approach draws on an intuition found in much other work on metaphor (e.g. Steen et al, 2010) that metaphors are used to express ideas in a more concrete form. The use of *imageability* scores allows us to operationalize concreteness, and the *topic chain* and *semantic cluster* filter out false positives to improve the precision of the final system.

Acknowledgment. This research is supported by the Intelligence Advanced Research Projects Activity (IARPA) via Department of Defense US Army Research Laboratory contract number W911NF-12-C-0024. The U.S. Government is authorized to reproduce and distribute reprints for Governmental purposes notwithstanding any copyright annotation thereon. Disclaimer: The views and conclusions contained herein are those of the authors and should not be interpreted as necessarily representing the official policies or endorsements, either expressed or implied, of IARPA, DoD/ARL, or the U.S. Government.

References

1. Balota, D.A., Yap, M.J., Cortese, M.J., Hutchison, K.A., Kessler, B., Loftis, B., Treiman, R.: The English lexicon project. Behavior Research Methods 39, 445–459 (2007)
2. Baayen, R.H., Feldman, L.F., Schreuder, R.: Morphological influences on the recognition of monosyllabic monomorphemic words. Journal of Memory and Language 53, 496–512 (2006)
3. Bleasdale, F.A.: Concreteness-dependent associative priming:Separate lexical organization for concrete and abstract words. Journal of Experimental Psychology: Learning, Memory, & Cognition 13, 582–594 (1987)
4. Broadwell, G., Stromer-Galley, J., Strzalkowski, T., Shaikh, S., Talor, S., Liu, T., Boz, U., Elia, A., Jiao, L., Webb, N.: Modeling Sociocultural Phenomena in Discourse. Journal of Natural Language Engineering (2012)
5. Cameron, L.: Operationalising 'metaphor' for applied linguistic research. In: Cameron, L., Low, G. (eds.) Researching and Applying Metaphor, pp. 3–28. Cambridge University Press, Cambridge (1999)
6. Cho, K.W., Webb, N., Feldman, L.B., Strzalkowski, T., Shaikh, S., Broadwell, G.A., Liu, T., Boz, U., Taylor, S.: Automatic Expansion of the MRC Psycholinguistic Database Imageability Ratings (forthcoming)
7. Coltheart, M.: The *MRC* Psycholinguistic Database. Quarterly Journal of Experimental Psychology 33A, 497–505 (1981)
8. Crisp, P., Gibbs, R., Deignan, A., Low, G., Steen, G., Cameron, L., Semino, E., Grady, J., Cienki, A., Kövecses, Z.: The Pragglejaz Group: MIP: A method for identifying metaphorically used words in discourse. Metaphor and Symbol 22, 1–39 (2007)
9. Douglas, A.J.: The case for bureaucracy. Government is Good: An Unapologetic Defense of a Vital Institution (2007),
   http://www.governmentisgood.com/articles.php?aid=20&print=1
10. Green, M.C., Brock, T.C.: In the mind's eye: Transportation-imagery model of narrative persuasion. In: Green, M.C., Strange, J.J., Brock, T.C. (eds.) Narrative Impact: Social and Cognitive Foundations, pp. 315–341. Lawrence Erlbaum Associates Publishers, New Jersey (2002)
11. Gilhooly, K.J., Logie, R.H.: Age of acquisition, imagery, concreteness, familiarity and ambiguity measures for 1944 words. Behaviour Research Methods and Instrumentation 12, 395–427 (1980)

12. Hargreaves, I.S., Pexman, P.M.: Does richness lose its luster? Effects of extensive practice on semantic richness in visual word recognition. Frontiers in Human Neuroscience 6, 234 (2012), doi:10.3389/fnhum.2012.00234
13. Koller, V., Hardie, A., Rayson, P., Semino, E.: Using a semantic annotation tool for the analysis of metaphor in discourse. Metaphorik.de 15, 141–160 (2008)
14. Martin, J.H.: A computational theory of metaphor. University of California, Berkeley, Computer Science Division. Rep. No. UCB/CSD, 88-465 (1988)
15. Miller, G.A.: WordNet: A Lexical Database for English. Communications of the ACM 38, 39–41 (1995)
16. Nelson, D.L., Schreiber, T.A.: Word concreteness and word structure as independent determinants of recall. Journal of Memory and Language 31, 237–260 (1992)
17. Paivio, A., Yuille, J.C., Madigan, S.A.: Concreteness, imagery, and meaningfulness: Values for 925 nouns. Journal of Experimental Psychology 76, 1–25 (1968)
18. Paivio, A., O'Neill, B.J.: Visual recognition thresholds and dimensions of word meaning. Perception and Psychophysics 8, 273–275 (1970)
19. Reilly, J., Kean, J.: Formal Distinctiveness of High- and Low-Imageability Nouns: Analyses and Theoretical Implications. Cognitive Science 31(1), 157–168 (2007)
20. Steen, G.J., Dorst, A.G., Herrmann, J.B., Kaal, A.A., Krennmayr, T., Pasma, T.: A Method for Linguistic Metaphor Identification: From MIP to MIPVU. John Benjamins Publishing, Amsterdam (2010)
21. Strain, E., Patterson, K., Seidenberg, M.S.: Semantic effects in single-word naming. Journal of Experimental Psychology: Learning, Memory, and Cognition 21, 1140–1154 (1995)
22. Wilson, M.D.: The MRC Psycholinguistic Database: Machine Readable Dictionary, Version 2. Behavioural Research Methods, Instruments and Computers 20, 6–11 (1988)
23. Winnick, W.A., Kressel, K.: Tachistoscopic recognition thresholds, paired-associate learning, and immediate recall as a function of abstractness-concreteness and word frequency. Journal of Experimental Psychology 70, 163–168 (1965)

Automated Trading in Prediction Markets

Anamaria Berea and Charles Twardy

C4I Center of Excellence
Volgenau School of Engineering
George Mason University

Abstract. This research presents the ongoing results of trading experiments that have been performed on the DAGGRE prediction market. DAGGRE is a research project that aims to improve the forecasting methods of world events using prediction markets, crowdsourcing and Delphi groups. The DAGGRE prediction market aggregates estimates from hundreds of participants to forecast the outcome of these events. On the prediction market that involves a few thousand human traders, during a time period of a year and a half, we introduced 3 trading algorithms that have been trading live on the market, based on different rules and trading policies. While all the Autotraders improved the overall market participation and activity and outperform most of the human traders, one of them is adaptive to the new information that continuously comes from the market. This paper presents the comparative analysis of the forecasting accuracy and market performance of these 3 Autotraders and discusses the preliminary results of these experiments.

Keywords: prediction market, combinatorial, autotrading algorithm, Bayes Net.

1 Introduction

Decomposition-Based Information Elicitation and Aggregation (DAGGRE) is a program sponsored by Intelligence Advanced Research Projects Activity (IARPA) and executed at George Mason University (GMU). Research has demonstrated that opinion pools and prediction markets outperform single expert opinion when forecasting the outcome of complicated aggregated events. DAGGRE is finding methods to improve over the unweighted average or plain-vanilla prediction markets (Hanson 2007). One of these methods involves automatic traders (Autotraders), using algorithms to determine a desirable estimate for trading without human intervention.

Autotraders are designed to examine a set of questions that appear on the DAGGRE prediction market website, and make trades on selected world events. These world events can be any type of macro level social phenomena, such as elections, riots, international agreements, epidemics, a.s.o. (Cameron 1963).

In the second year of research, DAGGRE became the first world generalized combinatorial market. In a combinatorial market, participants can make trades

A.M. Greenberg, W.G. Kennedy, and N.D. Bos (Eds.): SBP 2013, LNCS 7812, pp. 111–122, 2013.

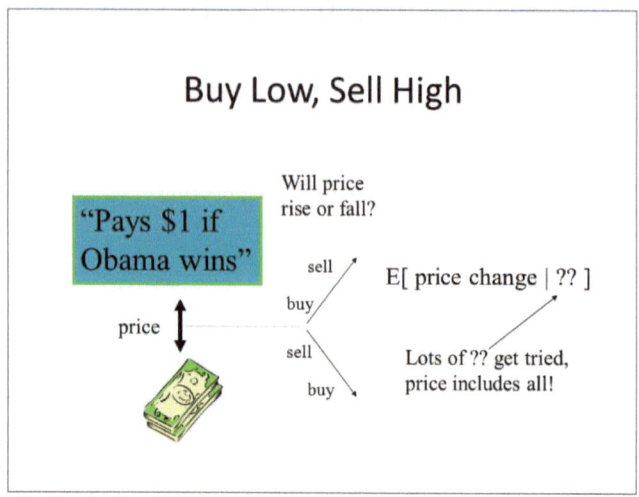

Fig. 1. A prediction market. *(Figure courtesy Robin Hanson)*

on combinations of events: i.e. "Will there be an uprising in Region A" assuming the truth or false value for "Will there be an uprising in Region B".

Besides a few thousand human users, DAGGRE also employs automated traders (Autotraders), using simple algorithms to determine a desirable estimate for trading, without human intervention (or, while in testing, with limited human intervention).

2   Methodology

Most multi-agent systems and simulations involve algorithms that are designed to "interact" with each other. In the case of DAGGRE, the algorithms are interacting with humans and are behaving adaptively to the aggregate human judgment, or the "wisdom of the crowds" (Surowiecki 2005).

Reasoning about trades for a given question is accomplished according to one of the 3 aforementioned algorithms. These algorithms consider the current market estimate, and then derive a new estimate to submit to DAGGRE as a trade.

The Autotrader code base is developed in Java. The Autotrader system consists of four packages: "Autotrader, "communication, "reasoning, and "unbbayes. The Autotrader package is responsible for initialization. The communication package is responsible for logging into the DAGGRE server, for issuing requests to the server, and for receiving responses. The reasoning package is responsible for deciding whether and how much to trade on a specific question. The unbbayes package is responsible for BAYES NET specific reasoning and trading (Sun et al., 2012; Matsumoto et al., 2011).

The DEFAULT algorithm traded in Year 1 based on a temporal rule and the assumption that, as the settlement date approaches, it is less and less likely that the status-quo of the event would change. The STATUS QUO algorithm attempts to improve the overall performance of a question by moving the market estimate to a range around 85% (or 15%), immediately after the question becomes live on the market. The BAYES NET algorithm traded in Year 2 and uses a knowledge-based approach; it uses the current market estimate of a set of supporting questions as 'soft evidence' (Jeffreys 1946) to influence the likelihood of the hypothesis question (Pearl 1988).

Currently, we only have 2 classes of Autotraders.

The STATUS QUO algorithm simply asserts the status quo, compensating for human change bias. In Year 1, IARPA found that a simple status quo heuristic remained competitive. We adapted the heuristic to a market: STATUS QUO tries to maintain 85% probability on the status quo answer, specified at startup. As the resolution date approaches, it gradually becomes more certain.

The BAYES NET algorithm uses a knowledge-based approach. Experts specify a Bayes net and conditional probabilities. It then reads the current market estimates for the set of supporting questions as 'soft evidence' to influence the likelihood of the hypothesis question, and trades on the hypothesis question.

In order to assess the performance and forecasting accuracy, either on the market or for the Autotraders presented in this paper, we calculate the Brier score. The Brier score (Brier 1950) is a measurement of the accuracy of probabilistic predictions. As a distance metric, lower is the better. The Brier score ranges from 0..2, and is the sum of the squared differences between the forecast and the outcome averaged over the number of forecasts. For example, on a binary (Yes/No) question, simply guessing 50% all the time yields a score of 0.5. The closer to 0, the better the forecasting accuracy.

All Autotraders receive points in order to make trades. As human traders, they gain points if they invest "well" and lose points in they disinvest. In other words, they are rewarded and punished equally as a human trader would if they made the same trading decisions.

The Autotraders are communicating with the market and the human trades daily, by reading the last values of the estimates and by placing their bids.

3   Results in Year 1

In Year 1m we used 2 simple algorithms (Status Quo and Default). They relied mostly on the initial values given by the human judgment with respect to the final outcome of the event.

The DEFAULT Autotrader has been trading since March 2012, and the STATUS QUO since August 2012. From the starting point of trading until the end of the first testing period – September 2012 – we resolved 11 questions that these Autotraders invested points in and we were able to compare their performance against the overall prediction market on these particular questions. In general the Autotraders had better average Brier scores than the market. On 8 out of

11 questions, the scores were below 0.05 and in 9 of the 11 questions, the scores were smaller than the market for the same period of time. K3 is the DEFAULT Autotrader, while K4 and K5 are the STATUS QUO Autotraders.

From some preliminary analysis and from their algorithms, we were able to infer early on that the STATUS QUO Autotrader was more active, but the DEFAULT one had a better Brier score overall. Their trades are represented by the blue dots.

Figure 2 shows us the differences in the average Brier Scores between the market, in red, and the Autotraders, in blue. On only two of the 11 questions, the Autotraders had worse performance than the overall market. For all the Autotraders, the average Brier Score was 0.07283.

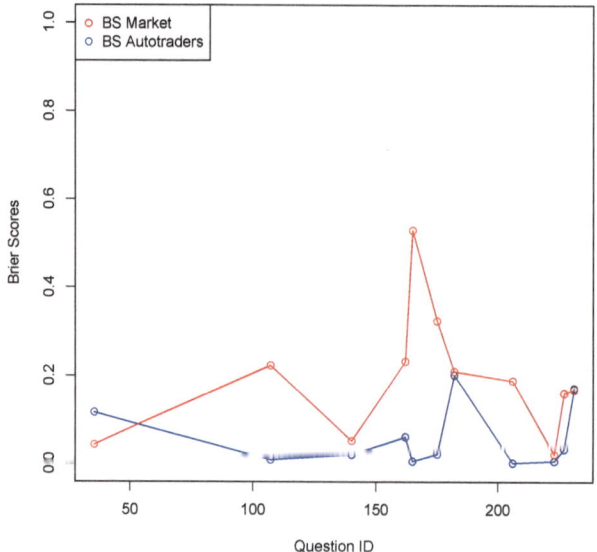

Fig. 2. The Brier Scores of the Autotraders comparatively to the prediction market

Figure 3 shows that the Brier score was improved for the rest of the questions and overall, the Autotraders improved the forecasting accuracy. The improvement here represents the difference between the Brier score of the Autotraders' values and the Brier score of the DAGGRE market before the autotrade.

The Brier Score improvement for Year 1 was very mild for all the Autotraders (Status Quo and Default). Since the first day of trading, the Autotraders made 1464 trades ("autotrades") and the average Brier score of all Autotraders, at all times, is of 0.3. We also calculated the average Brier Score of the market in the absence of the autotrades and for the same questions and trading period, the overall Brier score was 0.31. This means that the overall improvement of the Brier score was of only 0.01.

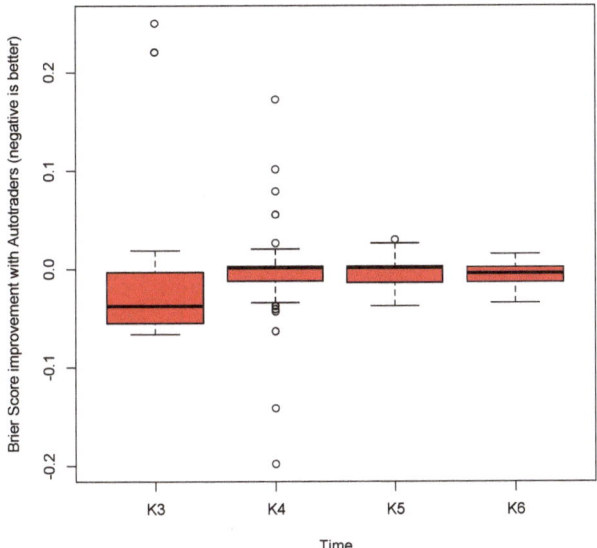

Fig. 3. The Brier score improvements per Autotrader. While K3 had most outliers, it also had most improvements (negative is better).

Figure 4 shows two examples of short-lived questions where the Autotraders traded in the market and pushed the estimates against the final outcome. In these examples, the autotrades were pushing the estimates away from what was eventually the resolution value for the questions. In the Spanish bonds question, both the DEAFULT and the STATUS QUO algorithms were trading, while in the Phelps question only the STATUS QUO algorithm. Both questions were live only for a few days.

Another interesting result of Year 1 was the individual performance of the Autotraders in the competition with the human users. Overall, they have consistently scored among the first 30 users out of a total of more than 1000 forecasters.

In general, the Autotraders are among the first 100 – 150 traders in the DAGGRE market on the leaderboard. In the examples above, the STATUS QUO was performing better and was on places 18 and 66 overall, above the DEFAULT one. We updated this analysis and, as of November 15, 2012, the K3 Autotrader, which combines the DEFAULT and the BAYES NET algorithms, was 28th on the market and 2 of the STATUS QUOs were ranked 43 and 165.

This analysis shows us that all the algorithms combined, although it improved the overall BS of the market, it only improved it marginally. On another hand, they did improve market activity and performed better than most human users.

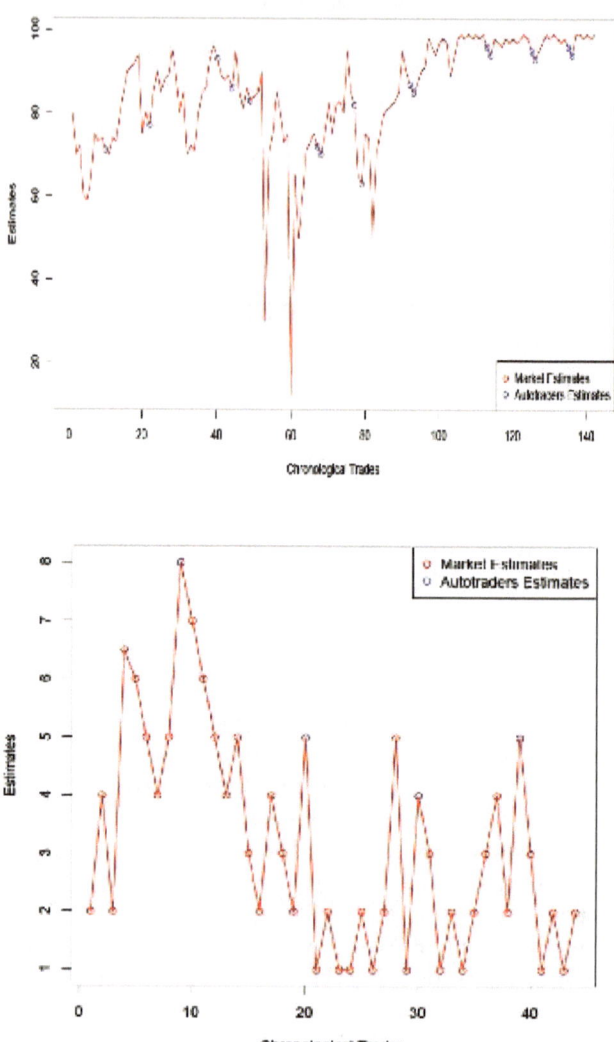

Fig. 4. The autotrades in the live market for 2 questions. Top: Will Spanish government generic 10-year bond yields equal or exceed 7% at any point before 1 September 2012?; Bottom: Will Michael Phelps win more than 4 gold medals at the London 2012 Olympics?

But each question on DAGGRE represents a discrete event and at any given point, we have around 100 live questions on the market. This means that we needed a more adaptive and specialized autotrading algorithm.

Autotrader	Algorithm	Rank	Points
autotraderK4	Status Quo	18	1829.03
autotraderK5	Status Quo	66	1450.33
autotraderK6	Status Quo	123	1310.3
autotraderK3	Default	155	1251.07

Fig. 5. The Leaderboard of Autotraders on the DAGGRE Market in Year 1

4   Results in Year 2

The results in Year 1 with respect to the simple algorithms as well as the launch of the DAGGRE combinatorial prediction market gave us the opportunity to launch an adaptive and more substantive Autotrader, the Bayes Net.

In Year 2, we retired the Default Autotrader and implemented 2 algorithms: a new Status Quo and a Bayes Net. The first one is a combination of the original 2 algorithms from year 1; the second one is a more substantive Autotrader.

The New Status Quo Autotrader is designed to test the forecasting accuracy and to improve market activity *(see Figure 6)*.

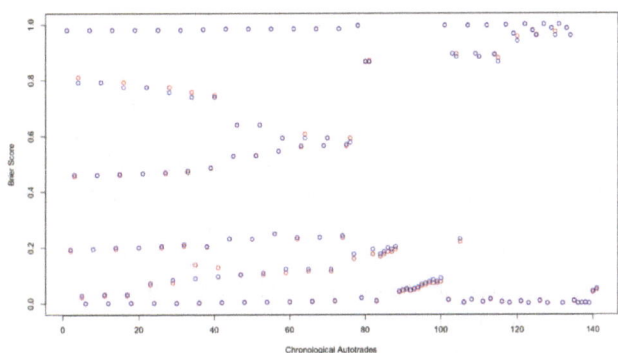

Fig. 6. The Brier Score of the new Status Quo Autotrader on the DAGGRE Market in Year 2

Over all the questions it has been trading, it did not improve the forecasting accuracy; after approximately 710 trades it showed the same average Brier Score as the market (0.248). The plot above shows that the Brier Scores of the market (red) and of the Autotrader (blue) mostly overlap.

Leaderboard Performance. Comparatively to the human users, in Year 2 the Status Quo autototrader has performed similarly to Year 1, being constantly among the first 100 users.

4.1   The Bayes Net Autotrader

In the BAYES NET algorithm, the conditional probabilities are endogenously informed by the market and are not kept constant as originally defined by human judgment.

The Bayes Net Autotrader was implemented to trade on the target question of a Bayes Net model with respect to the Failed States Index. The Failed States Index is a quantitative indicator published by Foreign Policy Magazine and the Fund for Peace each year in June.

The first case study adapts the Failed States Index model to Sudan *(see Figure 7)*. The target question in this model is: '' Will Sudan score less than 100 in the 2013 Failed States Index?''.

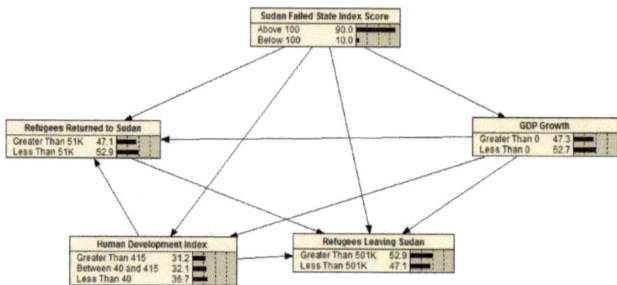

Fig. 7. The underlying Bayes Net model of the Failed States Index for Sudan in the combinatorial market

The conditional questions (assumptions) that are linked to this question in the market are those with respect to the Human Development Index of Sudan, mobile subscriptions, number of refugees originating from Sudan, number of refugees returning to Sudan and the percentage of agricultural production in the GDP. The choice for these assumptions relied on the analysis of the past Sudan Failed States Indexes. We chose the indicators that were more likely to change within a year, after the split of South Sudan from Sudan in 2012, and asked the market about all the questions in the model *(see Figure 8)*.

If this is the Bayes Net model, the Bayes Net Autotrader traded only on the targeted question while reading the market values with respect to the conditional questions. This effort is aimed at testing the following hypothesis:

Hypothesis: Does the Bayes Net Autotrader improve forecasting accuracy of the market?

In order to do so, we calculate the expected Brier Score of the forecasting accuracy in 2 cases: outcome No (0) and outcome Yes (1) *(see Figure 9)*.. These are questions still live on the market and we will know the final outcome in June 2013.

Fig. 8. The DAGGRE market trades on the hypothesis question in the Failed States Index of Sudan model

The market forecasted 6% at the stop time of the analysis. The score improvement would be of 28% in case the event does not happen and -3% in case the event happens (would hurt accuracy). But an outcome 1 has a very small probability, according to the market. This simple analysis of the expected Brier Score of the Autotrader versus the market in both outcomes shows that, overall, the Autotrader improves more than hurts accuracy.

5   Ongoing Experiments and Future Work

There are 3 major research objectives of the Autotraders:

- Improve forecasting;
- Improve market activity;
- Model interconnected events (which may be counterintuitive to human users).

We are now using a paired-question design. Pairs are designed to be similar on a set of factors hypothesized to affect performance on the question, but not highly correlated *(see Figure 10)*.

We are currently testing the Bayes Net Autotrader in a number of experiments that compare the performance of this algorithm on these pairs, where one set of pairs is the control group.*Examples: Failed States Index, European Union Exits, Eurozone Breakup.*

As the DAGGRE market is live, we are continuously receiving new data and updating our analyses and experiments. DAGGRE is currently a combinatorial market (Hanson 2003) that has many Bayes Nets under the hood. Over the following few months, our campaign of experiments will include multiple sets of Bayesian Networks interacting with the market to address longer term questions. Additionally, we expect to continue the work on model elicitation techniques, continuing to seek more efficient approaches to developing the initial models for placement in the market.

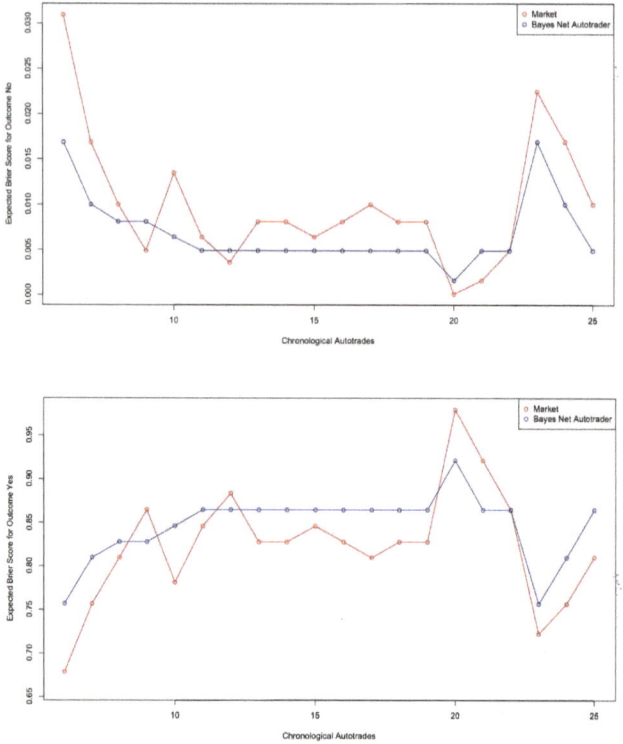

Fig. 9. The Expected Brier Score for Sudan FSI in the case the Index stays above 100 (above) and falls below 100 (below). The blue represents the Brier Score of the Bayes Net Autotrader and the red represents the Brier Score of the market.

Over the longer term, our goal is to explore how to best integrate a BN-based algorithm with the other classes of automated agents and techniques being developed by DAGGRE as well as their interaction with the human traders. While in financial markets there are already several automated traders that are involved in investing billions of dollars, the case of the markets for information – the prediction markets – poses a different set of issues with respect to algorithm design and analysis. Our overall goal is to maximize the accuracy, timeliness, and value of forecasting macro social phenomena and world events.

pair	category	Randomi zation (test or control)	name	settlement_at	default	starting prob 4-Jan	distance from goal
1	Middle East	0	Will a foreign or multinational military for	2013-01-21	0	0.01	0.01
1	Russia	1	Will the Russian military deploy* addition	2013-02-01	0	0.05	0.05
3	European Union	0	Will Mariano Rajoy resign or otherwise ve	2013-02-01	0	0.1	0.1
3	European Union	1	Will the United Kingdoms Liberal Democr	2013-03-31	1	0.91	0.09
4	European Union	0	What change will the European Union Co	2013-02-01	1	0.96	0.04
4	European Union	1	Will UK officially announces its intention t	2013-02-28	0	0.01	0.01
8	Iran	0	Before 1 April 2013, will substantial* evid	2013-03-31	0	0.35	0.35
8	Iran	1	Will Iran sign an IAEA Structured Approac	2013-03-31	0	0.34	0.34
11	Middle East	1	Will Mohammed Morsi cease to be Presic	2013-03-31	0	0.14	0.14
11	Middle East	0	Will Benjamin Netanyahu resign or other\	2013-03-31	0	0.02	0.02
12	European Union	1	Will the Republic of Macedonia* be a NA'	2013-04-01	0	0.1	0.1
12	European Union	0	Will any country officially announce its int	2013-04-01	0	0.08	0.08
14	International Politi	0	When will Japan officially become a mem	2013-04-01	1	0.89	0.11
14	International Politi	1	When will South Korea and Japan sign a n	2013-04-01	1	0.86	0.14
15	International Politi	0	When will the UN announce that Iran has	2013-04-01	1	0.8	0.2
15	North Korea	1	When will North Korea successfully deton	2013-04-01	1	0.97	0.03
16	European Union	1	Will Moodys issue a new downgrade of tr	2013-04-01	0	0.05	0.05
16	Financial Markets	0	Will Standard and Poor's downgrade the I	2013-04-01	0	0.10	0.1
17	European Union	0	When will Viktor Orban resign or otherwi	2013-04-01	1	0.92	0.08
17	Elections	1	When will Nouri al-Maliki resign, lose con	2013-04-01	1	0.9	0.1
19	European Union	0	Will the Greek government propose to wi	2013-04-01	1	0.99	0.01
19	Middle East	1	Will Israel officially announce that it recoj	2013-04-01	0	0.09	0.09
20	Nuclear Weapons	1	Will a UN nuclear inspector visits Iran's Pi	2013-04-01	0	0.26	0.26
20	International Politi	0	Will the Canadian consulate in Tehran offi	2013-04-01	0	0.17	0.17

Fig. 10. The DAGGRE market pairs of questions for controlled Autotrading experiments

Acknowledgments. The authors are very grateful for the software and coding support provided by Rob Alexander and Dan Maxwell from KaDSci. Supported by the Intelligence Advanced Research Projects Activity (IARPA) via Department of Interior National Business Center contract number D11PC20062. The U.S. Government is authorized to reproduce and distribute reprints for Governmental purposes notwithstanding any copyright annotation thereon. Disclaimer: The views and conclusions contained herein are those of the authors and should not be interpreted as necessarily representing the official policies or endorsements, either expressed or implied, of IARPA, DoI/NBC, or the U.S. Government.

References

Brier, G.W.: Verification of forecasts expressed in terms of probability. Monthly Weather Review 75, 1–3 (1950)

Cameron, W.B.: Informal Sociology: A Casual Introduction to Sociological Thinking. Random House (1963)

Hanson, R.: Combinatorial information market design. Information Systems Frontiers 5(1), 107–119 (2003)

Hanson, R.: Logarithmic market scoring rules for modular combinatorial information aggregation. The Journal of Prediction Markets 1(1), 3–15 (2007)

Jeffreys, H.: An Invariant Form for the Prior Probability in Estimation Problems. Proceedings of the Royal Society of London. Series A, Mathematical and Physical Sciences 186(1007), 453–461 (1946)

Matsumoto, S., Carvalho, R.N., Ladeira, M., da Costa, P.C.G., Santos, L.L., Silva, D., Onishi, M., Machado, E., Cai, K.: UnBBayes: a Java Framework for Probabilistic Models in AI. Java in Academia and Research (2011),
http://unbbayes.sourceforge.net

Pearl, J.: Probabilistic Reasoning in Intelligent Systems. Morgan Kaufmann, San Mateo (1988)

Sun, W., Hanson, R., Laskey, K.B., Twardy, C.R.: Probability and Asset Updating Using Bayesian Networks for Combinatorial Prediction Markets. In: Proceedings of the 28th Conference on Uncertainty in Artificial Intelligence (UAI 2012), Catalina, CA (2012)

Surowiecki, J.: The Wisdom of Crowds. Anchor Books (2005)

Social Network Analysis of Peer Effects on Binge Drinking among U.S. Adolescents

Marlon P. Mundt

Department of Family Medicine, University of Wisconsin School of Medicine and Public Health, Madison, Wisconsin, USA
Marlon.mundt@fammed.wisc.edu

Abstract. Adolescent binge drinking is a public health challenge. The study analyzes data from Add Health, a longitudinal survey of seventh through eleventh grade students enrolled between 1995 and 1996. A stochastic actor-based model simulates the co-evolution of binge drinking and friendship connections. Selection effects play a significant role in the creation of peer clusters with similar binge drinking. Friendship nominations between two students with similar binge drinking frequency were 3.46 (95% CI: 2.38-5.01) times more likely than between otherwise identical students with differing alcohol use frequency. An adolescent who nominated binge drinkers as friends was 14% more likely to begin binge drinking than adolescents with non-binge drinking friends. The data demonstrate that strong family ties reduced the odds of adolescent binge drinking by 7%.

Keywords: adolescent, alcohol, agent-based modeling, social influence, social selection, Markov model, systems science.

1 Introduction

Early adolescent alcohol abuse is a major public health challenge. According to data from the 2004 National Survey on Drug Use and Health, 6% of 14-year-olds, 18 % of 16-year-olds, and 33% of 18-year-olds had engaged in binge drinking, defined as 5 or more drinks in a single occasion, within the past 30 days [1]. Adolescent binge drinking is associated with increased risk of motor vehicle fatality, suicide, homicide, and violence [2-4].

A wide body of literature indicates that adolescents and their friends exhibit more similar binge drinking behavior than would be expected by chance alone [5,6]. An adolescent whose best friend binge drinks is twice as likely to be a binge drinker [6]. However, the causal pathway between binge drinking among friends is poorly understood. Similarities may occur because of peer influence, or the spread of behaviors and behavioral norms through social ties, where an individual's binge drinking behavior adjusts toward the average behavior of one's friends over time. Another pathway to homogeneity within friendships may be that friends are similar due to social selection, or homophily, the tendency for similar people to form friendships among one another.

A.M. Greenberg, W.G. Kennedy, and N.D. Bos (Eds.): SBP 2013, LNCS 7812, pp. 123–134, 2013.

Understanding the mechanism by which similarity in binge drinking among adolescent friends occurs is an important clinical matter. For example, influence-driven contagion in adolescent groups would lend itself to peer-to-peer methods of intervention. It would allow for possible multiplier or spillover effects from targeted individuals to a larger network of friends. On the other hand, selection-driven binge behavior patterns would suggest interventions that target adolescent friendship groups.

Previous studies have used structural equation modeling [7], latent growth models [8-11], and instrumental variables [12] in an attempt to separate selection from influence effects in adolescent binge drinking. These studies have generally relied on lagged indicators of binge drinking or friendship connections in an attempt to isolate selection effects from influence effects. The results indicate that both selection and influence are occurring, but cannot determine the relative contribution of the two factors.

A major limitation in prior studies on selection and influence effects is the failure to account for the dynamic interplay (i.e., co-evolution) of friendship ties and binge drinking. Assuming friendship connections are fixed while estimating changes in binge drinking would produce biased parameter estimates. In a similar manner, modeling binge drinking as constant when estimating friendship creation could lead to systematic error in results. In addition, previous studies are limited in their control for peer effects beyond the unidirectional dyadic relationship (e.g., *reciprocity*, the likelihood to reply to friendship with friendship, or *transitive closure*, the likelihood of friends of friends to become friends). The complexity of longitudinal social peer effects on binge drinking among adolescents necessitates more advanced statistical methods than have been used in previous studies of adolescent binge drinking.

A new analytical approach to the analysis of the co-evolution of friendship or social network ties and alcohol binge behaviors is stochastic actor-based modeling [13,14]. The primary assumption of the actor-based model is that individuals choose their friendship ties and their behaviors in one step-at-a-time micro-steps. At each micro-step, an agent maximizes a personal utility function for surrounding network and relative friend behavior. At that time point, the agent only considers the current network characteristics in deciding whether a change in behavior or network tie is preferable to the current state. In this manner, the process of co-evolution of network and behaviors from one wave of data to another is simulated as a result of a potentially large number of individually unobserved micro-step changes, and the network and behavior preferences parameters can be estimated. The actor-based model can disentangle selection and influence and determine their relative contribution to similarities in binge drinking behavior among friends.

Several studies have used actor-based modeling to examine selection and influence effects on adolescent alcohol use [15-18]. In one study, a sample of 1,204 7th graders in Finland was followed for 30 months to determine the degree to which the school children selected or were influenced by alcohol using friends [18]. The results indicated that alcohol use similarities among friends were a result of both the selection and influence processes. Another study followed cohorts of 4th graders, 7th graders, and 10th graders in Sweden for 2 years [17]. In this study, young adolescents selected friends based on similarity in alcohol use, but influence was not significant. Both selection and influence effects contributed to alcohol use similarities during later adolescence.

The present study will investigate the selection and influence processes as they relate to binge drinking in the larger context of adolescent friendship networks among adolescents in the U.S. Specifically, the study will address the following research questions:

Research Question #1 (*Selection*): Do adolescents select friends with similar binge drinking frequency?

Research Question #2 (*Influence*): Do adolescents adjust their binge drinking in correspondence with the binge drinking of their friends?

I hypothesize that both selection and influence effects will be present in the network model of adolescent binge drinking.

2   Methods

2.1   Data Source

This study analyzes data from the National Longitudinal Study of Adolescent Health (Add Health). Add Health used stratified sampling to choose high schools and middle schools which were representative of US schools nationwide based on region of the country, urbanicity, school funding, and racial composition [19].

All 7th through 12th graders at the 132 participating schools were invited to complete an In-School Survey (n=90,118). A random sample of 20,745 students from the In-School Survey respondents then completed a Wave I in-home survey, administered between April and December, 1995. Approximately one year later, Wave I participants who had not yet graduated were recontacted for a Wave II in-home survey. The Wave II in-home survey (n=14,738) took place between April and December, 1996.

The Add Health sampling design included 14 saturated sample schools. In the saturated sample schools, all students in attendance at the time of the In-School Survey were included in the sampling frame for the Wave I in-home survey. The inclusion of all students from a school allows the investigation of complete peer network structures and their influence on behaviors.

2.2   Analysis Sample

The study sample for the present analysis includes 2,563 Add Health subjects from saturated sample schools. Subjects who completed the Wave I in-home interview but were non-responders at Wave II were included in the model using imputation [20]. In this approach, outgoing friendship nominations of non-responders are imputed using last observation carry forward and treated as non-informative for statistical calculations while incoming nominations are allowed to vary and to contribute to the estimation procedures. The choice to impute wave II outgoing nominations was based on Wave II non-responders being eligible for nominations from other students. As such, their treatment as structural zeros for outgoing ties would have been inconsistent with the treatment for incoming ties. A structural zero designation would imply that the student was absent from the network. The choice of imputation for outgoing ties preserves the ability to include incoming friendship nominations in the analysis. The sample excluded subjects who did not send or receive any friendship nominations at either Wave I or Wave II.

Low network stability between study waves could lead to convergence failure in the iterative estimation process. A Jaccard Index of 0.20 or higher for the school was required for inclusion in the analysis [21]. One school from the saturated subsample was excluded from the analysis based on a Jaccard index of less than 0.20. A possible explanation for low network stability is that friendships change when students transition from middle school to a high school setting. A total of 13 schools were included in the analysis.

3   Measures

3.1   Binge Drinking

Students answered the question: "How often did you drink 5 or more drinks in a single setting in the past year?" on the in-home survey. Categorical responses included never, 1 or 2 times, 3 to 12 times, monthly but not weekly, weekly, and more than once a week.

3.2   Social Networks

At both Wave I and Wave II, students provided responses to: "Name your 5 best male and 5 best female friends from your school roster." Social networks within each school resulted from the formation of a friendship matrix based on the directed friendship designations.

3.3   Demographics

Students provided age, grade, and race/ethnicity data at the in-home interview. Age was calculated to the nearest month.

3.4   Family Characteristics

Study participants answered: "On a scale of 1 to 5, how often do you and your family have fun together?" In addition, parents of the sample students indicated how often they drank alcohol in the past year.

4   Statistical Analysis

4.1   Stochastic Actor-Based Modeling

The analysis uses stochastic actor-based modeling to assess the co-evolution of binge drinking and friendship ties from Wave I to Wave II of the study [20]. The stochastic actor-based model assumes changes in the network take place according to a continuous-time Markov chain with stationary transition distribution. Changes from Wave I to Wave II occur through mini-steps where the future state of the network is dependent only on the present state. Each mini-step is evaluated by choosing a random student i among all network members and either a potential friendship change or behavior change. In a potential friendship change, the student i might change an outgoing tie to student j so as to maximize the objective function for network structure and a random unexplained influence: $f_i^X(\beta, x(i \mapsto j), z) + U_i^X(t, x, j)$, where β is the parameter set, $x(i{\to}j)$ is the network changes that would occur if the tie between individual i and individual j in the network were changed, z is the state of behaviors within the network, and U is an independent random component. For a given actor i (the actor who takes the micro-step), the objective function is maximized over all potential alters j. The β parameters are estimated by Method of Moments (MoM). The MoM algorithm compares the observed network (obtained from the data) to hypothetical networks generated through repeated Monte Carlo simulations [13].

Similarly, the individual i might make a one-step change in binge drinking behavior based on the objective function for the parameterized binge drinking outcome and a random influence: $f_i^Z(\beta, x, z(i)) + U_i^Z(t, z, i)$, where β is the parameter set for behavior changes, x is the current state of network ties, z(i) is the next potential behavior state that would occur after a micro-step change, and U is an independent random component. By estimating the β parameters in the model, the simulated analysis seeks to define what tendencies and trends influence changes in friendship ties and in binge drinking behavior. The study model simultaneously evaluates the evolution of the adolescent friendship network and adolescent binge drinking while controlling for age, gender, race/ethnicity, parental drinking, and family bonding. The study uses the statistical program RSIENA (Simulated Investigation for Empirical Network Analysis) [20], originally designed by Snijders and van Duijn [22] and programmed by Ruth Ripley and Krists Boitmanis.

4.2   Model Specification

The model specification consists of two parts, friendship network evolution and binge drinking behavior evolution. First, the friendship network evolution portion of the model identifies the preferred choices in friendship ties depending on a list of friendship choice variables. Three binge drinking variables are included in the friendship evolution part of the model: (1) the effect of an adolescent's binge drinking on number of friends chosen (binge drinking-sent nominations), (2) the effect of an adolescent's binge drinking on the probability of being chosen as a friend by others (binge drinking-nominations received), and (3) the effect of similar binge drinking frequency on friendship selection (binge drinking similarity).

Several characteristics of the friendship network structure are included as control variables [13]. These take account of the overall density of friendship ties within the network (*density*), the likelihood to reciprocate friendship nominations (*reciprocity*), the tendency for friends of friends to be friends (*transitive triplets*), the propensity for closure in three-person friendships (*3-cycles*), the propensity for individuals with more in-coming friendship nominations to attract further friendship nominations (*in-degree popularity*), the tendency for individuals who name many others as friends to attract friendship nominations (*out-degree popularity*), and the inclination for students who name more friends to generate even more out-going friendship ties over time (*out-degree activity*). Control variables also include age, gender, and race/ethnicity effects on the number of friends chosen, on the probability of being chosen as a friend, and on the likelihood of friends being of similar age, gender, and race/ethnicity. Family bonding was considered for inclusion into the selection model, but was excluded for parsimony and lack of significance reasons.

Second, the binge drinking behavior evolution portion of the model specifies a list of variables that could influence potential binge drinking behavior changes. The model contains one main friendship-related influence component: the tendency for binge drinking to change based on the average binge drinking frequency of immediate friends. Control variables include age, gender, race/ethnicity, parental drinking, family bonding and linear and quadratic shape effects modeling average binge drinking frequency across the network. A secondary analysis added an interaction term (average friend binge drinking x reciprocity) to the model to determine if influence effects differed in reciprocal compared to non-reciprocal friendships.

The analysis employs the Snijders-Baerveldt meta-analysis (two-sided) test [23] to test overall significance of the primary and control variables across schools. The Snijders-Baerveldt test makes inference about parameters in the population of schools from which the studied schools are considered to be a sample.

5   Results

The sample consisted of 2,563 adolescents in grades 7 through 11 at Wave I of Add Health (Table 1).

Table 1. Descriptive Statistics (n=2,563)

Male (%)	50.8
Age, mean (sd)	15.8 (1.3)
Age, range (years)	12-18
Grade Level (%)	
7th grade	7.0
8th grade	7.3
9th grade	17.5
10th grade	34.7
11th grade	33.4
Race (%)	
Non-Hispanic white	60.9
Black	16.4
Native American	1.6
Asian	3.4
White Hispanic	17.7

The participants included 1,301 (51%) boys and 1262 (49%) girls. Mean age was 15.8 years. Forty-four percent of the adolescents' parents reported drinking alcohol in the past year. Over 70% of the adolescents indicated that they had fun with their family quite a bit or very much. Wave II respondents included 2,299 adolescents, 89.6 percent of the original sample. Subjects lost to follow-up between Wave I and Wave II were more likely than Wave II respondents to be weekly binge drinkers at Wave I (12% vs 7%, p=0.008). There were no significant differences between respondents and non-respondents in age, grade, gender, race/ethnicity, or family bonding.

Characteristics of the 13 schools in the study and binge drinking within the schools are reported in Table 2.

Table 2. School Network Statistics (n=13)

	Wave 1 (n=2,563)	Wave 2 (n=2,299)
Adolescents per school (range)	197.2 (48-987)	176.8 (41-889)
Friendship nominations per student	2.04	1.85
Reciprocal friendships (%)	22.9	25.3
Binge drinking, past 12 months (%)		
None	70.5	69.5
1-2 times	10.4	9.7
3-12 times	7.0	6.9
More than monthly, less than weekly	5.0	5.3
Weekly or more often	7.1	8.7

The mean number of students per school was 197. At Wave I, the average number of friendship nominations to other students in the same school was just over two per student. Thirty percent of students reported binge drinking in the past 12 months at Wave I. The percentage of binge drinkers increased to 31 percent at Wave II. The proportion of weekly binge drinkers increased between waves, from 7 percent at Wave I to 9 percent at Wave II. Thirty five percent of the students changed binge drinking frequency from Wave I to Wave II. Twenty percent of respondents increased binge drinking frequency while 15 percent decreased frequency.

Table 3 presents the friendship network evolution from Wave I to Wave II. Binge drinkers received significantly more nominations (p=.021), but they did not send more nominations (p=.293). Friendship formation was associated with similarity in binge drinking (p<.001). Other significant factors contributing to friendship creation included reciprocity (p<.001), transitive triplets (p<.001), 3-cycles (p=.009), in-degree popularity (p <.001) and out-degree popularity (p<.001). Students were more likely to choose as friends other students of similar age (p=.003) and gender (p<.001).

Table 3. Stochastic Actor-Based Model Results for Network Selection

	β	SE(β)	p-value*
Binge drinking (sent nominations)	0.03	0.03	.293
Binge drinking (received nominations)	0.07	0.03	.021
Binge drinking similarity	1.24	0.19	<.001
Density	-3.34	0.35	<.001
Reciprocity	2.49	0.20	<.001
Transitive triplets	0.82	0.09	<.001
3-cycles	-0.41	0.13	.009
In-degree popularity	0.09	0.01	<.001
Out-degree popularity	-0.15	0.02	<.001
Out-degree activity	-0.03	0.02	.144
Age (sent nominations)	-0.10	0.05	.052
Age (received nominations)	0.04	0.03	.253
Age similarity	1.28	0.37	.003
Male (sent nominations)	0.03	0.08	.677
Male (received nominations)	0.09	0.04	.020
Gender same	0.35	0.04	<.001
Minority (sent nominations)	0.05	0.07	.518
Minority (received nominations)	-0.13	0.25	.635
Same race	0.59	0.45	.240

*Parameter estimates β and standard error for stochastic actor-based model of the evolution of school friendships in the Add Health study. Coefficients correspond to the change in log-odds of a friendship nomination being present.

The beta coefficient in the network selection part of the model is comparable to a log-odds ratio of friendship formation in a logistic regression analysis theoretical framework. Exponentiation of the beta coefficient produces an odds-ratio. As such, binge drinkers were 7% (95% CI for Odds Ratio: 1.01-1.14) more likely to receive friendship nominations than otherwise similar students who did not binge drink. For binge drinking similarity (β=1.24), friendship formation between two students who shared the same binge drinking frequency was 3.46 times (95% CI: 2.38-5.01) more likely to occur than an otherwise identical friendship between two students who were maximally different with respect to binge drinking. Adolescents were more likely to nominate as friends others who binge drink similarly to themselves. These results are consistent with a selection effect.

Table 4 presents the results of the binge drinking evolution portion of the model from Wave I to Wave II. The stochastic actor-based model results support a significant influence effect on binge drinking from the binge drinking of friends. More frequent binge drinking by immediate friends was associated with increased binge drinking by the adolescent (p=.002). Students who were not binge drinkers at Wave I were 14% (95% CI: 1.07-1.21) more likely to begin binge drinking if their friends were binge drinkers. Non-binge drinking students were 68% (95% CI: 1.33-2.13) more likely to begin binge drinking if their friends were weekly binge drinkers. In terms of the other covariates included in the model, family bonding was a significant protective factor for binge drinking (p=.006), males were more frequent binge drinkers (p=.013), and more frequent parental alcohol use was significantly associated with more binge drinking by the adolescent (p=.013). Adolescents were 7% (95% CI: 0.90-0.97) less likely to binge drink if they had stronger family bonds. Males were 15% (95% CI: 1.09-1.21) more likely to binge drink than females. Reciprocity did not moderate the influence effects of friends on binge drinking. The addition of a (friend influence x reciprocity) interaction was not significant (β=-0.03, p=0.682).

Table 4. Stochastic Actor-Based Model Results for Influence Effects on Binge Drinking

	β	SE(β)	p-value*
Linear shape parameter	-1.00	0.13	<.001
Quadratic shape parameter	0.15	0.02	<.001
Average friend binge drinking	0.13	0.03	.002
Age	0.02	0.03	.508
Gender (male)	0.14	0.05	.013
Race (non-white)	0.03	0.04	.452
Parental drinking	0.04	0.02	.013
Family bonding	-0.07	0.02	.006

*Parameter estimates β and standard error for stochastic actor-based model of change in binge drinking in the Add Health study. Binge drinking frequency measured on a 0 (never) to 4 (weekly or more often) scale. Coefficients correspond to the change in log-odds of increased binge drinking frequency given a one-unit increase in the independent variable.

6   Discussion

The main objective of this investigation is to disentangle the selection and influence processes governing peer relationship's effect on adolescent binge drinking. The study uses a stochastic actor-based model to analyze the dynamic interplay of friendship formation and binge drinking changes as they co-evolve over time. Specifically, the study evaluates if similar binge drinking frequency among friends is more likely a result of a tendency for adolescents to choose friends with similar binge drinking behavior, or as a result of teen influence on each other's binge drinking.

The results demonstrate that selection effects in adolescent friendships are based, in part, around commonalities in binge drinking behavior. Homophily in friendship formation is also based on age, gender, and race/ethnicity. Our results show that reciprocity (i.e. tendency to have reciprocal friendships), transitivity (i.e. tendency to become a friend of a friends' friend), and degree effects (i.e. number of in-coming or out-going friendship nominations) contribute to friendship formation. Our findings are in line with previous studies showing selection effects to be a strong factor in alcohol use similarity within adolescent friendships [15-18].

The study findings also offer support for influence effects among teens after the social ties with their peers are in place. Our findings demonstrate that adolescents adjust to the binge drinking of their friends. An adolescent with binge drinking friends was 14% more likely to begin binge drinking than adolescents with non-binge drinking friends. The results are similar to two European studies that identified both selection and influence as significant in alcohol use similarities among adolescent friends [17,18].

Additionally, reciprocal friends do not exert influence on binge drinking frequency. Adolescents who nominate binge drinkers, but do not receive friendship nominations in return, change their binge drinking behavior to be in line in their attempt to make friends.

Interestingly, there is evidence that close family bonds, defined here as having family fun, are protective against binge drinking. These data demonstrate that strong family ties reduced the odds of adolescent binge drinking by 7%.

This study is the first to employ agent-based modeling for disentangling peer selection and influence effects on binge drinking in a sample of U.S. middle and high school students. The strength of the study lies in its innovative methodology, wealth of friendship variables, prospective design and large study sample size. The study also has several limitations. First, adolescents were limited to nominate up to 10 friends which may have obscured the friendship formation parameters in the model. However, other studies show that students on average report having four friends [15,18]. Second, the analysis, by limiting itself to friendships within a school, may not have captured all peers in the adolescent social network. The average number of friends reported in the sample was approximately two, fewer than in other studies [15,18]. This potentially means that only some of the influential peers were nominated, which limits the conclusions that can be drawn. However, school-based networks may be most relevant for intervention efforts. Third, the exclusion of isolates may influence the results; however, the impact is likely to be small. The co-evolution model of friendships and binge drinking is, in essence, a model of change in friendship ties and binge drinking between Wave I and Wave II conditional on network ties and binge drinking at Wave I. It is possible that isolates were more likely to binge drink, but the primary effect on the model would be driven by isolates who increased binge drinking dramatically between waves, while still remaining isolated. This is likely to be a small percentage of the student population. Fourth, binge drinking in the study was self-reported. However, self-reported alcohol use is generally considered to be a valid measure among adolescents. Add Health used a computer assisted data entry process for sensitive questions such as alcohol and drug use to protect confidentiality and enhance full reporting. Finally, the selection and influence processes may depend on contextual and cultural aspects of the schools analyzed, which limits generalizability to different school contexts and cultural settings.

6.1   Conclusions

This investigation demonstrates that both network selection and influence play a prominent role in adolescent binge drinking similarities among friends. The presence of both selection and influence in binge drinking similarities suggests multi-faceted prevention efforts to curb adolescent binge drinking. On the one hand, selection effects suggest that prevention interventions focus on protective factors, such as family bonding, to discourage binge drinking. Binge drinkers who select each other as friends may form cliques and propagate social settings which further encourage binging, but the initiation into binge drinking behavior may depend on factors outside of the school relationship. On the other hand, influence effects on binge drinking support interventions targeting peer leaders to promote "anti-binging" messages. Our findings may be of interest to parents, health care professionals, school administrators, law enforcement and community leaders who focus on alcohol prevention efforts.

Acknowledgements. This work was supported by NIAAA 1K01 AA018410-01. This research uses data from Add Health, a program project directed by Kathleen Mullan Harris and designed by J. Richard Udry, Peter S. Bearman, and Kathleen Mullan Harris at the University of North Carolina at Chapel Hill, and funded by grant P01-HD31921 from the Eunice Kennedy Shriver National Institute of Child Health and Human Development, with cooperative funding from 23 other federal agencies and foundations. Special acknowledgment is due Ronald R. Rindfuss and Barbara Entwisle for assistance in the original design. Information on how to obtain the Add Health data files is available on the Add Health website (http://www.cpc.unc.edu/addhealth). No direct support was received from grant P01-HD31921 for this analysis.

References

1. Substance Abuse and Mental Health Services Administration, Results from the, National Survey on Drug Use and Health: National Findings (Office of Applied Studies, NSDUH Series H-28, DHHS Publication No. SMA 05-4062). Rockville, MD (2005)
2. National Highway Traffic Safety Administration, Traffic Safety Facts 2002: Alcohol. U.S. Department of Transportation, DOT HS 809 606 (2003)
3. Youth Risk Behavior Survey, Centers for Disease Control. Atlanta, GA (2004)
4. Smith, G., Brandings, K., Miller, T.: Fatal non-traffic injuries involving alcohol: a meta-analysis. Annals of Emergency Medicine 33, 699–702 (1999)
5. Jaccard, J., Blanton, H., Dodge, T.: Peer influences on risky behavior: An analysis of the effects of a close friend. Developmental Psychology 41, 135–147 (2005)
6. Bahr, S.J., Hoffmann, J.P., Yang, X.: Parental and Peer Influences on the Risk of Adolescent Drug Use. The Journal of Primary Prevention 26(6), 529–551 (2005)
7. Schulenberg, J., Maggs, J.L., Dielman, T.E., Leech, S.L., Kloska, D.D., Shope, J.T., Laetz, V.B.: On peer influences to get drunk: A panel study of young adolescents. Merrill Palmer Q 45, 108–142 (1999)
8. Van Der Vorst, H., Vermulst, A.A., Meeus, W.H.J., Dekovic, M.: Engels RCME. Identification and prediction of drinking trajectories in early and mid-adolescence. J. Clin. Child Adolesc. Psychol. 38, 329–341 (2009)
9. Simons-Morton, B.: Social influences on adolescent substance use. Am. J. Health Behav. 29, 299–309 (2004)
10. Bray, J.H., Adams, G.J., Getz, J.G., McQueen, A.: Individuation, peers, and adolescent alcohol use: A latent growth analysis. J. Consult. Clin Psychol. 71, 553–564 (2003)

11. Li, F., Barrera, M., Hops, H., Fisher, K.J.: The longitudinal influence of peers on the development of alcohol use in late adolescence: A growth mixture analysis. J. Behav. Med. 25, 293–315 (2002)

12. Ali, M.M., Dwyer, D.S.: Social network effects in alcohol consumption among adolescents. Addict. Behav. 35, 337–342 (2010)

13. Snijders, T.A.B., van de Bunt, G.G., Steglich, C.E.G.: Introduction to actor-based models for network dynamics. Soc. Networks 32, 44–60 (2010)

14. Steglich, C.E.G., Snijders, T.A.B., Pearson, M.: Dynamic networks and behavior: Separating selection from influence. Sociol Methodol. 40, 329–393 (2010)

15. Knecht, A.B., Burk, W.J., Weesie, J., Steglich, C.E.G.: Friendship and alcohol use in early adolescence: A multilevel social network approach. J. Res. Adolesc. 21, 475–487 (2010)

16. Kiuru, N., Burk, W.J., Laursen, B., Salmela-Aro, K., Nurmi, J.E.: Pressure to drink but not to smoke: Disentangling selection and socialization in adolescent peer networks and peer groups. J. Adolescence 33, 801–812 (2010)

17. Burk, W.J., Van der Vorst, H., Kerr, M., Stattin, H.: Alcohol use and friendship dynamics: Selection and socialization in early-, middle-, and late-adolescent peer networks. J. Stud. Alc. Drugs 73, 89–98 (2012)

18. Mercken, L., Steglich, C., Knibbe, R., Hein de Vries, H.: Dynamics of friendship networks and alcohol use in early and mid-adolescence. J. Stud. Alc. Drugs 73, 99–110 (2012)

19. Harris, K.M., Halpern, C.T., Whitsel, E., Hussey, J., Tabor, J., Entzel, P., Udry, J.R.: The National Longitudinal Study of Adolescent Health: Research Design (2012), http://www.cpc.unc.edu/projects/addhealth/design

20. Ripley, R.M., Snijders, T.A.B., Lopez, P.P.: Manual for RSIENA 4.0 (version May 1, 2011). University of Oxford, Department of Statistics; Nuffield College, Oxford (2011), http://www.stats.ox.ac.uk/siena/

21. Real, R., Vargas, J.M.: The probabilistic basis of Jaccard's index of similarity. Syst. Biol. 45, 380–385 (1996)

22. Snijders, T.A.B., van Duijn, M.A.J.: Simulation of statistical inference in dynamic network models. In: Conte, R., Hegselmann, R., Terna, P. (eds.) Simulating Social Phenomena, pp. 493–512. Springer, Berlin (1997)

23. Snijders, T.A.B., Baerveldt, C.. A multilevel network study of the effects of delinquent behavior on friendship evolution. J. Math. Sociol 27, 123–151 (2003)

Feedback Dynamic between Emotional Reinforcement and Healthy Eating: An Application of the Reciprocal Markov Model

Edward H. Ip[1], Qiang Zhang[1], Ji Lu[2], Patricia L. Mabry[3], and Laurette Dube[4]

[1] Wake Forest University School of Medicine,
Medical Center Blvd., WC23, NC 27157, USA
{eip,qizhang}@wakehealth.edu
http://www.phs.wfubmc.edu/public/bios/home.cfm
[2] Dalhousie University Faculty of Agriculture, Truro, Canada
ji.lu@dal.ca
[3] Office of Behavioral and Social Sciences Research, Office of the Director, National
Institutes of Health, 31 Center Drive, Building 31, Room B1-C19;
MSC 2027, Bethesda, MD 20892, USA
mabryp@od.nih.gov
[4] Desautels Faculty of Management, McGill Univeristy,
1001 Sherbrooke Street West, Montreal, QC, Canada
laurette.dube@mcgill.ca

1 Introduction

Increasingly, the relationship between emotion and eating is recognized as an important issue in the study of obesity. Emotions influence food choice behaviors in human beings, but the evidence for the impact of negative and positive emotion on food intake appears to be asymmetric. Negative emotions have been thoroughly studied and there is consensus that they increase food consumption [1]. In addition, there is some evidence that among obese persons characterized as having a negative affect, there is a greater tendency to overeat in response to negative mood induction compared to obese persons characterized as low in negative affect or normal weight control subjects [2]. Positive emotions have also been associated with increased food intake; however, this finding is not as conclusive [1]. Emotion and food choice appear to have a reciprocal relationship; not only does emotional state influence food choice, but food choice impacts subsequent emotion in both positive and negative directions. In this vein, one psychosomatic theory of obesity proposes that eating may reduce anxiety, and that the obese overeat in order to alleviate their emotional discomfort. Intake of snacks such as cookies can also be used, purposefully or otherwise, as a mood-regulating tool [3]. Mediators of food intake and emotion, such as the context under which food is consumed-e.g., in a home setting or away from home (AFH)-also need to be taken into account in shaping the relationship between food choice and emotion. Some previous work has indicated that the nutritional quality of food is healthier at home than away, [4] but this trend has been questioned recently due to the general increase in high-fat and high-sugar foods in packaged meals typically

A.M. Greenberg, W.G. Kennedy, and N.D. Bos (Eds.): SBP 2013, LNCS 7812, pp. 135–143, 2013.
© Springer-Verlag Berlin Heidelberg 2013

consumed at home. Regardless of whether the food at home is healthier, it appears that there is strong evidence that the setting in which food is consumed contributes to the relationship between emotion and the nutritional quality of food.

Empirical research on emotion and food choice often encounters design and methodological challenges. Emotions of individuals tend to fluctuate throughout the day, and food intake is notoriously difficult to measure with great accuracy. Recent advances in data collection methods such as Experience Sampling Method (ESM) [5] provide new tool sets for collecting momentary emotion and food intake data. As a novel psychometric technique, the ESM captures emotions and behaviors "in the moment" and in the ecologically valid context in which they occur (as opposed to a laboratory setting). Now, ESM data are most often collected via personal devices, such as mobile phones, but in the past were recorded with paper and pencil via journals. More advanced tools include modern sensing and computational technologies that automatically trigger sampling based on the occurrence of specific events are now beginning to emerge as well [6]. An advantage of ESM is that intensively collected data could be used to pinpoint the moment when a behavior-for example food consumption-occurs, as well as the emotion that immediately precede or follows the behavior of interest.

In this paper, our goal is to examine, using ESM data, the reciprocal relationship between food intake and emotion; both before and after meal consumption. The analytical framework thus goes beyond existing approaches which tend to study uni-directional relationships between food and emotion. We look at both the emotion that precedes food consumption and the emotion that follows it to better understand the effect that each variable has on the other. While the ESM has the promise to gather highly granular data, the analysis of such data also poses important methodological and computational challenges. First, ESM-collected intensive longitudinal data may not be amenable to standard statistical models that assume independence between observations. Second, because the study focuses on reciprocity between behavior and emotion, the feedback mechanism between them needs to be captured within in the social context. For example, it is important to capture whether food is consumed at home or AFH because emotion and behavior may vary by context. A further challenge is that the dynamic interaction between food choice and emotion is subject to substantial day-to-day, if not meal-to-meal, fluctuation, and possibly both in direction and magnitude. For example, there could be abrupt changes of emotion over a short sequence of food intake occasions; such episodic changes of direction are not amenable to differential equation-based feedback models, such as system dynamics modeling, which are better suited to modeling smooth trajectories of change. To tackle these methodological challenges, this paper proposes using two discrete-time "intertwining" Markov chains for modeling and computing the dynamic interaction between emotional states and food intake statuses. In other words, emotion and food consumption assume a reciprocal relationship in which one influences the other over time as a Markov process. Pre-meal emotional state affects the status of the nutritional value of the meal, and the meal in turn affects post-meal emotion. A salient feature of the model is that no smooth relationship between either emotion or food variables and time is assumed. We call the model a Reciprocal Markov Model (RMM). Methodologically, the RMM is an extension of the hidden Markov model (HMM) [7] [8] [9]. By harnessing the power of modern computational methods, the RMM allows social behavior (eating) to be closely studied in conjunction with real-time, in-context psychological (emotion) data. Specifically, the role of social variables, including eating at home or AFH, within which the behavior occurs will be ascertained.

2 Data and Measure

Data used for computing the RMM were collected from N=160 white adult nonobese women in a large North American city. Participants were recruited through local advertisement. All participants signed an informed consent form before engaging in the study and each participant received a small incentive for participating. The study protocol was approved by the human-subjects ethics committee of McGill University, Canada.

The ESM method was used to assess individual food choice patterns and both concurrent (after meal) and lagged (before meal) emotional experiences on repeated episodes. Participants received instructions about the ESM protocol in a one-on-one training session conducted at a laboratory. During the first 10 observational days, a beeper would prompt a participant 6 times/day to fill out a short paper-and-pencil questionnaire at the next available moment. Data were collected over a period of 20 days, with systemic alternation between observational and nonobservational days. Participants were beeped every 2 hours during the typical waking hours (0800 to 2100) of a day. Thus, ideally there would be a total of 60 episodes collected for each participant. At each episode, a participant was asked to report on their momentary emotional states and on their food choice behavior, including meals and snacks, in the preceding 2 hours. For reporting episodes that entailed a meal, each participant was asked to report on the nutritional quality of the meal-i.e., whether it is healthier (H), the same (S), or less healthy (LH) than their baseline corresponding meal for breakfast, lunch, dinner, or snack. The baseline dietary habits were established in a face-to-face interviews, and the protocol was described in detail in [11]. By providing a summary measure of the perceived nutritional quality of each meal, the relative dietary measurement approach has the advantage of being highly efficient. The measure has been validated and used in the consumer and food research literature [10]. On the other hand, the ESM also asked participants to report the social setting of a meal-whether the meal was consumed alone or with others, whether the meal is consumed at home or AFH, and whether the day on which the meal was consumed was a weekend day. For measure of emotion, we followed [11] and used the following components: Positive Emotion (PE)- general; PE-peacefulness, Negative Emotion (NE) -general, NE-shame, NE-worry. See also [12]. Two other demographic and anthropometric measures that were deemed relevant-age and BMI-which were also included in the analysis.

3   Model

The setup for the RMM is rather general in that the model can accommodate multiple chains of both observed and hidden Markov processes. For the sake of illustration, the current application uses two chains, of which one (food choice) is directly observed and the other (emotion) is not. Consequently, the momentary food choice measure X_t at time t and the emotional states Z_t were modeled as two intertwining reciprocal Markov chains. An important feature of the RMM is that it captures the feedback mechanism inherent within the human emotion-behavior system. In other words, a variable X in the first chain at time t is modeled as having an effect on the variable of the second chain Z at time $t + 1$, whereas the variable in the second chain at time $t+1$ has an effect on the variable of the first chain at $t+2$, and so on. The interacting chains thus allow a feedback loop to form for each variable-for example, $X_t \to Z_{t+1} \to X_{t+2}$. Note that by the Markov assumption, $X_t \to X_{t+1} \to X_{t+2}$, and similarly, $Z_t \to Z_{t+1} \to Z_{t+2}$.

Generally, the RMM does not impose any restriction on the variable type of the respective Markov chains-i.e., the data type of an individual Markov chain could be discrete or continuous; the states could be directly observed or hidden (latent). In the current application, emotional state is modeled as an HMM of which the hidden (discrete) variable is indicated by several continuous observable outcomes (PE-general, PE-peaceful, and so on), whereas food choice pattern is modeled as an observed Markov chain of discrete outcome variables. To facilitate reading, we describe the RMM model with one hidden and one observed chain as follows.

Denote the hidden variable in HMM at time t by Z_t and the observed variable at time t by X_t. Not directly observed, the hidden variable Z_t is measured by multiple indicator variables $\{Y_{jt}, j = 1, \ldots, J\}$, which are assumed to be conditionally independent of each other given Z_t [9]. Figure 1 shows schematically the RMM of one hidden and one observed chain.

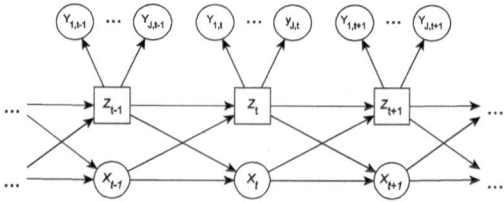

Fig. 1. A Reciprocal Markov Model (RMM) with one observed and one hidden chain

To model the observed indicator variables Y_{jt} for the hidden variables Z_{jt}, we assume a time-homogeneous Gaussian mixture model-i.e.,

$$P(Y_{jt}|Z_t = s) \sim N(\mu_{sj}, \sigma_{sj}^2). \tag{1}$$

In the study, the number of emotional states collected using ESM was not necessarily identical to the number of meals consumed. Indeed, the number of reported meals varied substantially across individuals. In this analysis, we fixed the number of daily reported time points to be identical to the number of ESM reported emotions-i.e., 6 times per day. When the emotion variable does not have a corresponding meal variable, we used last-value-carry-forward (LVCF) for the meal variable.

The RMM defines two sets of conditional models for the respective outcomes: $P(Z_t|Z_{t-1}, X_{t-1})$, and $P(X_t|X_{t-1}, Z_{t-1})$. Both sets of models are assumed to be homogeneous. For example, the probability of being in a specific emotional state given previous status is assumed to not vary across time points. The same could be said of food choice states. Logistic regression models were also fitted to model the transition between states by incorporating a set of relevant covariates. The contextual covariates included in the RMM were (1) the social variable, eating alone or with others; (2) the location variable, eating at home or AFH; and (3) the weekend variable (weekday or weekend). Because of the reciprocality structure of the RMM, a collection of conditional models needs to be specified. In general, if there are K and M states respectively for emotion Z and food choice X, there will be $K \times M$ different combinations of conditions. As an example, if there are two states of emotion k_1 and k_2, and three states of food choice status m_1, m_2, and m_3, then there are $3 \times 2 = 6$ regression models for each of the outcomes of emotion and food choice. In the interest of space, we present as an example one conditional model as follows. Given emotion status $Z_{t-1} = k_1$ and food choice status $X_{t-1} = m_1$ at the previous time point, the emotion and food choice statuses at the current time point t follows the following logistic formulation:

$$\log \left(\frac{p(Z_t = k_1)}{p(Z_t = k_2)} |. \right) = \beta_{01} + w_1^T \beta_1, \tag{2}$$

$$\log \left(\frac{p(X_t = m_1)}{p(X_t = m_3)} |. \right) = \beta_{02} + w_2^T \beta_2, \tag{3}$$

where w_1 and w_2 denote the vectors of covariates for the respective conditional models, β_{01} and β_{02} denote the respective intercept terms, and β_1, and β_2 denote the respective sets of regression coefficients. The notation $|.$ denotes the short-hand form of conditioning on the given values of X_{t-1} and Z_{t-1}. In Equation 3, the state m_3 is assumed to be a reference state. We used ordinal logistic regression when the states were deemed ordered. More generally, the RMM uses multinomial logistic regression for unordered categorical outcomes. Maximum likelihood method was used to estimate the parameters using specialized graphical model software written in MATLAB [8],[9].

4   Results

The mean age of the sample was 44.9 (range 18 to 83). In total, $9,365$ observations were collected from $n = 160$ participants and an average of 24.7 meals episodes per participant were reported. The HMM delineated two emotional states that can be visualized in Fig. 2. We labeled the states Positive Emotion (PE) and Negative Emotion (NE). The observed food choice variable contained three categories: Less Healthy (LH), Same (S), and Healthier (H). Tables 1 and 2 respectively show the conditional transition probabilities to different states for all combinations of values of the conditional emotional and food choice variables.

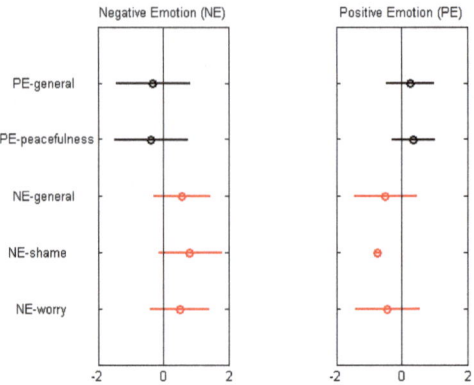

Fig. 2. Hidden state profiles for emotion

Table 1. Estimates of conditional tables for emotion, $p(Z_t|Z_{t-1}, X_{t-1})$

	S at t-1		H at t-1		LH at t-1	
	NE	PE	NE	PE	NE	PE
NE at t-1	0.77	0.23	0.83	0.17	0.72	0.28
PE at t-1	0.15	0.85	0.22	0.78	0.31	0.69

Table 2. Estimates of conditional tables for food choice, $p(X_t|X_{t-1}, Z_{t-1})$

	NE at t-1			PE at t-1		
	S	H	LH	S	H	LH
S at t-1	0.70	0.14	0.16	0.81	0.08	0.10
H at t-1	0.43	0.47	0.10	0.41	0.48	0.12
LH at t-1	0.41	0.14	0.45	0.42	0.07	0.51

The tables reveal some interesting patterns. For example, the probability of staying in PE is rather high (at 85%) if the preceding meal was reported the same compared to baseline (Table 1). Alternately, the probability of PE transitioning to NE is higher when the last meal is LH than when it is S or H. Table 2 also shows evidence that when pre-meal emotion is PE, there is a higher tendency to stay the same in eating. Surprisingly, when the last meal is LH and emotion is NE-i.e., the conditional combination (NE,LH), there is a higher probability of transitioning to a healthy meal (14%) than when emotion is positive (7%).

In our regression analysis, only the following conditions gave significant positive intercept terms: (PE,S) for eating the same (S), and (NE,S) for eating the same. The reference category is LH. In other words, for a participant who is in PE and eats the same quality meal compared to baseline, in her next meal the odds of her eating the same (S) is significantly higher than eating LH after controlling for other factors; so is when she is in NE. Because there are 2 (outcomes) $times 6$ (conditions) $= 12$ regression models, it is not possible to report all the results due to space limitations. We only highlight the cases in which the social variable AFH showed p-values < 0.01. With a more conservative criterion, we did not use the Bonferonni adjustment for multiple comparisons. We found that when eating AFH, participants were less likely to consume S and H food under the two preceding conditions: (PE, H) and (PE, S). This suggests that AFH does have an effect on nutritional quality under specific conditions. Finally, it is also noteworthy that age is highly significant for the following conditions: for (PE, LH), older adults are more likely to consume healthier meals ($p < 0.01$); and for (PE, S) or (NE, S), they are more likely than their younger counterparts to become emotionally more positive at the following meal ($p < 0.01$).

5   Conclusion

Using ESM data collected from a sample of women, this article presents an analysis of socio-behavioral interaction which incorporates feedback mechanisms between emotion and behavior. The analysis shows different transitional probabilities according to preceding emotional state and meal quality. We also demonstrate that social setting is an important factor in specific conditions, particularly eating AFH is more likely to be less healthy when prior emotion state is positive.

This paper offers two contributions. First, we propose a general Markov model that can accommodate both hidden and observed states. More importantly, we focus on a reciprocal mechanism that allows the examination of a possible feedback loop between emotion and eating behavior. Conceptually, the approach expands the more traditional unidirectional approach for analyzing the relation between emotion and eating behavior. As an methodologic extension of single-chain Markov models, we view the proposed RMM model as an important addition to the statistical toolkit for analyzing social-behavioral data. With a few exceptions in methodologies developed by social scientists [13], traditional statistical models are not generally well-equipped to analyze a system of variables that contain feedback loops. The lack of capacity to incorporate feedback mechanisms has long been viewed as a barrier for applying formal statistical models to complex systems. The feedback RMM proposed in this paper is an attempt to mend the barrier. A second contribution is in the application of the RMM to an important social-behavioral problem related to obesity. The RMM was used to analyze intensively collected ESM data, which are becoming more prevalent as mobile devices are increasingly used for momentary data collection. The RMM has the potential to be applied to other socio-behavioral situations in which multiple streams of momentary data are collected. One example is the collection of data related to alcohol consumption. The reciprocal pathways between influence of both positive and negative emotion on alcohol use and vice versa have been studied in depth (e.g., [14]) and ESM data arising from such situation could be analyzed using RMM.

There are also limitations to this study. First, the data used in this study was part of a broader study of age differences in affects, emotions, and lifestyle behaviors in women. The rather homogeneous sample in terms of demographics by design makes generalization of our findings difficult. Second, the method has not distinguished possible within-day and between-day variations. Finally, the LVCF assumption in treating meals that did not occur may not apply to all individuals. Finally, as pointed out by one reviewer, sustaining ESM data collection is challenging. There may exist a diminishing motivation amongst participants over time but we did not examine such potential bias in the data.

To summarize, despite its limitations, this article represents a first-step to offer a formal statistical Markov model that captures the interaction between two socio-behavioral processes. It also illustrates the complexity and challenges of modeling feedback loops that are not necessarily smooth.

Acknowledgment. The study is supported by NHLBI grant 1U01HL101066-01 (PI: Edward Ip). The content of this paper is solely the responsibility of the authors and does not necessarily represent the official views of the Office of Behavioral and Social Sciences Research or the National Institutes of Health.

References

1. Canetti, L., Bachar, E., Berry, E.M.: Food and emotion. Behavioral Processes 60, 157–164 (2002)
2. Jansen, A., Vanreyten, A., van Balveren, T., Nederkoorn, C., Havermans, R.: Negative affect and cue-induced overeating in non-eating disordered obesity. Appetite 51, 556–562 (2008)
3. Baumeister, R.F., Heatherton, T.F., Tice, D.M.: Losing control: why and how people fail at self-regulation. Academic Press (1994)
4. Lin, B.-H., Frazao, E., Guthrie, J.F.: Away-from-home foods increasingly important to quality of American diet. Agricultural Information Bulletin 749, 1–22 (1999)
5. Wheeler, L., Reis, H.T.: Self-recording of everyday life events: Origin, types, and uses. Journal of Personality 59, 339–354 (1991)
6. Context-aware Experience Sampling Project. MIT,
   http://web.mit.edu/caesproject/index.htm
   (last retrieved October 31, 2012)
7. MacDonald, I.L., Zucchini, W.: Hidden Markov and other models for discrete-valued time series. CRC Press (1997)
8. Ip, E.H., Snow-Jones, A., Heckert, D.A., Zhang, Q., Gondolf, E.: Latent markov model for analyzing temporal configuration for violence profiles and trajectories in a sample of batterers. Sociological Methods and Research 39, 222–255 (2010)

9. Zhang, Q., Snow Jones, A., Rijmen, F., Ip, E.H.: Multivariate discrete hidden markov models for domain-based measurements and assessment of risk factors in child development. Journal of Computational and Graphical Statistics 19(3), 746–765 (2010)

10. Block, G.: Foods contributing to energy intake in the US: data from NHANES III and NHANES 1999-2000. Journal of Food Consumption Analysis 17, 439–447 (2004)

11. Lu, J., Dube, L.: Emotional reinforcement as a protective factor for healthy eating in home settings. Journal of Clinical Nutrition 94, 254–261 (2011)

12. Richins, M.L.: Measuring emotions in the consumption experience. Journal of Consmuer Research 24, 127–146 (1997)

13. Paxton, P.M., Hipp, J.R., Marquart-Pyatt, S.: Nonrecursive models: Endogeneity, reciprocal relationships, and feedback loops. Sage Publication Inc. (2011)

14. Cooper, M.L., Frone, M.R., Russell, M., Mudar, P.: Drinking to regulate positive and negative emotions: A motivational model of alcohol use. Journal of Personality and Social Psychology 69(5), 990–1005 (1995)

Testing the Foundations of Quantal Response Equilibrium

Mathew D. McCubbins[1], Mark Turner[2], and Nicholas Weller[3]

[1] University of Southern California, Marshall School of Business,
Gould School of Law and Department of Political Science
[2] Case Western Reserve University, Department of Cognitive Science
[3] University of Southern California, Department of Political Science

Abstract. Quantal response equilibrium (QRE) has become a popular alternative to the standard Nash equilibrium concept in game theoretic applications. It is well known that human subjects do not regularly choose Nash equilibrium strategies. It has been hypothesized that subjects are limited by strategic uncertainty or that subjects have broader social preferences over the outcome of games. These two factors, among others, make subjects boundedly-rational. QRE, in essence, adds a logistic error function to the strict, knife-edge predictions of Nash equilibria. What makes QRE appealing, however, also makes it very difficult to test, because almost any observed behavior may be consistent with different parameterizations of the error function. We present the first steps of a research program designed to strip away the underlying causes of the strategic errors thought to be modeled by QRE. If these causes of strategic error are correct explanations for the deviations, then their removal should enable subjects to choose Nash equilibrium strategies. We find, however, that subjects continue to deviate from predictions even when the reasons presumed by QRE are removed. Moreover, the deviations are different for each and every game, and thus QRE would require the same subjects to have different error parameterizations. While we need more expansive testing of the various causes of strategic error, in our judgment, therefore, QRE is not useful at *predicting* human behavior, and is of limited use in *explaining* human behavior across even a small range of similar decisions.

Keywords: bounded rationality, human behavior, Nash equilibrium, behavioral game theory, strategic uncertainty, social preferences, Quantal Response Equilibrium.

1   Game Theoretic Models of Behavior

A common approach to predicting strategic human behavior across many different situations utilizes the theory of non-cooperative games. At the same time, decades worth of human subject experiments now demonstrate that the basic predictions of game theory are inaccurate [1, 2, 3]. As a result, scholars have developed a variety of modifications of the standard Nash equilibrium concept that relax the strict behavioral and cognitive assumptions contained in Nash equilibrium.

A.M. Greenberg, W.G. Kennedy, and N.D. Bos (Eds.): SBP 2013, LNCS 7812, pp. 144–153, 2013.
© Springer-Verlag Berlin Heidelberg 2013

One of the most well-known models that relaxes the predictions of Nash equilibrium is the Quantal Response Equilibrium (QRE) [4]. Some argue that QRE "almost always explains the direction of deviations from Nash and should replace Nash as the static benchmark to which other models are routinely compared." [5] The QRE model maintains the assumption that individuals have beliefs that are supported in equilibrium by the strategies that players choose, but that players make systematic "mistakes" or deviations in their choices. The individual deviations from Nash equilibrium strategies and outcomes can come from a variety of different sources, but two reasons seem to predominate discussions. The first reason is strategic uncertainty or bounded rationality [6]. The second reason is that individual utility functions may involve something other than the individual subject's payoffs, typically thought of as a "social preference," in which subjects appear altruistic or fair or seek to reciprocate fairness or seek to limit inequality in payoffs [7, 8].

In the QRE model the deviations between subjects' actual behavior and the Nash equilibrium of a game are represented by a parameter, λ, and as λ approaches zero, behavior is essentially random, while as λ approaches infinity, behavior is consistent with the strategies predicted by Nash equilibrium. The parameter in QRE is a free parameter that is fitted via maximum likelihood estimation and is used to fit the model to the observed behavior. The source or interpretation of the parameter, however, is not determined *a priori* but is left to the analyst. The QRE model predicts that deviations from Nash predictions will be less likely as behavior becomes more costly in terms of the losses a player suffers from non-equilibrium (non-rational) behavior.

The QRE model has been widely demonstrated to provide a better fit to observed behavior than Nash equilibrium, and has become popular in fields outside of economics, such as computer science. The most rigorous statistical examinations of QRE estimate λ from a subset of experimental data and then examine how the estimated λ "predicts" the remaining, unused or out-of-sample data (basically a form of cross-validation or training/test data) [9]. While clever and exciting, this approach falls short as a test of the QRE model for two main reasons.

First, this approach does not make specific predictions about expected behavior prior to the design of the study. That is, the QRE approach does not rule out much, if any, behavior as inconsistent with the theory, data is collected and used to estimate a free parameter that can accommodate almost any observations. Despite the empirical failure of the Nash-equilibrium model, it has the virtue that it does make clear predictions about the expected pattern of observed behavior. To put it another way, is there a distribution of observed behavior that would suggest QRE is an incorrect model of behavior?

Second, statistical approaches to QRE take experimental data and then identify which model best "fits" (by some standard) that data. It is unspecified how to reject QRE in this approach. We could certainly find that some other model better fits the data or that combining QRE with another model leads to a better overall fit to the distribution of the data [9]; however, this approach does not reject the model.

An alternative approach to testing QRE is to identify the underlying source of the deviations from rationality (Nash equilibrium strategies) that give rise to λ, and then design experiments that remove these sources and observe whether deviations still occur. This is an ambitious research program, as there are at present about four dozen causes of bias or error identified in the literature. It is this approach that we pursue here. We discuss in the next section what is required to test a behavioral theory.

2   Testing a Behavioral Theory

We begin by identifying the basic requirements to test a theory. A simple representation is given by the following two statements, in which c = a condition or set of conditions, t = a treatment, and y = an outcome of interest.

$$\text{if } c + t, \text{ then } y$$
$$\text{if } c + \sim t, \text{ then } \sim y$$

In laboratory economic experiments, we can think of the t as the particular game being played (i.e. a prisoner's dilemma, chicken, ultimatum, matching pennies, etc.). The conditions, c, represent other parameters of the experimental design (i.e. elements of the protocol, the number of repetitions, possibilities for learning, etc.), and the outcome, y, is how subjects play a given game.

The outcome, y, is a function of t and c; therefore, to test whether the treatment causes the outcome requires that we hold constant the conditions under which the treatment is implemented; otherwise, it may be the changes in the conditions that lead to the presence or absence of a particular outcome. Overall, the framework implies that if we change some element of t (without changing anything else) and the outcome changes, then we can infer that the change in the outcome is a result of the change of t. This framework implies that we must be able to specify, in advance of the research, what are c, t and y. We will know that t is the cause of the outcome if its removal (or change) leads to a change in the outcome. On the other hand, if we remove (or change) t and y still occurs, then we know that it is not the t that is causing y to be observed.

This framework implies a straightforward way to test various theories of behavior, although it can be quite difficult to implement. In a nutshell: 1) identify the treatment and how it causes the outcome of interest; 2) identify a way to remove the aspect of the treatment that causes the behavior, and 3) determine whether the behavior persists. If the behavior persists, then we can rule out that aspect of the treatment as the cause.

3   Testing the QRE Model

QRE has gained popularity in behavioral economics, political science, and computer science, with some researchers claiming it should be used as the baseline model for explaining subject behavior [5]. Despite these claims, the QRE model proves quite difficult to test – that is, it is difficult to design a test of QRE that would allow us to reject the approach [10].

One direct way to examine whether QRE can explain behavior is to create situations where the reasons for subjects to deviate from predictions are minimized, ideally to the point of non-existence. To do so requires us to be clear about the sources of the observed errors. In the QRE framework, deviations are thought to come from strategic uncertainty and/or some form of social preference. Extending our previous experiments [11, 12] we employ two different experimental approaches to study whether removing the causes of strategic error actually eliminates those deviations in experimental settings.

Our first approach is to examine subjects in situations where they do not have to make a strategic decision. In these tasks an individual subject makes a decision that by itself completely determines her payoff. Most often this is either a single-play task or it is the last stage of a sequential game of complete and perfect information, thus establishing the conditions for subjects to be sequentially rational. If strategic uncertainty leads to error, then the removal of any strategic element of the task should eliminate the deviations.

The second approach we take combines removing strategic uncertainty with reducing the likelihood that we misunderstand the subjects' utility function (that is we do not observe how social are their preferences). Although we cannot observe utility functions directly, a common reason offered for why the utility function is different than that assumed by classical game theory is that subjects have some form of social preferences in their utility function [7]. In general, these approaches explain deviations from Nash equilibrium by arguing that individuals receive some utility from the payoffs received by other actors. To eliminate this rationale for behavioral deviations, we have subjects play economic games against an algorithmic agent (computer player). We describe these experiments in more detail in the next section. While we have received yarns spun to (post-hoc) explain why subjects taking private and anonymous actions against a computer would act socially with the computer, we do not see any sequentially rational reason why subjects would be concerned with the payoffs a computer receives and would thus deviate from Nash equilibrium strategies. Together, these two approaches are good first steps to gaining leverage over the sources of deviations in game theoretic experiments.

4   Experimental Design

We describe our general experimental design in this section. One of the most important attributes of our experimental design is that we have subjects complete a battery of different, sometimes related, tasks over the course of an experimental session. This will allow us to compare behavior in one task to behavior in another task. As a result, we can examine whether the behavior we observe happens only in one game or if it happens across a variety of games. This is crucial to determining if the behavior is anomalous or reflective of a larger pattern.

The first step we take in our experimental design is to control the conditions of the experiment in a way that makes it unlikely that they are the cause of the deviations in behavior. Subjects were recruited using flyers and email messages distributed across a large public California university and were not compelled to participate in the experiment, although they were given $5 in cash when they showed up. The experiment lasted approximately two hours. The subjects in our experiment completed the tasks using pen and paper in a controlled classroom environment.

In all of the tasks we discuss below subjects played against an anonymous opponents (human or computer) and in all settings we used a double blind protocol to ensure their anonymity from the experimenters as well [13]. In between tasks subjects were also randomly paired with a new subject who was in another room. Subjects also only completed each task one time to ensure that they were not learning about the other subjects in the experiment, which might also cause changes in behavior. To ensure that subjects understood the tasks facing them we utilize very simply experimental tasks and we also quiz subjects before each task to ensure that they understand what they are to do. Subjects are paid for correct answers on the quizzes and if they get an answer wrong we describe the correct answer to them. Overall, subjects very rarely got wrong answers on the quizzes, which gives us confidence that their behavior is not due to a lack of understanding the task. We take great care in our control of the experimental conditions to ensure that we remove the aspects of the experiment that are typically argued to be the causes of Nash equilibrium.

We report on a portion of our battery of tasks here that involve either individual decisions or decisions by subjects acting in the role as the second player of a sequential task. These tasks do not involve strategic uncertainty about the actions of another subject. Players in all of these games know they are randomly paired with another subject in a different room. We report on the choices made by subjects when they are Player 2 in the Trust Game [14]. In this game, Player 1 and Player 2 each begin with a $5 endowment. Player 1 chooses how many dollars to send to Player 2 (ranging from $0 to $5). That amount is tripled by the experimenters and given to Player 2, who then has the sum of the tripled amount received and the initial $5 endowment. Player 2 then chooses how many of those dollars (possibly 0) to transfer to Player 1. Player 2 keeps whatever she does not transfer. In another game, called the Dictator game, The Dictator (Player 1) and the Recipient have endowments equal to exactly what Player 2 and Player 1 had in the Trust Game when the present Dictator played it as Player 2. Accordingly, the Dictator game is identical right down to the specific endowments to the second half of the Trust game: the Dictator is in exactly the position he or she was in as Player 2 in Trust. In effect, each subject replays the second half of the Trust game, but now without the reciprocity frame. The Donation game is identical to The Dictator game, except that each player begins with a $5 endowment and the amount Player 1 chooses to send is quadrupled before it is given to Player 2. In both of these games, the subject keeps all the money he does not send to the other player and there is no action required by the other player.

In what is called the Sequential Chicken Game, players maximize their payoffs if they choose the opposite of the action chosen by the other player. We implement a sequential version in which the 1st player chooses STOP or GO, and then the 2nd player observes the 1st player's choice before also choosing between STOP or GO. The payoffs in the game are shown in Table 1, where Player 1's payoffs are listed first and the grey cells represent deviations from Nash equilibrium.

Table 1. Payoffs in Sequential Chicken Game where Player 1 moves first and play is revealed to Player 2

		Player 2	
		GO	STOP
Player 1	GO	$0, $0	$5, $3
	STOP	$3, $5	$4, $4

At the end of the experiment, we present the subjects with tasks that would allow them to learn something about the choices made by subjects in the other room. They are asked to make a choice as Player 2 in the Trust game as one of the final tasks. In this last task, we have no choice but to provide subjects with information about what other subjects have done – Player 2 in Trust must know what Player 1 chose to send.

We also conducted the same experimental games reported above but had players play against computer opponents. We informed the subjects that the computers (their opponents) would always take the action that maximized the computers financial gain for the present task– that is, the computers played as perfect "Nash players." Playing a game against a computer represents a change in the treatment: it removes exactly the elements associated with playing against an anonymous person.

5   Behavior without Opportunity for Strategic Error

The experiments we designed—including those in which people play each other—explicitly remove the opportunity for subjects to make miscalculations about the actions of others, because in all of the settings, subjects' payoffs are determined solely by their own actions.   We go to great lengths to ensure subject anonymity from each other and the experimenters, and we believe that we have done as well as possible in that regard. This does not remove the possibility that subjects' utility is not related solely to their monetary payoffs, which remains a source of deviations.

In Table 2 we report the results for the different tasks in which subjects had to decide how much, if any, money to pass to another player. Recall that in all of these settings the choices of players determine directly and solely what is kept or passed there was no need for strategic calculation. Therefore, the possibility of deviations caused by the inability of players to understand and predict the strategic interaction is eliminated by the experimental design. Yet, subjects send money to another anonymous player, which contrasts with the Nash equilibrium. In all three tasks, many of the players (the number of subjects is 180 in these tasks) send money to the other player. Furthermore, the assumption within QRE that costly mistakes will be less likely is inconsistent with these data. The incidence of deviations does not decline in a consistent pattern as we move away from the choice to send $0.

We turn now to the sequential chicken game and in particular to examining the behavior of Player 2, who already knows the move of Player 1 before making her own choice in this task. There are 17 subjects in the role of player 2 that choose GO even after the 1st player already chose "GO." These players knowingly made a choice that leads both subjects to get a payoff of $0 instead of the second player's choosing STOP and guaranteeing $3 for herself and $5 for the first player. In addition, there are 36 players in the role of Player 2 who choose STOP even after Player 1 also chose STOP. This choice gives Player 2 a lower payoff than if she chooses GO in this situation. Player 2 faces zero strategic uncertainty, because, before choosing, Player 2 is informed of Player 1's choice.   Yet, overall, 53 subjects (29.4%) chose the action that leads to the lower payoff.

Across a wide number of tasks, we observe that subjects take actions that are inconsistent with standard game theoretic expectations and that cannot be explained by strategic uncertainty. However, it may be that these deviations come not from strategic uncertainty but from the fact that the subjects' utility functions differ from our assumptions about them, which is another source of behavioral deviations in the QRE approach. We turn now to a set of experiments designed to address that question.

Table 2. Choices in three different experimental tasks when playing with other humans. Trust results include the 80 players who received $0 from Player 1. If those are excluded the number who sent $0 drops to 38.

	Amount passed by subjects in each experimental task										
	$0	$1	$2	$3	$4	$5	$6	$7	$8	$9	$10
Donation	87	32	23	11	6	21	n/a	n/a	n/a	n/a	n/a
Dictator	132	11	18	8	1	7	0	1	0	1	1
Trust Player 2	118	10	20	8	8	8	0	1	0	1	6

6   Misunderstood Utility Functions

The results in the prior section demonstrate that subjects still deviate from basic Nash equilibrium even when we remove the strategic elements; however, it is still possible that the behavior we observe is a result of the fact that we misunderstand their utility functions. In this setting, that will usually mean that individual's utility functions include something other than their own monetary payoffs. This could include a concern for reducing inequality, maximizing social welfare, punishing others for the choices they made, or anything else that makes them take actions in which they knowingly earn less money because their utility depends in some way on what others earn as well. This is not a very specific argument, and the unobservability of utility functions means that if we allow ourselves to utilize post-hoc rationalizations, we can justify nearly any behavior via utility function modification.

We turn now to examining the actions our human subjects take when matched to a computer that they know will take actions to maximize its payoffs – a perfect Nash player. In Table 3, we display the humans' behavior (for 40 subjects) across a range of different experimental tasks. This time we focus on the decisions humans make as the first player in Trust, choosing how much of $5 to pass the computer, rather than the second player, because computer players never pass money to the other player and the computer will never return money as Player 2 in the Trust game. Despite the fact that there is no strategic uncertainty and players know they are playing a computer that will maximize its own payoffs, we still observe that players choose to pass money to the computer player in all three situations. In both Donation and Trust, at least 25% of subjects in the role of Player 1 pass money to a computer opponent.

In the sequential chicken game, the 2nd player is always faced with a computer who chose "GO," which is the Nash equilibrium. In this setting, the human players chose "STOP" 39 out of the 40 times.

Table 3. Choices when subjects play against computers who play as perfect Nash players

	Amount passed by subjects in each experimental task					
	$0	$1	$2	$3	$4	$5
Donation	30	3	5	2	0	0
Dictator	38	0	1	0	0	1
Trust Player 1	28	3	5	1	1	2

The overall pattern of behavior when subjects are playing against computer players is much closer to Nash behavior. By combining tasks where our subjects' actions directly determine their payoffs with computer opponents who will always take the payoff maximizing option, we remove all of the strategic uncertainty that often faces players in game theoretic experiments. In addition, play against a computer eliminates much (though perhaps not all) of the ways that a subject's utility function can depend on the payoffs to another subject. Despite removing the explanations for deviations that underpin QRE, we still observe behavior that is inconsistent with Nash equilibrium. In the Donation and Trust games, 25% of subjects deviate from the Nash predictions, whereas in the sequential chicken game, only 2.5% of subjects deviate.

These results suggest that there may be something about the games themselves that makes them more amenable to non-Nash behavior even after we remove the standard rationales that underpin the QRE model.

7   Discussion

Prior research has established that the quantal response equilibrium provides a better fit to experimental behavior than Nash equilibrium, but it is also clear from the results we report that subjects' behavioral deviations do not occur simply from the standard explanations that underpin QRE. We show via a variety of experiments that even when we design the experimental conditions and treatments to remove the reasons why people might deviate from Nash equilibrium we still observe such deviations. The deviations also differ across games, but we lack an explanation for why that occurs. We could still estimate the lambda parameter of the QRE model for that data we discuss in this paper, but the parameter would have no substantive meaning given that the causes of the deviations do not exist in these experiments.

Additionally, as we report elsewhere, the decision to send money in one of these tasks is not predictive of the decision to send money in one of the other tasks [11]. That is, players may send money in one setting but then not send money in another very similar setting. In the context of the QRE model, this implies that the lambda parameter is not consistent within a subject across very similar, albeit not identical, tasks. The QRE model does not assume that the parameter will be consistent, but it makes it difficult to use the QRE approach to predict behavior in new tasks if individual deviations are inconsistent across similar tasks.

The results in this paper suggest that even though QRE improves on Nash equilibrium in its ability to fit observed behavior, we cannot be sure that the observed deviations come from the sources assumed in QRE, because deviations persist even when subjects are placed in environments that remove the sources of the deviations. Therefore, these results suggest that we need to revise the basic cognitive and behavioral assumptions of game theory in order to build models that more accurately reflect human capabilities and actions.

Acknowledgments. McCubbins acknowledges the support of the National Science Foundation under Grant Number 0905645. Any opinions, findings, and conclusions or recommendations expressed in this material are those of the author(s) and do not necessarily reflect the views of the National Science Foundation. Turner acknowledges the support of the Centre for Advanced Study at the Norwegian Academy of Science and Letters.

References

1. Smith, V.: Bargaining and Market Behavior: Essays in Experimental Economics. Cambridge University Press (2000)
2. Camerer, C.. Behavioral Game Theory. Cambridge University Press (2003)
3. Gigerenzer, G.: Rationality for Mortals: How People Cope with Uncertainty. Oxford University Press, New York (2008)
4. McKelvey, R., Palfrey, T.: Quantal Response Equilibria for Normal Form Games. Games and Economic Behavior 10, 6–38 (1995)
5. Camerer, C., Ho, T.-H., Chong, J.-K.: Behavioral Game Theory: Thinking, Learning and Teaching. Paper Presented at the Nobel Prize Symposium (2001)
6. Simon, H.: Rational Choice and the Structure of Environments. Psych. Rev. 63, 129–138 (1956)
7. Rabin, M.: Incorporating Fairness Into Game Theory and Economics. The American Economic Review 83, 1281–1302 (1993)
8. Souten, J., DeCremer, D., van Dijk, E.: Violating Equality in Social Dilemmas: Emotional and Retributive Reactions as a Function of Trust, Attribution, and Honesty. Personality & Social Psychology Bulletin 32, 894–906 (2006)
9. Wright, J., Leyton-Brown, K.: Beyond Equilibrium: Predicting Human Behavior in Normal Form Games. In: Conference of the Association for the Advancement of Artificial Intelligence, AAAI 2010 (2012)
10. Haile, P.A., Hortaçsu, A., Kosenok, G.: On the empirical content of quantal response equilibrium. Cowles Foundation Discussion Paper no. 1227 (August 2008)

11. McCubbins, M.D., Turner, M., Weller, N.: The Mythology of Game Theory. In: Yang, S.J., Greenberg, A.M., Endsley, M. (eds.) SBP 2012. LNCS, vol. 7227, pp. 27–34. Springer, Heidelberg (2012a)
12. McCubbins, M.D., Turner, M., Weller, N.: The Challenge of Flexible Intelligence for Models of Human Behavior. Association for Advancement of Artificial Intelligence Spring Symposium on Game Theory for Security, Sustainability and Health. AAI Technical Report (2012b)
13. Hoffman, E., McCabe, K., Shachat, K., Smith, V.: Preferences, Property Rights and Anonymity in Bargaining Games. Games and Economic Behavior 7, 346–380 (1994)
14. Berg, J., Dickhaut, J., McCabe, K.: Trust, Reciprocity and Social History. Games and Economic Behavior 10, 122–142 (1995)

Modeling the Social Response
to a Disease Outbreak

Jane Evans[1,2], Shannon Fast[1,2], and Natasha Markuzon[1]

[1] Draper Labratory, Cambridge MA 02139, USA
{sfast,nmarkuzon}@draper.com
[2] Massachusetts Institute of Technology, Cambridge MA 02139, USA

Abstract. With the globalization of travel and economic trade, disease
can spread rapidly across the globe, sometimes causing panic, population
flight and other forms of social disorder. These responses often herald a
significant change in the epidemiological pattern or etiology of an infectious disease event. It is therefore increasingly important not only to
detect outbreaks of infectious disease early, but also to anticipate and describe the social response to the disease. We use social network analysis
to model situations in which a society exhibits social strain in connection
with a disease. We model negative social response (NSR) by coupling disease spread and opinion diffusion and verify the results against real-world
scenarios. This model captures the complex interaction between disease
and culturally determined social responses, providing insights that may
help operational analysts and policy makers better respond to sudden
disease outbreaks.

Keywords: social network, epidemic, opinion dynamics.

1 Introduction

The public response to a disease outbreak is usually calm and orderly [1]. In
rare cases, however, the outbreak of disease can trigger social and economic
disturbances, including panic or suspicion [2,3,4]. Severe behavioral responses
observed in response to disease include rioting, hoarding of medical supplies,
flight, and violence against members of groups believed to have or carry the
disease [5]. Analysis of 15 months of historical data obtained from Ascel Bio [6],
a biosurveillance company that gathers near-real-time data on infectious disease
outbreaks, revealed that negative social responses (NSR), including panic and
anxiety as well as more severe behavioral responses, are relatively uncorrelated
with the severity of disease. Instead, the region where the outbreak occurred and
whether the disease in question is endemic to that region were much stronger
predictors of whether the society exhibits NSR. We present a model of the spread
of NSR associated with disease spread. This model incorporates the inherent
interdependence of the disease and social processes, but is also flexible enough
to capture cultural differences in the degree to which NSR can spread through
the population.

A.M. Greenberg, W.G. Kennedy, and N.D. Bos (Eds.): SBP 2013, LNCS 7812, pp. 154–163, 2013.

Although there are many models predicting disease spread [7] and many models for the spread of information over networks [8], models examining the interaction between disease spread and information spread are comparatively limited. Nevertheless, as more researchers recognize the importance of individuals' beliefs and actions for determining the progression of epidemics, such models are becoming more common [10]. Epstein et al. [12] explore the effects of fear-induced behavioral changes on the spread of epidemics. They find that self-isolation behaviors decrease the scale of the epidemic. If, on the other hand, frightened individuals flee the area, the scale of the epidemic can dramatically increase. Meloni et al. [11] also consider the effects of population mixing. They conclude that real-time information on a disease spread can reduce the efficacy of disease containment and mitigation measures. Funk et al. [13] take a different view. They model the spread of awareness over a network with individuals taking greater self-protection measures in proportion to their disease awareness. They find that awareness can reduce the size of the outbreak, and, under certain assumptions, stop it altogether.

While a growing number of models address the interplay between disease spread and human behavior, they rarely take into account cultural factors that contribute to differences in the social response to disease. The proposed model predicts NSR while addressing this issue. Namely, the agent-based model has parameters controlling the rate of spread of opinions regarding the disease. The rate of opinion diffusion is independent of the rate of disease spread but is related to the structure of the society. The readiness with which members of a society communicate disease-related concerns to their social connections, as well as, the size and interconnectedness of the social network determine the degree to which behavioral or non-behavioral NSR can propagate. These features of NSR spread were incorporated into the model's NSR parameters.

We demonstrate that the model is able to capture the disease and NSR spread reported in two real-world outbreaks of dengue fever, one in India and another in Argentina. Dengue fever is endemic to India, and the outbreak signature pattern examined in this study did not cause any NSR there. The outbreak signature pattern in Argentina, in contrast, was characterized by readily observable NSR. We will discuss how the proposed model can be used in predicting future unusual disease activity and associated NSR, and in identifying actions that will enable policy makers to make proactive decisions to mitigate both NSR and disease spread.

2   Methods

The progression of disease and social response on the social network are simulated using agent-based modeling [9]. We consider individuals in a population who are connected to each other through social ties. These agents repeatedly interact according to behavioral rules, which govern the spread of disease and social response. The disease and NSR are assumed to spread independently, coupled by an interaction rule – when an agent is infected, her NSR level is maximized. The model's interaction rules imply that agents' likelihood of experiencing heightened NSR will increase when they interact either with the sick or with agents who already have heightened NSR.

The model is based on the spread of misinformation model proposed by Acemoglu et al. [14]. Interactions between agents occur in a pair-wise manner according to a rate 1 Poisson process, independent of all other agents. Thus, time can be discretized into interactions, with the probability that agent i interacts at interaction k equal to $\frac{1}{N}$.

2.1   Disease Process

We employ a variation of the susceptible-infected-removed disease model first proposed by Kermack and McKendrick [15]. This model partitions the population into three classes: susceptible individuals (S) who are not infected but could become infected if exposed, infected individuals (I) who currently have the disease, and recovered individuals (R) who have either recovered from the disease and have immunity or have died. An agent's disease status is estimated as a discrete random variable $D_i(k) \in \{S, I, R\}$. At each interaction k, an agent i is randomly selected for interaction. Agent i then selects an interaction partner j uniformly from among his neighbors. Given that i and j interact, the following interaction types can occur:

1. *Infection.* Given that agent i is infected and agent j is not, with probability ν, agent i infects agent j:

$$D_i(k+1) = D_j(k+1) = D_i(k) \ . \tag{1}$$

   Given that agent j is infected and agent i is not, with probability ν, agent j infects agent i:

$$D_j(k+1) = D_i(k+1) = D_j(k) \ . \tag{2}$$

2. *Recovery.* Given that agent i is infected, with probability κ, agent i recovers:

$$\begin{aligned} D_i(k+1) &= R \\ D_j(k+1) &= D_j(k) \ . \end{aligned} \tag{3}$$

3. *Identity.* If neither infection nor recovery occurs, there is no change in infection state for either i or j.

2.2   Negative Social Response Process

Coupling Process. In addition to his disease state, each agent has a value associated with negative social response, indicating his emotional state related to the disease. Negative social response is treated as a continuous random variable $NSR_i(k) \in [0, 1]$ where 0 indicates no anxiety and 1 indicates extreme panic. NSR spread is modeled under separate rules, but is coupled with disease spread. Specifically, the coupling process maximizes the NSR of newly infected agents:

$$\text{If } D_i(k) = S \text{ and } D_i(k+1) = I, \text{ then } NSR_i(k+1) = 1, \forall i \in N \ . \tag{4}$$

Initially, $NSR_i(0) = 0 \ \forall i \in \{N : D_i(0) \neq I\}$ and $R_j(0) = 1 \ \forall j \in \{N : D_i(0) = I\}$.

Opinion Exchange Dynamics. In addition to the coupling rule, the NSR states update according to interaction rules, which are described below. At each interaction k, agent i is randomly selected to interact with probability $\frac{1}{N}$. Agent i then selects an interaction partner j uniformly at random from among his neighbors. Conditioned on i and j meeting, the following pair-wise interactions can occur:

1. *Forceful.* If $NSR_i(k) > 0.5$, agent j adopts agent i's NSR level with probability α_{High}. If $NSR_i(k) \leq 0.5$, with probability α_{Low}, agent j adopts agent i's NSR level:

$$NSR_i(k+1) = NSR_j(k+1) = NSR_i(k) \ . \tag{5}$$

2. *Averaging.* With probability β, agents i and j average their NSR levels:

$$NSR_i(k+1) = NSR_j(k+1) = \frac{NSR_i(k) + NSR_j(k)}{2} \ . \tag{6}$$

3. *Decay.* With probability δ, agent i's NSR level decreases by a fixed decay parameter, Δ:

$$NSR_i(k+1) = \Delta \times NSR_i(k) \ . \tag{7}$$

4. *Identity.* If none of the above interaction types has occurred, neither i nor j changes his NSR level.

Although there are several interaction rules, the interpretation of the NSR process is straightforward. Forceful interactions capture the idea that NSR is contagious. In particular, when $\alpha_{High} > \alpha_{Low}$, the model represents the realistic situation in which fear and anxiety, which can potentially induce behavioral changes, are more easily spread than is calm. Averaging reflects agents arriving at a consensus about their social response to the situation. Finally, decay reflects the eventual cease of NSR panic as the disease spread tapers off. The degree to which NSR can successfully spread is reflected in the number of connections available to each person and the relative probability of an α_{High} interaction, compared with other interactions. These parameters can be set to reflect particular cultures, diseases and environmental factors. In the current version of the model, we do not explicitly consider the translation of anxiety into behavioral NSR. It is assumed that as a society experiences increasing social strain the likelihood and severity of behavioral NSR increases. In future models, we will provide more detail on how behavioral negative social responses, such as flight, deleterious effects on the economic system, and protests or riots, emerge.

3   Results

The model was implemented on a Waxman random graph $G(N, E)$ with 100 nodes and on average 12 connections per node [16]. For each set of model parameters examined, we simulated 50 realizations of the model, where each realization consisted of 5000 interactions on the test network. At each interaction,

k, both the number of infected agents and the average NSR level of the agents were computed. These values were averaged over all realizations of the model to obtain the expected number of infected agents and the expected average NSR level at interaction k.

3.1   Simulation Results

The behavior of the model is controlled by adjusting the interaction probabilities. A shorter, more severe disease outbreak is achieved by increasing ν and κ and decreasing the difference between them. To achieve a more severe negative social response, we increase the probability of forcefully spreading NSR (α_{High}) or increase the number of connections between agents. We performed sensitivity analysis to determine the ranges of NSR parameters that can produce a given NSR spread. The parameters can effectively model both societies in which individuals do not readily communicate NSR and societies in which NSR spreads readily.

Parameter variation allows us to independently control the spread of disease and the spread of NSR. Even with limited disease spread, one can model a significant negative social response. This finding is consistent with observations from the 2009 H1N1 pandemic. The morbidity and mortality rates associated with H1N1 were not substantially higher than those of seasonal influenza; however, public concern over H1N1 infection was much greater [17].

3.2   Application to Two Outbreaks of Dengue Fever

We validate the model against data from two real-world outbreaks of dengue fever. Detailed information on the outbreaks and associated NSR was obtained from Ascel Bio [6]. The data are a historical combination of multi-source biosurveillance reports, covering more than 200 countries and 300 infectious diseases affecting primarily humans and animals.

In September and October of 2008, New Delhi, India experienced an outbreak of dengue fever [18]. Dengue has been known to be endemic to India for over 200 years, and outbreaks in the New Delhi region are common [19]. The 2008 outbreak lasted for 62 days with 1070 cases reported, most confirmed. Since New Delhi routinely experiences cases of dengue fever, the public health infrastructure was sufficient to deal with the increased caseload. No serious NSR was reported.

A year later, an outbreak of dengue fever occurred in Charata, Argentina, resulting in a severe negative social response [20]. Dengue fever is a relatively novel disease in Argentina. It had been eradicated for 82 years before it resurfaced in 1998. The scale of the 2009 epidemic was nearly unprecedented in Argentina [21]. Severe negative social response occurred, including reports of residents fleeing the city, hospitals being overrun with people seeking treatment, and citizens protesting what they perceived as an insufficient response from the government. There was widespread disagreement regarding the extent of the epidemic. Reported total case counts ranged from 3000 to over 11,000, with 1200 confirmed cases. The extent of the social disorder in Charata made it difficult for officials to arrive at reasonable estimates of the number of cases, and it seems plausible that as people became increasingly anxious in response to the disease, reports became exaggerated and spread through word of mouth.

We modeled the disease and NSR spread for both cities, reflecting scaled disease spread and the reported NSR. Although dengue fever is transmitted by mosquitoes and therefore does not directly spread through a social network, we assume that individuals living in close physical proximity are likely to have similar exposure to mosquitoes carrying the virus and are also likely to have social connections with one another. We therefore consider social transmission of disease to be a sufficient proxy for the true disease process. Figure 1 shows the expected number of cases and expected average NSR for both New Delhi and Charata. The spread of infection in New Delhi is mild and is accompanied by only a negligible social response, while the spread of infection in Charata is more severe and is accompanied by a severe negative social response. To demonstrate the effect of society and outbreak-specific factors that influence NSR spread, we examined the effects of replacing the disease spread parameters of New Delhi with those of Charata while keeping New Delhi's NSR parameters constant. The resulting expected average NSR is displayed in the right-most column of Fig. 2. The expected average NSR in New Delhi remains low even when the disease spread is made more severe, indicating that in our model, societal factors – including the extent to which the society is familiar with and equipped to cope with the disease – have a comparable influence to the disease itself on the extent of NSR.

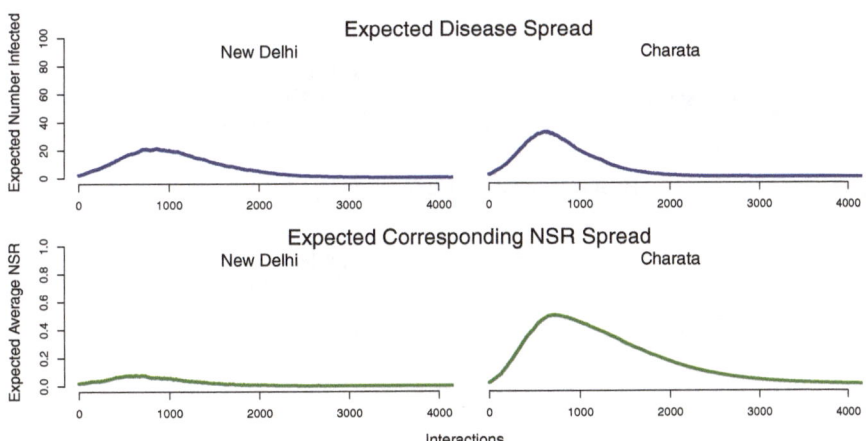

Fig. 1. Although the disease spread (uppermost plots) is more severe in Charata than in New Delhi, Charata's heightened NSR (lowermost plots) is much more severe than can be accounted for by disease severity alone. Factors specific to the society and outbreak, which we incorporated into our NSR spread parameters, added to the increased level of NSR in the simulation of Charata.

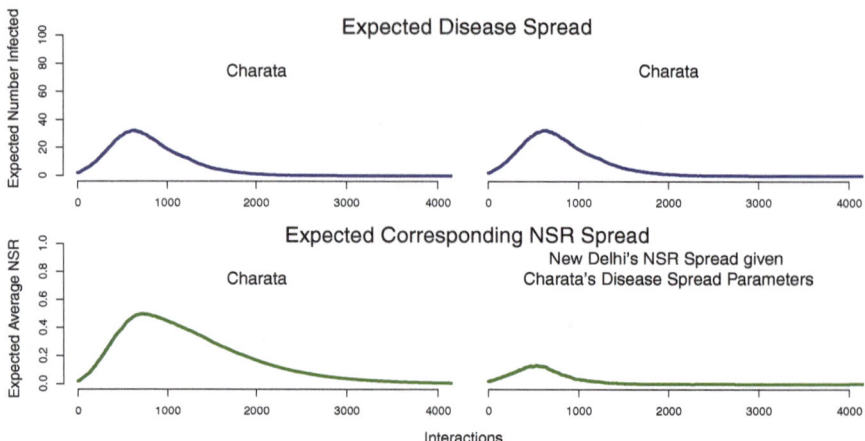

Fig. 2. The flexibility of the model to incorporate outbreak and culture-specific factors in order to explain differences in NSR spread is demonstrated by simulating NSR spread in both New Dehli and Charata, given disease spread modeled using the parameters set for Charata (uppermost graphs). Even with the same disease parameters, the NSR spread is dramatically different in the two cities (lowermost graphs). The NSR parameters set for New Delhi prevent even a more severe disease outbreak from causing severe NSR.

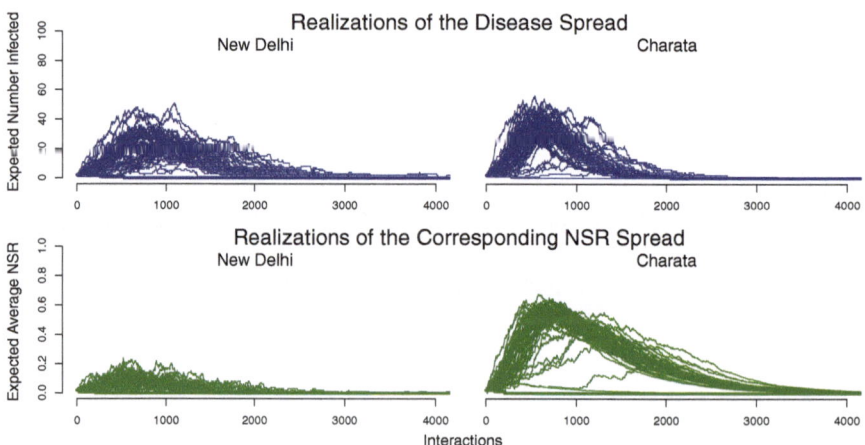

Fig. 3. The number of cases and average NSR in New Delhi and Charata are shown for each of the 50 realizations of the model. In general, different realizations of disease spread produced similar NSR responses, with a few exceptions for Charata, where in several realizations disease spread failed to cause heightened NSR.

Figure 3 describes the variation across the realizations of the outbreak. New Delhi displays limited variability in both the disease and NSR processes. Charata shows a consistently sharp spike in the number of infections, though some realizations take more interactions for the disease spread to accelerate than do others. There are interesting outliers in the realizations of Charata's NSR. In most realizations, Charata's average NSR rises sharply and then tapers off, but in a few realizations heightened NSR does not spread through the population.

4   Discussion

We have shown that a society's negative social response to a disease outbreak can be successfully modeled as a series of interactions on a social network. The framework provides unique insight into separate nature of disease and NSR spread. While it is reasonable to assume that the driving mechanism behind NSR is the severity of the disease, the data suggest that cultural and situational factors play a more significant role. As mentioned previously, the 2009 H1N1 outbreak is the most recent evidence that a disease perceived to be non-routine can provoke anxiety that is disproportionate with the severity of the disease. In our model, we simulated cultural differences by varying the number of connections for agents and the degree to which agents communicate their NSR to others, and we have demonstrated that this variation plays an important role in increasing the level of NSR in the network. Further work will aim to better incorporate information about the novelty of the disease in the area, as analysis of Ascel Bio data suggests that when people are faced with an uncommon situation – a new disease or an unusually high fatality rate – their socio-economic response is more dramatic, as we have seen in Argentina where dengue fever outbreaks were relatively novel.

The simulations validated the assumption that the complex behavior of an outbreak can be explained by a system of basic probability rules. As the simulation of dengue fever outbreaks showed, the model can accurately capture real-world bio-events. By adjusting the probability of infection and the probability of recovery, we can control the severity and duration of the disease process, matching a variety of types of infectious disease outbreaks. By adjusting the probability of agents with elevated NSR diffusing their NSR into the wider society, we can model either mild or severe NSR.

Further research will take the model a step further and investigate the possible feedback of NSR on disease spread. Specifically, the presence of NSR in the network could make the disease less likely to spread if people with high anxiety stay home thus reducing the number of connections and the probability of spreading the disease. Alternatively, if people with high anxiety take actions that disrupt health care delivery, or if they flee to other regions, NSR could make the disease spread more quickly.

Agent-based modeling design allows for monitoring and analyzing the temporal patterns of disease and corresponding NSR spread. The current simulations demonstrated the explanatory power of the model; in the future, we will concentrate on the predictive power and monitoring power. We will evaluate the approach for its ability to anticipate elevated NSR at the beginning of disease spread. Furthermore, when NSR is detected, we will temporally model NSR spread and the effect of containment measures.

The model is extremely flexible, allowing us to simulate a variety of real-world situations with just a few parameters. These parameters are best estimated by incorporating and modeling historical data. In order to estimate NSR parameters, experts can match disease outbreaks to past outbreaks in countries with similar cultures and similar histories with the disease in question. As we become more aware of the causes of NSR and cultural differences in its expression, we will be able to better hone the parameters to fit any number of diseases and cultures, allowing for better prediction of if and how negative social response will spread through a given society faced with a disease outbreak.

Acknowledgments. We thank Ascel Bio for the use of their biosurveillance data, and especially thank James Wilson for his assistance and advice. We also thank Professor Marta Gonzalez for her invaluable advice. This work was partially supported by an internal Charles Stark Draper Laboratory grant.

References

1. Glass, T.: Bioterrorism and the people: How to vaccinate a city against panic. Clin. Infect. Dis. 34, 217–223 (2002)
2. Mob stones Chilean bus amid flu fears. Sydney Morning Herald (2009), http://news.smh.com.au
3. Desvarieux, J.: At the heart of Haiti's cholera riots, anger at the U.N. Time (2010), http://www.time.com
4. Biryabarema, E.: Uganda says avoid handshakes as ebola returns. Reuters (2012), http://www.reuters.com
5. Strong, P.: Epidemic psychology: A model. Sociol. Health Ill. 12, 249–259 (1990)
6. Ascel Bio, http://www.ascelbio.com/
7. Keeling, M.J., Eames, K.T.D.: Networks and epidemic models. J. R. Soc. Interface 2, 295–307 (2005)
8. Acemoglu, D., Ozdaglar, A.: Opinion dynamics and learning in social networks. Dyn. Games Appl. 1, 3–49 (2011)
9. Bonabeau, E.: Agent-based modeling: Methods and techniques for simulating human systems. PNAS 99, 7280–7287 (2002)
10. Funk, S., Salathe, M., Jansen, V.A.A.: Modelling the influence of human behavior on the spread of infectious diseases: A review. J. R. Soc. Interface 7, 1247–1256 (2010)
11. Meloni, S., Perra, N., Arenas, A., Gomez, S., Moreno, Y., Vespignani, A.: Modeling human mobility responses to the large-scale spreading of infectious diseases. Sci. Rep. 1, 1–7 (2011)
12. Epstein, J.M., Parker, J., Cummings, D., Hammond, R.A.: Coupled contagion dynamics of fear and disease. Mathematical and Computational Explorations. PLOS One 3, 1–11 (2008)
13. Funk, S., Gilad, E., Watkins, C., Jansen, V.A.: The spread of awareness and its impact on epidemic outbreaks. Proc. Natl Acad. Sci. USA 106, 6872–6877 (2009)

14. Acemoglu, D., Ozdaglar, A., ParandehGheibi, A.: Spread of misinformation in social networks. GEB 70, 194–227 (2010)
15. Kermack, W., McKendrick, A.: A contribution to the mathematical theory of epidemics. Proc. R. Soc. A 115, 700–721 (1927)
16. Waxman, B.M.: Routing of multipoint connections. IEEE J. Sel. Areas Commun. 6, 1617–1622 (1998)
17. Gilman, S.: Moral panic and pandemics. Lancet 375, 1866–1867 (2010)
18. Dengue cases touch 478. Hindustan Times (2008),
http://www.hindustantimes.com
19. Gupta, E., Dar, L., Kapoor, G., Broor, S.: The changing epidemiology of dengue in Delhi, India. Virol. J. 3 (2006)
20. Argentina admits dengue epidemics in northern provinces. MercoPress (2009),
http://en.mercopress.com
21. Masuh, H.: Re-emergence of dengue in Argentina: Historical development and future challenges. Dengue Bulletin 32, 44–54 (2008)

Discovering Consumer Health Expressions from Consumer-Contributed Content

Ling Jiang, Christopher C. Yang, and Jiexun Li

College of Information Science and Technology, Drexel University

Abstract. It has long been recognized that health consumers and professionals use different vocabularies to express health related concepts. Consumers often find it difficult to understand medical terminologies. If consumers misinterpret the health information they received and rely on it for decision making, this language gap would cause severe consequences. Many efforts have been taken to build Consumer Health Vocabulary (CHV) to bridge the gap and facilitate health information consuming. Extracting vocabularies used by consumers to express health concepts is a significant as well as challenging subtask in developing CHV. However, few studies have focused on developing methods for extracting consumer health expressions. In this work, we proposed a semi-automatic method that employs Principal Components Analysis (PCA) and Logistic Regression for identifying consumer health expressions from consumer-contributed content in social media. The experiment results showed that the proposed method is effective in identifying consumer health expressions from consumer-contributed content. These identified expressions can help to extend CHV and to enhance the performance of Adverse Drug Reactions (ADRs) signals detection.

Keywords: Consumer Health Vocabulary, Principal Components Analysis, Logistic Regression.

1 Introduction

There are an increasing number of health consumers using the Internet to search for health related information. According to a recent Pew Internet Survey, 74% of American adults use the internet, of which 80% have looked online for information about any of 15 health topics, and this translates to 59% of all American adults [1]. Health consumers actively seek health information online so that they could be well informed of healthcare knowledge and involved in personal healthcare decision making. Nowadays, the relationship between consumers and healthcare professionals is transforming from a "doctor says/patient does model" to a "partnership model". Consumers expect a more active role in their own healthcare.

To satisfy the increasing consumer needs, many online healthcare social media sites are emerging. Not only do these sites provide healthcare information sources, but also build platforms for consumer interactions such as discussion forums and online social groups, which meanwhile generate an enormous stockpile of

A.M. Greenberg, W.G. Kennedy, and N.D. Bos (Eds.): SBP 2013, LNCS 7812, pp. 164–174, 2013.

consumer-contributed healthcare content. Healthcare research could benefit from taking advantage of this rich information resource.

Several researchers have realized the potential of consumer-contributed content in detecting signals of Adverse Drug Reactions (ADRs) [2-4]. In a previous study [2], we proposed a lexicon-based association mining technique for detecting signals of ADRs in online health social website, and proved consumer-contributed content to be reliable source for ADR detection. Since this technique is based on matching the customer-contributed content with the ADR lexicon, the lexicon is crucial for the detecting accuracy. It has been long recognized that consumers and professionals use different vocabularies to express health related information [5-8]. In order to mine signals of ADRs from consumer-contributed content efficiently, the Consumer Health Vocabulary (CHV) [9] was used to generate the ADR lexicon. However, being still in the progress to perfection, the CHV is not complete itself. By relying on only this one external source to generate the lexicon, we could miss many other expressions used by consumers. Therefore, the lexicon needs to be extended to enhance the performance of ADR detection.

Expanding the ADR lexicon would also be beneficial to CHV development. The ADR lexicon expansion task here is close to Consumer-Friendly Display (CFD) names identification, which is an essential subtask in CHV development [10]. "Consumer vocabulary problem" has long been regarded as a fundamental issue in health information provision [6]. Health consumers and healthcare professionals usually express health concepts in different ways. Consumers often find it difficult to understand medical terminology due to their lack of professional knowledge. This language gap obstructs effective communications between consumers and professionals, and it is also a barrier to successful health information retrieval. The situation could get even worse for the fact that consumers actively explore health information by themselves. If consumers misinterpreted the health information they received and rely on it for decision making, this vocabulary issue could cause serious consequences.

One solution to consumer vocabulary problem is to link common health-related language to professional medical concepts through consumer health vocabularies (CHVs) [10]. To achieve this, one important task is to identify expressions used by most consumers. In this work, we presented a semi-automatic method to identify health expressions from consumer-contributed content on the Web.

In recent years, consumer health vocabulary has been an important topic in health information study. McCray et al. first identified terminological problems in user queries submitted to the National Library of Medicine website [5]. Patrick et al. called the mismatch between consumer vocabulary and the professional vocabulary as the "consumer vocabulary problem", and recognized it as a fundamental issue in health information study [6]. Zeng et al. discovered the difference between patient and clinician terminology by analyzing the information retrieval performance resulting from these terms, and the results showed that patient terms lead to poorer performance [11]. Zielstorff discussed about the health consumerism movement and explained the significance of consumer health vocabulary issues in this movement [12].

Currently, studies on solutions for consumer vocabulary problem mainly focus on mapping consumer vocabularies to professional medical concepts through Consumer Health Vocabularies (CHVs) [10, 12]. Zeng et al. have been devoting themselves to the development of open access, collaborative Consumer Health Vocabulary (CHV) [7-11]. The first-generation CHV was developed by Zeng et al. and it is defined as "a collection of forms used in health-oriented communication for a particular task or need (e.g., information retrieval) by a substantial percentage of consumers from a specific discourse group and the relationship of the forms to professional concepts" [8].

The development of first-generation CHV means a huge step in consumer health vocabulary study. However, the CHV still needs improvement and perfection. One significant challenge is to identify health expressions used by consumers. An effective approach is to extract consumer health vocabularies from existing text. Some researchers used e-mail messages as a source for identifying medical terms used by consumers [6, 13]. Tse and Soergel identified consumer health expressions from online discussion postings, health related articles from popular magazines and newspapers, commercial ads, government publications, and patient pamphlets [14, 15]. Zeng et al. generated candidate Consumer-Friendly Display (CFD) names from the National Library of Medicine MedlinePlus query logs[10].

So far, very few researches have been focused on automatic methods for consumer health expressions identification. Given the massive health information on web, which is still increasing every day, identifying consumer health expressions manually is very time-consuming and onerous. Hence, an automatic method to support consumer health expressions extraction is desirable. Since consumer-contributed content in online health social websites is generated by consumers, it should be a reliable source for consumer health expressions extraction.

2   Methods

The purpose of this work is to explore a semi-automatic method for expanding consumer health vocabularies; in particular, we focus on the ADRs lexicon. First, a statistic method was used to automatically identify candidate terms for consumer health expressions from consumer-contributed content. Next, the candidate terms were reviewed to form into medical expressions used by consumers and added into the lexicon, which would be used to enhance the performance of ADRs signals detection, and contribute to the development of CHV as well. **Fig. 1** shows the method flowchart of this study. Although this work only focused on consumer vocabularies of adverse drug reactions, the same techniques could be applied to other health vocabulary problems to assist developing CHV.

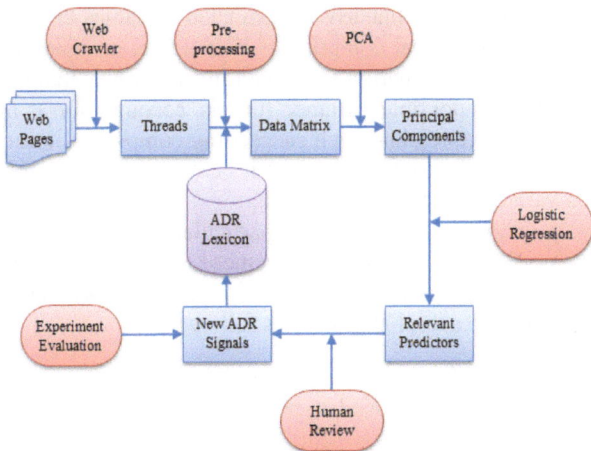

Fig. 1. Method Flowchart

2.1   Data Preparation

The data could be obtained from online health social websites, such as MedHelp and PatientsLikeMe, where consumers talked about drugs, side effects, and treatments in discussion threads, which contain a large amount of health vocabularies used by consumers to express health concepts. Moreover, this kind of data is public and easy to get.

After we collect the data, some natural language processing techniques, including removing stop words and stemming, would be applied to pre-process the data. Then, the data would be represented into a term by thread matrix, each cell of which is the TF-IDF value of the term. The TF-IDF weight (term frequency–inverse document frequency) is a numerical statistic which reflects how important a word is to a document in a collection or corpus. In this case, it reflects how important a term is to a thread. The TF-IDF weight was used here instead of the term frequency because most of the terms related to ADRs were relatively infrequent in the whole dataset but would appear frequently in those threads that were talking about the side effects. Using TF-IDF would increase the predicting ability of this kind of terms while reducing the predicting ability of the terms that appeared in almost every thread.

In preparation for running the logistic regression on the data, all the threads would be divided into two groups based on the original ADRs lexicon. A method introduced in [2] can be used to label these threads. If a thread contains any ADRs signals, it would be labeled as 1; otherwise, it would be labeled as 0.

2.2   Principal Components Analysis (PCA)

As the consumer corpus contains hundreds of thousands of terms, we need to address the high dimensionality. We adopted PCA to deal with this problem.

Principal component analysis (PCA)[16] is a mathematical procedure that uses an orthogonal transformation to convert a set of observations of possibly correlated variables into a set of values of linearly uncorrelated variables called principal components. The number of principal components is less than or equal to the number of original variables. This transformation is defined in such a way that the first principal component has the largest possible variance (that is, accounts for as much of the variability in the data as possible), and each succeeding component in turn has the highest variance possible under the constraint that it be orthogonal to the preceding components. So it is a variable reduction procedure.

Algebraically, principal components are particular linear combinations of the p random variables $X_1, X_2, \ldots, X_p$. Geometrically, these linear combinations represent the selection of new coordinate systems obtained by rotating the original system with X_1, $X_2, \ldots, X_p$ as the coordinate axes[17], shown as follows:

$$
\begin{aligned}
Y_1 &= a_{11}X_1 + a_{12}X_2 + \cdots + a_{1p}X_p \\
Y_2 &= a_{21}X_1 + a_{22}X_2 + \cdots + a_{2p}X_p \\
&\;\;\vdots \qquad\qquad\qquad\qquad \vdots \\
Y_p &= a_{p1}X_1 + a_{p2}X_2 + \cdots + a_{pp}X_p
\end{aligned}
$$

the new axes represent the directions with maximum variance and provide a simpler and more parsimonious description of the covariance structure.

2.3   Logistic Regression

After conducting PCA, we used logistic regression to discover which components contribute the most to the discriminant process. Logistic regression determines the impact of multiple independent variables presented simultaneously to predict membership of one or other of the two dependent variable categories.

Let the dependent variable Y be either 0 or 1, and $X = \{X_1, X_2, \cdots X_n\}$ be the independent variables, the goal of logistic regression is to determine the probability P $(Y|X) \in [0, 1]$ of Y given independent variables X [17]. The logistic regression model is as follows:

$$
\ln(odds) = \ln\frac{p}{1-p} = \beta_0 + \beta_1 X_1 + \beta_2 X_2 + \cdots + \beta_n X_n
$$

where the odds ratio

$$
odds = \frac{p}{1-p}
$$

denotes the ratio of probability of 1 to the probability of 0. Because we were trying to detect consumer vocabularies of adverse drug reactions here, we needed to discriminate threads that include the ADRs we were interested in from those did not. Therefore, all the threads have been classified into two groups according to if consumers discussed about certain ADRs in the threads [2]. By running logistic regression, we can find out the predictive power of each component for differentiating threads related to ADRs from others. Furthermore, from the components identified as relevant predictors, we can identify terms with high loadings as potentially candidates for new ADR expressions.

2.4   Human Review

After PCA and Logistic Regression, we could find a number of potential ADRs expressions from consumer-contributed content. However, not all of these terms are necessarily related to ADRs. At this step, we utilize the human judgment based on domain expertise to further review and screen out the real ADRs-related terms. In addition, human annotators need to review the threads in which these detected terms occur to explore different ways that consumers were using them. The commonly used expressions involving these terms will be identified as new ADRs signals.

2.5   Experiment Evaluation

After human review, these newly discovered expressions about ADRs will be eva-luated in terms of their ability to actually detect ADRs in consumer-contributed content. For this purpose, we need a test set of data points labeled as positive and negative instances (i.e., threads related to and not related to ADRs). We can compare the extended lexicon with the original lexicon for detecting ADRs related threads based on metrics such as precision, recall, and F-1 measure. It is worth noting that our proposed approach for ADRs signal extension is an iterative process. Once shown effective for improving ADRs detection, the extended lexicon will replace the existing lexicon for future iteration of detection and extension.

3   Experiment

3.1   Dataset Preparation

In this study, we conducted an experiment to evaluate our proposed approach for ADRs signal extension. We collected online discussion threads of three drugs, Biaxin, Lansoprazole, and Luvox, from MedHelp, one of the most popular health social net-working websites. MedHelp provides a platform for people to share needs for better medical information and support. In the Drugs section of MedHelp, users can start a thread of a certain drug with a post, on which all users can comment. There are up to thousands of threads under each drug. We selected the drugs with more than 500 threads of discussion, and collected all the original posts and comments of these drugs. A highly parallelized automatic web crawler was used to collect the data.

Table 1. Data Summary

Drug Name	# Threads	# Threads Labeled 1	# Threads Labeled 0	# Terms
Biaxin	686	222	464	1272
Lansoprazole	592	214	378	1245
Luvox	570	362	208	1265

Online discussion threads were composed of natural language. Natural language processing techniques, including stop-words removing and stemming, were used to preprocess the data. Then for each drug, the data was represented into a term by thread matrix. All the threads were divided into two groups according to if the threads included signals of any of the three adverse reactions we are interested in, including diarrhea, heart disease, and depression. If one thread included signals of ADRs, it was labeled as 1; otherwise, it was labeled as 0. This step was achieved by matching the original ADR lexicon with the threads to identify signals of ADRs. **Error! Reference source not found.** summarizes basic statistics of the data, including the total number of threads and terms for each drug, and number of threads in each group. The value of each cell in the matrix was the TF-IDF of the term in the corresponding thread. This step was done using a Java program which imported a WEKA filter package.

3.2   Consumer Expressions Extraction

We use SPSS 19 software package to perform PCA. Taking Lansoprazole as an example, 231 components with eigenvalues greater than one were extracted out from 1245 unique terms, which explained 87.78% of the variance.

Then the logistic regression was performed in SPSS. There were 231 components in total extracted in the PCA, so we need to decide which variables to choose for processing before generating the logistic regression model. This can be done by selecting certain logistic regression variable selection method in SPSS. In this case, the "Forward: Conditional" method was used to include variables based on their significance.

By running logistic regression in SPSS, 52 components were selected out of the 231 components to generate the model. Component 1 was included in the first step, then Component 83, followed by Component 9 etc. As more components were added into the function, the accuracy increased. After 52 components were all included, the categorizing accuracy reached 89.4%. **Table 2** shows top 5 Variables in the Equation table in terms of the Exp(B) score. The Exp(B) column represents the odds ratios for the predictors. For example, with all other variables constant, a thread with one unit greater value in component 1 than another group is 51.407 times more likely to be categorized into group 1, which means the thread contains signals of ADRs.

Table 2. Top 5 Varibales in the Equation Table

	B	S.E.	Wald	df	Sig.	Exp(B)
FAC1_1	3.940	.471	69.820	1	.000	51.407
FAC19_1	1.024	.274	13.992	1	.000	2.785
FAC59_1	.921	.182	25.710	1	.000	2.513
FAC6_1	.904	.229	15.597	1	.000	2.470
FAC10_1	.823	.198	17.349	1	.000	2.277

By looking at the above table, we can conclude that component 1 has the highest predictive capability, which is almost as 25 times as that of component 19, in separating the threads into the two different groups, and the value of component 1depends on the terms with high loadings in this component. Using the component matrix produced in PCA, we can identify the terms that affect the value of component 1. In this step, human review is required to check the terms with high loadings in component 1 to select candidate terms of adverse reactions used by most consumers. Two human annotators were recruited to review the terms. For example, terms such as *feel*, *down*, *heart*, *attack*, *anxiety*, and *problem* were selected from component 1. Then human annotators went back to the threads where these terms were found to examine how consumers used these terms and identified expressions like *feel down*, *heart attack*, *heart problem*, and *anxiety disorders* were used by consumers when they were talking about side effects of heart disease and depression. After the human review and annotation process, these expressions were added into the lexicon.

3.3   Evaluation

In order to evaluate the effectiveness of the methods, we compared the ability of identifying ADRs related threads between original ADRs lexicon and extended lexicon. The original lexicon was generated by searching the adverse reaction terms in CHV Wiki[1] to get the consumer expressions in CHV. We used the same dataset of discussion threads of the three drugs, Biaxin, Lansoprazole, and Luvox for evaluation. For each drug, we detect if one thread was talking about any one of the three adverse reactions, including diarrhea, heart disease, and depression, by matching the threads with the lexicon.

Gold Standard and Metrics

The goal of evaluation is to compare the original and extended lexicons, so we needed to set up a gold standard. The same two human annotators set up a gold standard for the evaluation. First, we randomly selected 165 threads for Biaxin, 185 threads for both Lansoprazole and Luvox from the data. Next, the two human annotators manually went through all the selected threads to review whether each thread talked about any one of the three ADRs. Finally, weighted Kappa measure was computed to examine the reliability of the generated gold standard. Weighted Kappa measure is a statistical measure extended from Kappa measure for computing inter-rater agreement or inter-annotator agreement for qualitative (categorical) items [18]. It has a maximum value of 1 which indicates a perfect agreement between two raters and a value of 0 representing total disagreement. In general, a weighted kappa measure value larger than 0.8 is considered to be a very good agreement between the two raters. In this experiment, the weighted Kappa measure value was 0.86, which means the two annotators had a very good agreement and the gold standard was reliable. When disagreement happened, the two human annotators would discuss about their opinions until they reached a consensus.

In this work, we use precision, recall, and F-1 measures as the metrics in our experiment:

$$\text{Precision} = TP / (TP + FP), \quad \text{Recall} = TP / (TP + FN)$$
$$\text{F-1 measure} = 2 \times \text{Precision} \times \text{Recall} / (\text{Precision} + \text{Recall})$$

---

[1] http://consumerhealthvocab.chpc.utah.edu/CHVwiki/

where TP, FP, FN are the number of true positives, false positives, and false negatives, respectively.

Results and Discussion

Table 3 shows the evaluations results. As we can see in the table, most of the precision decreased while the recall increased, which reflected the trade-off relationship between precision and recall. This can be explained by the expansion of the lexicon. When useful adverse reaction expressions were added into the lexicon, some noise might also be included. In terms of F-1 measure, the value increased a lot for all the ADRs except diarrhea.

For the ADR diarrhea, it turned out that the lexicon expansion did not improve the performance of detecting ADRs threads, but rather led to a poorer performance. However, this is not hard to understand after closely examining the results. For diarrhea, the algorithm using the original lexicon has already produced quite satisfactory results. The values of precision, recall, and F-1 measure are nearly all over 0.8. This means that the original lexicon for adverse reactions was already quite good for detecting diarrhea. Including any other terms into the lexicon could increase the possibility of having false positives. For example, we identified that *dizziness* was described by some consumers as one of the symptoms of diarrhea. However, it turned out that dizziness might happen with diarrhea, but being dizzy did not necessarily mean having diarrhea. Therefore, the extended lexicon may not be desirable for diarrhea. However, currently there is no benchmark to determine whether a lexicon is good enough so that there is no need for extension. This is one limitation of current study, but this could be a part of future work. By determining when to stop extending a lexicon, we can avoid such performance drop by adding noise into the lexicon.

Table 3. Comparison Between Original and Extended Lexicon

ADR	Drug Name	Precision		Recall		F-1 Measure	
		Original	Extended	Original	Extended	Original	Extended
Diarrhea	Biaxin	**0.79**	0.55	0.83	**0.91**	**0.81**	0.69
	Lansoprazole	**0.84**	0.59	0.90	**0.93**	**0.87**	0.72
	Luvox	**0.80**	0.40	**1.0**	**1.0**	**0.89**	0.57
	Average	*0.81*	*0.51*	*0.91*	*0.95*	*0.86*	*0.66*
Heart Disease	Biaxin	0.50	**0.60**	0.06	**0.81**	0.11	**0.68**
	Lansoprazole	**0.75**	0.47	0.09	**0.47**	0.17	**0.47**
	Luvox	0.09	**0.56**	0.04	**1.0**	0.06	**0.71**
	Average	*0.45*	*0.54*	*0.06*	*0.76*	*0.11*	*0.62*
Depression	Biaxin	0.29	**0.50**	0.19	**0.95**	0.23	**0.66**
	Lansoprazole	**0.75**	0.51	0.46	**0.85**	0.57	**0.64**
	Luvox	**0.59**	0.58	0.67	**0.94**	0.63	**0.72**
	Average	*0.54*	*0.53*	*0.44*	*0.91*	*0.48*	*0.67*

For heart disease and depression, the performance of all three drugs was enhanced a lot after extending the lexicon. When extracting consumer vocabularies, we found that there were diverse expressions used by consumers, and most of them were very different from professional vocabularies. When talking about heart disease, consumers may use expressions such as *heart issues, heart problems* rather than *cardiac disease.* Sometimes they would describe detailed incidents as *heat attack* or *irregular heartbeats.* The discovery of all these terms reflects how consumers' language in expressing health concepts is different from professional vocabularies. By extracting these expressions used by most consumers, we could not only detect signals of ADRs more accurately but also better understand the consumer health language and therefore contribute to the development of CHV.

Overall, the proposed method is effective in identifying ADR expressions from consumer-contributed content. Not only can this method be useful for ADRs lexicon expansion to promote the ADRs signals detection performance, but also be applied to consumer health vocabulary extraction.

4   Conclusion

As consumers take increasing responsibility for their own health care, the language gap existing between consumers and healthcare professional presents a serious issue. Many efforts have been taken to build CHV to bridge the gap and facilitate health information consuming. Extracting health vocabularies used by consumer to express health concepts is a significant as well as challenging subtask in developing CHV. In this paper, we propose a semi-automatic method for identifying ADR expressions from consumer-contributed content. Not only can the extracted consumer health expressions be used for lexicon expansion to enhance the performance of ADRs signals detection, but also assist in developing CHV.

In our future work, we will further extend the corpus by including more drugs and ADRs to extract more consumer health expressions. Moreover, although the key part of the task was finished automatically, the method still involved human efforts. Hence, another direction for future work is to develop techniques to automatically extract consumer health expressions from consumer corpus.

References

1. Fox, S.: The Social Life of Health Information. Pew Internet & American Life Project (May 12, 2011), http://pewinternet.org/Reports/2011/Social-Life-of-Health-Info/Summary-of-Findings.aspx (accessed October 17, 2012)
2. Yang, C.C., Jiang, L., Yang, H., Tang, X.: Detecting Signals of Adverse Drug Reactions from Health Consumer Contributed Content in Social Media. In: Proceedings of ACM SIGKDD Workshop on Health Informatics (August 12, 2012)
3. Chee, B.W., Berlin, R., Schatz, B.: Predicting Adverse Drug Events from Personal Health Messages. In: Annual Symposium Proceedings, pp. 217–226 (2011)
4. Leaman, R., Wojtulewicz, L., Sullivan, R., Skariah, A., Yang, J., Gonzalez, G.: Towards internet-age pharmacovigilance: extracting adverse drug reactions from user posts to health-related social networks. In: Proceedings of the 2010 Workshop on Biomedical Natural Language Processing, pp. 117–125 (2012)
5. McCray, A.T., Loane, R.F., Browne, A.C., Bangalore, A.K.: Terminology issues in user access to Web-based medical information. In: AMIA Annual Symposium, pp. 107–111 (1999)

6. Patrick, T.B., Monga, H.K., Sievert, M.E., Houston Hall, J., Longo, D.R.: Evaluation of controlled vocabulary resources for development of a consumer entry vocabulary for diabetes. Journal of Medical Internet Research 3(3), E24 (2001)
7. Zeng, Q., Kogan, S., Ash, N., Greenes, R.A., Boxwala, A.A.: Characteristics of consumer terminology for health information retrieval. Methods Inf. Med. 41(4), 289–298 (2002)
8. Zeng, Q.T., Tse, T.: Exploring and developing consumer health vocabularies. Journal of the American Medical Informatics Association: JAMIA 13(1), 24–29 (2006)
9. Zeng, Q.T., Tse, T., Divita, G., Keselman, A., Crowell, J., Browne, A.C.: Exploring Lexical Forms: First-Generation Consumer Health Vocabularies. In: AMIA Annual Symposium Proceedings, pp. 1155–1155 (2006)
10. Zeng, Q.T., Tse, T., Crowell, J., Divita, G., Roth, L., Browne, A.C.: Identifying Consumer-Friendly Display (CFD) Names for Health Concepts. In: AMIA Annual Symposium Proceedings, pp. 859–863 (2005)
11. Zeng, Q., Kogan, S., Ash, N., Greenes, R.A.: Patient and clinician vocabulary: how different are they? Stud. Health Technol. Inform, 399–403 (2001)
12. Zielstorff, R.D.: Controlled vocabularies for consumer health. Journal of biomedical informatics 36(4-5), 326–333 (2003)
13. Smith, C.A., Stavri, P.Z., Chapman, W.W.: In their own words? A terminological analysis of e-mail to a cancer information service. In: AMIA Annual Symposium, pp. 697–701 (2002)
14. Tse, T., Soergel, D.: Exploring Medical Expressions Used by Consumers and the Media: An Emerging View of Consumer Health Vocabularies. In: AMIA Annual Symposium Proceedings, pp. 674–678 (2003)
15. Tse, T., Soergel, D.: Procedures for Mapping Vocabularies from Non-Professional Discourse A Case Study: "Consumer Medical Vocabulary". In: Proceedings of the ASIST Annual Meeting, vol. 40(1), pp. 174–183 (2003)
16. Wikipedia. Principal Components Analysis (October 12, 2012), http://en.wikipedia.org/wiki/Principal_component_analysis (accessed on October 17, 2012)
17. Johnson, R.A., Wichern, D.W.: Applied Multivariate Statistical Analysis. Pearson Prentice Hall (2007)
18. Wikipedia. Principal Components Analysis (July 20, 2012), http://en.wikipedia.org/wiki/Cohen%27s_kappa#Weighted_kappa (accessed on October 17, 2012)

Patient-Centered Information Extraction
for Effective Search on Healthcare Forum

Yunzhong Liu and Yi Chen

School of Computing, Informatics, and Decision Systems Engineering
Arizona State University, Tempe, AZ
{liuyz,yi}@asu.edu

Abstract. Online healthcare forums are one of the major social media in Health 2.0 for patients and caregivers to share personal experience and to help each other. However, current forums do not support effective information search and thus users are unable to fully leverage the rich information in the forums. In this work, we propose patient-centered information extraction to better organize the information in the forum and have developed a patient-centered medical information database extracted from a forum. In this system, the patients discussed on the forum are identified and their shared medical information is aggregated and associated with the corresponding patients. The experimental evaluation shows that our system can provide better information search results than traditional approaches.

1 Introduction

Nowadays, Health 2.0, the web-based applications and services for healthcare, has become very popular. In Health 2.0, forums are one of the major social media where patients or their caregivers share personal experience, support and encourage each other, and form patient communities. In a forum, a user, or a post *author*, may publish a *post*, the smallest information unit in a forum. An initial post and the replying posts submitted by the same or different authors compose a *thread*, or a topic.

Online healthcare forums provide valuable information for patients, caregivers, doctors and researchers. There is a large and increasing volume of user cases, evidences, and facts shared by patients, which may provide insights to the research on diseases and treatments. It is also an important resource for patients and caregivers to seek for other patients with similar symptoms and to check what treatments have been taken by or suggested for those patients for self-education on their diseases and treatments.

However, currently the rich information on healthcare forums has not been fully leveraged. While it is easy to share information by posts, and to browse and read the posts shared by other patients, current technology does not provide effective ways for a user to easily *discover* information that she is interested in, in a large repository of posts. Let us look at two examples, both of which are observed in questions issued by real users to the epilepsy discussion forum[1].

---

[1] http://epilepsyfoundation.ning.com/forum

A.M. Greenberg, W.G. Kennedy, and N.D. Bos (Eds.): SBP 2013, LNCS 7812, pp. 175–183, 2013.
© Springer-Verlag Berlin Heidelberg 2013

Consider a user who wants to check whether Vitamin can be used to alleviate aggression, and would like to search the epilepsy forum for other patients' experience to gain more knowledge. She would issue a keyword query "Vitamin, aggression" on the forum. One approach commonly used in forums to support information search is to consider each post as an information unit (like a document) and to return a post if it contains the query keywords, referred as *post-based search* in this paper. The Patientslikeme forum[2] and WebMD forum[3] are mainly based on this method.

Adopting the post-based search, the information shown in Table 1 will be missed from the result since there is no single post in this thread containing both query keywords. For space reason, only post fragments are included in the table. PostID is a post's sequence number in a thread. For privacy concern, we replaced the real AuthorID of the forum participants with C1, C2, and C3. The ParentPostID is the PostID of the post that the current post replies to. For example, the 4*th* post with PostID 4 replies to the first post with PostID 1. However, when we read through posts 4 and 6 in this thread, we can see that even though they are from different authors, they are closely connected and collectively show that Vitamin B6 can help aggression, which could be caused by Keppra.

As we can see, post-based search tends to put too strict criteria on search and thus misses relevant results, that is, it suffers *low recall*. To improve recall, an intuitive approach is to search forums using *thread-based search*: take each thread as an information unit and consider a thread as relevant if all the posts in the thread collectively contain the query keywords. This approach and its variants have been used in the Healthboards message boards[4] and the Epilepsy forum. Thread-based search can identify the thread in Table 1 as relevant to query "Vitamin, aggression". However, it suffers other problems, as shown in the following example.

Suppose a patient suffers seizure due to weaning, and would like to search the epilepsy forum to learn how to cope with her problems from similar patients. She would issue a keyword query "seizure, wean". Using the thread-based search method, a thread with title "B6 wondering" will be returned, where some fragments are shown in Table 2.

As we can see, this thread is returned as a query result since it contains a post with PostID 6 and a post with PostID 11, which together contain both query keywords. However, after reading these two posts, we find that keyword "seizure" and "wean" are associated with different patients: C4's mom and C5's son. There is no relationship between "seizure" and "wean" described in the thread, and thus this thread is not useful for the user who searches for the information about the seizure disease caused by weaning. As illustrated in this example, the thread-based search tends to return some results that are not relevant to the user, that is, it suffers *low precision*.

---

[2] http://www.patientslikeme.com
[3] http://exchanges.webmd.com
[4] http://www.healthboards.com/boards

From the above examples, we observe that existing approaches do not perform well for a user query with multiple keywords. We analyze those queries and find that when a query contains multiple keywords, these keywords are expected to have close relationships between each other. For instance, a query may involve the relationship between a symptom and a disease, the relationship among several symptoms, the relationship among multiple diseases, the relationship between a disease and treatments, or the relationship between a treatment and side effects. To correctly find such relationships, it is critical that the matches to query keywords refer to the *same* patient. However, post-based or thread-based search does not consider *who* a keyword is associated with. They only check syntactic information units, either a post or a thread. It is common to see multiple posts refer to the same patient, and a thread contains information of multiple patients. Therefore the root cause of the low-quality results generated by existing approaches is the mis-alignment between the syntactic information unit (a post or a thread) that existing methods are based on and the semantic information unit (a patient) that the query user refers to.

Table 1. Samples from one thread for the query "Vitamin, aggression"

Thread link: http://epilepsyfoundation.ning.com/forum/topics/katies-temper			
PostID	AuthorID	*Content*	Parent PostID
1	C1	Katie woke up a swinging her arms this morning and hitting things,very combative. I just wished I knew if it was the Keppra she is taking that is making her do this. She is VERY tempermental,alot of times I don't know what to do with her. ... She has in home therapys and her therapists the other day was telling me it seems she has a sensory integration disorder.	Null
4	C2	Yes, Keppra can cause **aggression**.	1
6	C3	Have you tried giving her **Vitamin** B6 with the keppra?? It is supposed to help with the Keppra-rage.	1

Table 2. Samples from one thread for the query "seizure, wean"

Thread link: http://epilepsyfoundation.ning.com/forum/topics/b6-wondering			
PostID	AuthorID	*Content*	Parent PostID
6	C4	My mom is 59 and she takes keppra as well says that she gets tired very very early, usually around 7 she's just about ready for bed. She swears by keppra for controlling both her **Seizure**s and the auras.	1
11	C5	My son is **wean**ing off keppra, but he's still taking 250mgtwo times a day. (He was on something like 1000mg and life was hell). He gets angry really fast- right after taking his meds.	1

Table 3. Simplified records from the patient-centered medical information database

PatientID	Note	Medical Info ID	Medical Information
1	Katie	1	Katie woke up...
1	Katie	2	Yes, Keppra can cause ...
1	Katie	3	Have you tried giving her ...
2	C4's mother	1	My mom is ...
3	C5's son	1	My son is ...

In light of this observation, we propose to mine the semantic information unit - each individual patient and the associated information - from the posts. Then a user query is processed with respect to the semantic information unit, finding out the patients whose experience is related to query keywords and therefore can bring insights about the relationships among the keywords. We developed an information extraction system, which takes the original forum data as the input, identifies the patients and the information associated with each individual, and outputs a patient-centered medical information database. Table 3 shows a simplified version of our database records extracted from the information shown in Table 1 and Table 2. In Table 3, multiple pieces of information from three different posts in Table 1 are identified to be associated with the same patient. On the other hand, the information from post 6 and 11 in Table 2 is extracted and associated with two different patients with different PatientIDs. With such a patient-centered database, it is easy to find which patients are relevant to a user query, thus improving the search quality achieved by post-based or thread-based search. In our example, patient 1 corresponds to a relevant result to user query "Vitamin, aggression", while none of them is relevant to user query "seizure, wean".

Related Work: There are existing studies [8,4] on improving the traditional thread or post-based search on other types of forums, such as technical forum. However, these types of forums are different from healthcare forum since their focus is topics rather than individuals described in the posts. Although the thread structure information, such as the reply relationship, has been exploited in these studies, they do not make deep NLP analysis to mine the semantic information unit in the posts, like each individual patient in a health forum. Therefore, they would have similar problems as post-based or thread-based search for the two example queries discussed earlier.

2   System Overview

To build the patient-centered medical information database, we need to identify the patient mentions that refer to the same person and to associate and aggregate the medical information with the corresponding patients. Our system includes four major components. In *person identification* module, we discover all the person mentions to find the potential patient mentions. Since it is difficult to identify a patient from some individual person mentions, we apply *person resolution* to group all the person mentions into clusters such that all the mentions in the same cluster refer to the same person. Then we make *patient identification* based on all the information in each cluster of person mentions. At last, we make *medical information association* for each identified patient from the posts. State-of-art natural language processing (NLP) techniques and MetaMap tool [2] in the Unified Medical Language System (UMLS) [6] have been integrated into our system.

2.1   Person Identification

This component takes sentences in posts as input and outputs person mentions. Our method is based on the Stanford NLP [1] and MetaMap tool in UMLS. First, all the person names identified by Named Entity Recognition (NER) and pronouns (except "it") identified by Part of Speech (POS) tagger are identified as person mentions. For example, "Katie" and all "she" and "her" in post 1 in Table 1 are identified as person mentions. Second, all the phrases extracted by MetaMap with their semantic types belonging to "living beings" semantic group will be identified as person mentions. For example, "son" in post 11 in Table 2 can be identified as a person mention since its semantic type is "family group", which belongs to "living beings" semantic group.

2.2   Person Resolution

This component groups the person mentions within a thread into clusters such that each cluster includes all the mentions that refer to the same person. Stanford deterministic co-reference resolution system [5], which was the top ranked system at the CoNLL-2011 shared task, is used for generating the co-reference resolution results. Since co-reference resolution is more general than person resolution, we can easily extract the person resolution results from the co-reference resolution results. For post 1 in Table 1, "Katie" and all "she" and "her" in this post will be identified as co-referent.

In addition to person resolution within a post, we also incorporate the author information and the reply relationship between posts for inter-post person resolution. First, we assume the same role with the same relationship with the same author in the same thread refers to the same person. For example, if "my son" has been mentioned by the same author in two different posts in the same thread, we consider them as co-referent. Second, we transform one thread into multiple multi-person conversation documents based on the reply relationship, in which a post author is a speaker and the post content is analogous to the utterance. In this way, the person mention in the replying post that refers to the person in its parent post can be identified.

2.3   Patient Identification

This component identifies the patient mentions from the identified person mentions. We assume a person mentioned in a thread is either a patient or a non-patient. Then we propose to combine the semantic role labeling (SRL) [3], MetaMap, and a few patient identification patterns. We identify patients mainly using SRL with Propbank [7] annotation. In addition, we also used 12 patient identification patterns based on a sample data set. As shown in experimental evaluation later, this small number of patterns, such as "take *pharmacologic substance*", "have *disease* or *syndrome*", have a very high coverage in identifying patients and scale well in a large dataset. Here "*pharmacologic substance*" and "*disease* or *syndrome*" are two semantic types for medical phrase, which can be extracted from post content by MetaMap. For example, in post 1 in Table 1, from "she has a sensory integration disorder" we can identify "she" is a patient since "sensory integration disorder" has the semantic type *"disease or syndrome"*. Note that all the co-referent person mentions will be identified as patient mentions if at least one of them has been identified as a patient mention. Therefore, "Katie" and all "she" and "her" in this post will be identified as patient mentions.

2.4   Medical Information Association

This component associates the medical information with the closest patient or person if no patient has been identified at all. Note that the medical information in a replying post can also be associated with the patient mentioned in its parent post if that replying post does not introduce a new patient that is closer to the information. In Table 1, "aggression" in post 4 and "Vitamin B6" in post 6 are both associated with "Katie" in post 1. Also note that no information should be associated with "you" in post 6 or "I" in post 1 since they refer to the caregiver "C1", rather than a patient.

3   Experiments

To evaluate our system, we use the publicly available data in the epilepsy foundation discussion forum, which is initiated and maintained by National Institute of Neurological Disorders and Stroke (NINDS). We collected 9210 posts included in 911 threads (topics) published on the "Patient help patient" sub-forum by Nov. 2011. In this forum, the explicit quotation information has been used to identify the reply relationship between two posts. Otherwise, by default, we consider all the following posts in a thread reply to the first post in this thread, which follows the assumption in the feature used in [9] that the following posts tend to reply to the first one.

3.1   Query Set

Our query set includes ten multiple-keyword queries. In order to leverage real user queries without introducing bias, we follow the method used in [4] to randomly select queries. First, we find all the thread titles in the forum that end with a question mark. Since such a title indicates that a user, the thread initiator, is looking for answers to a question, it naturally represents as a user query. We then extract keywords from these thread titles. Instead of using a stopword list to filter out unimportant words, we choose MetaMap tool to extract phrases as the query keywords. The reason is that we want to identify each medical phrase containing multiple words and treat it as a unit in query processing. We randomly chose ten such thread titles with each corresponding to one query. We only tested ten queries because it is extremely labor-intensive to generate the ground truth for each query, especially since some queries may involve a large number of threads, which may include an enormous number of posts. Table 4 shows the chosen questions (thread titles) and the extracted keywords for each query.

3.2   Ground Truth

We manually find the ground truth of relevant results for each query, based on the analyzed user expectation as discussed in Section 1. We assume AND semantics among all the keywords in a query. To generate the ground truth for a query, we first define a relevant thread as a thread that contains all the query keywords. Consider the intensive human labor, we randomly choose 30 relevant threads for manual checking if a query involves more than 30 relevant threads. Since a patient is a semantic unit, we find the relevant patients whose associated information contains all the query keywords from the relevant threads. Then we consider the posts that are associated with such patients and contain at least one query keyword in the associated information as ground truth.

Table 4. Ten randomly chosen questions and keywords extracted from them

	Query questions (Keywords are underlined)
1	can **sz**[5] **types** change?
2	**New Seizures**...What does this mean?
3	Has your **temporal lobe epilepsy** become **worse** over **time**?
4	What is the **difference** in recordings between an **ambulatory EEG** and nonambulatory **EEG** (without **stimulus**)?
5	Anyone have a **child** with **Alternating Hemiplegia**?
6	Has anyone tried **Stiripentol** with their **kids**?
7	**Growth** Spurt - **Breakthrough seizures**?
8	**Vitamins** to help with **aggression**???
9	**Seizure** due to **weaning**?
10	**tonic** clonic after **flu virus**?

Table 5. Evaluation for ten randomly chosen queries

Query	Post-based			Thread-based			Patient-based		
	Precision	Recall	F1	Precision	Recall	F1	Precision	Recall	F1
1	0.978	0.379	0.547	0.509	1.0	0.674	0.764	0.698	0.73
2	0.973	0.379	0.545	0.477	1.0	0.646	0.702	0.695	0.698
3	1.0	0.059	0.111	0.378	1.0	0.548	0.923	0.706	0.8
4	1.0	0.333	0.5	0.5	1.0	0.667	1.0	1.0	1.0
5	1.0	0.286	0.444	0.636	1.0	0.778	1.0	0.714	0.833
6	1.0	0.5	0.667	0.5	1.0	0.667	1.0	0.5	0.667
7	1.0	0.063	0.118	0.087	1.0	0.16	0.8	0.75	0.774
8	1.0	0.429	0.6	0.28	1.0	0.438	1.0	0.714	0.833
9	1.0	0.534	0.696	0.349	1.0	0.518	0.716	0.658	0.686
10	1.0	0.3	0.462	0.455	1.0	0.625	0.6	0.3	0.4
Overall	0.995	0.326	0.491	0.417	1.0	0.589	0.851	0.674	0.752

---

[5] "sz" is identified as "seizure" using the acronym list in
http://epilepsyfoundation.ning.com/forum/topics/acronym-thread.

3.3   Comparison Systems

We compare the ground truth with post-based search, thread-based search, and our approach, referred as patient-based search. Post-based search returns all the posts each containing all the query keywords. Thread-based search returns all the posts each containing at least one query keyword in a relevant thread. Our patient-based search returns all the posts each containing at least one query keyword associated with a relevant patient. Note that our approach shares the same intuition as the ground truth, but automatically identifies patients and automatically associate information to each patient. The quality of these automated processes has been evaluated.

3.4   Evaluation Metrics

We use standard evaluation metrics in information retrieval: precision (P), recall (R), and f-measure $(F1)$. Precision is the ratio of the number of correctly returned posts to the total number of returned posts. The recall is the ratio of the number of correctly returned posts to the total number of posts that should be returned according to the ground truth. f-measure is defined as the harmonic mean of precision and recall: $F1 = \frac{2*P*R}{P+R}$.

3.5   Evaluation Results

The experimental results are shown in Table 5. It shows that post-based approach has almost perfect precision as in most cases keywords in the same post refer to the same patient and have close relationship, but it has very low recall. On the other hand, thread-based search achieves perfect recall since we do not consider the relationships of keywords in different threads, but it has a very low procision. In contrast, our patient-based search has good precision and recall in general, and achieves a much higher f-measure than the other two approaches.

We also analyzed the major reasons that affect our system performance. First, some forum acronyms cannot be recognized, like "my DD" cannot be identified as "my daughter". Second, some patients cannot be identified by our system due to informal language used in a forum and the limited context. Third, some assumed reply relationships between posts are incorrect. We plan to leverage the method proposed in [9] to extract more accurate reply relationships between posts. Fourth, the performance of the current NLP tools, especially the co-reference resolution tool, is not perfect.

4   Conclusion and Future Work

To the best of our knowledge, this is the first work that makes patient-centered information extraction on healthcare forum. By building a database of patient information, we can process user search on the semantic units in the forum (patients) rather than the syntactic units (posts or threads) and thus achieve high quality in information search. Our experimental evaluation verifies the effectiveness of our approach.

In future, besides addressing the several problems that we analyzed in experimental evaluation discussed earlier, we will also investigate the following issues to further improve our system. First, we will relax the AND semantics and develop a ranking model that ranks the results based on the relevance of the patients discussed in the results. Second, since obtaining a ground truth in this application is extremely labor-intensive, we will also investigate obtaining ground truth through crowdsourcing, where the challenge is how to design tasks for the crowd and how to consolidate their opinions to obtain ground truth.

Acknowledgments. This material is based on work partially supported by NSF CAREER Award IIS-0845647, IIS-0915438, an IBM Faculty Award and a Google Research Award.

References

1. Stanford core NLP tools, http://nlp.stanford.edu/software/corenlp.shtml
2. Aronson, A.R.: Metamap: Mapping text to the UMLS metathesaurus (2006), http://skr.nlm.nih.gov/papers/references/metamap06.pdf
3. Collobert, R., Weston, J., Bottou, L., Karlen, M., Kavukcuoglu, K., Kuksa, P.: Natural language processing (almost) from scratch. Journal of Machine Learning Research 12, 2493–2537 (2011)
4. Duan, H., Zhai, C.: Exploiting Thread Structures to Improve Smoothing of Language Models for Forum Post Retrieval. In: Clough, P., Foley, C., Gurrin, C., Jones, G.J.F., Kraaij, W., Lee, H., Mudoch, V. (eds.) ECIR 2011. LNCS, vol. 6611, pp. 350–361. Springer, Heidelberg (2011)
5. Lee, H., Peirsman, Y., Chang, A., Chambers, N., Surdeanu, M., Jurafsky, D.: Stanford's multi-pass sieve coreference resolution system at the CoNLL 2011 shared task. In: Proceedings of ACL CoNLL 2011 Shared Task (2011)
6. Lindberg, D., Humphreys, B., McCray, A.: The unified medical language system. Methods of Inf. Med. 32(4), 281–291 (1993)
7. Palmer, M., Gildea, D., Kingsbury, P.: The proposition bank: A corpus annotated with semantic roles. Computational Linguistics 31(1) (2005)
8. Seo, J., Croft, W.B., Smith, D.A.: Online community search using thread structure. In: Proceedings of the 18th ACM Conference on Information and Knowledge Management (CIKM) (2009)
9. Wang, H., Wang, C., Zhai, C., Han, J.: Learning online discussion structures by conditional random fields. In: Proceedings of the 34th International ACM SIGIR Conference on Research and Development in Information Retrieval (2011)

Controlling for Population Variances in Health and Exposure Risk Using Randomized Matrix Based Mathematical Modeling[*]

Brian M. Gurbaxani, Troy D. Querec, and Elizabeth R. Unger

Chronic Viral Diseases Branch, Division of High Consequence Pathogens and Pathology, Centers for Disease Control and Prevention, Bldg 24, MS A-30, 1600 Clifton Rd NE, Atlanta, GA 30329
buw8@cdc.gov

Abstract. In a previous work, we analyzed the co-occurrence of HPV types in 6 large studies with cervicovaginal samples, representing >32,000 women, to ascertain if associations exist among HPV types and to guide policies on HPV vaccination and vaccine development. The data showed that more women either were uninfected by HPV or had multiple concurrent infections than could be explained by independent assortment, which could result from variance in health and exposure risk factors. Modeling exposure and immune competence proved unstable, so we used a randomized matrix based approach that obviated the need to understand the underlying risk factors. We randomized our source data while preserving increasing levels of fidelity to the original data structures to discover the type associations for HPV infection. We offer that this could be a generally useful technique for studying any type of association in biosocial science, e.g. between demographic, socioeconomic, or other variables.

Keywords: statistical modeling, HPV, vaccine, risk, immune competence, randomized matrix.

1 Introduction

In a previous work (Querec, et al. in revision), we took on a problem which had been extensively studied in the literature, but with conflicting results. The problem was to ascertain whether there is evidence of HPV type associations in human infections. It quickly became apparent that the null model that had been used, i.e. types assorting in samples independently, was severely flawed [1]. By this null model, most 2 type combinations occur much less often than expected by chance, but combinations of 4 or more types are seen much more often than expected, and p-values exceeding the nominal significance threshold of 0.05 by hundreds of orders of magnitude were being reported. Later papers incorporated some confounding factors into more complicated models, but were limited by not having data on all possible confounders [2,3]. These papers also had mixed results.

---

[*] This work was performed by employees of the U.S. Government. The rights of this work are governed by title 17 U.S.C. 105.

A.M. Greenberg, W.G. Kennedy, and N.D. Bos (Eds.): SBP 2013, LNCS 7812, pp. 184–192, 2013.
© Springer-Verlag Berlin Heidelberg 2013

We adapted methods for modeling species distribution among archipelago islands [4] into a statistically powerful and robust set of tools with potential use beyond analysis of viral associations. One way to improve upon a null model is to reproduce the observed data with either mathematically closed form (e.g. ordinary differential equation based) or simulation based (e.g. agent based) models. These can have the advantage of providing insight into the nature of how the data is generated. On the other hand, parameter fits to such models can be unstable because the parameters are unidentifiable from the data, or the fits can be poor in key parts of the domain of independent variables because the model dimension is too small, so that the model's predictive value is in question. Furthermore, data on all the confounding variables may not be available and/or a realistic model for generating the observed data may not exist. Moreover, building such complicated models of the phenomenon might become so much of an endeavor in themselves that they detract from the ultimate objective of the study. We used an "end run" around the modeling issues to build statistically powerful tools for analyzing certain questions in the biological, physical, and social sciences – wherever the association of many variables of interest needs to be studied. The goal of this paper is to focus on the computational methodology while the original paper focuses on the biology. We have leveraged the strengths of both Monte Carlo based and analytical statistics to build computationally efficient and robust algorithms for testing statistical significance.

2   Statistical Modeling

2.1   Null Model

Using an aggregate dataset of 6 different study populations for a total of over 30,000 subjects with HPV results for 37 genotypes, we observed that multiple types do not co-occur with anything close to Poisson joint distributions assuming independent assortment, either in the data as a whole or in the individual HPV studies that we considered. Rather, multiple types occur much more frequently than would be expected at random (Figure 1 – expected data from Monte Carlo simulations).

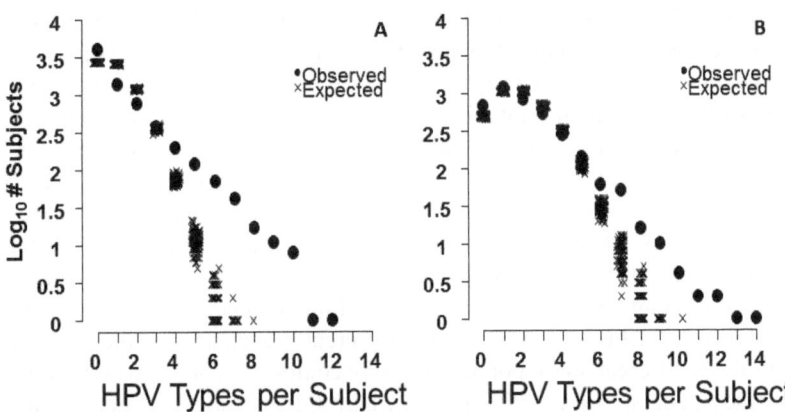

Fig. 1. Observed (•) vs. expected (×) distributions of multiple HPV types for (A) a national population based study (7012 subjects), and (B) a clinic based study (3828 subjects)

2.2   Exposure and Immune Competence Risk Model

It is plausible to consider that the greater than expected frequency of multiple HPV types in the data is due to subgroups within each sample that accumulate multiple types because of increased risk of infection, poor viral clearance due to immune compromise, or both. A simple 4 parameter model was attempted which modeled both the increased risk (2 parameters: fraction of the population at increased risk and a coefficient of increased exposure to viral types) and the immune compromise (2 parameters: fraction of the population immune compromised and a coefficient of decreased ability to clear infection). The parameters were then adjusted to match the observed frequency of type infections and co-infections. While these models replicated the original data much better than simple Poisson random assortment, the model was shown to be unstable in that increased fidelity to observed type combination frequencies often yielded unstable solutions with physically impossible parameter estimates (e.g. greater than 100% of the at risk population having no immune protection at all). Other than difficulties with the parameter estimates, it would have been difficult to convince an audience of experts to believe the model, and hence the model would compromise the credibility of our results.

2.3   Randomized Matrix Based Statistical Model

Rather than building more complex models to try and control for the confounding factors within each sample population, we were able to achieve our original goal of determining association between HPV types using a randomized matrix based approach adapted from ecological studies [4]. In this approach, we randomized the original matrix of observed type frequency data while preserving different amounts of the original data structure (Figure 2). Curve fits to this randomized data help define the level of statistical significance with which various types are observed to co occur.

Figure 2 shows an example of the data randomization procedure. In all of the procedures, column totals (representing the frequency of each variable in the population) are preserved. The simplest model is to swap 1's and 0's in the original binary matrix preserving column totals only, shown at the top right of Figure 2. This model was not pursued as it did not yield realistic results, as is shown in Figure 1. The next model preserves row and column totals for the data matrix as a whole, but not within any subset or "strata" of the matrix (middle figure), and was dubbed the "non-strata" model. Finally, we can preserve row and column totals not just for the matrix as a whole, but within various strata and sub-strata of the data matrix. We chose to randomize the matrix while preserving two types of strata: "study-strata" where each strata represents a different study in the aggregate data and so called "k-strata" where each strata has a unique integer number of multiple co-occurring types. Stratification was performed, for instance, because the different prevalences of the 37 HPV genotypes in the 6 different study populations may confound analysis with randomization across study populations. Finally, the most high fidelity model we employed was the "study-k strata" model, where we randomized the data by preserving the k-strata row

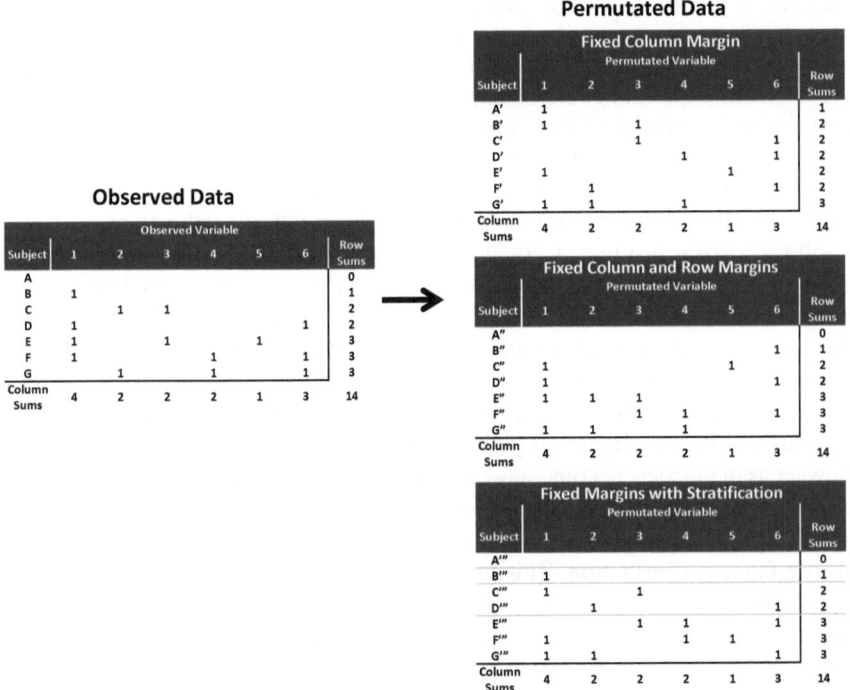

Fig. 2. Matrix randomization procedure starting from the observed data

and column totals within each of the contributing studies. Thus a total of 5 different models, with increasing levels of complexity and fidelity to the original dataset, were employed.

For the matrix randomizations themselves, we used the Vegan package [5] in the R programming language [6]. For all but the most trivial model, we used the 'permatswap' function to preserve higher order data structure. We used the trial-swap method to avoid the potential for bias in other fixed column/row sum algorithms. While the number of randomization steps were set to 1×10^7 for the 'thin' parameter, the 'burnin' parameter was set to 5×10^7, as this value was found to maximize Bray-Curtis dissimilarity values, indicating thorough matrix randomization (data not shown).

2.4   Analysis of the Randomized Matrices

We generated 1000 randomized matrices for each of the 5 models, and these served as the raw material for the method of assessing statistical significance in the observed data, which we now describe. Co-occurrences of HPV types were counted using a Perl script (Active Perl 5.8). The core feature of the script is that it counts all occurrence of types $\{x_1, x_2, \ldots x_k\}$, where k varies from 1 to n (the total number of interacting variables being studied), whether or not other types are present in a given sample. The speed of the counting is greatly enhanced by an adaptation of Knuth's "n choose k" algorithm [7] modified for this purpose. Due to resource limitations on the workstation being used, matrices were processed in batches of 200 at a time and the results were combined using a different Perl script. Resources also limited the number of type combinations that could be exported and stored in separate files (in our study $> 1 \times 10^{11}$ type combinations were possible), so that we only output the Monte Carlo data for $k \leq 4$ types. Our computational studies of the data showed that we could see nothing statistically significant in $k \geq 5$ anyway (data not shown).

Processing the counts generated in Perl was done with a combination of the JMP 9.0 software (SAS Institute, Cary NC) and Mathematica 8.0 (Wolfram Research, Champaign IL). First cubic equations were fit to the raw Monte Carlo data as a function of simple frequency based counts (these are counts derived from the frequencies of the $\{x_1, x_2, ...x_k\}$ only, i.e. assuming independent assortment). These are shown in Figure 3. Cubic equations fit the best and were the most parsimonious. Note that a different curve must be fit for each value of k and for each type of model. The fits are fairly snug and become tighter as you increase the number of input matrices, but fray somewhat as the models become more sophisticated and constrained.

Analyzing the Monte Carlo data, it was discovered that the number of times a given type combination $\{x_1, x_2, ...x_k\}$ was observed in each Monte Carlo run fit very well to a Poisson distribution. This greatly simplified the remainder of the analysis, because now we had a distribution upon which to base our computations of statistical significance, and we had an estimate of the lone parameter needed to specify that distribution from the Monte Carlo runs.

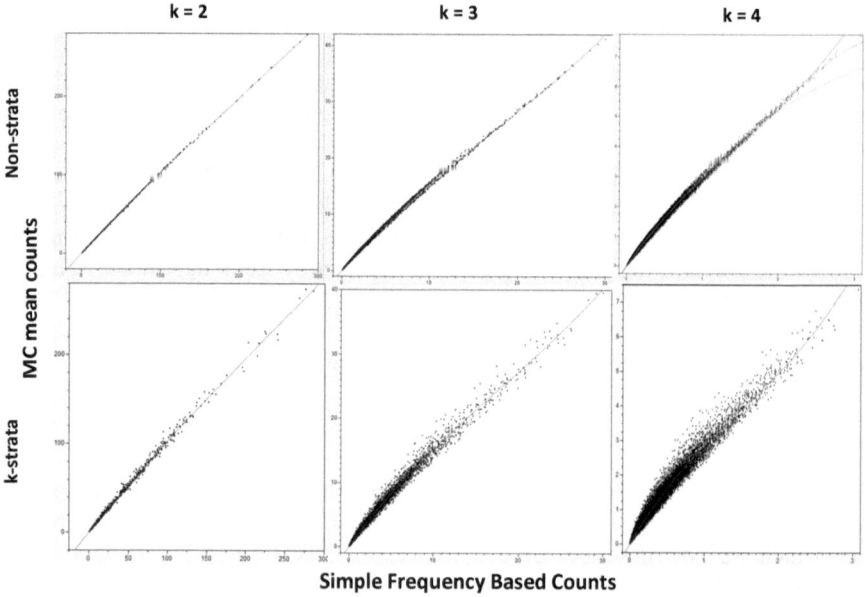

Fig. 3. Curve fits to data from 1000 matrix randomizations. Each black dot represents a given type combination for 2, 3, and 4 types. A cubic equation (red) always fit best. Quadratic and quartic (green, blue in the upper right) were tried in each case but not used.

To save ourselves from computing the cumulative density function (cdf) for the Poisson distribution potentially billions of times, we again used a curve fitting technique as shown in Figure 4. Significance boundaries for evenly log spaced values of the mean parameter are computed and then curve fit so they could be interpolated. Notice that the left tail boundaries (the counts below which you know that a combination is under-represented in the data) bottom out at zero for a fairly high mean parameter. For example in Figure 4, the graph shows that if the mean for the type combination in the Monte Carlo runs is 16 observations per randomized matrix, then even 0 observations in the real data does not meet the significance standard of $p \leq 10^{-4}$. An example of the curve which interpolates the points in Figure 4 is shown in equation 1. Although equation 1 looks complicated, it is a power law with only 3 fitted parameters, and it fit as well as shown to all of the Poisson generated cdf values.

$$y = \sqrt{x} * (6.17 + 8.88 \cdot 10^{-3} * (Log_2(x) + 1)^{3.34})$$ (1)

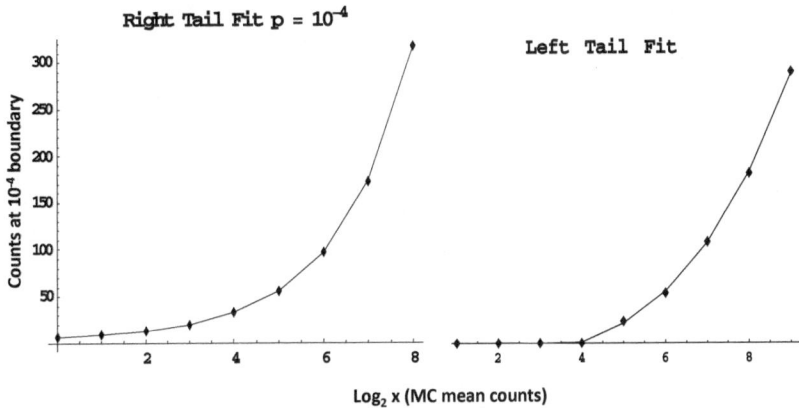

Fig. 4. Curve fits to pre-defined p-value boundaries given a Poisson distribution mean "x". Y-axis values indicate the number of counts needed for an event to be significantly over (right tail) or under (left tail) represented with a p-value of 0.0001.

3   Results

We started with a binary matrix of data indicating that certain types or variables would co-occur at certain rates, and set out to determine if those rates were more or less than would happen independently, at random. We paired each of the billions of possible type combinations $\{x_1, x_2, \ldots x_k\}$ with a simple frequency based count – roughly the counts expected only by multiplying the type frequencies together. The simple frequency based counts were transformed into Monte Carlo mean observed counts (as in Fig. 3), and those in turn were further transformed into p-values (Fig. 4). A typical result for a given model (non-strata, in this case) is shown in Fig. 5.

As one can see in Fig. 5, in our study we have several results that exceed a significance boundary of 10^{-4}, both over and under observed. Further analysis revealed that only one type combination exceeded 10^{-8} (56 and 66, shown), and none were under observed at that level. We also observed that with increasing model complexity, the statistical significance of the observed outliers decreases (data not shown). The results could signify potential synergy or antagonism among types and provide input to both policy makers and next generation vaccine designers.

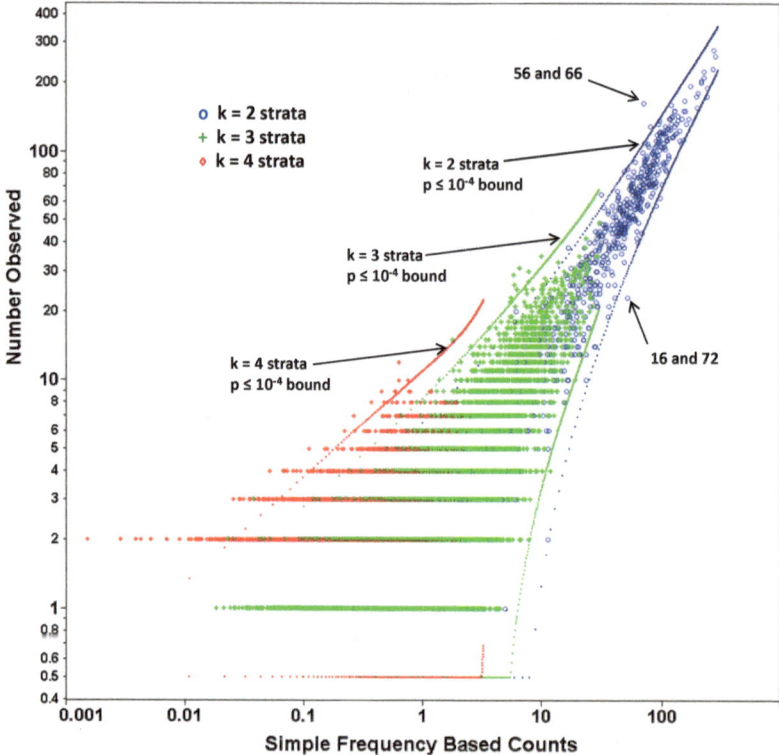

Fig. 5. Volcano plot for observed HPV multiple type data. Only 3 strata are shown for clarity (2, 3, and 4 type combinations). The model used to compute the p-value boundaries only preserved the row and column sums for the entire dataset and not for sub-strata (non-strata model).

4   Discussion

The method presented here works well with a large number of interacting variables including those at low frequency, and so could be a better alternative to popular data mining approaches like market basket analysis. In fact, market basket analysis was tried on this data, but most of the association rules did not have enough support or confidence to meet minimum thresholds. The method was applied to matrices of binary association data (where each variable is either present or absent in each subject), but there is nothing which prevents the method from being adapted for non-binary data.

The strength of the curve fitting is in reducing the number of Monte Carlo runs, which can be computationally very expensive, especially if the input matrix is large. The interpolation methods mean that not all observed variable associations need to be observed in the Monte Carlo runs for the method to work.

Another benefit of the method is in the power of the statistics themselves. A common rule of thumb for Monte Carlo analysis is that the p-values computed can be no better than the inverse of the number of Monte Carlo runs analyzed. For example, in the HPV multiple types analysis, we randomized our input matrix 1000 times for each of the 5 null models, potentially limiting $p \geq .001$. But this limit assumes non-parametric statistics when, in fact, we do estimate parameters with our approach. Because our mean observed values in the type-associations data collapse to a well defined curve, and because the Monte Carlo data fit a Poisson distribution for each type combination, p-values $\leq .001$ can reasonably be computed.

Besides the analytical strengths of the method outlined here, there are other benefits to medical, social, or behavioral study which might employ it. Other than the technical difficulties (e.g. of parameter estimation, dimensionality, etc.) involved with mathematical or simulation based modeling, models can be based on hidden assumptions or could suggest conclusions about the population that would be difficult to justify unless they could be proved with complete rigor. More specifically, modeling confounding factors to explain the association of population variables could involve making assumptions about how different subsets of the population associate, assumptions which could be controversial, extremely difficult to prove, and ultimately would detract from the analysis. Such difficult discussions about the nature and accuracy of the proposed model can be avoided and still allow one to answer central research questions using this method.

Disclaimer. The findings and conclusions in this report are those of the authors and do not necessarily represent the views of the CDC.

References

1. Mendez, F., Munoz, N., Posso, H., Molano, M., Moreno, V., van den Brule, A.J., Ronderos, M., Meijer, C., Munoz, A.: Cervical coinfection with human papillomavirus (HPV) types and possible implications for the prevention of cervical cancer by HPV vaccines. J. Infect. Dis. 192, 1158–1165 (2005)
2. Chaturvedi, A.K., Myers, L., Hammons, A.F., Clark, R.A., Dunlap, K., Kissinger, P.J., Hagensee, M.E.: Prevalence and clustering patterns of human papillomavirus genotypes in multiple infections. Cancer Epidemiol. Biomarkers Prev. 14, 2439–2445 (2005)
3. Vaccarella, S., Franceschi, S., Snijders, P.J., Herrero, R., Meijer, C.J., Plummer, M.: Concurrent infection with multiple human papillomavirus types: pooled analysis of the IARC HPV Prevalence Surveys. Cancer Epidemiol. Biomarkers Prev. 19, 503–510 (2010)

4. Connor, E.F., Simberloff, D.: The Assembly of Species Communities: Chance or Competition? Ecology 60, 1132–1140 (1979)
5. Oksanen, J., Blanchet, F.G., Kindt, R., Legendre, P., Minchin, P.R., O'Hara, R.B., Simpson, G.L., Solymos, P., Stevens, M.H., Wagner, H.: Vegan: Community Ecology Package, R package version 2.0-1 (2011), http://CRAN.R-project.org/package=vegan
6. R Core Team: R: A Language and Environment for Statistical Computing, Ver. 2.14.1. R Foundation for Statisctical Computing. Vienna, Austria (2011)
7. Knuth, D.E.: Generating all combinations and partitions. The art of computer programming, vol. 4, Fascicule 3. Addison-Wesley, Upper Saddle River (2005)

How Do E-Patients Connect Online?
A Study of Social Support Roles in Health Social Networking

Katherine Y. Chuang and Christopher C. Yang

Department of Information Science and Technology
Drexel University
{katychuang,chris.yang}@drexel.edu

Abstract. E-patients use online support communities as a way to meet other patients who have experienced or are currently undergoing similar health issues. In these communities they either seek social support or provide social support. This study incorporates the blockmodeling research technique for graphically representing the community member interactions as social positions in order to provide a clear picture of the network's social structure. This analysis views network data from six lenses to compare three different computer-mediated communication formats (forum information, journal informational, notes informational, forum nurturant, journal nurturant, notes nurturant). Results show that forum users are more likely to only be in the position of receiving or offering support, while journal users are often straddling both roles. Notes users tend to be recipients of messages.

Keywords: health informatics, social media, social support, computer-mediated communication, social network analysis, blockmodeling.

1   Introduction

Positional metrics in a social network structure refers to a person's social position in a group; a concept that comes from sociology describing the sets of actors with similar ties to others, which can be found using structural equivalence [1-3]. This metric refers to the extent to which nodes have a common set of linkages to other nodes. However, nodes do not need to have ties to others in the same cluster to be considered structurally equivalent. This perspective of looking for social positions using the structural equivalence measure is a type 'positional analysis'. In the social psychology field, positioning theory suggests that people use actions and speech when interacting with others, which develops into social roles [4]. For example a teacher may say something to imply herself as knowledgeable.

Structural equivalence is valuable in measuring online support group data because not all relationships are equal, and categorizing individuals into social roles helps us to better understand factors in social support exchanges [5]. There are very few studies that use positional analysis to study these behaviors. In fact, previous research

A.M. Greenberg, W.G. Kennedy, and N.D. Bos (Eds.): SBP 2013, LNCS 7812, pp. 193–200, 2013.
© Springer-Verlag Berlin Heidelberg 2013

studying communication patterns of online communities limit measurement to distance between various actors such as centrality and density and location measures such as in degrees and out degrees [6, 7]. The position of users in a social network has only been studied in a face-to-face environment [8].

Two nodes in a network structure are considered equivalent (aka same position) when they have the same ties or communicate with the same nodes. Position is the collection of individuals similarly embedded in networks of relations (actors in similar social activity, ties or interactions) and represented as a block in a blockmodel. The role shows patterns of relations, which obtain data between actors or between positions (association among relations). The positions are typically displayed in a reduced matrix, which can tangibly show the active clusters as different from the inactive clusters. In the following sections, an introduction to positional analysis in online support groups will be given, followed by the approach for producing blockmodels, and finally the results are presented with a discussion of these findings.

2   Social Support Networks

Treating social support occurrences as a variable that may occur as a resource exchanged between users in a social network rather than as given allows study of the social network as the subject and social support as the object of study [9]. In addition to this perspective, researchers have measured *social capital*, which are resources characterized by norms of reciprocity and social trust; *social influence*, where thoughts and actions are changed by actions or words of others; *social undermining*, a process by which others express negative affect or criticism or hinder one's attainment of goals, *companionship*, sharing leisure or other activities with network members; and finally *social support*, which is aid and assistance exchanged through social relationships and interpersonal transactions.

This study measures the transfer of social support, or the aid exchanged through interpersonal interactions in an online community. Understanding how it is transferred through multiple computer-mediated communication (CMC) formats could help reveal insights to how the user interface design impacts interpersonal communication. These niche social networking websites are valuable resources because of the benefits of support groups, such as inter-patient discourse and self-decision-making, improved health, and the betterment of the general quality of life. Users who participate are often motivated by the sense of community and empowerment [10, 11].

When measuring relationships among actors, the network approach can be used to consider (1) supportive ties anywhere in the network, (2) content, strength and symmetry of ties within a network, (3) structure of social support, and (4) characteristics of either network or components of the network [12]. Positional analysis shows roles and positions based on actor's structural similarities and patterns of relations in multiple relational networks [3]. This paper describes the approach for measuring user activity based on the actor's patterns of communication with multiple individuals in the network.

3   Methods

Three data samples were used in this study, one from each of the CMC formats (forum, journal, notes) in the Alcoholism community from a three-month period. Inactive users who posted less than two messages in each format were removed from analysis. The remaining data samples include 102 active users in forum networks, 62 active users in journal networks, 40 active users in notes informational network and 52 active users in notes nurturant network.

Blockmodeling is a process of identifying social positions within a social network by representing them in a matrix format or reduced graph. A block is a section of the matrix indicating a cluster of individuals from the network that has similar ties to others in the network. A block can indicate one of a few types of positions depending on the number of 'choices' available and the number present in the block [13]. The procedure for constructing a blockmodel begins by selecting a number of partitions to cluster the nodes in a social network. Related studies used four partitions [13-14].

Once partitioned, a blockmodel is represented in a matrix, with values either fractional value for a *density table* (a matrix with blocks of densities, fractional values between the range of 0 to 1) or binary values in an *image graph* (a matrix coded with 0 or 1). The difference between these two is the value presented. The advantage of using an image matrix is that it provides a simpler view, whereas a density table allows a more custom selection of nodes present in a reduced graph based on the density criterion (δ). A density table can be used to construct the image matrix, with a standard density threshold of $\delta = 0.5$, however one can select a different threshold to account for the non-perfect nature of structural equivalence in networks [15]. The social network data rarely contains (perfectly) structurally equivalent actors, so the block models based on structural equivalence are rarely perfect oneblock or perfect zeroblocks. By using the threshold density, the observed block would be coded as one block if is greater than or equal to the threshold (alpha) or coded as zeroblock when less than the threshold. Interpreting the blockmodels can be done for both density table and image graphs. The advantage of using image matrices is to show the type of position each node belongs to, whereas the advantage of using a density table can show the strength of relationship ties between position blocks.

Matrices can also be represented graphically in a reduced graph indicated by nodes and ties, to highlight the typology of positions [13]. These typologies can be identified as *Transmitters*, *Receivers*, or *Ordinary* nodes [3]. Positions can also be labeled as *Primary*, *Broker/Liaisons*, or *syncopates*, or *isolates* [13-14]. In the results, nodes are referred to as *Transmitters*, *Receivers*, *Carriers*, or *Isolates*.

4   Results

The following six subsections report the results for each of the six networks (forum informational, forum nurturant, journal informational, journal nurturant, notes informational, notes nurturant). Each user in the six networks was clustered into four and eight partitions to compare the two selections and reported as image graphs. The rows in each image graph represent the senders and the columns represent target nodes.

Table 1. Summary of Block model positions (4-partitions), informational network

	Total	Isolates	Transmit	Receiver	Carriers
Forum	102	29	21	51	1
Journal	62	47	4	6	5
Notes	40	24	2	14	0

Table 2. Summary of Block model positions (4-partitions), nurturant network

	Total	Isolates	Transmit	Receiver	Carriers
Forum	77	44	21	9	3
Journal	102	75	10	13	4
Notes	52	32	18	3	0

4.1   Forum Format

The forum network encompasses 102 active users over a three-month period. In the informational network, two edges in the reduced graph (Figure 1) show communication activity in a chain direction (cluster 1 to 2, cluster 2 to 4). This shows that the distribution of users either post a lot of messages and to a small subset or do not post to the community (isolates). The users in cluster 3 and 4 are most likely to be recipients of support, which suggests lack of history as a member of the community. In contrast, users in cluster 1 tend provide support, which would indicate the veterans of the community. Cluster 2 contains one user, who speaks with both new and old members, which is indicated by its carrier position. This user is a gatekeeper helps keep the conversation going for the entire community. Cluster 1 may be more selective about who they would they would message and topics they would contribute to, since the forum is a public space. There are more individuals in receiving position than transmitters, which make sense because the conversation in a group is in a one-to-many direction.

In the nurturant network, two edges remain in a chain pattern (Figure 4). Cluster 3 is a transmitter block connected to the carrier block (cluster 1), which is connected to the receiver block (cluster 2). The network clusters in this forum nurturant network have similar pattern to the forum informational network in the chaining sequence. The direction of communication here similarly suggests that some users are may be selective about thread topics they want to participate in about but also the pattern of communication fits the idea that the CMC is a one-to-many space. Along with the idea that users here tend to provide support rather than ask for support, the triangular pattern of out-degrees suggest that users are comfortable with reaching out to strangers with this CMC. A cluster that is a carrier is more likely to be commenting than a receiver, who starts a thread. This finding is consistent with the idea that people who start a thread in the forum are seeking answers. Conversations show that information support and nurturant support is exchanged in one direction, which suggests that social positions are either 'seeking information' or 'providing information'.

4.2   Journal Format

The journal network contained 62 active users for the three month time period. In the informational network, a tight circle of communication appeared among three clusters (Figure 2). Three edges remain with one group of users composing a majority of the messages, another that is mostly on the receiving end, and another that engages in back and forth exchanges (cluster 2). The triangular pattern rather than chain suggests that a user that comments on a friend's journal will very likely comment on multiple friends' journals. And the same group of 'friends' will comment on the same journals.

The results in the journal nurturant network show that blocks are active in small groups, similar to the journal informational network because of the triangular structure. There is one transmitter block (cluster 1), one receiver block (cluster 3), and one carrier block (cluster 2). Cluster 2 has a self-loop, which again indicates a majority of users communicating explicitly within their group. There are 4 individuals in this cluster, which makes it a small subset. The journal nurturant network blockmodels support the idea that journals are for friends who have conversations in small groups. The reduced graph demonstrates that cross-group links stay in small groups. There is not much 'reaching out' to random members of the entire network. The journal network is more densely interrelated than the forum network, which is consistent with the idea that friends read each other's journals and use that space for conversations.

4.3   Notes Format

The notes informational network contained 40 active users in a three-month period that posted notes to at least two friends. There is only one active block (cluster 3) acting as a transmitter, which only contained only two individuals. The lack of communication between clusters for this network is not evident. This lack of cross-communication suggests that friends usually post a note in a one-to-one manner rather than one to many. In contrast to the journal and forum informational networks, groups in this notes informational network tend pair up with another group in a one-way communication with an equal amount of receiver and transmitter nodes. There are also a higher proportion of isolate clusters (clusters 2, 4). This 'paired' communication patterns show that users will transmit information to a specific neighboring cluster. This communication tends to be unidirectional, suggesting the 'checking in' type of behavior that friends will use to maintain an intimate relationship. Users also tend to be grouped into the social role of either providing or seeking support, or neither. This evidence supports the idea of users maintaining relationships with periodic check-ins.

In contrast, the notes nurturant network had 52 users across the three-month period that posted to at least two different friends' walls. The reduced model has two edges to show two different transmitting blocks (cluster 1, 2) connected to receiving blocks (cluster 3). This nurturant network is different from the nurturant informational network in the number of receiving and transmitting blocks. When viewed as a whole, it most of the users will post to a specific group or receive notes from users of limited clusters. This supports the idea that friends are likely to communicate through the notes format in a targeted manner. There are more transmitter nodes in this network compared to the notes informational network, which can suggest that in this format more users tend to compose emotional expressions rather than informational support.

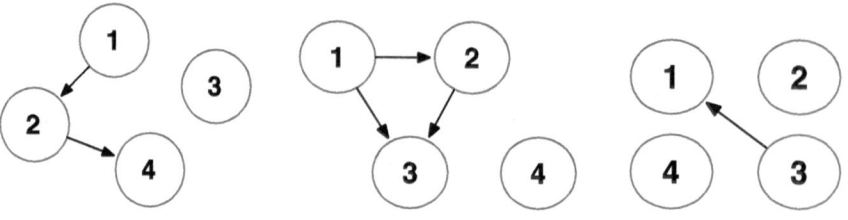

Fig. 1. Forum Informational   **Fig. 2.** Journal Informational   **Fig. 3.** Notes Informational

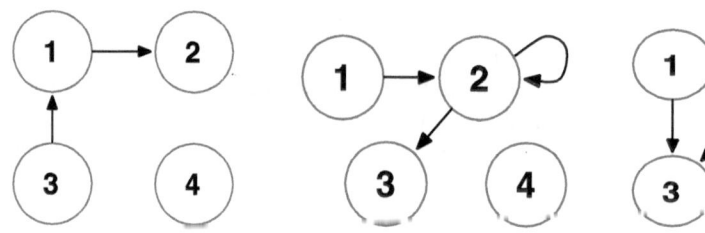

Fig. 4. Forum Nurturant   **Fig. 5.** Journal Nurturant   **Fig. 6.** Notes Nurturant

5   Discussion

This study measured the transfer of social support, or the aid exchanged through interpersonal interactions in an online community. Understanding how supports are transferred through multiple CMC formats could help reveal insights to how the user interface design impacts interpersonal communication. Results show varying levels of behaviors through multiple CMC formats in regards to social support exchanged for each type [16]. Positioning theory suggests that these differences can result because of the social role that results through repeated use of particular combinations of actions and speech when interacting with others, which develops into social roles [4]. Additionally, these differences in individual characteristics in a social network structure can be attributed to the design of the community software [7].

Forum users are more likely to only be one position of either receiving or offering support, while journal users are often straddling both roles. Notes users tend to be recipients of social support. The journal network is more densely interrelated than the forum network, which is consistent with the idea that friends read each other's journals and use that space for conversations. Notes users post to a specific group, which indicates that friends are likely to communicate through the notes format in a targeted manner. This evidence supports the idea of users maintaining relationships with periodic check-ins. Users of notes format tend to compose emotional expressions rather than informational support. Users who participate in this online support community seem to be motivated by the sense of community, a connection of two concepts suggested by other research studies [10, 11]. Future work can analyze specific roles of certain individuals to compare users operating under different CMC formats, in order to validate whether roles are attributed to the CMC format or the individual's personality.

6   Conclusion

This paper reports findings on blockmodeling to compare social positions of users across multiple computer-mediated communication formats of an online health support community. Positional analysis is a metric based on who each individual communicates with regularly rather than activeness in the community based on a mathematical measure, *'structural equivalence'* to find *social positions* within a group. Results show that different social positions exist for each of the CMC formats, however a user tends to be in a role of providing informational support in the forum format, mixed support types in the journal format and receiving emotional support in the notes format. The technique can be used in other communities but the findings are only for this community using this approach and data.

References

1. Garton, L., Haythornthwaite, C., Wellman, B.: Studying Online Social Networks. JCMC 3(1) (1997), http://jcmc.indiana.edu/vol3/issue1/garton.html
2. Granovetter, M.: The Strength of Weak Ties. American Journal of Sociology 78(6), 1360–1380 (1972)
3. Wasserman, S., Faust, K.: Social Network Analysis: Methods and Applications. Cambridge University Press, Cambridge (1994)
4. Harré, R., van Lagenhove, L.: Positioning theory: Moral contexts of intentional action. Blackwell, Oxford (1999)
5. Gilbert, E., Karahalios, K.: Predicting tie strength with social media. Paper Presented at the Proceedings of the 27th International Conference on Human Factors in Computing Systems (2009), http://dx.doi.org/10.1145/1518701.1518736
6. Chang, H.: Online supportive interactions: Using a network approach to examine communication patterns within a psychosis social support group in taiwan. Journal of the American Society for Information Science and Technology 60(7), 1504–1518 (2009), doi:10.1002/asi.21070/pdf

7. Pfeil, U., Zaphiris, P.: Investigating social network patterns within an empathic online community for older people. Computers & Human Behavior 25(5), 1139–1155 (2009)
8. Wellman, B.: Applying Network Analysis to the Study of Support. In: In, B. (ed.) Social Networks and Social Support. Sage, Beverly Hills (1981)
9. Walker, M.E., Wasserman, S., Wellman, B.: Statistical Models for Social Support Networks. Sociological Methods and Research 22, 71–98 (1993)
10. Takahashi, Y., Uchida, C., Miyaki, K., Sakai, M., Shimbo, T., Nakayama, T.: Potential Benefits and Harms of a Peer Support Social Network Service on the Internet for People With Depressive Tendencies: Qualitative Content Analysis and Social Network Analysis. Journal of Medical Internet Research 11, e29 (2009),
http://www.jmir.org/2009/3/e29/
11. Wright, K.B., Bell, S.B.: Health-related Support Groups on the Internet: Linking Empirical Findings to Social Support and Computer-mediated Communication Theory. Journal of Health Psychology 8(1), 39–54 (2003), doi:10.1177/1359105303008001429
12. Levy, J.A., Pescosolido, B.A.: Social Networks and Health 8 (2002)
13. Burt, R.S.: Models of Network Structure. Annual Review of Sociology 6, 79–141 (1980), doi:10.1146/annurev.so.06.080180.000455
14. Bambina, A.D.: Online Social Support: The Interplay of Social Networks and Computer-Mediated Communication: Cambria Press (2007)
15. Arabie, P., Boorman, S.A., Levitt, P.R.: Constructing Blockmodels: How and Why. Journal of Mathematical Psychology 17, 21–63 (1978)
16. Chuang, K., Yang, C.C.: Interaction Patterns of Nurturant Support Exchanged in Online Health Social Networking. Journal of Medical Internet Research 14(3), e54 (2012) PMID: 22555303, http://www.jmir.org/2012/3/e54/, doi:10.2196/jmir.1824

Dynamic Stochastic Blockmodels: Statistical Models for Time-Evolving Networks

Kevin S. Xu* and Alfred O. Hero III

Department of Electrical Engineering and Computer Science,
University of Michigan, Ann Arbor, MI, USA
{xukevin,hero}@umich.edu

Abstract. Significant efforts have gone into the development of statistical models for analyzing data in the form of networks, such as social networks. Most existing work has focused on modeling static networks, which represent either a single time snapshot or an aggregate view over time. There has been recent interest in statistical modeling of *dynamic networks*, which are observed at multiple points in time and offer a richer representation of many complex phenomena. In this paper, we propose a state-space model for dynamic networks that extends the well-known *stochastic blockmodel* for static networks to the dynamic setting. We then propose a procedure to fit the model using a modification of the extended Kalman filter augmented with a local search. We apply the procedure to analyze a dynamic social network of email communication.

Keywords: dynamic network, stochastic blockmodel, state-space model.

1 Introduction

Many complex physical, biological, and social phenomena are naturally represented by networks. Tremendous efforts have been dedicated to analyzing network data, which has led to the development of many formal statistical models for networks. Most research has focused on static networks, which either represent a single time snapshot of the phenomenon being investigated or an aggregate view over time. As such, statistical models for static networks have a long history in statistics and sociology among other fields [2]. However, most complex phenomena, including social behavior, are time-varying, which has led researchers to consider dynamic, time-evolving networks.

In this paper, we consider dynamic networks represented by a sequence of snapshots of the network at discrete time steps. We characterize such networks using a set of unobserved *time-varying states* from which the observed snapshots are derived. We propose a state-space model for dynamic networks that combines two types of statistical models: a static model for the individual snapshots and a temporal model for the evolution of the states. The network snapshots are modeled using the stochastic blockmodel [5], a simple parametric model commonly

---

* Current affiliation: 3M Corporate Research Laboratory, St. Paul, MN, USA.

A.M. Greenberg, W.G. Kennedy, and N.D. Bos (Eds.): SBP 2013, LNCS 7812, pp. 201–210, 2013.
© Springer-Verlag Berlin Heidelberg 2013

used in the analysis of static social networks. The state evolution is modeled by a stochastic dynamic system. Using a Central Limit Theorem approximation, we develop a near-optimal procedure for fitting the proposed model in the on-line setting where only past and present network snapshots are available. The inference procedure involves a modification of the extended Kalman filter, which is used for state tracking in many applications [3], augmented with a local search strategy. We apply the proposed procedure to analyze a dynamic social network of email communication and predict future email activity.

2   Related Work

Several statistical models for dynamic networks have previously been proposed by extending a static model to the dynamic setting in a similar fashion to our proposed model [2]. Two such models include temporal extensions of the exponential random graph model [1] and latent space model [13]. More closely related to the state-space model we propose are several temporal extensions of stochastic blockmodels (SBMs). SBMs divide nodes in the network into multiple classes and generate edges independently with probabilities θ_{ab} dependent on the class memberships a, b of the nodes [5]. Yang et al. [15] propose a dynamic SBM involving a transition matrix that specifies the probability that a node in class i at time t switches to class j at time $t + 1$ for all i, j, t and fit the model using Gibbs sampling and simulated annealing. Ho et al. [4] propose a temporal extension of a mixed-membership version of the SBM using linear state-space models for the class membership vectors of node clusters. One major difference between [4, 15] and this paper is that we treat the edge probabilities θ_{ab} as *time-varying states*, while [4, 15] treat them as time-invariant parameters. In addition, our model allows for a simpler inference procedure using a Central Limit Theorem approximation. We demonstrate the importance of the time varying states for analysis of a dynamic social network in Section 5.

3   Static Stochastic Blockmodels

We first introduce notation and summarize the static stochastic blockmodel (SSBM), which we use as the static model for the individual network snapshots. We represent a dynamic network by a time-indexed sequence of graphs, with $W^t = [w_{ij}^t]$ denoting the adjacency matrix of the graph observed at time step t. $w_{ij}^t = 1$ if there is an edge from node i to node j at time t, and $w_{ij}^t = 0$ otherwise. We assume that the graphs are directed, i.e. $w_{ij}^t \neq w_{ji}^t$ in general, and that there are no self-edges, i.e. $w_{ii}^t = 0$. $W^{(s)}$ denotes the set of all snapshots up to time s, $\{W^s, W^{s-1}, \ldots, W^1\}$. The notation $i \in a$ indicates that node i is a member of class a. $|a|$ denotes the number of nodes in class a. The classes of all nodes at time t is given by a vector c^t with $c_i^t = a$ if $i \in a$ at time t. We denote the submatrix of W^t corresponding to the relations between nodes in class a and class b by $W_{[a][b]}^t$. We denote the vectorized equivalent of a matrix X, i.e. the vector obtained by simply stacking columns of X on top of one another, by x. Doubly-indexed subscripts such as x_{ij} denote entries of matrix X, while singly-indexed subscripts such as x_i denote entries of the vectorized equivalent x.

Consider a snapshot at an arbitrary time step t. An SSBM is parameterized by a $k \times k$ matrix $\Theta^t = [\theta^t_{ab}]$, where θ^t_{ab} denotes the probability of forming an edge between a node in class a and a node in class b, and k denotes the number of classes. The SSBM decomposes the adjacency matrix into k^2 blocks, where each block is associated with relations between nodes in two classes a and b. Each block corresponds to a submatrix $W^t_{[a][b]}$ of the adjacency matrix W^t. Thus, given the class membership vector c^t, each entry of W^t is an independent realization of a Bernoulli random variable with a block-dependent parameter; that is, $w^t_{ij} \sim \text{Bernoulli}\left(\theta^t_{c_i c_j}\right)$.

SBMs are used in two settings:

1. The *a priori* blockmodeling setting, where class memberships are known or assumed, and the objective is to estimate the matrix of *edge probabilities* Θ^t.
2. The *a posteriori* blockmodeling setting, where the objective is to simultaneously estimate Θ^t and the class membership vector c^t.

Since each entry of W^t is independent, the likelihood for the SBM is given by

$$f\left(W^t; \Phi^t\right) = \prod_{i \neq j} \left(\theta^t_{c_i c_j}\right)^{w^t_{ij}} \left(1 - \theta^t_{c_i c_j}\right)^{1 - w^t_{ij}}$$

$$= \exp\left\{\sum_{a=1}^{k}\sum_{b=1}^{k}\left[m^t_{ab}\log\left(\theta^t_{ab}\right) + \left(n^t_{ab} - m^t_{ab}\right)\log\left(1 - \theta^t_{ab}\right)\right]\right\}, \quad (1)$$

where $m^t_{ab} = \sum_{i \in a}\sum_{j \in b} w^t_{ij}$ denotes the number of *observed* edges in block (a, b), and

$$n^t_{ab} = \begin{cases} |a||b| & a \neq b \\ |a|(|a| - 1) & a = b \end{cases} \quad (2)$$

denotes the number of *possible* edges in block (a, b) [6]. The parameters are given by $\Phi^t = \Theta^t$ in the a priori setting, and $\Phi^t = \{\Theta^t, c^t\}$ in the a posteriori setting. In the a priori setting, a sufficient statistic for estimating Θ^t is the matrix Y^t of *block densities* (ratio of observed edges to possible edges within a block) with entries $y^t_{ab} = m^t_{ab}/n^t_{ab}$. Y^t also happens to be the maximum-likelihood estimate of Θ^t, which can be shown [6] by setting the derivative of the logarithm of (1) to 0.

Estimation in the a posteriori setting is more involved, and many methods have been proposed, including Gibbs sampling [8], label-switching [6, 16], and spectral clustering [12]. The label-switching methods use a heuristic for solving the combinatorial optimization problem of maximizing the likelihood (1) over the set of possible class memberships, which is too large to perform an exhaustive search.

4   Dynamic Stochastic Blockmodels

We propose a state-space model for dynamic networks that consists of a temporal extension of the static stochastic blockmodel. First we present the model and inference procedure for a priori blockmodeling, and then we discuss the additional steps necessary for a posteriori blockmodeling. The inference procedure is on-line, i.e. the state estimate at time t is formed using only observations from time t and earlier.

4.1   A Priori Blockmodels

In the a priori SSBM setting, Y^t is a sufficient statistic for estimating Θ^t as discussed in Section 3. Thus in the a priori dynamic SBM setting, we can equivalently treat Y^t as the observation rather than W^t. The entries of $W^t_{[a][b]}$ are independent and identically distributed (iid) Bernoulli (θ^t_{ab}); thus by the Central Limit Theorem, the sample mean y^t_{ab} is approximately Gaussian with mean θ^t_{ab} and variance $(\sigma^t_{ab})^2 = \theta^t_{ab}(1 - \theta^t_{ab})/n^t_{ab}$, where n^t_{ab} was defined in (2). We assume that y^t_{ab} is indeed Gaussian for all (a, b) and posit the linear observation model

$$Y^t = \Theta^t + Z^t,$$

where Z^t is a zero-mean iid Gaussian noise matrix with variance $(\sigma^t_{ab})^2$ for the (a,b)th entry.

In the dynamic setting where past snapshots are available, the observations would be given by the set $Y^{(t)}$. The set $\Theta^{(t)}$ can then be viewed as states of a dynamic system that is generating the noisy observation sequence. We complete the model by specifying a model for the state evolution over time. Since θ^t_{ab} is a probability and must be bounded between 0 and 1, we instead work with the matrix $\Psi^t = [\psi^t_{ab}]$ where $\psi^t_{ab} = \log(\theta^t_{ab}) - \log(1 - \theta^t_{ab})$, the logit of θ^t_{ab}. A simple model for the state evolution is the random walk

$$\boldsymbol{\psi}^t = \boldsymbol{\psi}^{t-1} + \boldsymbol{v}^t,$$

where $\boldsymbol{\psi}^t$ is the vector representation of the matrix Ψ^t, and $\boldsymbol{v}^t$ is a random vector of zero-mean Gaussian entries, commonly referred to as process noise, with covariance matrix Γ^t. The entries of the process noise vector are not necessarily independent or identically distributed (unlike the entries of Z^t) to allow for states to evolve in a correlated manner. The observation model can then be written in terms of $\boldsymbol{\psi}^t$ as[1]

$$\boldsymbol{y}^t = h\left(\boldsymbol{\psi}^t\right) + \boldsymbol{z}^t, \tag{3}$$

where the function $h : \mathbb{R}^{k^2} \to \mathbb{R}^{k^2}$ is defined by $h_i(\boldsymbol{x}) = 1/(1 + e^{-x_i})$, i.e. the logistic function applied to each entry of $\boldsymbol{x}$. We denote the covariance matrix of $\boldsymbol{z}^t$ by Σ^t, which is a diagonal matrix[2] with entries given by $(\sigma^t_{ab})^2$. A graphical representation of the proposed model for the dynamic network is shown in Fig. 1.

---

[1] Note that we have converted the block densities Y^t and observation noise Z^t to their respective vector representations $\boldsymbol{y}^t$ and $\boldsymbol{z}^t$.

[2] The indices (a, b) for $(\sigma^t_{ab})^2$ are converted into a single index i corresponding to the vector representation $\boldsymbol{z}^t$.

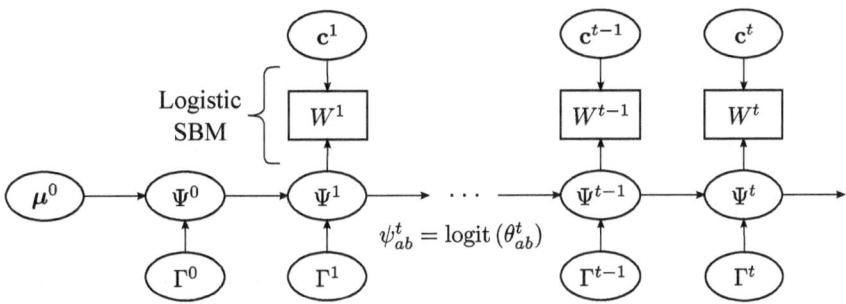

Fig. 1. Graphical representation of the proposed model. The rectangular boxes denote observed quantities, and the ovals denote unobserved quantities. The logistic SBM refers to applying the logistic function to each entry of Ψ^t to obtain Θ^t then generating W^t using Θ^t and c^t.

To perform inference on this model, we assume the initial state is Gaussian distributed, i.e. $\psi^0 \sim \mathcal{N}\left(\mu^0, \Gamma^0\right)$, and that $\{\psi^0, v^1, \ldots, v^t, z^1, \ldots, z^t\}$ are mutually independent. If (3) was linear in ψ^t, then the optimal estimate of ψ^t in terms of minimum mean-squared error would be given by the Kalman filter [3]. Due to the non-linearity, we apply the extended Kalman filter (EKF), which linearizes the dynamics about the predicted state and provides an near-optimal estimate of ψ^t. The predicted state under the random walk model is simply $\hat{\psi}^{t|t-1} = \hat{\psi}^{t-1|t-1}$ with covariance $R^{t|t-1} = R^{t-1|t-1} + \Gamma^t$. Let J^t denote the Jacobian of h evaluated at the predicted state $\hat{\psi}^{t|t-1}$. The EKF update equations are as follows [3]:

Near-optimal Kalman gain: $\quad K^t = R^{t|t-1}\left(J^t\right)^T \left[J^t R^{t|t-1} \left(J^t\right)^T + \Sigma^t\right]^{-1}$

Posterior state estimate: $\quad \hat{\psi}^{t|t} = \hat{\psi}^{t|t-1} + K^t \left[y^t - h\left(\hat{\psi}^{t|t-1}\right)\right]$

Posterior estimate covariance: $\quad R^{t|t} = \left(I - K^t J^t\right) R^{t|t-1}$

The posterior state estimate $\hat{\psi}^{t|t}$ provides a near-optimal fit to the model at time t given the observed sequence $W^{(t)}$. How to choose the hyperparameters $\left(\mu^0, \Gamma^0, \Sigma^t, \Gamma^t\right)$ in an optimal manner is beyond the scope of this paper and is discussed in [14, chap. 5].

4.2   A Posteriori Blockmodels

In many applications, the class memberships c^t are not known a priori and must be estimated along with Ψ^t. This can be done using label-switching methods [6, 16], but rather than maximizing the likelihood, we maximize the posterior state density given the entire sequence of observations $W^{(t)}$ up to time t to account for the prior information. This is done by alternating between label-switching and applying the EKF.

The posterior state density is given by

$$f\left(\boldsymbol{\psi}^t \mid W^{(t)}\right) \propto f\left(W^t \mid \boldsymbol{\psi}^t, W^{(t-1)}\right) f\left(\boldsymbol{\psi}^t \mid W^{(t-1)}\right). \tag{4}$$

By the conditional independence of current and past observations given the current state, $W^{(t-1)}$ drops out of the first term in (4). It can thus be obtained simply by substituting $h(\boldsymbol{\psi}^t)$ for $\boldsymbol{\theta}^t$ in (1). The second term in (4) is equivalent to $f\left(\boldsymbol{\psi}^t \mid \boldsymbol{y}^{(t-1)}\right)$ because the class memberships at all previous time steps have already been estimated. By applying the Kalman filter to the linearized temporal model [3], $f\left(\boldsymbol{\psi}^t \mid \boldsymbol{y}^{(t-1)}\right) \sim \mathcal{N}\left(\hat{\boldsymbol{\psi}}^{t\mid t-1}, R^{t\mid t-1}\right)$. Thus the logarithm of the posterior density is given by

$$\log f\left(\boldsymbol{\psi}^t \mid W^{(t)}\right) = c - \frac{1}{2}\left(\boldsymbol{\psi}^t - \hat{\boldsymbol{\psi}}^{t\mid t-1}\right)^T \left(R^{t\mid t-1}\right)^{-1}\left(\boldsymbol{\psi}^t - \hat{\boldsymbol{\psi}}^{t\mid t-1}\right)$$
$$+ \sum_{a=1}^{k}\sum_{b=1}^{k}\left\{m_{ab}^t \log\left[h\left(\psi_{ab}^t\right)\right] + \left(n_{ab}^t - m_{ab}^t\right)\log\left[1 - h\left(\psi_{ab}^t\right)\right]\right\}, \tag{5}$$

where c is a constant term independent of $\boldsymbol{\psi}^t$ that can be ignored[3].

We use the log-posterior (5) as the objective function for label-switching. We find that a simple local search (hill climbing) algorithm [11] initialized using the estimated class memberships at the previous time step suffices, because only a small fraction of nodes change classes between time steps in most applications. At the initial time step, we employ the spectral clustering algorithm of Sussman et al. [12] for the SSBM as the initialization.

5   Application to Enron Email Network

We demonstrate the proposed procedure on a dynamic social network constructed from the Enron corpus [9, 10], which consists of about 0.5 million email messages between 184 Enron employees from 1998 to 2002. We place directed edges between employees i and j at time t if i sends at least one email to j during week t. Each time step corresponds to a 1-week interval. We make no distinction between emails sent "to", "cc", or "bcc". In addition to the email data, the roles of most of the employees within the company (e.g. CEO, president, manager, etc.) are available, which we use as classes for a priori blockmodeling. Employees with unknown roles are placed in an "others" class.

5.1   State Tracking

We begin by examining the temporal variation of the states, which we refer to as *state tracking*. Recall that the states $\boldsymbol{\Psi}^t$ correspond to the logit of the edge probabilities $\boldsymbol{\Theta}^t$. We first apply the a priori EKF to obtain the state estimates $\hat{\boldsymbol{\psi}}^{t\mid t}$ and their variances (the diagonal of $R^{t\mid t}$). Applying the logistic function, we can then obtain the estimated edge probabilities $\hat{\boldsymbol{\Theta}}^{t\mid t}$ with confidence intervals.

---

[3] At the initial time step, $\hat{\boldsymbol{\psi}}^{1\mid 0} = \boldsymbol{\mu}^0$ and $R^{1\mid 0} = \Gamma^0 + \Gamma^1$.

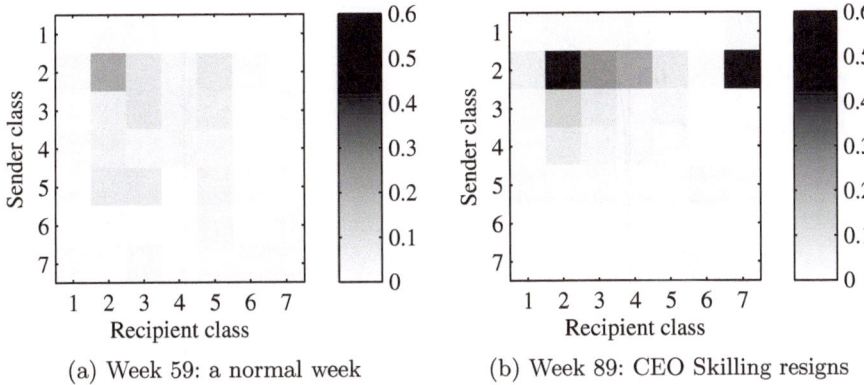

(a) Week 59: a normal week   (b) Week 89: CEO Skilling resigns

Fig. 2. Estimated edge probability matrices for two selected weeks. Entry (i, j) denotes the estimated probability of an edge from class i to class j. Classes are as follows: (1) directors, (2) CEOs, (3) presidents, (4) vice-presidents, (5) managers, (6) traders, and (7) others. Notice the increase in the probability of edges from CEOs during the week of Skilling's resignation.

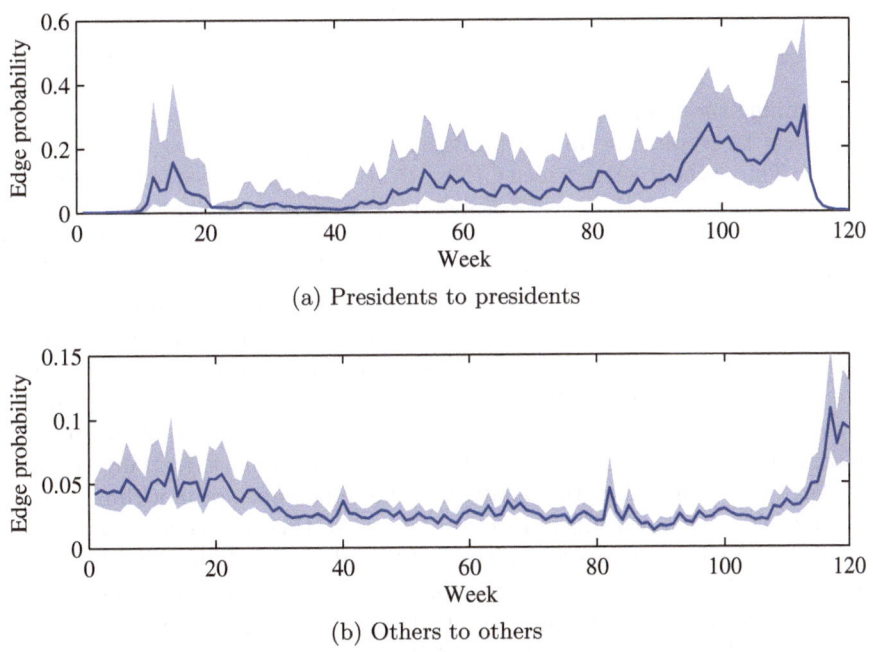

(a) Presidents to presidents

(b) Others to others

Fig. 3. A priori EKF estimated edge probabilities $\hat{\theta}_{ab}^{t|t}$ (solid lines) with 95% confidence intervals (shaded region) for selected a, b by week. An increase in edge probabilities between Enron presidents (a) occurs prior to a similar increase between those in other roles (b) suggesting insider knowledge.

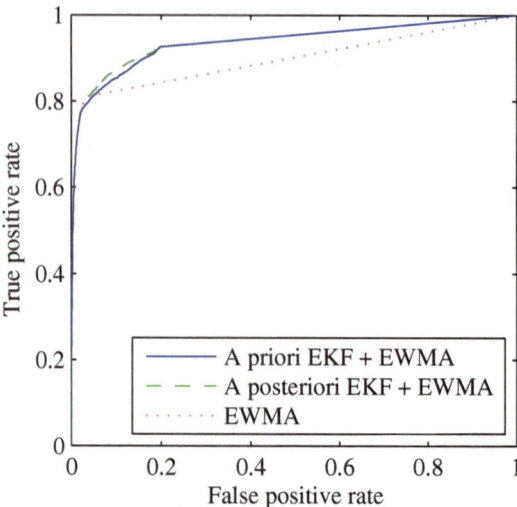

Fig. 4. Comparison of ROC curves for link prediction on Enron data. True positive rate denotes the fraction of actual edges that are correctly predicted, and false positive rate denotes the fraction of non-edges that are predicted to be edges. The convex combination of either EKF with the EWMA outperforms the EWMA alone by accounting for block-level characteristics.

Examining the temporal variation of $\hat{\Theta}^{t|t}$ reveals some interesting trends. For example, a large increase in the probabilities of edges from CEOs is found at week 89. This is the week in which CEO Jeffrey Skilling resigned and is confirmed to be the cause of the increased probabilities by examining the content of the emails. Fig. 2 shows a comparison of the matrix $\hat{\Theta}^{t|t}$ during a normal week and during the week Skilling resigned.

Another interesting trend is highlighted in Fig. 3, where the temporal variation of two selected edge probabilities over the entire data trace with 95% confidence intervals is shown. Edge probabilities between Enron presidents show a steady increase as Enron's financial situation worsens, hinting at more frequent and widespread insider discussions, while emails between others (not of one of the six known roles) begin to increase only after Enron falls under federal investigation.

A key observation from this analysis is the importance of modeling the edge probabilities as time-varying states, as opposed to time-invariant parameters as in [4, 15]. Indeed the *temporal variation* of the edge probabilities is what reveals the internal dynamics of this time-evolving social network. Furthermore, the temporal model provides estimates with less uncertainty than the static SBM, with 95% confidence intervals that are 24% narrower on average.

5.2 Dynamic Link Prediction

Next we turn to the task of dynamic link prediction, which differs from static link prediction [7] because the link predictor must simultaneously predict the new edges that will be formed at time $t + 1$, as well as the current edges (as of time t) that will disappear at time $t + 1$, from the observations $W^{(t)}$. The latter task is not addressed by most static link prediction methods in the literature.

Since the SBM assumes stochastic equivalence between nodes in the same class, the EKF alone is only a good predictor of the block densities Y^t, not the edges themselves. However, the EKF can be combined with a predictor that operates on individual edges to form a link predictor. A simple individual-level predictor is the exponentially-weighted moving average (EWMA) given by $\hat{W}^{t+1} = \lambda \hat{W}^t + (1 - \lambda)W^t$. Using a convex combination of the EKF and EWMA predictors, we obtain a better link predictor that incorporates both block-level characteristics (through the EKF) and individual-level characteristics (through the EWMA). This can be seen from the receiver operating characteristic (ROC) curves in Fig. 4. The a posteriori EKF slightly outperforms the a priori EKF because the a posteriori EKF finds a better fit to the dynamic SBM via a better assignment of nodes to classes than the a priori (assumed) assignment.

6   Conclusion

This paper proposes a statistical model for dynamic networks that utilizes a set of unobserved time-varying states to characterize the dynamics of the network. The proposed model extends the well-known stochastic blockmodel for static networks to the dynamic setting can be used for either a priori or a posteriori blockmodeling. The main contribution of the paper is a near-optimal on-line inference procedure for the proposed model using a modification of the extended Kalman filter, augmented with a local search. We applied the proposed inference procedure to the Enron email network and discovered some interesting trends when we examined the estimated states. One such trend was a steady increase in emails between Enron presidents as Enron's financial situation worsened, while emails between other employees remained at their baseline levels until Enron fell under federal investigation. In addition, the proposed procedure showed promising results for predicting future email activity. We believe the proposed model and inference procedure can be applied to reveal the internal dynamics of many other dynamic networks.

Acknowledgments. This work was partially supported by the Army Research Office grant W911NF-12-1-0443. Kevin Xu was partially supported by an award from the Natural Sciences and Engineering Research Council of Canada.

References

[1] Ahmed, A., Xing, E.P.: Recovering time-varying networks of dependencies in social and biological studies. Proc. Nat. Acad. Sci. 106(29), 11878–11883 (2009)
[2] Goldenberg, A., Zheng, A.X., Fienberg, S.E., Airoldi, E.M.: A survey of statistical network models. Found. Trends Mach. Learn. 2(2), 129–233 (2010)
[3] Haykin, S.: Kalman filtering and neural networks. Wiley-Interscience (2001)

[4] Ho, Q., Song, L., Xing, E.P.: Evolving cluster mixed-membership blockmodel for time-varying networks. In: Proc. 14th Int. Conf. Artif. Intell. Statist. (2011)

[5] Holland, P.W., Laskey, K.B., Leinhardt, S.: Stochastic blockmodels: First steps. Soc. Netw. 5(2), 109–137 (1983)

[6] Karrer, B., Newman, M.E.J.: Stochastic blockmodels and community structure in networks. Phys. Rev. E 83, 016107 (2011)

[7] Liben-Nowell, D., Kleinberg, J.: The link-prediction problem for social networks. J. Am. Soc. Inf. Sci. 58(7), 1019–1031 (2007)

[8] Nowicki, K., Snijders, T.A.B.: Estimation and prediction for stochastic blockstructures. J. Am. Stat. Assoc. 96(455), 1077–1087 (2001)

[9] Priebe, C.E., Conroy, J.M., Marchette, D.J., Park, Y.: Scan statistics on Enron graphs. Comput. Math. Organ. Theory 11(3), 229–247 (2005)

[10] Priebe, C.E., Conroy, J.M., Marchette, D.J., Park, Y.: Scan statistics on Enron graphs (2009), http://cis.jhu.edu/~parky/Enron/enron.html

[11] Russell, S.J., Norvig, P.: Artificial intelligence: A modern approach, 2nd edn. Prentice Hall (2003)

[12] Sussman, D.L., Tang, M., Fishkind, D.E., Priebe, C.E.: A consistent adjacency spectral embedding for stochastic blockmodel graphs (2012), arXiv:1108.2228v3 [stat.ML]

[13] Westveld, A.H., Hoff, P.D.: A mixed effects model for longitudinal relational and network data, with applications to international trade and conflict. Ann. Appl. Stat. 5(2A), 843–872 (2011)

[14] Xu, K.S.: Computational methods for learning and inference on dynamic networks. Ph.D. thesis, University of Michigan (2012)

[15] Yang, T., Chi, Y., Zhu, S., Gong, Y., Jin, R.: Detecting communities and their evolutions in dynamic social networks—a Bayesian approach. Mach. Learn. 82(2), 157–189 (2011)

[16] Zhao, Y., Levina, E., Zhu, J.: Consistency of community detection in networks under degree-corrected stochastic block models. Ann. Stat. (2012) (in press)

LA-LDA: A Limited Attention Topic Model for Social Recommendation

Jeon-Hyung Kang[1], Kristina Lerman[1], and Lise Getoor[2]

[1] USC Information Sciences Institute, 4676 Admiralty Way, Marina del Rey, CA
jeonhyuk@usc.edu, lerman@isi.edu
[2] University of Maryland, Computer Science Department, College Park, MD
getoor@cs.umd.edu

Abstract. Social media users have finite attention which limits the number of incoming messages from friends they can process. Moreover, they pay more attention to opinions and recommendations of some friends more than others. In this paper, we propose $\mathcal{L}A$-LDA, a latent topic model which incorporates limited, non-uniformly divided attention in the diffusion process by which opinions and information spread on the social network. We show that our proposed model is able to learn more accurate user models from users' social network and item adoption behavior than models which do not take limited attention into account. We analyze voting on news items on the social news aggregator Digg and show that our proposed model is better able to predict held out votes than alternative models. Our study demonstrates that psycho-socially motivated models have better ability to describe and predict observed behavior than models which only consider topics.

Keywords: social media, diffusion, link analysis, influence.

1 Introduction

Information overload has been drastically exacerbated by social media. On sites such as Twitter, YouTube and Facebook, more videos and images are uploaded, blog posts written, and new messages posted than people are able to process. Social media sites attempt to mitigate this problem by allowing users to subscribe to, or follow, updates from specific users only. However, as the number of friends people follow grows, and the amount of information shared expands, the information overload problem returns.

Though social media contributes to the information overload problem; however it also creates opportunities for solutions. We can apply statistical techniques to social media data to learn user preferences and interests from observations of their behavior. The learned preferences could then be used to more accurately filter and personalize streams of new information. Consider social recommendation: when a user shares an item, e.g., by posting a link to a news story on Digg or Twitter, he broadcasts it to all his followers. Those followers may in turn share the item with their own followers, and so on, creating a cascade through which information and ideas diffuse through the social network.

A.M. Greenberg, W.G. Kennedy, and N.D. Bos (Eds.): SBP 2013, LNCS 7812, pp. 211–220, 2013.
© Springer-Verlag Berlin Heidelberg 2013

By analyzing these cascades, who shares what items and when, we can learn what users are interested in and use this knowledge to filter and rank incoming information.

The generic diffusion process described above ignores two important elements: (*i*) users have finite attention, which limits their ability to process recommended items, and (*ii*) users divide their attention non-uniformly over their friends and interests. Attention is the psychological mechanism that integrates perceptual and cognitive factors to select the small fraction of input to be processed in real time [8,12]. Attention has been shown to be an important factor in explaining online interactions [17,7]. Attentive acts, e.g., reading a tweet, browsing the web, or responding to email, require mental effort, and since the brain's capacity for mental effort is limited, so is attention. Attention has been shown to impact the popularity of memes [18,17], what people retweet [3,7] and the number of meaningful conversations they can have [5]. Attention is important, because most sites, including Digg and Twitter, display items from friends as a chronologically sorted list, with the newest items at the top of the list. The more friends a user follows, the longer the list, in average. A user scans the list, beginning at the top, and if he finds an item interesting, he may share it with his followers. He will continue scanning the list until he gets bored or distracted, which is likely to happen before he had a chance to inspect all new items. While a user must divide his limited attention among his friends, he does not divide it uniformly. Some friends are closer or more influential [6,4]; therefore, their recommendations may receive more attention, making them more likely to be adopted. Users may also preferentially pay more attention to each friend depending on topic.

In next section we describe a diffusion mechanism that takes into consideration the limited, non-uniformly divided attention of social media users. We use this mechanism to motivate $\mathcal{L}A$-LDA, a probabilistic topic model we introduce. Next, we analyze voting on news items on the social news aggregator Digg and show that our model is better able to predict held out votes than alternative models that do not take limited attention into account. Our study demonstrates that psycho-socially motivated models are better able to describe and predict observed user behavior in social media, and may lead to better tools for solving the information overload problem.

2   LA-LDA

Social Recommendation Setting. We begin by describing the social recommendation scenario we are modeling. We assume an idealized social media setting, with U users who recommend to each other and adopt items A. Users have interests X, and items have topics Z, with users more likely to adopt items whose topics match their interests. In addition, each user u has $N_{frds(u)}$ friends and can see the items friends adopted.

The social recommendation model we propose is dynamic, and describes a number of user actions. A user u can share an item i at time t. An item could be a link to an online resource that a user shares by tweeting it on Twitter or submitting for it on Digg. We assume that when an item is shared by u, the recommendation is broadcast of all of u's followers. A user u can share a recommended item i at time t, for example, by retweeting the link on Twitter or voting for it on Digg.

We also introduce the notion of a *seed*, the user who introduced the item into the social network. For any item i, there is a set of seed users whose adoptions diffuse through the social network along follower links, based on users' interests.

Finally, what sets our model apart from previous models for social recommendations is that we also model user's attention. Users have limited attention and may not attend to all the items their friends recommend. After attending to an item, they may decide to adopt and share it. Once an item is shared, the limited attention diffusion process continues to unfold.

In summary, in the context of social recommendation, limited attention implies that users may process all items their friends recommend. How they limit their attention depends on both their interests and their social network.

Probabilistic Model. We now introduce a topic model $\mathcal{L}A$-LDA that captures the salient elements, including the limited attention of users, of social recommendation. Our model consists of four key components which describe user's interests $(\theta_{(u)})$, item's topics $(\psi_{(i)})$, user's attention to friends on different interests $(\tau_{(u)})$, and user's limited attention $(\phi_{(u)})$. We assume there are N_u users, N_i items, and each user u follows $N_{frds(u)}$ friends. Moreover, each user has N_x interests, and each item has N_z topics.

The $\mathcal{L}A$-LDA model is presented in graphical form in Figure 1(a). There are four parts to the model representation: user level (θ, τ, ϕ), item level (ψ), interest $\times$ topic level (π), and global hyperparameters $(\alpha, \beta, \rho, \text{and } \eta)$. Each adoption of an item i by a user u has an associated item topic z, and user interest x; Y denotes the friend(s) whose recommendations for i were adopted by u. Variables A and Y are observed, while X and Z are hidden. User u's interest profile $\theta_{(u)}$ is a distribution over N_x interests. Similarly, item i's topic profile $\psi_{(i)}$ is a distribution over N_z topics. Each user pays attention to different friends depending on interests, so that for user u and interest x, there is an interest-specific distribution $\tau_{(u,x)}$ over $frds(u)$. The distribution of user u's attention over both N_x interests and $frds(u)$ is captured by $\phi_{(u)}$. Finally, each interest x and topic z pair has an adoption probability $\pi_{(x,z)}$ for items. The generative process for item adoption through a social network is shown in Figure 1(b).

Inference. The inference procedure for our model follows the derivation of the equations for collapsed Gibbs sampling, since we cannot compute posterior distribution directly because of the summation in the denominator. By constructing a Markov chain, we can sample sequentially until the sampled parameters approach the target posterior distributions. In particular, we sample all variables from their distribution by conditioning on the currently assigned values of all other variables. To apply this algorithm, we need the full conditional distribution and it can be obtained by a probabilistic argument.

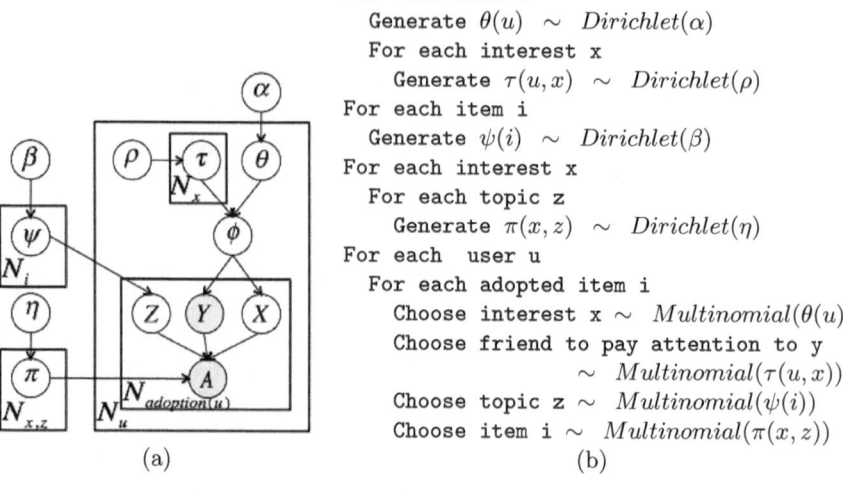

For each user u
  Generate $\theta(u) \sim Dirichlet(\alpha)$
  For each interest x
    Generate $\tau(u,x) \sim Dirichlet(\rho)$
For each item i
  Generate $\psi(i) \sim Dirichlet(\beta)$
For each interest x
  For each topic z
    Generate $\pi(x,z) \sim Dirichlet(\eta)$
For each user u
  For each adopted item i
    Choose interest $x \sim Multinomial(\theta(u))$
    Choose friend to pay attention to y
                $\sim Multinomial(\tau(u,x))$
    Choose topic $z \sim Multinomial(\psi(i))$
    Choose item $i \sim Multinomial(\pi(x,z))$

(a)   (b)

Fig. 1. The $\mathcal{L}A$-LDA model (user interest profiles(θ), interest-specific attention profiles(τ), item topic profiles(ψ), and adoption probabilities(π))

The Gibbs sampling formulas for the variables are:

$$P(Z_{(u,v)} = k | Z_{-(u,v)}, X, Y, A_u) \propto \frac{n^k_{-(u,v)} + \beta}{n^{(\cdot)}_{-(u,v)} + \beta \times N_z} \frac{n^{x,k}_{-(u,v)} + \eta}{n^{x,k}_{-(\cdot,\cdot)} + \eta \times N_i}$$

$$P(X_{(u,v)} = j | X_{-(u,v)}, Y, Z, A_u) \propto \tag{1}$$

$$\frac{n^j_{-(u,\cdot)} + \alpha}{n^{(\cdot)}_{-(u,\cdot)} + \alpha \times N_x} \frac{n^y_{-(u,j)} + \rho}{n^{(\cdot)}_{-(u,j)} + \rho \times N_{(frds(u))}} \frac{n^{j,u}_{-(u,v)} + \eta}{n^{j,z}_{-(\cdot,\cdot)} + \eta \times N_i}$$

where $n^k_{-(u,v)}$ is the number of times topic k is assigned on item (u,v) excluding the current assignment of $Z_{(u,v)}$, $n^{x,k}_{-(u,v)}$ is the number of adoptions of item (u,v) under item topic assignment k and user interest assignment of x, excluding the current item topic assignment of $Z_{(u,v)}$, A_u is the set of items adopted by user u, and v ranges over the items in A_u. (u,v) denotes the index of the vth item adopted by user u. The first ratio expresses the probability of topic k for item (u,v), and the second ratio expresses the probability of item (u,v)'s adoption under the item topic assignment k and user interest assignment x. In the second equation, $n^j_{-(u,\cdot)}$ is the number of times user u pays attention to interest j excluding the current assignment of $X_{(u,v)}$ and $n^y_{-(u,j)}$ is the number of times user u pays attention to friend y on interest x excluding the current assignment of $X_{(u,v)}$. The first ratio expresses the probability of user u paying attention to interest j and the second ratio expresses the probability that user u pays attention to friend y on interest j. Our model allows the algorithm learn each user's interests by taking into account the limited attention on friends for certain interests from local perspective, while adopting is given by user's interest and item's topic assignment from global perspective. To make the model simple we use symmetric Dirichlet priors. We estimate θ, ψ, π, and ϕ with sampled values in the standard manner.

3   Evaluation on Synthetic Data

Our first set of experiments illustrate the properties of the $\mathcal{L}A$-LDA model used in conjunction with synthetic data. We used social network links among top 5,000 most active users in 2009 dataset, who are followed by in average 81.8 other users (max 984 and median 11). We begin generating synthetic data by creating N_i items and N_u users according to the generative model.

We model the propagation of items through the social network over a period of N_{day} days. We first choose a set of seeders ($S\%$) from N_u users. Seeders will be able to introduce new items into the network. We introduce a special source node, which contains all of the items. Seeders will have the source node as one of their friends. Every user u is assigned a fixed attention budget V_u, which determines the total number of items from friends that u can attend to in a day. For simplicity, we represent V_u as a function of a global attention limit parameter v_g and the number of friends user has. This is motivated by the observation that, at least on Digg, user activity is correlated with the number of friends they follow (the correlation coefficient is 0.1626–0.1701). Intuitively, the number of items a user adopts is some fraction of the number of stories to which a user attends; here, to simplify matters, we assume that user's attention budget is simply proportional to the number of friends she follows.

function GENERATE SYNTHETIC DATA
    for $day = 1 \rightarrow N_{day}$ **do**
        for $u = 1 \rightarrow N_u$ **do**
            for $attention = 1 \rightarrow V_u$ **do**
                choose interest $x \sim Mult(\theta_{(u)})$
                choose friend $y \sim Mult(\tau_{(u,x)})$
                choose a item i from y
                choose topic $z \sim Mult(\psi_{(i)})$
                Adopt and share item with probability $\pi_{(x,z)}$
            end for
        end for
    end for
  end function

Synthetic cascades are generated as follows. Each day, every user within her allotted attention budget, will check to see whether her friends have any items that match her interests. Initially, when the cascade starts, the source node is the only friend, which has items, so only seed nodes will be able to adopt and share items. However, as time progresses, and items begin flowing through the network. Eventually users will exhaust their attention budget, without being able to attend to all the items that their friends shared with them. When user chooses to attend to an item i that has been shared by a friend y, they choose without replacement, so that an item will only be attended to once from a particular friend y. However, we do allow a user to attend the same item from different friends. Once an item has been chosen, the user will adopt (and share) the item with probability $\pi_{x,z}$.

By varying parameters (S and v_g) and hyperparameters (α, β, η, and ρ) we can create different synthetic datasets and we investigate how well we are able to recover the user interests from the generated data using $\mathcal{L}A$-LDA (or LDA) model. We evaluate the performance of models by measuring the similarity of the learned and the actual distributions by the average deviation between the Jensen-Shannon divergence of their vectors. The average deviation is small when two vectors are similar without considering the indexing of the interests.

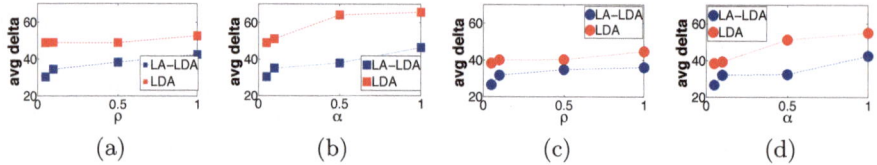

Fig. 2. The average deviation of user interest (θ) and item topic (ψ) with different limited attention values (ρ and α) on synthetic. The top two figures show average deviation between learned and actual θ when (a) α=0.05 and ρ=0.05, 0.1, 0.5, and 1.0 and (b) ρ=0.05 and α=0.05, 0.1, 0.5, and 1.0. The bottom two figures show average deviation between learned and actual ψ when (c) $\alpha = 0.05$ and (d) $\rho = 0.05$.

For comparison, we learned two different LDA models, one for user interests and one for item topics. We learn the LDA for interest distributions of users θ by viewing a user as a document and items as terms in a document, and we learn the LDA for topic distributions of items ψ by setting item as a document and users as terms in a document. We also ran $\mathcal{L}A$-LDA to learn both θ and ψ in accordance with that model. For generating the synthetic data, we set v_g=2, β=0.1, η=0.1 and S=30%) and varied α (0.05, 0.1, 0.5, and 1.0) and ρ (0.05, 0.1, 0.5, and 1.0). We applied the same hyperparameters used to generate the synthetic data in the models.

The average deviation between learned and actual interests and topics of items in the synthetic datasets are shown in Fig. 2. With large values of α, users allocate their attention uniformly over interests, so users are more likely to adopt items on a variety of interests. Because of this adoption tendency, it is hard to distinguish their interests. For small values of α, users pay attention to a limited number of interests and more can be learned from their adoption behavior. That is why both LDA and $\mathcal{L}A$-LDA perform better for small α values. Similarly, large values of ρ cause users to pay attention to their friends uniformly, while small values focuses users' attention to a smaller subset of their friends. With large ρ values, average deviations of both models are high, whereas for lower values both models perform better. In all four cases, $\mathcal{L}A$-LDA is superior to LDA in learning interests distribution of users and topics distribution of items for all α and ρ values.

4   Evaluation on Digg

We evaluate $\mathcal{L}A$-LDA on real-world data from the social news aggregator Digg, which allows users to submit links to news stories and other users to vote for (or "digg") stories they find interesting. Digg also allows users to follow the activity of other users to see the stories they submitted or dugg recently. When a user votes for a story, this recommendation is broadcast to all his followers. At the time data was collected, users were submitting many thousands of stories, from which Digg selected a handful to promote to its popular front page.

We evaluated two datasets The 2009 dataset [9] contains information about the voting history of 70K active users (with 1.7M social links) on 3.5K stories promoted to Digg front page in June, and contains 2.1M votes. At the time, Digg assigned stories to one of eight topics (Entertainment, Lifestyle, Science, Technology, World & Business, Sports, Offbeat, and Gaming). The 2010 dataset [15] contains information about voting histories of 12K users (with 1.3M social links) over a 6 months period (Jul – Dec). It includes 48K stories with 1.9M votes. At the time data was collected, Digg assigned stories to 10 topics, replacing the "World & Business" topic with "World News," "Business," and "Politics".

Before a story is promoted to the front page, it is visible on the upcoming stories queue and to the submitter's followers. With each new vote, the story becomes visible to that voter's followers. We examine only the votes that the story accrued before promotion to the front page, during which time it propagated mainly via friends' recommendations. In the 2009 dataset, 28K users voted for 3K stories and in the 2010 dataset, 4K users voted for 36K stories before promotion. We focused the data further by selecting those users who voted at least 10 times, resulting in 2,390 users (who voted for 3,553 stories) in the 2009 dataset and 2,330 users (who voted on 22,483 stories) in the 2010 dataset.

$\mathcal{L}A$-LDA has six parameters: the number of interests (N_x) and topics (N_z) and hyperparameters α, β, η, and ρ. The choice of hyperparameters can have implications inference results. While our algorithm can be extended to learn hyperparameters, here we fix them (0.1) and focus on the consequences of varying the number of topics and interests (from 5 to 800). We estimate the performance of model by computing the likelihood of the training set given the model for different combinations of parameters. We took samples at a lag of 100 iterations after discarding the first 1000 iterations and both algorithms stabilize within 2000 iterations. The best performance is obtained for $N_x = 10$ interests and $N_z = 200$ topics in the 2009 dataset and $N_x = 30$ interests and $N_z = 200$ topics in the 2010 dataset for both ITM and $\mathcal{L}A$-LDA. LDA results in best performance for 200 interests in the 2009 and 500 interests in the 2010 dataset.

Evaluation of Learned User Interests. The topics assigned to stories by Digg provide useful evidence for evaluating topic models. We represent user u's preferences by constructing an empirical interest vector that gives the fraction of votes made by u on each topic. The empirical interest vector serves as gold standard for evaluating user interests learned by different topic models. We measure the similarity of the distributions using average Jensen-Shannon divergence. In both datasets, $\mathcal{L}$A-LDA (2009 dataset: 15.11 & 2010 dataset: 28.71) outperforms ITM [11] (36.38 & 36.01) and LDA [1] (37.72 & 55.43) models by learning user interests that are closer to the gold standard.

Evaluation on Vote Prediction. We evaluate our proposed topic models by measuring how well they allow us to predict individual votes. There are 257K pre-promotion votes in the 2009 dataset and 1.5M votes in the 2010 dataset, with 72.34 and 68.20 average votes per story, respectively. For our evaluation, we randomly split the data into training and test sets, and performed five-fold cross validation. To generate the test set, we use the held-out votes (positive examples) and augment it with stories that friends of users shared but that were not adopted by user. Depending on a user's and their friends' activities, there are different numbers of positive (N_{pos}^u) in the test set. The average percentage of N_{pos}^u in the test set is 0.73% (max 18%, min 0.02%, and median 0.13%), suggesting that friends share many stories that users do not end up not voting for. This makes the prediction task extremely challenging, with less than one in a hundred chance of successfully predicting votes if stories are picked randomly.

We train the models on the data in the training set. Then, for each story i in the test set, we compute the probability user u votes for it, given training data $\mathcal{D}$. For LDA, the probability of the vote on i is the probability of adopting a_i:

$$P(a_i|\mathcal{D}) = \int_\theta \sum_x P(a_i|x)P(x|\theta)P(\theta|\mathcal{D})\,\mathrm{d}\theta \qquad (2)$$

For ITM, the probability that user u votes for story i is obtained by integrating over the posterior Dirichlet distributions of θ and ψ:

$$P(a_i|\mathcal{D}) = \int_\psi \int_\theta \sum_{x,z} P(a_i|z,x)P(z|\psi)P(x|\theta)P(\psi|\mathcal{D})P(\theta|\mathcal{D})\,\mathrm{d}\theta\mathrm{d}\psi \qquad (3)$$

Finally, in the $\mathcal{L}$A-LDA model, the probability user u votes for story i is:

$$P(a_i|\mathcal{D}) = \int_\psi \int_\phi \sum_{x,y,z} P(a_i|x,z)P(z|\psi)P(x,y|\phi)P(\psi|\mathcal{D})P(\phi|\mathcal{D})\,\mathrm{d}\phi\mathrm{d}\psi \qquad (4)$$

where the probability of a user's vote is decided by the distribution of the user's limited attention over friends and interests ϕ and story's topic profile ψ. We evaluate performance of the models on the prediction task using average precision. Average precision at N_{pos}^u for each user is $\sum_{k=1,n} Prec(k)/(N_{pos}^{user})$, where $Prec(k)$ is the precision at cut-off k in the list of votes ordered by their likelihood.

We divide users into categories based on their activity in the training set. The first category includes *all users* and the remaining categories include users who voted for at least 7.5%, 15%, and 25% of the stories in the training set. While $\mathcal{L}A$-LDA outperforms baseline methods in all cases, its comparative advantage improves with user activity. When there is little information about user interests, the precision of all methods is ranges from 1%–3%. As the amount of information about user interests, as expressed through the votes they make, grows, performance of all models improves, but that of $\mathcal{L}A$-LDA improves much faster. $\mathcal{L}A$-LDA correctly predicts more than 30% of the votes made by the most active users, as compared to 11% of the randomly guess.

Average Precision	2009 Data				2010 Data			
	All users	$\geq$7.5%	$\geq$15%	$\geq$25%	All users	$\geq$7.5%	$\geq$15%	$\geq$25%
random	0.0192	0.0477	0.0617	0.1092	0.0111	0.03619	0.0557	0.1054
LDA	0.0209	0.0440	0.0621	0.1107	0.0182	0.0415	0.0562	0.1117
ITM	0.0220	0.1100	0.1526	0.2693	0.0244	0.1363	0.1763	0.2370
$\mathcal{L}A$-LDA	0.0224	**0.1164**	**0.1677**	**0.3204**	0.0376	**0.1368**	**0.1881**	**0.3154**
Submitter	0.0379	0.0873	0.1138	0.1517	0.0283	0.0483	0.0746	0.1257
Max	**0.0789**	0.0964	0.1240	0.1707	**0.0702**	0.0733	0.1080	0.1616
ITM+Submitter	0.0241	0.0904	0.1311	0.1889	0.0381	0.0845	0.1121	0.1816
ITM+Max	0.0257	0.0977	0.1471	0.2365	0.0482	0.1243	0.1645	0.2436

One may ask whether a simple attention allocation heuristic could predict votes as well as $\mathcal{L}A$-LDA, but at a reduced computational cost. We answer this question by presenting results of four experiments studying the effect of the influence heuristic on the prediction task. In the first experiment, predicted votes for each user are sorted based the influence of the *submitter*, the first user to post the story on Digg. In the second experiment, they are sorted based on the influence of the most influential (*max*) voter. The third experiment investigates the effect of including either influence heuristic into the ITM model. In this case, the vote probability given by Eq. 3 is multiplied by relative influence (with respect to the most influential user in the network) of the *submitter* or *max* voter. When there is little information to learn user interests, using a simple heuristic that a user votes for a story if a very influential user recommended it, works well to predict votes, three to four times better than random guess. However, as $\mathcal{L}A$-LDA receives more data about user interests, it is able to learn a model that outperforms the simpler influence-based models.

5   Conclusion

Traditional topic models have been extended to a networked setting to model hyperlinks between documents [10], and the varying vocabularies and styles of different authors [13]. Collaborative filtering methods examine item recommendations made by many users to discover their preferences and recommend new items that were liked by similar users ([14],[2]) and improve the explanatory power of recommendations by extending LDA [16].

We introduced $\mathcal{L}A$-LDA, a novel hidden topic model that takes into account social media users' limited attention. Our work demonstrates the importance of modeling psychological factors, such as attention, in social media analysis. These results may apply beyond social media and point to the fundamental role that psychosocial and cognitive factors play in social communication. People do not have infinite time and patience to read all status updates or scientific articles on topics they are interested in, see all the movies or read all the books. Attention acts as an "information bottleneck," selecting a small fraction of available input for further processing. Since human attention is finite, the mechanisms that guide it become ever more important. Uncovering the factors that guide attention will be the focus of our future work.

References

1. Blei, D., Ng, A., Jordan, M.: Latent dirichlet allocation. The Journal of Machine Learning Research 3, 993–1022 (2003)
2. Chua, F.C.T., Lauw, H.W., Lim, E.-P.: Generative models for item adoptions using social correlation. In: TKDE (2012)
3. Counts, S., Fisher, K.: Taking it all in? visual attention in microblog consumption. In: ICWSM (2011)
4. Gilbert, E., Karahalios, K.: Predicting tie strength with social media. In: CHI (2009)
5. Goncalves, B., Perra, N., Vespignani, A.: Validation of Dunbar's number in Twitter conversations (2011), arXiv.org
6. Granovetter, M.S.: The Strength of Weak Ties. American Journal of Sociology 78(6), 1360–1380 (1973)
7. Hodas, N., Lerman, K.: How limited visibility and divided attention constrain social contagion. In: SocialCom (2012)
8. Kahneman, D.: Attention and effort. Prentice Hall (1973)
9. Lerman, K., Ghosh, R.: Information contagion: an empirical study of spread of news on digg and twitter social networks. In: ICWSM (2010)
10. Nallapati, R., Cohen, W.: Link-PLSA-LDA: A new unsupervised model for topics and influence of blogs. In: ICWSM (2008)
11. Plangprasopchok, A., Lerman, K.: Modeling social annotation: a bayesian approach. ACM Transactions on Knowledge Discovery from Data 5(1), 4 (2010)
12. Rensink, R., O'Regan, J., Clark, J.: To see or not to see: The need for attention to perceive changes in scenes. Psychological Science 8(5), 368 (1997)
13. Rosen-Zvi, M., Griffiths, T., Steyvers, M., Smyth, P.: The author-topic model for authors and documents. In: UAI (2004)
14. Sarwar, B., Karypis, G., Konstan, J., Riedl, J.: Itembased collaborative filtering recommendation algorithms. In: WWW (2001)
15. Sharara, H., Rand, W., Getoor, L.: Differential adaptive diffusion: Understanding diversity and learning whom to trust in viral marketing. In: ICWSM (2011)
16. Wang, C., Blei, D.M.: Collaborative topic modeling for recommending scientific articles. In: KDD (2011)
17. Weng, L., Flammini, A., Vespignani, A., Menczer, F.: Competition among memes in a world with limited attention. Scientific Reports 2 (2012)
18. Wu, F., Huberman, B.A.: Novelty and collective attention. Proc. the National Academy of Sciences 104(45), 17599–17601 (2007)

Graph Formation Effects on Social Welfare and Inequality in a Networked Resource Game

Zhuoshu Li[1,2], Yu-Han Chang[2], and Rajiv Maheswaran[2]

[1] Beihang University, Beijing, China
[2] Information Sciences Institute, University of Southern California, CA, USA
zslibuaa@gmail.com, {ychang,maheswar}@isi.edu

Abstract. We introduce the Networked Resource Game, a graphical game where players' actions are a set of resources that they can apply over links in a graph to form partnerships that yield rewards. This introduces a new constraint on actions over multiple links. We investigate several network formation algorithms and find bilateral coalition-proof equilibria for these games. We analyze the outcomes in terms of social welfare and inequality, as measured by the Gini coefficient, and show how graph formation affects these aspects of a networked economy.

1   Introduction

Graphical games that model social phenomena have been an emerging research area applied to group consensus making, networked bargaining and trading strategies. Here, we investigate the interactions of a society where actions are resource-bounded, i.e., agents have limits on how they are able to act across their network. We model the notion that people have a finite number of resources and their network affects how those resources can be coupled with others' resources in order to produce rewards. One example of this is in professional networks where agents need to form partnerships and the payoffs of the partnerships are a function of the capabilities that each bring to the table.

In this paper, we introduce the *Networked Resource Game*, and study how the structure of the network and its dynamics affect social welfare and inequality, measured by the Gini coefficient, of the resulting equilibria. For network formation, we utilize Erdos-Renyi [13] and preferential attachment [1] models and introduce several new algorithms as well. We introduce an algorithm to find bilateral coalition-proof equilibria as Nash equilibria do not lead to reasonable outcomes in this domain. In this context, we study how the various algorithms affect social welfare and inequality and the impact of network properties on performance.

2   Related Work

Graphical games [9] provide compact representation of multi-agent interaction when players' payoffs depend only on actions of agents in their neighborhood.

A.M. Greenberg, W.G. Kennedy, and N.D. Bos (Eds.): SBP 2013, LNCS 7812, pp. 221–230, 2013.

It is known that finding Nash equilibria for graphical games is difficult even for restricted structures [4]. Local heuristic techniques are commonly employed [7,3]. A seminal work in using agent-based simulation to study human interaction was Axelrod's tournament for Prisoner's Dilemma [2]. Prisoner's Dilemma has also been studied in a graphical setting with simulated agents [11]. Dynamic networked games based on the Ultimatum Game have also been investigated [10] Research on identification and development of networks includes analyzing event-driven growth [14] and inferring social situations by interaction geometry [6]. Other work has described algorithmic methods to discover temporal patterns in networked interaction data [8]. Researchers have formulated efficient solution methods for games with special structures, such as limited degree of interactions between players linked in a network, or limited influence of their action choices on overall payoffs for all players [12,15]. In terms of these work, our model takes the networked interaction into a completely different domain, as we focus on the influence of the structure and topology of the network, on the dynamics of resource allocation in the network.

3   Networked Resource Game Model

The Networked Resource Game is characterized by a set of N players $\{p_i\}_{i=1}^{N}$, a card distribution C, a graph G and a reward function R. Each player p_i has a set of cards $C_i = \{c_{i,1}, \ldots, c_{i,N_i^C}\}$ where N_i^C is the number of cards for that player. The cards represent a skill or resource that the player can play on a link. Each card has a type which comes from a predetermined type set T, i.e., $c_{i,j} \in T \ \forall i, j$. For simplicity, given a discrete type set, we can think of the type as a color and that each card has a color. The graph is a set of edges over N nodes, i.e., $G = \{e_{ij}\}$ where e_{ij} refers to a *link* between players p_i and p_j. It is possible that some players have no links associated with them.

The graph specifies the links over which players may play their cards. Here, we include the restriction that a player may play at most one card on a link. Thus,

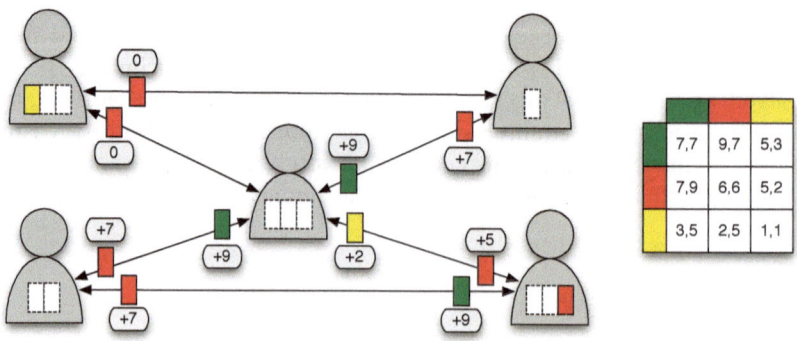

Fig. 1. The Networked Resource Game

the number of cards indicate a players ability to have multiple simultaneous partnerships. It is possible that a player has more links than cards and also more cards than links. Based on what cards are played on a link, each player gets a reward specified by the function $R(a, \bar{a})$ which is the reward to a player for performing action a on a link where the other player performed action $\bar{a}$. The action space for player p_i on link e_{ij} is $A_{ij} = C_i \cup 0$, where 0 indicates that the player chose not to play one of their cards on that link.

The reward function has $(|T| + 1)^2$ inputs representing every combination of actions, i.e., all card types and not playing a card, for each player. The Networked Resource Game is similar to a standard graphical game, however, the action space has restrictions over multiple links whereas in standard graphical games, actions on link are independent. Here, we have the restriction that $\cup_j a_{ij} \subset C_i$ where a_{ij} is player p_i's action on link e_{ij}. This states that a player cannot play more cards than they have, which introduces a coupling over links.

An illustration of the game is shown in Figure 1. It shows a game involving three card types (green, red and yellow). One can imagine that these cards represent assets of value in an economy that yield different outcomes to each contributor in partnerships. For example, green could represent capital, red could represent skilled labor and yellow could represent unskilled labor.

4   Network Formation and Finding Equilibria

Network Formation. Here, we describe the algorithms that we use to create our social network graphs and find equilibria for a given graph. Network formation is determined by various growth processes that describe how a link is added to an existing graph. We describe four such models:

- **Erdos-Renyi (ER)**: This is a baseline process where we add a link chosen uniformly from those links that do not already exist in the graph.
- **Preferential Attachment (PA)**: If the input graph has zero or one link, we use the ER process. Thus, the network is seeded with two random links. After this, to add a link, we choose a node randomly and consider the links it could add to the graph, i.e., the set of links connected to the chosen node that are not already in the graph. Each such link is given a weight equal to the degree of the target node it connects to, and a link is chosen in proportion to these weights. Preferential attachment models have been proposed as a model that reflects how social networks are formed, particularly online.
- **Most Free Cards (MFC)**: Each node is given an MFC score: the number of cards it has minus the number of links it has, i.e., a measure of the number of free cards for that player. The process selects a node uniformly from those that have the highest MFC score. This node then chooses a link uniformly from other nodes that have the highest MFC score. When the MFC scores are all zero, the algorithm becomes ER.

- **Poor-to-Rich Chain (PRC)**: We first associate each player with a wealth calculated as the sum of the value of their cards, where the value of each card is the maximum reward obtainable from applying that card:

$$w_i = \sum_{c \in C_i} \max_{a \in T \cup 0} R(c, a)$$

We first create a chain, where agents are ordered by wealth with ties broken randomly. Then, a player chosen uniformly from those with the highest MFC score adds a link. The target node is the closest node in the chain with a free card, i.e., an MFC score greater than zero. Again, ties are broken randomly. When all MFC scores are zero, the algorithm becomes ER.

The various processes described above capture various degrees of control that players may have over the network on which they play. In the ER and PA models, players have no control over links. One may consider PA as player driven, but the game properties (card, rewards) do not affect the formation of the links so the processes are not strategic. The MFC model is a decentralized strategic model where agents have partial information about the state of the world, namely the number of cards and links for each player. The PRC model is a centralized model that takes game parameters into account when making the graph and incorporates a social structure onto the world where people with similar wealth are more likely to be connected to each other.

Finding Equilibria. Given a game structure (cards, rewards, and a graph), we would like to determine an appropriate outcome. Nash equilibria are often considered as a solution concept for games and graphical games as well, however, it has issues in the context of the Networked Resource Game. Consider the simple example of four players in a fully connected graph where each player has one card. Two players have a single red card and two players have a single green card. Let the rewards for having two cards with same color on a link be 100 for each player, two cards with different colors on the same link be 10, and all links with one or zero links be worth nothing. Consider the situation where we have two red-green links. Each player receives 10 and has no incentive to deviate, i.e, move their one card to another link, because that would cause a loss of 10, even though each player has a link to a player with the same color card.

Thus, in the Networked Resource Game, Nash equilibria lead to artificially poorer results than one would expect if one was playing this game assuming players could communicate over the links that they have. Thus, we consider equilibria where players can make bilateral deviations. An equilibrium in this context is a state where no player would choose to make a unilateral deviation and no two players would choose to make a bilateral deviation. We use the procedure below to discover such equilibria for a given game structure.

Each player first assigns cards randomly to available links. We then perform action updates in a series of rounds. In each round, we order the set of links. For each link, the players iterate back and forth on card choices for the link. On the first iteration, the first player assumes that the other player plays one of their cards, chosen from all cards that player has, i.e., not necessarily the card being played on the link currently. The first player then plays their best response on all links given the cards that are played on all the links that they have. In the second iteration and all following iterations, the acting player chooses their best response to the cards that are being played on their link. This procedure continues until an equilibrium is reached for that link or we reach a preset limit of interactions. We continue this procedure for all links in each round. The procedure terminates, when at the end of a round, the joint actions are the same as the joint actions in the previous round. The procedure continues for a preset number of rounds. Finding equilibria in graphical games is a challenging problem. The algorithm presented is sound in that if it terminates before reaching the preset number of rounds, we know that the resulting joint action is an equilibrium for the game, however, we may not find all equilibria.

Abstract algorithm FINDING-EQUILIBRIA for computing bilateral coalition-proof equilibria

Algorithm **FINDING-EQUILIBRIA**
Inputs: one game structure(cards,rewards and a graph)
Outputs: bilateral coalition-proof equilibria
  Each player first assigns cards randomly to available links
  equilibria ← 0
  for *round* ← 1 to n1
    do order the set of links
    num ← 1
    repeat
      (1)one player P_i assumes that the other player P_j
        which links with P_i plays one of its cards
      (2)P_i plays their best response on all links given the
        cards that are played on all the links that they have
      (3)P_j chooses their best response to the cards
        that are being played on their link
      (4)*num* ← *num* + 1
    until an equilibrium is reached for that link
        or num = *n*2
    if the joint actions = joint actions in the previous round
    then return equilibria
    else return -1

5   Experiments

We considered societies of 12 players. In each scenario, each player was given a number of cards chosen uniformly from one to five: $|C_i| \sim U(1,5)$. We had three card types: green, red, and yellow. Card colors were selected independently for each card using the following probabilities: $P([\text{green red yellow}]) = [0.20\ 0.40\ 0.40]$. There were two methods for selecting reward functions. In the baseline method, each reward for links with two cards on them were chosen randomly: $R(c_1, c_2) \sim U(1, 1000)$ for $c_1, c_2 \in T$. Links with one or zero cards gave zero reward to both players. In the alternate method, the reward for a green-green link is replaced with 100 times the value of the maximum reward of all the rewards in the baseline method. The latter is to investigate a society where there is a significantly outlying reward available to a small number of people if they make the right connections. It is for this reason that the green cards occur at lower likelihood than the others. For a given game card and reward structure, we would run our various network formation algorithms and generate graphs of increasing size. Each network formation algorithm was run 10 times, thus generating 10 graphs with the same number of edges for each process. For each game structure (cards, rewards and graph) that resulted, we would find the set of equilibria. For each graph, the equilibrium-finding algorithm was run 40 times and each run was ended if the algorithm didn't terminate in 15 rounds.

For any single equilibrium, we calculated the *social welfare* as the sum of all the rewards to all players and the *Gini coefficient*, a measure of income disparity [5,16]. The Gini coefficient measures the gap in the cumulative distribution function (CDF) of total share of wealth as a function of percentile income between a uniformly wealthy society which would have a linear CDF and the CDF of the society being investigated. Larger Gini coefficients indicate greater income disparity. For each game structure, we calculated an associated social welfare with the weighted average of social welfares of equilibria of that game structure, where weights were the number of times the equilibrium was discovered. We calculated associated Gini coefficients for each game structure similarly.

The Gini coefficient is normalized between zero (everyone has equal wealth) and one (one person has all the wealth), but social welfare for each game is a function of the reward matrix. We first solve the following integer program:

$$\max \sum_{(c_1, c_2) \in C_2} n_{c_1, c_2} \left(R(c_1, c_2) + R(c_2, c_1) \right)$$

$$\text{such that } \sum_{\tilde{c} \in T} n_{c, \tilde{c}} \leq n_c \quad \forall c \in T, \quad n_{c, \tilde{c}} \geq 0, \quad \forall c, \tilde{c}$$

This considers all possible combinations of cards on a link $(c_1, c_2) \in C_2$ and maximizes the reward obtained for having a particular number of card combinations on the graph (n_{c_1, c_2}) with the rewards obtained for that card combination $(R(c_1, c_2) + R(c_2, c_1))$, such that the number of card combinations of the graph does not violate the card constraints, i.e., the number of cards of a particular type (n_c) and non-negativity of the number of combinations. This yields an upper bound on the social welfare because it allows multiple links between players and links between cards of the same player. We use this to normalize social welfares across different card and reward structures.

6   Results

Figure 2 shows how social welfare changes as a function of network formation algorithm and graph size. We did not show the error bars for clarity in presentation but we discuss significance below. We see that social welfare improves as the society gets more connected for all algorithms. MFC and PRC are significantly better than ER and PA. ER is slightly better than PA but the result is not statistically significant. These results hold in both reward scenarios. For baseline rewards, MFC and PRC both reach about 0.9 efficiency in social welfare at about 18 links and do not improve much beyond that. We also see the impact of network structure as the 28-link ER and PA graphs are less efficient that MFC and PRC graphs that are half the size. We noticed that ER and PA graphs are not easy to reach equilibria when the graph is larger than 30 links. For alternate rewards, the efficiency is significantly smaller than the baseline word, this could be the result of two factors: there are green-green links that are not being formed, and our normalization could be overcounting the number of potential green-green links.

Figure 3(a) shows how Gini coefficients change as a function of network formation algorithm and graph size. Inequality decreases as the network sizes increase. For the baseline reward structure, MFC, PRC and ER are significantly better than PA. The key change is that ER has jumped from the PA equivalence class to the MFC/PRC equivalence class. We note that the Gini coefficient is relatively flat after about 18 links. For the alternate reward structure, all the algorithms are in the same equivalence class. This is because once a few green-green links are formed, it is difficult to change the inequality of the world.

We then investigated the number of wasted cards in equilibrium, i.e., the number of cards that did not yield any reward to the player holding it. Figure 3(b) shows the number of wasted cards as a percentage of the total number of cards in a society. We see that wasted cards explains a lot of the phenomena in social welfare. The MFC and PRC algorithm, which has an MFC component, waste the fewest cards because that is part of their process. The others form links that are not as useful in allowing players to use their cards. ER performs slightly better that PA because it does not overload particular users with large numbers of links. Thus as fewer cards are wasted, social welfare improves. This similarly explains the Gini coefficient because as more cards are used, we have fewer users with low or no rewards. Nevertheless, it is interesting to note that while ER

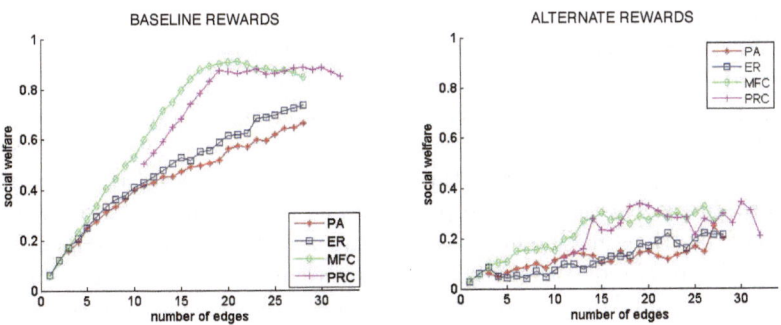

Fig. 2. Social Welfare by Algorithm and Graph Size

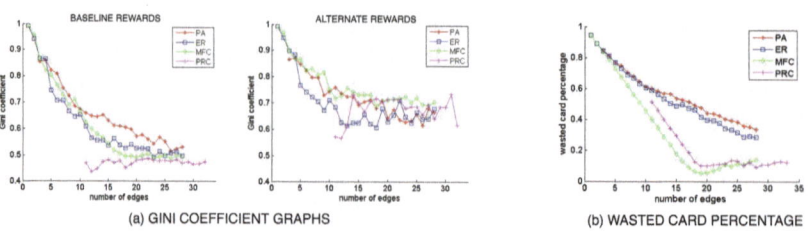

Fig. 3. (a) Gini Coefficient and (b) Wasted Card Percentage by Algorithm, Graph Size

wastes more cards than MFC and PRC, it does not perform worse in terms of inequality. This remains an open question. Interestingly, with half the possible links (33), we still have about 10% of cards being wasted.

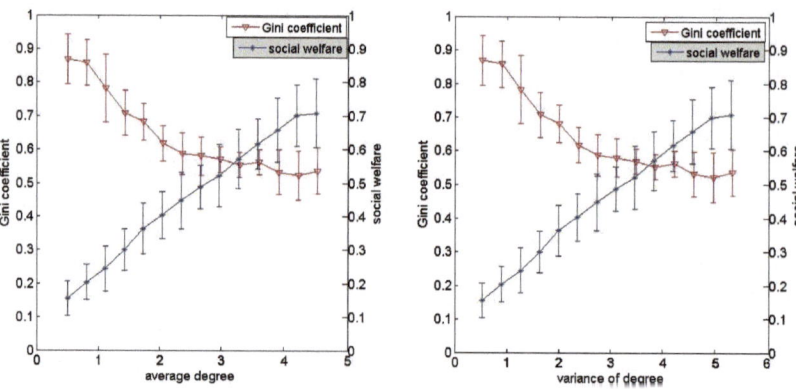

Fig. 4. Social Welfare and Gini Coefficient by Average and Variance of Degree for Baseline Rewards

We also looked at the impact of network properties on outcomes. Figure 5 shows social welfare and Gini coefficient as a function of the average and variance of the degrees of the nodes in the graph. Clearly, this will depend on the card and reward structure. In our case, both average and variance of degree showed similar curves in increasing social welfare and decreasing inequality. The inequality curves are similar in both reward structures and the social welfare curves are close to the best performing algorithms as a function of graph size. One potential future direction is using these properties as part of the network formation process because they may be more easily estimated than the requirements of the processes we presented. We also plan on investigating games where more than two players can collaborate. It is also a challenge to investigate appropriate outcomes for graphs as the scale of the society grows as equilibrium discovery will become more computationally demanding. We believe the Networked Resource

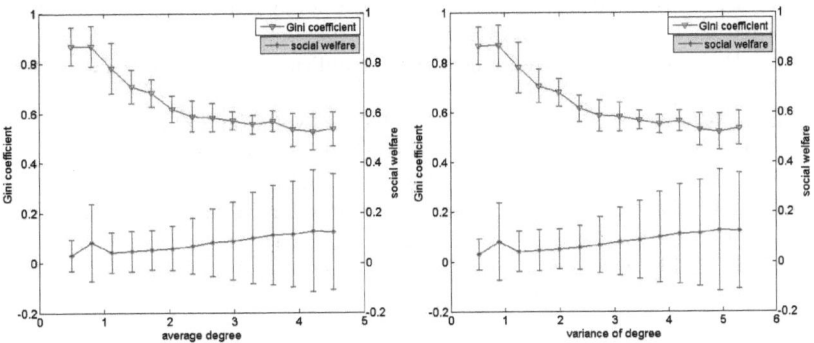

Fig. 5. Social Welfare and Gini Coefficient by Average and Variance of Degree for Alternate Rewards

Game is a good starting point for modeling and investigating the complexities and design of economies of resource-bounded and socially networked agents.

References

1. Albert, R., Barabasi, A.-L.: Statistical mechanics of complex networks. Reviews of Modern Physics 74, 47–97 (2002)
2. Axelrod, R., Hamilton, W.D.: The evolution of cooperation. Science 211, 1390–1396 (1981)
3. Duong, Q., Vorobeychik, Y., Singh, S., Wellman, M.P.: Learning graphical game models. In: IJCAI (2011)
4. Elkind, E., Goldberg, L., Goldberg, P.: Nash equilibria in graphical games on trees revisited. In: Proceedings of the 7th ACM Conference on Electronic Commerce, pp. 100–109. ACM (2006)
5. Gini, C.: On the measure of concentration with special reference to income and statistics. Colorado College Publication (1936)
6. Groh, G., Lehmann, A., Reimers, J., Friess, M., Schwarz, L.: Detecting social situations from interaction geometry. In: 2010 IEEE Second International Conference on Social Computing (SocialCom), pp. 1–8 (August 2010)
7. Heckerman, D., Geiger, D., Chickering, D.M.: Learning bayesian networks: The combination of knowledge and statistical data. In: Machine Learning, pp. 197–243 (1995)
8. Hsieh, H.-P., Li, C.-T.: Mining temporal subgraph patterns in heterogeneous information networks. In: 2010 IEEE Second International Conference on Social Computing (SocialCom), pp. 282–287 (August 2010)
9. Kearns, M., Littman, M., Singh, S.: Graphical models for game theory. In: Conference on Uncertainty in Artificial Intelligence, pp. 253–260 (2001)
10. Kim, E., Chi, L., Maheswaran, R., Chang, Y.-H.: Dynamics of behavior in a network game. In: IEEE International Conference on Social Computation (2011)
11. Luo, L., Chakraborty, N., Sycara, K.: Prisoner's dilemma in graphs with heterogeneous agents. In: 2010 IEEE Second International Conference on Social Computing (SocialCom), pp. 145–152 (August 2010)

12. Ortiz, L., Kearns, M.: Nash propagation for loopy graphical games. In: Neural Information Processing Systems (2003)
13. Erdös, P., Rényi, A.: On random graphs. i. Publicationes Mathematicae 6, 290–297 (1959)
14. Qiu, B., Ivanova, K., Yen, J., Liu, P.: Behavior evolution and event-driven growth dynamics in social networks. In: 2010 IEEE Second International Conference on Social Computing (SocialCom), pp. 217–224 (August 2010)
15. Vickrey, D., Koller, D.: Multi-agent algorithms for solving graphical games. In: National Conference on Artificial Intelligence, AAAI (2002)
16. Yitzhaki, S.: More than a dozen alternative ways of spelling gini economic inequality. Economic Inequality (1998)

Recommendation in Reciprocal and Bipartite Social Networks–A Case Study of Online Dating

Mo Yu[1], Kang Zhao[2], John Yen[1], and Derek Kreager[1]

[1] The Pennsylvania State University,
University Park, PA 16802, USA
[2] Tippie College of Business, The University of Iowa,
Iowa City, IA 52242, USA
{muy145,jyen}@ist.psu.edu, kang-zhao@uiowa.edu, dkreager@psu.edu

Abstract. Many social networks in our daily life are bipartite networks that are built on reciprocity. How can we recommend users/friends to a user, so that the user is interested in and attractive to recommended users? In this research, we propose a new collaborative filtering model to improve user recommendations in reciprocal and bipartite social networks. The model considers a user's "taste" in picking others and "attractiveness" in being picked by others. A case study of an online dating network shows that the new model outperforms a baseline collaborative filtering model on recommending both initial contacts and reciprocal contacts.

Keywords: bipartite social network, reciprocity, online dating, user recommendation.

1 Introduction

Nowadays, online social networking and social media websites attract billions of people [1]. These services not only change how people search and spread information, but also have great impact on the way people get to know and interact with each other. Along with their popularity and importance, the business values of these websites are also well recognized. To improve users online experience and boost their levels of activities or engagement, most of these websites deploy recommender systems to help users find friends or other users of interest.

In the literature, many approaches for friend/user recommendation in social networks utilize network structures. One popular idea along this direction is that the more common neighbors two users share, the more likely the two users will be connected [2]. In addition, other approaches based on network structural features, such as the social balance in triad [3] and the multi-relational perspective of social networks [4], have also been used for link prediction, including networks with reciprocity [5]. However, these approaches do not directly apply to bipartite social networks, in which there are two types of nodes and an edge only exists between two nodes of different types. For example, in a heterosexual dating network, only a male user can be connected to a female user. No matter how many

A.M. Greenberg, W.G. Kennedy, and N.D. Bos (Eds.): SBP 2013, LNCS 7812, pp. 231–239, 2013.
© Springer-Verlag Berlin Heidelberg 2013

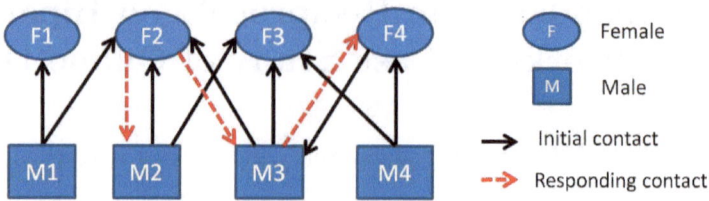

Fig. 1. An example bipartite dating network

female common neighbors two male users share, there will be no edge between the two male users.

Collaborative filtering (CF) has been used for recommendation in bipartite networks that include both people and items [6], such as the consumer-product network in Amazon.com. Recommendations are based on the similarity between people or items. However, links in a consumer-product network are built in a parasocial or unilateral way, because products can only be passively chosen by people in sale transactions. By contrast, social networks consist of autonomous people. Many social networks (e.g., Facebook and LinkedIn) are based on reciprocity: a link between two users can only be established if both users agree to be connected to each other. People in these reciprocal networks actively make decisions to initiate or accept connections based on their own preferences. Little research has tackled the problem of recommendation in bipartite reciprocal social networks. [7] recognized the importance of reciprocity in online dating recommendation and used data mining techniques for dating link prediction.

In this research, we propose a simple yet effective CF model to improve user recommendation in reciprocal bipartite social networks. We evaluate the new approach using metrics that reflect the effectiveness of recommending both unilateral and bilateral connections.

The social network in an online dating website is used as a case study. In addition to being a typical bipartite social network with strong reciprocity, online dating is chosen for two other reasons: First, online dating websites are very popular. In the U.S., 37% of all single Internet users looking for a partner have visited an online dating website [8]. In 2011, it was estimated that the online dating business is worth more than 2 billion British pounds [9]. Second, user/partner recommendation is especially important in online dating. While Facebook or LinkedIn users can directly search the names of their acquaintance, users of online dating websites usually do not know many people in the social network and rely more on recommender systems to find potential partners. Thus by improving user recommendation, an online dating website could potentially attract more users, improve their satisfaction, and generate more revenues.

2 The Proposed Approach

In a consumer-product network, products picked by a consumer collectively reflect the consumer's preference. Similarly, looking at whom a user approaches through initial contacts in a dating network, we can infer the user's personal taste in dating partners (i.e., what type of partners she/he is interested in). In addition to taste, another important factor in providing recommendations for dating is attractiveness, which measures a user's characteristics that make her/him attractive to others. For example, Mike (a male user) can certainly approach all the female users that match his taste, but sending out these initial contacts do not necessarily mean Mike will be happy at the end of the day. Mike would like to approach female users that he likes, and get responses from these females. It is Mike's taste that affects whom he may approach through initial contacts, and his attractiveness that determines whether he can get responses or not.

Taste and attractiveness are like two sides of the same coin in a dating relationship. For example, if Mike (male) and Mary (female) are in a relationship, it means Mary's attractiveness matches Mike's taste and vice versa. In other words, reciprocity in a dating network means that there is a match of both taste and attractiveness between two users. Leveraging reciprocity in online dating, our research tries to improve dating partner recommendations by boosting a user's chance of getting responses.

2.1   The Dataset

The dataset used in this study comes from a popular online dating website. It includes profiles and dating activities of 15,131 users in a U.S. city during a time span of 196 days. We can represent user activities in the dataset as a bipartite dating network, in which each user is represented as a node. Because the dataset only contains information of heterosexual dating, an edge in the dating network always connects a male and a female.

In this website, if user A is interested in user B, she/he could approach user B by sending B a message. We refer to this as an initial contact from A to B. If user B is also interested in A, she/he could respond by sending a responding message back to A, which constitutes a reciprocal contact between the two users. Figure 1 shows an example bipartite contact network among female and male users. The dataset contains more than 180,000 initial contacts, 70% of which are initiated by male users. Among all initial contacts, 24% of them eventually became reciprocal contacts.

2.2   The Baseline CF Model

Before introducing our approaches, we briefly describe the baseline CF model. The classic model has three components. The first component represents users' dating activities in an $N \times M$ contact matrix C. Among all users U, we define users we would like to provide recommendation to as service users S, where $S \subseteq U$. $N = |S|$ is the number of service users, and $M = |U|$ is the total number of users ($N \leq M$). The model separates service users from all users because this approach, as many CF approaches, relies on historical activities and works best for users with enough online activities. In the binary matrix C, $c_{i,j} = 1$ if user

	F1	F2	F3	F4
M1	1	1	0	0
M2	0	1	1	0
M3	0	1	1	0
M4	0	0	1	1

(a) Baseline CF model

	F1	F2	F3	F4
M1	0	0	0	0
M2	0	1	0	0
M3	0	1	0	1
M4	0	0	0	0

(b) Reciprocity-only model

	F1	F2	F3	F4
M1	<1,0>	<1,0>	<0,0>	<0,0>
M2	<0,0>	<1,1>	<1,0>	<0,0>
M3	<0,0>	<1,1>	<1,0>	<1,1>
M4	<0,0>	<0,0>	<1,0>	<1,0>

(c) The hybird model

Fig. 2. Contact matrices for the dating network in Figure 1. Only consider male users as service users.

i approached user j through initial contacts. It does not matter whether user j responded to user i or not. If user i did not sent any initial contacts to j, then $c_{i,j} = 0$. Thus, a row in the matrix represents a service user's all activities in initiating contacts and could reflect his/her taste. Considering only male users as service users, Figure 2a shows the contact matrix for the dating network in Figure 1.

The second component quantifies the similarity $s_{p,q}$ between service users p and q. In other words, the similarity between two row vectors p and q in C are measured. In this study, we use cosine similarity as the measure. High similarity between two row vectors in the matrix indicates that the two corresponding service users have similar tastes (i.e., approached similar users). Users of opposite genders will have a similarity score of zero because they do not approach any common users and their row vectors cannot have value 1 at the same column.

The third component finds potential dating partners based on user similarity scores and the contact matrix. To provide recommendations for service user p, the model iterates through each target user t ($t \in U$ and $t \neq p$), with whom user p has not interacted with, and calculates a success score between p and t as $E_{p,t} = \sum_{k=1}^{M} s_{p,k} \times c_{k,t}$. The higher the $E_{p,t}$ score is, the more likely user t will be recommended to user p. The model basically says that if user t is approached by users whose tastes are similar to those of user p, then user t could be a potential partner for user p.

2.3   The Reciprocity-Only Model

This model extends the baseline CF model simply by considering only reciprocal contacts, which reflect both parties' taste (or attractiveness). In its binary matrix C, $c_{i,j} = 1$ if and only if there is a reciprocal contact between user i and user j (it does not matter who initiates the contact); $c_{i,j} = 0$ otherwise, even if there is an initial contact between i and j. This matrix can never have more 1s than the matrix of the baseline CF model. As a result, a row in the contact matrix could reflect the corresponding service user's taste and attractiveness. An example of the contact matrix is shown in Figure 2b. Similarity between service users is calculated in the same way as the baseline CF model. High similarity between two rows in the matrix means that the two corresponding service users have similar tastes (i.e., approached similar users) and attractiveness (i.e., responded by similar users). The success score for recommending user t to user i is still calculated in the same way as the baseline CF model. According to this model, if user t is interested in and attractive to users whose tastes and attractiveness are similar to user p, then user t could be recommended to user p.

2.4   The Hybrid Model

While the reciprocity-only model captures users' tastes as well as attractiveness, it has 2 limitations: (1) It loses information on users' tastes and attractiveness when an initial contact is not responded. For the example in Figure 1, M1 will have an empty row vector and we cannot track his taste. (2) It fails to leverage users' preference reflected by un-responded contacts. For instance, when F2 chose not to respond to M1's initial contact, it suggests M1's attractiveness does not match F2's taste. Then for users whose attractiveness is similar to M1, F2 may not be a good candidate. The 2 limitations also have implications for calculating the similarity between service users. For example, M3 and M4 have common taste and attractiveness (albeit in a discouraging way) because both approached F3 yet both failed to get responses. Such similarity between M3 and M4 is not captured by the reciprocity-only model.

Thus we propose a hybrid model that considers both initial and responding contacts. While the $N \times M$ contact matrix C is essentially a 3 dimensional matrix, for the purpose of simplicity, we still denote it with a 2 dimensional one, whose elements are vectors. Formally, $c_{i,j}$ is a 1×2 vector, i.e., $c_{i,j} = < c_{i,j,1}, c_{i,j,2} >$. $c_{i,j,1} = 1$ if user i has sent a message (initial or responding) to user j (meaning i's taste matches j's attractiveness); $c_{i,j,1} = 0$ if user i is not interested in j. Similarly, $c_{i,j,2} = 1$ if user j has approached user i (meaning i's attractiveness matches j's taste); $c_{i,j,2} = 0$ if user j is not interested in i. Similar to the reciprocity-only model, we do not differentiate between initial and responding messages and it does not matter who initiates the contact. An example contact matrix for this hybrid model is shown in Figure 2c. Note that in the figure, if user M3 does not respond to F4's message, then $c_{3,4} = < 0, 1 >$.

Then how do we calculate the similarity between two service users (i.e., two row vectors in the matrix)? We would like to measure such similarity based on three factors: (1) two users are interested in similar users; (2) two users attract similar users; and (3) two users reject or are rejected by similar users. Thus we denote the similarity between service users p and q as $s_{p,q} = \sum_{k=1}^{M} f(c_{p,k}, c_{q,k})$. It is basically the aggregation of users p's and q's interaction with all other users. Considering similarities in both taste and attractiveness, we would like the function $f(c_{p,k}, c_{q,k})$ to meet the following criteria: $f(< 1,1 >, < 1,1 >) = f(< 1,0 >, < 1,0 >) = f(< 0,1 >, < 0,1 >) > f(< 1,1 >, < 1,0 >) = f(< 1,1 >, < 0,1 >) > f(\text{otherwise})$. In other words, users who share similar taste and attractiveness should have higher similarity scores than those with only similar taste (or attractiveness). Users whose taste and attractiveness are both different will have the lowest similarity score. Thus we pick $f(c_{p,k}, c_{q,k})$ as defined in Equation 1, where $\oplus$ represents exclusive OR operation. $dgr(i)$ is the degree centrality of user i in the undirected and unweighted dating network. Dividing by $dgr(p) + dgr(q)$ normalizes the similarity function.

$$f(< c_{p,k,1}, c_{p,k,2} >, < c_{q,k,1}, c_{q,k,2} >) =$$
$$\frac{[(\overline{c_{p,k,1} \oplus c_{q,k,1}}) + (\overline{c_{p,k,2} \oplus c_{q,k,2}})] * [(c_{p,k,1} + c_{p,k,2}) \cap (c_{q,k,1} + c_{q,k,2})]}{dgr(p) + dgr(q)} \tag{1}$$

To rank potential dating partners, we add a penalty factor when calculating the success score of recommending user t to p using $E_{p,t} = \sum_{k=1}^{M} s_{p,k} \times g(c_{k,t})$. The introduction of function $g(c_{k,t})$ (defined in Equation 2) is meant to reward a match of both taste and attractiveness while penalizing a partial match of either taste or attractiveness. In other words, potential partners who match the taste and attractiveness of a user similar to a service user will get high success scores. This is because these potential partners may be interested in and draw interests from the service user. Meanwhile, potential partners who match only the taste or attractiveness of users similar to a service user will get lower scores (by s) for potentially unilateral interests between them and the service user.

$$g(c_{k,t}) = \begin{cases} 1, & \text{if } c_{k,t} = < 1, 1 >; \\ 1 - s, (0 < s < 1) & \text{if } c_{k,t} = < 1, 0 > \text{ or } < 0, 1 >; \\ 0, & \text{otherwise.} \end{cases} \tag{2}$$

In sum, the hybrid model extends the baseline CF model in two ways. First, it considers both taste and attractiveness when calculating the similarity between service users. Users with similar taste and attractiveness will have higher similarity scores than those who only share common taste or attractiveness. Second, it considers the match of both taste and attractiveness when recommending dating partners. Those who match both a service user's taste and attractiveness are more likely to be recommended than those who may only ignite unilateral interests.

3   Evaluation and Discussions

We use two sets of metrics to evaluate the performance of our model. The first set is based on initial contacts (IC): IC Precision@K measures how many of the recommended K users were approached by the service user); IC Recall@K evaluates among all users approached by the service user, how many can be ranked within top K. The second set of metrics focuses on reciprocity–whether an initial contact is responded. Reciprocal-contact (RC) Precision@K evaluates how many of the recommended K users become a service user's reciprocal contacts; RC Recall@K measures among all users that responded to the service user's initial contact, how many can be ranked within top K by the recommender.

For the experiment, user activities from day 1 to day 100 are for training and those from day 101 to day 196 are for testing. We pick users who have sent 5 or more messages in both training and testing periods as service users in this case study. This step reduces N to 1,597. Figure 3 compares performance of the

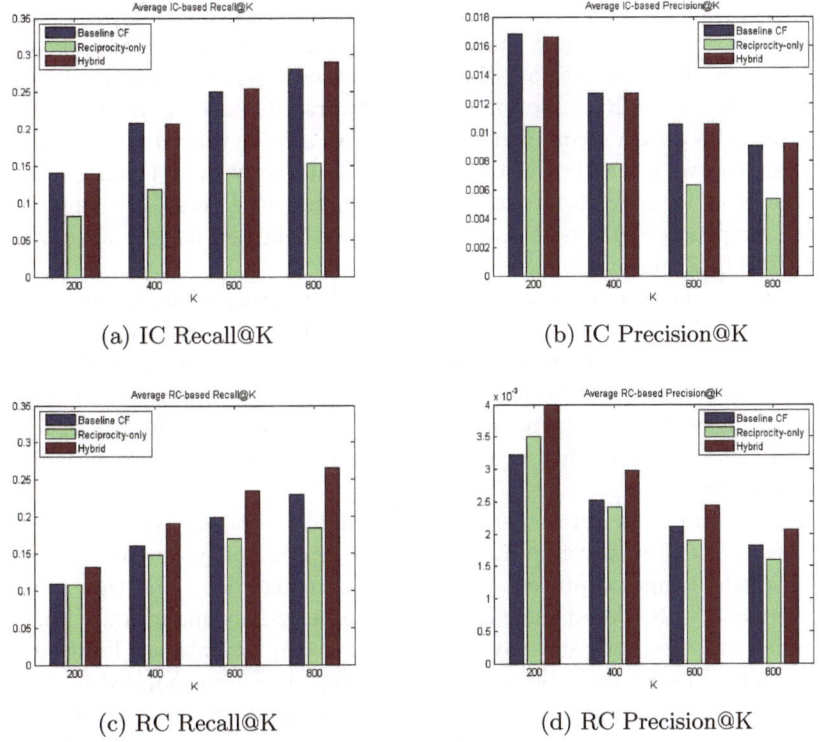

(a) IC Recall@K (b) IC Precision@K

(c) RC Recall@K (d) RC Precision@K

Fig. 3. Compare the performance of 3 models (the hybrid model with $s = 0.6$)

3 models on the four metrics. Overall, the hybrid model (with $s = 0.6$) has the dominant performance and the reciprocity-only model is the worst.

The performance of the reciprocity-only model on IC-based metrics is not surprising, because the model does not take advantage of all the information about users' tastes. However, it also fails to improve on RC-based metrics, as we expect the model can increase the chance that a service user gets responses from a potential partner. The performance of the reciprocity-only model illustrates that only considering reciprocal contacts is not enough for providing effective partner recommendations. A user's taste, even when it is unilateral, can still be valuable for recommending partners, no matter whether the goal is to recommend unilateral or reciprocal contacts.

The hybrid model with $s = 0.6$ dominates the baseline CF model, as well as the reciprocity-only model. On one hand, it outperforms the baseline on RC-based metrics: the hybrid model offers average improvements of 17.41% on RC precisions and 18.32% on RC recalls. The improvement is more obvious when K value is low, which is desirable as a recommender usually cannot make too many recommendations. On the other hand, the model's performance on IC-based metrics is as good as that of the baseline: the improvements on IC precisions and recalls are 0.26% and 1.03% respectively.

For the hybrid model, we try different values for the penalty factor s, ranging from 0.1 to 0.9. When the s value is low, the model gets slightly better IC-based metrics, albeit at the cost of lower RC-based metrics. Similarly, increasing s will lead to better RC-based metrics and worse IC-based performance. This makes sense because a recommender with higher s value will tend to recommend more partners with matches in both taste and attractiveness, leading to higher possibilities of forming reciprocal connections, and vice versa. For this specific dataset, we pick $s = 0.6$ as it provides the most balanced performance on both IC and RC-based metrics.

The hybrid model aims at recommending dating partners who are attractive to and interested in a service user. The results show that the model can accomplish this job. More specifically, if a service user approaches a partner recommended by the hybrid model, he/she will have a better chance of getting responses. In addition, this performance is achieved without compromising the model's ability to recommend partners that a user is unilaterally interested in (measured by IC-based metrics).

The hybrid model's better performance should be attributed to its consideration of one's taste as well as attractiveness. Based solely on taste, the baseline CF model cannot capture a service user's attractiveness and thus have bad performance on RC-based metrics. The reciprocity-only model only considers reciprocal contacts between two users but has two limitations: First, it loses information of user's taste because a large number of unresponded initial contacts are ignored. Second, although it does explicitly utilize one's attractiveness, it only captures one's partial attractiveness. In other words, a users attractiveness is reflected only when someone she/he approached is attracted by her/him. However, an initial contact that does not get any response also carries information of a users attractiveness (or strictly speaking, non-attractiveness to certain users). The new hybrid model does not waste such information on a users taste and attractiveness. Thus it could dominate the baseline and the reciprocity-only models in recommending potential partners.

4   Conclusions and Future Work

In this research, we propose a new CF model to improve user recommendation in bipartite social networks, where the formation of links depends on reciprocity. This model leverages users' historical activities in approaching others and getting responses. Different from traditional CF approaches, the new model focuses on users' taste and attractiveness in establishing bilateral connections. The model proposes new ways to calculate user similarity and to rank potential connections based on the match of both taste and attractiveness.

Using an online dating network as a case study, we illustrate that the new model can outperform the baseline CF model on recommending both unilateral and reciprocal contacts. In other words, the new model can recommend partners that matches a user's taste and attractiveness. The implications of this research are not limited to online dating only. It can also be incorporated into existing recommenders for other reciprocal bipartite networks, such as college admission network (high school students and universities as nodes), job-hunting network (applicants and employers as nodes), and so on.

There are also many possible directions for future research. We would like to further validate the effectiveness and robustness of the hybrid model through sensitivity analysis (e.g., change the pool of users, the testing/training periods, etc). It is also interesting to compare our approach to other model-based CF techniques. A unique challenge for recommending partners in a dating network is to find whether a user is still actively seeking partners. Once a user finds a good match and starts an offline relationship, she/he may not be active any more. Recommending such a user to others will mostly likely be unsuccessful. Thus we are interested in identifying inactive users and filtering them out in the recommendation.

Acknowledgements. This work was partially supported by the Pennsylvania State University and the National Science Foundation under an IGERT award # DEG-1144860, Big Data Social Science.

References

1. Zhao, K., Kumar, A.: Who blogs what: understanding the publishing behavior of bloggers. World Wide Web Online First, 1–24 (2012)
2. Newman, M.: Clustering and preferential attachment in growing networks. Physical Review E 64, 025102 (2001)
3. Leskovec, J., Huttenlocher, D., Kleinberg, J.: Predicting positive and negative links in online social networks. ACM, 641–650 (2010)
4. Zhao, K., Yen, J., Ngamassi, L.M., Maitland, C., Tapia, A.: Simulating inter-organizational collaboration network: a multi-relational and event-based approach. Simulation 88, 617–631 (2012)
5. Hopcroft, J., Lou, T., Tang, J.: Who will follow you back?: reciprocal relationship prediction, 2063740. ACM, 1137–1146 (2011)
6. Huang, Z., Zeng, D.D.: Why does collaborative filtering work? transaction-based recommendation model validation and selection by analyzing bipartite random graphs. Informs Journal on Computing 23, 138–152 (2011)
7. Cai, X., Bain, M., Krzywicki, A., Wobcke, W., Kim, Y.S., Compton, P., Mahidadia, A.: Reciprocal and Heterogeneous Link Prediction in Social Networks. In: Tan, P.-N., Chawla, S., Ho, C.K., Bailey, J. (eds.) PAKDD 2012, Part II. LNCS, vol. 7302, pp. 193–204. Springer, Heidelberg (2012)
8. Madden, M., Lenhart, A.: Online dating. Technical report, Pew Research Center (2006),
   http://www.pewinternet.org/~/media//
   Files/Reports/2006/PIP_Online_Dating.pdf.pdf
9. Abbott, M.: Internet dating defies economic gloom. BBC (December 2011)

In You We Follow: Determining the Group Leader in Dialogue

David B. Bracewell and Marc T. Tomlinson

Language Computer Corporation
Richardson, TX
{david,marc}@languagecomputer.com

Abstract. In this paper, we investigate whether the social roles of dialogue participants can be recognized through the social actions performed by the participant in their interactions with others in the group. Specifically we focus on determining if a participant is the leader of the group. We decompose the problem into identifying the social goals for participant discourse segments. These social goals are represented through a set of eleven psychologically-motivated social acts. We then model leadership using a sociological-inspired model called social rank which takes into account the social capital accumulated by the participant over the course of a single dialogue. We explore these models in task-oriented dialogues communicated in English, Arabic, and Chinese and show that the incorporation of social rank can improve precision of detecting the leader by 14% in English, 8% in Arabic, and 4% in Chinese.

1   Introduction

Leaders in task-oriented groups provide guidance, push for success, and facilitate discussion. The group affords a leader power granting the leader influence and status within and over the group. Identification of group leaders and leadership qualities is an important first-step in understanding the power structure of a group.

Much attention has been given to pure network-based approaches (e.g. PageRank [1]) to determining the power and influence an individual wields within an online community [2, 3]. Recently, work in natural language processing has begun addressing the question of power in dialogue through analysis of the messages exchanged between individuals [4–7].

In this paper we examine a method for determining who the leader of a group is using the interactions contained within a single dialogue. We determine the leader through examination of the social actions performed by the dialogue participants as evidenced through the linguistic expressions they employ. Using these social actions, we rank participants using a metric we call *Social Rank* which is roughly based on the sociological theories of social capital and group stratification. We define *Social Rank* as the honor and prestige afforded to an individual by others in the group which increases their status and power within and over the group. We explore how *Social Rank* can improve over baseline and

A.M. Greenberg, W.G. Kennedy, and N.D. Bos (Eds.): SBP 2013, LNCS 7812, pp. 240–248, 2013.
© Springer-Verlag Berlin Heidelberg 2013

machine learning approaches for determining the leader in dialogues communicated in English, Arabic, and Chinese.

2   Sociological Roots of Social Capital and Social Rank

The social rank of an individual is the honor and prestige afforded them by others in the group. We base our idea of social rank on the theories of group stratification formulated by Max Weber [8]. Weber's Three-component theory of stratification examines how wealth, power, and prestige work together in determining the social power of an individual [8]. Weber defines wealth as the economic resources available to an individual. Prestige is defined as the respect a person is afforded by their status or position. The final aspect of Weber's theory is power. Power can come via a variety of ways. Realizations of power come in two main types, formal power (power given to an individual by an authority) and informal power (power based on an individual's characteristics, e.g. expertise, skills; [9]). In the case of identifying the leader, we are looking for power over the task and discussion of the group.

One way in which social wealth and prestige can be determined is through an individual's social capital. Social capital originates from the field of Sociology. A variety of definitions for social capital can be found in literature. The definition of particular interest to task-oriented discourse and leadership revolves around the role that cooperation and trust play in the collective results of the group. We base our definition of social capital on Coleman [10], who describes social capital as an employable resource of individuals. He states that social capital is gained through the changing relations of individuals and in particular those changes that cause action. Coleman describes three forms of social capital: (1) *Obligation*, (2) *Information*, and (3) *Norms and Effective Sanctions*.

Obligation based social capital is gained by performing an action for another where there is an expectation and trust that the other person will repay the action. An individual is able to collect on the social capital they gained at some point in the future, i.e. "call in a favor." Information based social capital arises when one uses their network to gain information, or expert knowledge, about a topic. For example, someone interested in technology, but who does not have the time to keep up with the latest trends, might use their friends to gain information. Unlike the obligation form of social capital, information-based social capital has no expectation of being repaid in the future. Coleman's final form of social capital relates to the social norms and effective sanctions that govern an individual's actions. One of the more powerful forms of norms arises when individuals act in the interest of the group instead of their own self-interest. The rewards for adhering to this norm are often increased status, honor, and social support. In the case of task-oriented groups this may be an increase in power or even leadership.

3   Computational Model of Social Capital and Social Rank for Dialogue

An individual's desire to reach their goals drives their predisposition to belong to a group. A negotiation is performed between group members over the social identity they assume within the group, which in the process transforms individual collaborators into a collective and cooperative group [11]. Based on these social identities, individuals communicate and make social contacts within their negotiated role. The social contacts affect individual and group productivity [12]. Furthermore, individuals build social capital based on the amount and type of contact they make whilst in the group.

3.1   Social Actions

While the social contacts made by individuals in social networks are often explicit through friending, liking, disliking, etc., in dialogue they are more likely to be manifested through linguistic expressions of social intentions. We label the social intentions of an utterance using social acts. Social acts are pragmatic speech acts that signal a dialogue participant's social intentions.

For calculation of social capital and social rank we have defined three categories of social acts: *Cooperation, Support,* and *Hostility.* Cooperation and support are designed to roughly capture the three forms of social capital discussed in section 2. Cooperation is captured through the social acts of agreement, offering gratitude, mediation, and solidarity. Support is captured through agreement, group affordance, solidarity, and supportive behavior [13]. Finally, hostility is used to capture the lack of prestige, social wealth, and power and individual has within the group. Hostility is captured through social actions of undermining, disrespect, relationship conflict, task conflict, and challenges of credibility. We use the same definitions of and methodology for identification of these social acts in text as [14, 15].

3.2   Social Capital

Based on Coleman's [10] work, we have defined a measure of social capital based on the actions performed by individuals in a single dialogue. In particular, we posit that individuals who have higher levels of interactivity, cooperation, and support will have higher amounts of social capital. Furthermore, an individual's social capital is increased by the capital they can collect from others due to obligations, support, etc. More formally, we calculate social capital ($SCap$) for an individual P_i as:

$$SCap(P_i) = \alpha \cdot I(P_i) + \beta \cdot C(P_i) + \frac{\delta \cdot \sum_{P_j \in A(P_i)} \left(S(P_j, P_i) \cdot SCap(P_j) \right)}{N - 1} \quad (1)$$

where N is the total number of dialogue participants; $I(P_i)$ is the level of interactivity participant P_i had with the group; $C(P_i)$ is the amount of cooperation

P_i had toward the group; $S(P_J, P_i)$ is the amount of support participant P_j showed toward participant P_i; and $A(P_i)$ is the set of participants who made more affordances than detractions to P_i. The α, β, and δ parameters control the effect of the individual components (interactivity, cooperation, and support) on the social capital.

Increased interaction is one mechanism to build prestige and is needed to gain obligation and information based social capital. Moreover, we posit that a leader should have more interaction with group as they need to control the group toward their outcome. $I(P_i)$ is the interactivity of P_i which captures the breadth of a participant's interaction with the group and is calculated as:

$$I(P_i) = \frac{\sum_{P_j \neq P_i} Reply(P_i, P_j)}{N} \tag{2}$$

where $Reply(P_i, P_j)$ means that there exists a turn (t) in which participant P_i was the speaker and P_i was directing their message toward P_J (the target).

While groups require a small level of conflict to be there most productive [16], it is generally agreed that cooperative groups are more likely to reach their goals than non-cooperative ones [17]. As such, we posit that a leader is someone who shows more signs of being cooperative than not. $C(P_i)$ is the amount of cooperation P_i shows towards other group members and is calculated as:

$$C(P_i) = \frac{\sum_{t=1}^{T}(Speaker(P_i, t) \cdot C_{action}(t)}{T_{P_i}} \tag{3}$$

where $Speaker(P_i, t)$ returns 1 if P_i is the speaker of turn t, $C_{action}(t)$ is the number of social acts indicative of cooperation at turn t, T is the total number of turns in the dialogue, and T_{p_i} is the total number of turns by P_i the dialogue.

The affordance of power, or support of power, by group members to an individual is a sign of that individual's power. Moreover, with these affordances comes social capital with which an individual can borrow against. We capture these affordances through signs of support from a participant P_j to P_i for participants in the set $A(P_i)$. A participant P_j belongs to set $A(P_i)$ when the total number of times P_j employs a positive (cooperative or supportive) social act is more than they employ a negative (hostile) social act toward P_i, where cooperative, supportive, and hostile are defined in section 3. In other words, the participants in set A are those for whom P_i can gain social capital.

$S(P_j, P_i)$ is the amount of support a participant P_j (in set $A(P_I)$) shows towrd P_i and is calculated as:

$$S(P_j, P_i) = \frac{\sum_{t=1}^{T}(Speaker(P_j, t) \cdot Target(P_i, t) \cdot S_{action}(t))}{T_{P_j}} \tag{4}$$

where $Target(P_i, t)$ returns 1 if P_i is the target (i.e. who the utterance was directed toward) for turn t and $S_{action}(t)$ is the number of social acts indicative of support at turn t.

Calculating social capital is done using an iterative algorithm, similar to that used for PageRank [1] which is shown in figure 1.

Iterative Method for Calculating Social Capital

1: Initialize $SCap$ for each Person P_i at turn $t = 0$ to $\alpha \cdot I(P_i) + \beta \cdot C(P_i)$

2: $k = 0$

3: do

4:     $k = k + 1$

5:     $SCap(P_i; t = k) = \alpha \cdot I(P_i) + \beta \cdot C(P_i) + \dfrac{\delta \cdot \sum_{P_j \in A(P_i)} (S(P_j, P_j) \cdot SCap(P_i; t=k))}{N-1}$

6: $until \left(\sum_{P_i} SCap(P_i; t = k) - \sum_{P_i} (P_i; t = k - 1) \right) < \epsilon$

Fig. 1. The iterative method for calculating social capital

The calculation of social capital begins with each participant P_i in the dialogue having an initial value equal to their interactivity and cooperation with the group. The algorithm then performs a number of iterations updating the social capital value for each P_i based on the social capital of the other participants in set $A(P_i)$. The process completes when the total social capital for the group at iteration k changes less than ϵ from the total social capital at iteration $k - 1$.

3.3   Social Rank

Joining social capital with the concepts of Weber [8], we define social rank as a metric for determining power, prestige, and status within in a group. Within the confines of virtual online groups (excluding those taking part in a game, such as World of Warcraft), economic resources, as are traditionally discussed within the context of social rank and social class, are often of little consequence. Instead, it is the social capital of an individual that defines their wealth. An individual's social capital facilitates their ability to affect change or action that makes them wealthy in online settings. Thus, we formally define social rank ($SRank$) for a participant P_i as:

$$SRank(P_i) = \lambda \cdot SCap(P_i) - \frac{\gamma \cdot \sum_{P_j \notin A(P_i)} (H(P_j, P_i) \cdot SCap(P_j))}{N - 1} \tag{5}$$

We define the negative argument of the social rank calculation as the social detraction for participant P_i. We hypothesize that an individual who does not follow their negotiated social identity creates conflict and hostility within in the group. λ and γ are free parameters, which adjust the amount that the individual's social capital and social detraction play in determining their social rank.

Social detraction measures the amount of an individual's disregard or deviance from their agreed upon social identity throw the reactions of others in the group. The main component of the social detraction is $H(P_j, P_i)$ which is a measure of hostility, or conflict, toward P_i as exhibited by another participant P_j and is calculated as:

$$H(P_j, P_i) = \frac{\sum_{t=1}^{T} (Speaker(P_j, t) \cdot Target(P_i, t) \cdot H_{action}(t))}{T_{P_j}} \tag{6}$$

where $H_{action}(t)$ is the number of social acts indicative of hostility at turn t. The group of participants who contribute to the social detraction of P_i are those who are not in the set $A(P_i)$, i.e. those participants whose showed more negative actions than positive actions toward P_i.

4   Experimental Results

For evaluation, we gathered dialogues communicate in each language from Wikipedia discussions, web forums, blog comments, and chat transcripts. The number of dialogues used for training and testing are shown in table 1. The average number of participants across the entire dataset was 7.8 for English, 16.7 for Arabic, and 8.3 for Chinese with an average of 40.8 turns in English, 29.0 turns in Arabic, and 23.6 turns in Chinese. The training data consisted of dialogues in which a leader may not have been present. The testing data consisted only of dialogues in which there was a leader present.

Table 1. Number of dialogues used for training/development and used for testing per language

Language	# Training Dialogues	# Testing Dialogues
English	83	75
Arabic	135	25
Chinese	353	75

We set the parameters to equally weight each factor in the social capital (as $\alpha = 0.33$, $\beta = 0.33$, and $\delta = 0.33$) and social rank ($\lambda = 0.5$ and $\gamma = 0.5$) equations. We believe that these default set of parameters should provide decent performance on a wider range of data genres. However, in the future we will explore tuning the parameters to different genres (chat, forum, blog, etc.) of data using a development set and grid search.

We compared social rank against a random baseline model (randomly picking one participant as the leader) and a motif model which discovers patterns of social acts that are indicative of leadership. The motif model was used in [4] to capture pursuits of power in dialogue. The motif model determines whether or not a participant is exhibiting leadership qualities. The participant which the motif model has the highest confidence of exhibiting leadership qualities was then determined to be the leader of the dialogue. A motif model was constructed for each of the languages using a set of Yes/No annotations over the training data which was multiply annotated. The training data consisted of a total of 425 participants in English dialogues, 1,387 participants in Arabic, and 3,136 participants in Chinese with Yes/No annotations. The inter-annotator agreement rates were 72.8% for English, 83.4% for Arabic, and 88.0% for Chinese. The testing data consisted of an average of 8.3 participants in English, 4.2 participants in

Arabic, and 11.1 participants in Chinese (these averages differ than those found in the entire dataset).

We also examined the performance when incorporating the confidence score of the motif model into the social rank calculation. The equation for social rank ($SRank$) after incorporating the motif model is as follows:

$$SRank(P_i) = TL(P_i) \cdot \left(\lambda \cdot SCap(P_i) - \frac{\gamma \cdot \sum_{P_j \in D(P_i)} (H(P_j, P_i) \cdot SCap(P_j))}{N-1} \right) \quad (7)$$

where $TL(P_i)$ is the confidence that participant P_i exhibits leadership using the motif model. Table 2 lists the accuracy for determining the **one** leader in the dialogue for the random baseline, the motif model, social rank, and the combination of social rank and the motif model.

Table 2. Comparison of accuracy results in determining the leader of a dialogue between using and not using social rank

	Random	Motif	Social Rank	Motif + Social Rank
English	12%	26%	38%	40%
Arabic	24%	48%	52%	56%
Chinese	9%	48%	48%	52%

As illustrated in Table 2, social rank increased the accuracy in determining the leader by 26% over baseline and 12% over the motif model for English, 28% over baseline and 4% over the motif model for Arabic, and 39% over baseline for Chinese. Chinese was the only language which social rank did not improve the accuracy over the motif model. This most likely due to the large size of training data used that existed to train the motif model which resulted in the model generating better confidence scores. The incorporation of the confidence that a participant was exhibiting leadership qualities generated by the motif model into social rank resulted in improved accuracy for all three languages. English improved by 2%, Arabic by 4%, and Chinese by 4% of social rank alone.

5   Conclusion

In this paper we presented a computational model for determining the group leader in a dialogue using the sociological theories of social capital and social rank. We evaluated social rank on dialogues communicated in English, Arabic, and Chinese. We showed that the social rank can drastically increase performance over baseline (up to 39% for Chinese) and provides better accuracies than a machine learning classifier. The incorporation of the classifier into the social rank computation further boosted accuracies by 2% to 4%.

Acknowledgments. This research was funded by the Office of the Director of National Intelligence (ODNI), Intelligence Advanced Research Projects Activity (IARPA), and through the U.S. Army Research Lab. All statements of fact, opinion or conclusions contained herein are those of the authors and should not be construed as representing the official views or policies of IARPA, the ODNI or the U.S. Government.

References

1. Brin, S., Page, L.: The anatomy of a large-scale hypertextual Web search engine. Computer Networks and ISDN Systems 30(1-7), 107–117 (1998)
2. Brendel, R., Krawczyk, H.: Detection of Roles of Actors in Social Networks Using the Properties of Actors' Neighborhood Structure. In: 2008 Third International Conference on Dependability of Computer Systems DepCoS-RELCOMEX, pp. 163–170. IEEE (2008)
3. Fisher, D., Smith, M., Welser, H.T.: You Are Who You Talk To: Detecting Roles in Usenet Newsgroups. In: Proceedings of the 39th Annual Hawaii International Conference on System Sciences HICSS 2006, p. 59b (2006)
4. Bracewell, D., Tomlinson, M., Wang, H.: A motif approach for identifying pursuits of power in social discourse. In: 2012 IEEE Sixth International Conference on Semantic Computing (ICSC), pp. 1–8. IEEE (2012)
5. Bramsen, P., Escobar-Molano, M., Patel, A., Alonso, R.: Extracting social power relationships from natural language. In: Proceedings of the 49th Annual Meeting of the Association for Computational Linguistics: Human Language Technologies, HLT 2011, vol. 1, pp. 773–782. Association for Computational Linguistics, USA (2011)
6. Danescu-Niculescu-Mizil, C., Lee, L., Pang, B., Kleinberg, J.: Echoes of power: language effects and power differences in social interaction. In: Proceedings of the 21st International Conference on World Wide Web, WWW 2012, pp. 699–708. ACM, New York (2012)
7. Tomlinson, M., Bracewell, D.B., Draper, M., Almissour, Z., Shi, Y., Bensley, J.: Pursuing power in arabic on-line discussion forums. In: Proceedings of the Eighth Conference on International Language Resources and Evaluation (2012)
8. Weber, M.: The theory of social and economic organization. W. Hodge, London (1947)
9. French, J.R.P., Raven, B.: The Bases of Social Power. In: Cartwright, D. (ed.) Studies in Social Power. Studies in Social Power, vol. 35, pp. 150–167. Institute for Social Research (1959)
10. Coleman, J.S.: Social capital in the creation of human capital. American Journal of Sociology 94(S1), 95 (1988)
11. Swann Jr., W.B., Johnson, R.E., Bosson, J.K.: Identity negotiation at work. Research in organizational behavior 29, 81–109 (2009)
12. Putnam, R.: Bowling Alone: The Collapse and Revival of American Community. Touchstone Books by Simon & Schuster (2000)
13. Bracewell, D.B., Tomlinson, M., Shi, Y., Bensley, J., Draper, M.: Who's playing well with others: Determining collegiality in text. In: Proceedings of the 5th IEEE International Conference on Semantic Computing (ICSC 2011). IEEE Computer Society, Palo Alto (2011)

14. Bracewell, D.B., Tomlinson, M.T., Brunson, M., Plymale, J., Bracewell, J., Boerger, D.: Annotation of Adversarial and Collegial Social Actions in Discourse. In: 6th Linguistic Annotation Workshop, pp. 184–192 (July 2012)
15. Bracewell, D., Tomlinson, M., Wang, H.: Identification of social acts in dialogue. In: 24th International Conference on Computational Linguistics, COLING (2012)
16. Jehn, K.A., Mannix, E.A.: The Dynamic Nature of Conflict: A Longitudinal Study of Intragroup Conflict and Group Performance. Academy of Management Journal 44(2), 238 (2001)
17. Little, J.W.: Norms of Collegiality and Experimentation: Workplace Conditions of School Success. American Educational Research Journal 19(3), 325–340 (1982)

Pareto Distance for Multi-layer Network Analysis*

Matteo Magnani[1] and Luca Rossi[2]

[1] Dept. of Computer Science
Aarhus University, Denmark
magnanim@cs.au.dk
[2] Dept. of Communication Studies
University of Urbino, Italy
luca.rossi@uniurb.it

Abstract. Social Network Analysis has been historically applied to single networks, e.g., interaction networks between co-workers. However, the advent of on-line social network sites has emphasized the stratified structure of our social experience. Individuals usually spread their identities over multiple services, e.g., Facebook, Twitter, LinkedIn and Foursquare. As a result, the analysis of on-line social networks requires a wider scope and, more technically speaking, models for the representation of this fragmented scenario. The recent introduction of more realistic *layered* models has however determined new research problems related to the extension of traditional single-layer network measures. In this paper we take a step forward over existing approaches by defining a new concept of geodesic distance that includes heterogeneous networks and connections with very limited assumptions regarding the strength of the connections. This is achieved by exploiting the concept of Pareto efficiency to define a simple and at the same time powerful measure that we call *Pareto distance*, of which geodesic distance is a particular case when a single layer (or network) is analyzed. The limited assumptions on the nature of the connections required by the Pareto distance may in theory result in a large number of potential shortest paths between pairs of nodes. However, an experimental computation of distances on multi-layer networks of increasing size shows an interesting and non-trivial stable behavior.

Keywords: Multi-layer networks, Pareto efficiency, distance, social network sites.

1 Introduction

The last two decades have witnessed the proliferation of several on-line social network services (SNS) like Classmates.com, Friendster, followed by MySpace,

---

* This work has been supported in part by the Danish Council for Strategic Research, grant 10-092316, by the Italian Ministry of Education, Universities and Research PRIN project *Relazioni sociali ed identità in Rete: vissuti e narrazioni degli italiani nei siti di social network* and FIRB project RBFR107725.

A.M. Greenberg, W.G. Kennedy, and N.D. Bos (Eds.): SBP 2013, LNCS 7812, pp. 249–256, 2013.

Facebook, Twitter and Google+, just to name a few. While it is not clear whether only one or few big players will survive in the near future, or multiple specialized services will still exist separately, we can make the hypothesis that a single concept of *social connection* or *social network* is not sufficient to satisfy the *sociability requirements* of human beings.

Decades before the advent of SNSs this had already been described by Goffman [1] and other researchers, for which individuals (or *actors*) perform on multiple stages, creating a sort of sociologically fragmented personality whose different components relate to different audiences (and thus networks). When we move into the on-line context this scenario is, if possible, even harder to manage. On-line communication undoubtedly offers many opportunities to experience this multiplicity of identities and of heterogenous relationships.

As a consequence, the fact that a thorough analysis of human communities may require models that can capture more than just the existence of generic connections between specific people has become a popular topic in the research literature on SNSs. Recent works have defined the foundations of new multi-layer (or multi-dimensional) models that extend previous work (like multiplex networks, [2]) by clearly separating the multiple co-existing networks [3,4,5]. More in general, research on complex social networks has often dealt with several kinds of network able to represent different kinds of relationships within the same graph [6] or even relationships between users with different attributes [7]. The mutual influences between different co-existing networks inside the same SNS have also been empirically studied [8,9]. However, now that basic models of multiple networks are available, we have to face the challenge of extending traditional network analysis metrics to be applicable to these models.

Existing works end up by either considering different social connections separately [3], or treating all social connections in the same way [4,5]. Both approaches are not satisfactory, because the former does not take into consideration that real processes like information propagation happen traversing all these network layers and heterogeneous connections, while the latter does not distinguish between potentially very different connections, e.g., friendship or co-working.

In this paper we define a new concept of geodesic distance that explicitly treats different link types as heterogenous entities, but at the same time composes them in a homogeneous measure. This is achieved by exploiting the concept of Pareto efficiency to define a simple and at the same time powerful measure that we call *Pareto distance*, of which geodesic distance is a particular case. We consider this concept a first step toward the effective analysis of real multi-layer network dynamics using extended network measures well corroborated during many decades of social network analysis. In fact, our measure is a conservative extension of single-network distances that introduces as few assumptions as possible with regard to the multi-network layers. In addition, we show an interesting explosion of the complexity of multi-layer models when distances must be computed: the juxtaposition of N networks is not just N times more complex than a single network in general. However, we also provide preliminary experimental results showing an interesting stable behavior of our Pareto distance when the size of the network increases.

The paper is organized as follows. In the next section we briefly introduce the multi-layer model used in the following and show by example how the concept of *distance* becomes more complex when we switch to a multi-network perspective. Then we introduce the concept of Pareto distance, showing that it generalizes single-layer geodesic distances, and we discuss its properties. We then show how this concept may theoretically produce a large result, i.e., a large number of alternative distances, because of its generality and limited assumptions. However, preliminary experimental evidence highlights an interesting stable behavior of this measure. We conclude the paper with some final remarks.

2   Multi-layer Models and Distances

The main distinguishing feature of multiple social network models is the presence of heterogeneous information. Three main basic options are available to represent heterogeneous networks: models allowing multiple node types (also called heterogeneous or multi-type networks), represented in Figure 1(a) [10,11], models allowing multiple relationship types (also called multi-dimensional networks), represented in Figure 1(b) [3,4] and models explicitly representing the co-existence of multiple networks (also called multi-layer(ed) or multi-stratum networks), represented in Figure 1(c) [5].

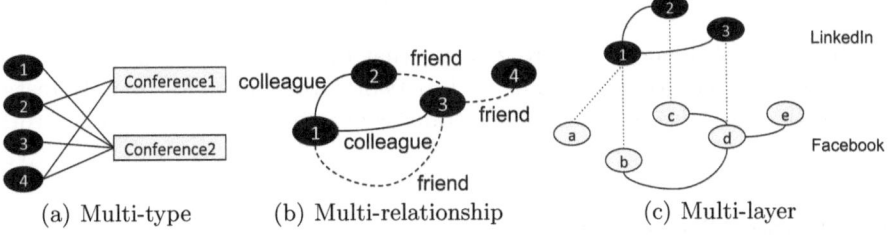

(a) Multi-type   (b) Multi-relationship   (c) Multi-layer

Fig. 1. Three models of heterogeneous networks. An example of multi-type network is an author-conference graph (a), indicating who (ellipses) presented at which conference (rectangles). Multi-relationship networks (b) allow different kinds of links, e.g., connecting people who are friend (dashed lines) and people who work together (continuous lines). Multi-layer networks (c) are made of multiple social graphs, with mappings to indicate that nodes in different graphs correspond to the same individual.

In the following we use a multi-layer network (Figure 1(c)) as a base model, because it only contains nodes related to individuals (therefore it is a social network strictly speaking) and is more expressive than multi-relationship models — in particular, a multi-relationship model can be represented using a multi-layer model with exactly the same number of nodes in each layer and with every node corresponding to exactly one node in each layer.

As it appears from Figure 1(c), the main constituents of this model are two or more network layers, not dissimilar from traditional networks, and *mappings* indicating which nodes in different layers correspond to the same individual.

2.1   Multi-layer Distance

Before introducing the concept of Pareto distance it is important to appreciate why distances may significantly change in a multi-layer model [5]. The left hand side of Figure 2 shows that two users A and D whose accounts are not connected to each other can be in fact connected through the path A → A' → B' → B → C → C' → D' → D. This information flow process involves some normal in-network propagations and the *choices* of some users that the information is worth propagating also in the other network.

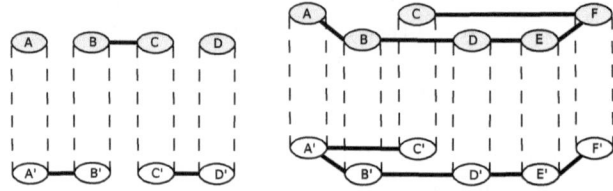

Fig. 2. Node reachability (left) and node distances (right) in a layered context

The right hand side of Figure 2 shows that even for users that are already connected to each other in one or both networks their distance may decrease. Also in this case this depends on the transfer of information from one network to the other.

2.2   Pareto Distance

The main problem with regard to distances in multi-layer models concerns the heterogeneity of the connections. Let us consider again the example in Figure 1(c). What is the distance between the user corresponding to nodes b and the user corresponding to node e? One option is to go from one to the other by traversing links b-d and d-e. Another option is to start from the top network (node 1, owned by the same individual owning b), go through 1-3, then have a network switch 3-d (corresponding to the forwarding of information from one network to the other) and finally go through the link d-e. The two corresponding complete multi-network paths are then:

1. b-d-e
2. b-1-3-d-e

Now we can continue this analysis making two assumptions. First, we will not consider network switches — this is only for the sake of simplicity of presentation, but does not limit the model because switches can be considered without any formal modifications to the model. The second assumption is that we consider the links inside the *same* network being *homogeneous*, meaning that we do not make any difference between, e.g., link b-d and link d-e being both *friend* connections on the same network. This is the same assumption done whenever we compute a geodesic distance on a single unweighted network: the distance between b and e on the lower network would in fact be just 2, hiding the explicit path b-d-e under the simple number of links.

Given these assumptions, we can thus summarize the two paths using the number of links traversed in each network. In our example, we indicate with F a link on the Facebook network and with L a link on the LinkedIn network. As a result the two paths become:

1. 0L + 2F
2. L + F

This can be formalized through the concept of multi-layer path length:

Definition 1 (Multi-layer path length). *The multi-layer path length of path p on n networks is an array $r_1 + \cdots + r_n$ where r_i indicates the number of connections traversed in the ith network inside p.*

As we mentioned previously, we may also include network switches in this definition by extending the array – however in the following we stick to this simpler definition to present more compact examples.

As within traditional single-networks, we may have several paths connecting two nodes, and *distance* is defined using a concept of *efficiency*, favoring the *shortest* among all possible paths. However, when multiple networks are involved it is more problematic to define efficiency. The key concepts that we use to define a multi-network distance without converting the heterogeneous links into a homogeneous unit are *Path dominance* and *Path incomparability*, extending the corresponding concepts in the Pareto efficiency theory:

Definition 2 (Path dominance). *Let r and s be two multi-layer path lengths on n networks. r dominates s iff $\forall l \in [1, n]\ r_l \leq s_l \wedge \exists i\ r_l < s_l$.*

Definition 3 (Path incomparability). *Let r and s be two multi-layer path lengths. r and s are incomparable iff r does not dominate s and s does not dominate r.*

If we compare our two paths we can see that they are incomparable. The meaning of this property is that both paths may be *shorter* than the other under a specific path evaluation function. As a simple example, consider the case where information on Facebook propagates 3 times less frequently than on LinkedIn. We would have:

$$0L + 2F = 0L + 2 \cdot (3L) = 6L > 4L = L + 3L = L + F$$

On the contrary, assume that a specific piece of information propagates on LinkedIn two times less frequently than on Facebook. We would have:

$$0L + 2F = 0 \cdot (2F) + F = F < 3F = 2F + F = L + F$$

From this example we can appreciate how different interpretations of the two kinds of links might lead to one or the other path being shorter. As a consequence, in absence of additional information we cannot discard any of the two options. These would both be alternative shortest paths, in a similar way as we may have multiple shortest paths of the same length in a traditional single-layer model.

This definition has the nice property that it returns all paths that can be shortest under some monotone path evaluation function. However, this is not enough: we do not want to return paths that cannot be shorter, and we can now use the concept of dominance for this. For example, consider the path b-1-2-c-d-e. This can be summarized as L + 2F. Notice that this time this path is dominated by 2F, because independently of the weight we put on L and F, L + 2F will always be longer than 2F (or equal when L has weight 0).

This brings us to the second property of our Pareto distance: we not only return all paths that are the shortest under some path evaluation function, but we do not return any path that cannot be the shortest one. This means that our Pareto distance operator is *tight* (sound and complete).

Finally, we can define the Pareto distance between two nodes n_1 and n_2 as follows:

Definition 4 (Pareto distance). *Let $ML(n_1, n_2)$ be set of all multi-layer path lengths between nodes n_1 and n_2. The Pareto distance between n_1 and n_2 is a set $P \subseteq MP$ such that $\forall p \in P \; \nexists p' \in ML : p'$ dominates p.*

2.3   Complexity and Stability of Pareto Distances

The fact that the Pareto distance set is tight (i.d., minimal) is clearly a feature of this definition. However, minimality does not necessarily mean that the set of potential shortest paths is small. As an example, consider Figure 3. In this specially constructed multi-layer network with only two networks there are five path lengths in the Pareto distance between a and e, corresponding to as many as $2 \cdot \binom{4}{2}$ real paths. For example, the path a-b-2-3-c-d-e, or L + 3F, is one of the shortest paths under the evaluation function: *a short path cannot have more than one link on LinkedIn and more than three links on Facebook.*

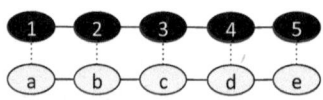

Fig. 3. A constructed multi-layer network with a large number of Pareto paths

One possible way to reduce the probability that this kind of situation happens in practice is to restrict the notion of Pareto efficiency. For example, it is possible to focus only on linear path evaluation functions, i.e., functions in the form

$\alpha_1 \cdot$(number of links in L) $+ \alpha_2 \cdot$(number of links in F). Another option consists in keeping the original concept of Pareto distance and select some *representative* path lengths among the potentially numerous choices.

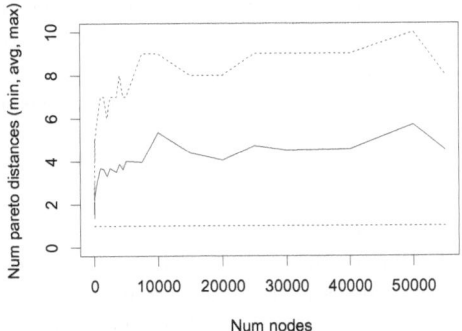

Fig. 4. Stability of Pareto distance (the continuous line indicates the average value, dashed lines indicate minimum and maximum)

However, an experimental evaluation of our measure suggests that the number of path lengths that are Pareto efficient maintains a stable behavior for increasing network sizes. To evaluate this behavior, we have first grown several multi-layer networks of increasing size, each made of two scale-free networks [12]. Then, we computed the Pareto distance between the first node inserted in the networks and all the other reachable nodes in its giant component. In Figure 4 we show the minimum, average and maximum number of alternative Pareto distances found in these networks depending on their size.

This experiment presents two interesting and non-trivial outcomes. The first, as said, is that the size of the result of our method quickly stabilizes. This is strictly connected to the fact that real social networks (including scale-free ones, like those generated in our tests) have a small diameter, therefore the theoretically large distances we could find in increasingly large networks cannot be found in practice, providing a bound also for multi-layer distances. The second result is that the maximum number of alternative paths also stabilizes. This could not be concluded analyzing only the average behavior of Pareto distances: in fact, on a large network we might have expected the existence of a few pairs of nodes with a large number of alternative Pareto distances not affecting the average. However, in our tests we could not observe this behavior, this resulting in an even stronger notion of stability.

3   Concluding Remarks

Recent works have claimed that the analysis of single SNSs may return a partial and sometimes wrong picture of many interesting phenomena, including information propagation and network centralities. The introduction of multi-layer social network models has partially provided an answer to these claims, but traditional network measures are now no longer applicable to these extended models.

In this paper we have defined a new concept of geodesic distance for heterogeneous networks based on very limited assumptions regarding the strength of the connections. This is achieved by exploiting the concept of Pareto efficiency: *Pareto distance* generalizes geodesic distance and provides an intuitive and powerful notion of proximity, which is a fundamental network measure being at the basis of widely used metrics such as closeness and betweenness. We have also highlighted how this simple definition may theoretically result in large and thus computationally hard sets of alternative distances. However, a preliminary experimental evaluation has shown the practical stability of our measure, both in its average behavior and its bounds.

The development of efficient algorithms for the computation of Pareto distances and their application to a wide range of synthetic and real networks will constitute future work to improve our understanding of this new measure for the analysis of multiple networks, a topic that is receiving more and more attention because of its theoretical interest but also its practical applicability to current real scenarios.

References

1. Goffman, E.: Frame analysis: an essay on the organization of experience. Harper & Row, New York (1974)
2. Wasserman, S., Faust, K.: Social Network Analysis. Cambridge University Press (1994)
3. Berlingerio, M., Coscia, M., Giannotti, F., Monreale, A., Pedreschi, D.: Foundations of Multidimensional Network Analysis. In: ASONAM Conference, pp. 485–489. IEEE Computer Society (2011)
4. Brodka, P., Stawiak, P., Kazienko, P.: Shortest Path Discovery in the Multi-layered Social Network. In: ASONAM Conference, pp. 497–501. IEEE Computer Society (2011)
5. Magnani, M., Rossi, L.: The ML-model for multi-layer social networks. In: ASONAM Conference, pp. 5–12. IEEE Computer Society (2011)
6. Kazienko, P., Bródka, P., Musial, K., Gaworecki, J.: Multi-layered social network creation based on bibliographic data. In: SocialCom/PASSAT, pp. 407–412. IEEE Computer Society (2010)
7. Zhao, P., Li, X., Xin, D., Han, J.: Graph cube: on warehousing and olap multidimensional networks. In: SIGMOD Conference, pp. 853–864. ACM (2011)
8. Magnani, M., Montesi, D., Rossi, L.: Information propagation analysis in a social network site. In: ASONAM Conference, pp. 296–300. IEEE Computer Society, Los Alamitos (2010)
9. Rossi, L., Magnani, M.: Conversation practices and network structure in twitter. In: ICWSM (2012)
10. Cai, D., Shao, Z., He, X., Yan, X., Han, J.: Community Mining from Multi-relational Networks. In: Jorge, A.M., Torgo, L., Brazdil, P.B., Camacho, R., Gama, J. (eds.) PKDD 2005. LNCS (LNAI), vol. 3721, pp. 445–452. Springer, Heidelberg (2005)
11. Sun, Y., Han, J., Zhao, P., Yin, Z., Cheng, H., Wu, T.: RankClus. In: EDBT, pp. 565–576. ACM Press (2009)
12. Magnani, M., Rossi, L.: Formation of Multiple Networks. In: Greenberg, A.M., Kennedy, W.G., Bos, N.D. (eds.) SBP 2013. LNCS, vol. 7812, pp. 257–264. Springer, Heidelberg (2013)

Formation of Multiple Networks[*]

Matteo Magnani[1] and Luca Rossi[2]

[1] Dept. of Computer Science
Aarhus University, Denmark
magnanim@cs.au.dk
[2] Dept. of Communication Studies
University of Urbino, Italy
luca.rossi@uniurb.it

Abstract. While most research in Social Network Analysis has focused on single networks, the availability of complex on-line data about individuals and their mutual heterogenous connections has recently determined a renewed interest in multi-layer network analysis. To the best of our knowledge, in this paper we introduce the first network formation model for multiple networks. Network formation models are among the most popular tools in traditional network studies, because of both their practical and theoretical impact. However, existing models are not sufficient to describe the generation of multiple networks. Our model, motivated by an empirical analysis of real multi-layered network data, is a conservative extension of single-network models and emphasizes the additional level of complexity that we experience when we move from a single- to a more complete and realistic multi-network context.

Keywords: Multi-layer networks, Network formation, Social Network Sites.

1   Introduction

Network formation models are among the most important tools in Network Science and Social Network Analysis (SNA). A typical application of artificially generated networks is to provide *null* models that can be used to test new measures and make comparisons with real networks, so that significant patterns can be highlighted in the real data. In addition, these models are useful to test hypotheses on the dynamics underlying network evolution.

However, existing generative models have been developed to describe the evolution of *single* networks. While this is very relevant, as most of the research in SNA has been devoted to single networks, recent empirical studies have emphasized how on-line social systems including Social Network Sites (SNS) are made of multiple stratified networks influencing each other [1,2]. Multi-network

---

[*] This work has been supported in part by the Italian Ministry of Education, Universities and Research PRIN project *Relazioni sociali ed identità in Rete: vissuti e narrazioni degli italiani nei siti di social network* and FIRB project RBFR107725.

A.M. Greenberg, W.G. Kennedy, and N.D. Bos (Eds.): SBP 2013, LNCS 7812, pp. 257–264, 2013.

models were discussed several years ago in the field of SNA (also known as multi-plex networks, [3]) and described as everyday experience by sociological research [4], but only recently the availability of real multi-network data has boosted the development of new models [5,6] and algorithmic approaches [7,8] based on the assumption that the analysis of the single networks may provide a distorted scenario if their multi-layered organization is not taken into consideration. As a simple example, on-line information propagation is typically characterized by the traversal of different networks [9].

In this paper we introduce the problem of multi-layer network generation. This is a challenging task, because models describing the formation of multiple networks should still generate network layers compatible with existing models and experimental observations of single networks, but should also consider the mutual relationships between different layers. Therefore, we propose a model where network evolution may be characterized both by *internal* dynamics, as described by existing single-network models like Preferential Attachment, and by *external* dynamics, where events like the creation of a new connection are influenced by the structure of other networks (here called *network layers*).

The paper is organized as follows. In the next two sections we briefly review the main theoretical basis of our work, namely network formation and multi-layer models. Then, in Section 4 we propose our approach. Our work is based on an analysis of real data that are used as a guideline for the definition of our model and also to test its ability to reproduce real observations. These data are presented in Section 5.

To the best of our knowledge our model is the first to deal with the generation of multi-network data. As such, it raises many new questions regarding the parameters and processes to be used to represent the dynamics underlying the formation of multiple networks. We devote our concluding remarks to these issues.

2   Network Formation Models: A Quick Review

Research on random network models, their definition and related algorithms, is at least as old as modern network science and it has always been characterized by a common goal: being able to reproduce networks as they are observed in social, biological or physical phenomena. Within this perspective, we provide an essential summary of the most popular network models.

The definition of more and more sophisticated models can be seen as a never-ending attempt to catch the true complexity and inner nature of networks [10]. Among the first attempts in this direction, the Erdős-Rènyi model [11], often notated as the $G(n,p)$ model, provides a simple but effective way to generate basic random networks. While this model has been historically useful to rise the interest on research topics such as edge probability and normal degree distribution, it fails at describing networks appearing in real-life phenomena. Its major caveats, i.e., the lack of scale-free degree distribution and the lack of high clustering values that are often observed in real-life contexts, have later been addressed by the Barabàsi-Albert and the Watts-Strogatz models [12,13]. The Barabàsi-Albert model is based on the concept known as Preferential Attachment, stating that important nodes in a network have higher probability than others to further increase their popularity. In addition, like in other more recents models [14,15] Preferential Attachment is not only used as a method to generate a network, but describes its formation step by step — in particular, the growing aspect is essential to obtain the required degree distribution.

While all these models provide a rather detailed level of description of several existing networks, and have been fruitfully used to simulate many real-world phenomena, none of them supports a multi-layer structure, therefore they cannot be directly used to describe the whole complexity of entangled multi-layered social phenomena.

3   Multi-layer Network Models

In the recent literature on multiple social network models we can find proposals allowing multiple node types [16,17], exemplified in Figure 1(a), models allowing multiple relationship types [5,8], represented in Figure 1(b), and models explicitly representing the co-existence of multiple networks (also called multi-layer networks) [6], represented in Figure 1(c).

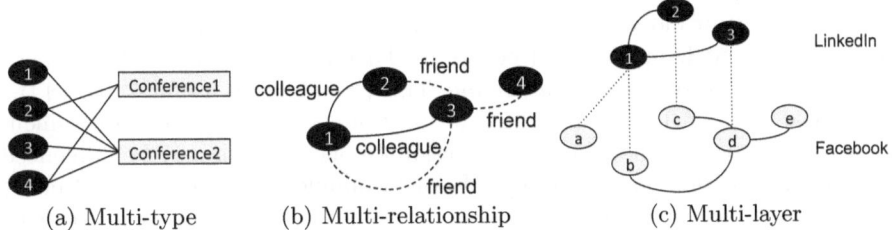

(a) Multi-type   (b) Multi-relationship   (c) Multi-layer

Fig. 1. Three examples of heterogeneous networks: an author-conference graph (a), a multi-relationship network (b) and a multi-layer network (c), made of multiple social graphs and mappings indicating that different nodes correspond to the same individual

In the following we use a multi-layer network containing only nodes related to individuals (therefore, we do not consider heterogeneous nodes). In addition, we only allow nodes to have a single correspondence with nodes in other layers — the more complex situation of a node corresponding to n different nodes in another layer, e.g., a Facebook user having multiple Twitter accounts, is left to future extensions. As it appears from Figure 1(c), the main constituents of this model are two or more network layers, not dissimilar from traditional networks, and *mappings* indicating which nodes in different layers correspond to the same individual.

4   From Single- to Multi-layer Generators

Figure 2 shows a generic process of network formation. We call this network N_1. At different timestamps $t_0, t_1, \ldots$ a new node (+n) or a new edge (+e) are added to the network[1]. The specific mechanisms regulating the creation of nodes and edges vary depending on the formation model. In the following, we will adopt a well known model using Preferential Attachment to generate directed networks [14]. In summary, this model chooses the nodes to be connected together either at random, or with a probability proportional to the in- and out-degrees of the nodes.

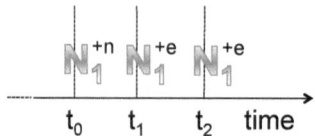

Fig. 2. Abstract view of the evolution of a network

Figure 3 extends the previous example to two networks N_1 and N_2. If we focus on a single horizontal layer, say N_1, we can observe the same dynamics of Figure 2. However, the whole process highlights two main new aspects. First, considering two or more networks we can no longer assume that at every timestamp t_i an event happens in all networks. Therefore, every network will have some associated probability of *no action*. This probability is useful to model the fact that different networks may grow at different speeds. The second fundamental aspect consists in the fact that an action on one network may be influenced by a previous action on another one. In our example, an edge is created in N_1 following the fact that the same two nodes were already linked in N_2. Practically speaking, if I already know someone, e.g., we are friends on Facebook, this may increase the probability that we will also connect on another on-line social network.

In summary, according to our model at every time t_i there are three possible events on each network:

1. **no-action**: nothing happens, i.e., the network remains unchanged.
2. **internal-growth**: the network grows according to internal dynamics, i.e., something happens independently of the other networks. For example, a Twitter user may find a tweet interesting and thus start following its author. In the following this event will be modeled as a Preferential Attachment process.

---

[1] As in all the aforementioned models, in this paper we only consider growing networks and not the deletion of nodes and edges, to keep this initial model simple and focus on the multi-layered aspects of network evolution. The extension of the model to deletion events will be object of future work.

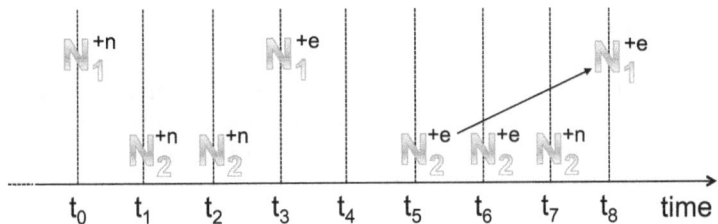

Fig. 3. Abstract view of the joint evolution of two networks

3. **external-growth**: the network grows according to external dynamics, i.e., something happens because of the configuration of an external network. For example, Dante and Beatrix are already "off-line" friends and after opening their accounts they also start following each other on Twitter (it is worth noting how in this example we do not limit our model to on-line social networks).

As said, in the following we will use Preferential Attachment as a network formation model in case of internal growth [14]. However, nothing prevents us to consider other models that can be seamlessly plugged into our approach — we do not further develop this idea because of space limitations and also to provide a well-defined first version of our model.

On the contrary, we need to discuss more the event of external growth. In this case, we can either add an edge coming from an other network, or a node. The first important aspect is that different networks may be more or less correlated, therefore the probability that the edge or node is "imported" from a specific other network is not uniform. The second important point regards the choice of the node to be imported and the corresponding creation of a link. We allow two possible actions:

1. The new node is just imported from the other network *at random*, meaning that the choice is not influenced by other nodes already in the target network[2].
2. The new node is chosen from the set of nodes connected to individuals already in the target network.

In practice, these actions can be exemplified as follows. Guido has an account on Twitter and an account on Facebook. At some point, another Facebook user creates an account on Twitter. Under option (1), this is just a random user that decided to join Twitter. Under option (2), this is a friend of Guido on Facebook who decided to join Twitter and start following Guido.

---

[2] By *target network* we indicate the network into which we are inserting the new node or edge.

5   Experimental Analysis

Our experimental analysis has two main objectives. The first is to highlight the presence of the theoretical features of our model in real multi-network data. The second objective is to test the ability of the model to replicate these data.

The data used in our experimental analysis of a real multi-layer network has been initially extracted from Friendfeed, a social media aggregator [18]. In this system while users can directly post messages and comment on other messages much like in Facebook and other similar SNSs, they can also register their accounts on other systems. In this way, using the Friendfeed API we could retrieve the multiple accounts of the same users for several social services.

As a result we obtained a Friendfeed network with 7 677 120 arcs, a Twitter network with 37 805 211 arcs and a YouTube network with 708 911 arcs. These networks have been used for the analysis of degree centrality correlations reported in the following. In addition, we also built three networks by keeping only those connections between users in our sample, with respectively 37 997, 67 123 and 1 185 arcs. The (not surprising) different sizes of these networks motivate the *no-action* steps in our network model.

The left hand side of Figure 4 shows the correlation between user rankings according to their degree centrality on the Twitter network and on the Friendfeed network, while on the right of Figure 4 we have shown the correlation between user rankings according to their degree centrality on the Twitter network and on the YouTube network. To interpret these figures consider that each point represents a user, and users with a high x or y coordinate are among the top users on the corresponding SNS according to their degree centrality (more precisely, x and y coordinates correspond to the ranking of the user, 0 for the user with lower degree centrality, up to 7 628 for the user with the highest degree centrality in that SNS).

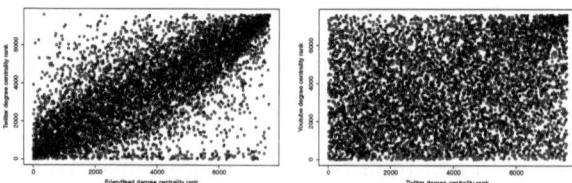

Fig. 4. User ranking (according to their degree centrality) in different networks: Friendfeed and Twitter (left) and Twitter and YouTube (right): Pearson correlation indexes are respectively .75 and .21

These figures show an interesting phenomenon corresponding to the varying probability of pairs of networks to be correlated that can be found in our network formation model. The high correlation between Friendfeed and Twitter means that users with a high degree centrality on Twitter tend to maintain it on the

Friendfeed network. On the opposite side, when we compare the degree centrality ranking on Twitter with the one on YouTube we are unable to detect a clear linear relationship.

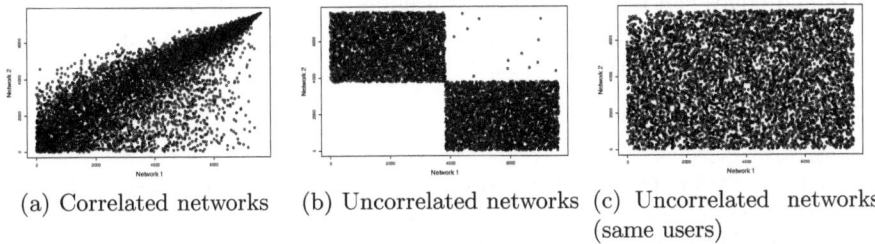

(a) Correlated networks   (b) Uncorrelated networks   (c) Uncorrelated networks
                                                               (same users)

Fig. 5. Degree rank correlation of network pairs generated using our model

In Figure 5 we have represented the corresponding rankings computed and compared on artificial networks generated using our model. In particular, Figures 5(a) and 5(b) correspond respectively to a pair of correlated networks and a pair of uncorrelated networks. While Figure 5(a) shows a behavior similar to the one observed on the Friendfeed and Twitter networks (Figure 4, left), the uncorrelated pair of networks (Figure 5(b)) presents a peculiar distribution of the degree rankings not observed in the real data. However, this can be explained by noticing that the way in which we collected our real data introduced a bias as we only selected individuals present in all the networks. On the contrary, our model does not enforce this controlled choice of users, and the uncorrelated growth of two networks results in different users joining either one network or the other, producing the well separated plot in Figure 5(b). We can make the hypothesis that this is what we would observe by comparing two unrelated real social networks, e.g., the QZone and Cloob networks[3]. In fact, by simply adding the constraint that users should be selected from a common basis, our model finally produces the networks corresponding to Figure 5(c).

6   Concluding Remarks

Multi-layer network data are everywhere in on-line social networks, but due to legal, privacy-related and technical issues they are still very hard to collect. This is also why research on topics such as the definition of new centrality metrics for these networks or the study of propagation patterns in multi-layer contexts is still at its early stages, although having been marked as very relevant for many years (e.g., in [3]). Therefore, the availability of our model could boost this area of research providing a tool to generate prototypical ML-networks for experimental researches [19]. At the same time, as far as new real data are collected our model and its variations can be used to test hypotheses regarding the evolution of multiple correlated networks.

---

[3] Respectively, the principal SNSs in China and Iran.

In our opinion, while it is still difficult to provide a thorough experimental analysis of our approach because of the limited availability of real data and the novelty of the topic, this paper draws many new research questions. A certainly non-comprehensive list includes the study of models for multiple networks where nodes and edges can be deleted, where a node in a network may correspond to multiple nodes in another, and where different networks evolve according to different internal formation models.

References

1. Szell, M., Stefan Thurner, R.L.: Multirelational organization of large-scale social networks in an online world. PNAS (107), 746–759 (2010)
2. Rossi, L., Magnani, M.: Conversation practices and network structure in twitter. In: ICWSM (2012)
3. Wasserman, S., Faust, K.: Social Network Analysis. Cambridge University Press (1994)
4. Goffman, E.: Frame analysis: an essay on the organization of experience. Harper & Row, New York (1974)
5. Berlingerio, M., Coscia, M., Giannotti, F., Monreale, A., Pedreschi, D.: Foundations of Multidimensional Network Analysis. In: ASONAM, pp. 485–489 (2011)
6. Magnani, M., Rossi, L.: The ML-model for multi-layer social networks. In: ASONAM, pp. 5–12. IEEE Computer Society (2011)
7. Rodriguez, M.A., Shinavier, J.: Exposing multi-relational networks to single-relational network analysis algorithms. J. Informetrics 4(1), 29–41 (2010)
8. Brodka, P., Stawiak, P., Kazienko, P.: Shortest Path Discovery in the Multi-layered Social Network. In: ASONAM, pp. 497–501. IEEE Computer Society (2011)
9. Magnani, M., Montesi, D., Rossi, L.: Information propagation analysis in a social network site. In: ASONAM, pp. 296–300. IEEE Computer Society (2010)
10. Newman, M.E.J.: Networks: an introduction. Oxford University Press (2010)
11. Erdős, P., Rényi, A.: On the evolution of random graphs. Magyar Tud. Akad. Mat. Kutató Int. Közl 5, 17–61 (1960)
12. Barabási, A., Albort, R.: Emeigence of scaling in random networks. Science 286(5439), 509–512 (1999)
13. Watts, D., Strogatz, S.: The small world problem. Collective Dynamics of Small-World Networks 393, 440–442 (1998)
14. Bollobás, B., Borgs, C., Chayes, J., Riordan, O.: Directed scale-free graphs. In: SODA, pp. 132–139 (2003)
15. Kumar, R., Raghavan, P., Rajagopalan, S., Sivakumar, D., Tomkins, A., Upfal, E.: Stochastic models for the web graph. In: FOCS, pp. 57–65 (2000)
16. Cai, D., Shao, Z., He, X., Yan, X., Han, J.: Community Mining from Multi-relational Networks. In: Jorge, A.M., Torgo, L., Brazdil, P.B., Camacho, R., Gama, J. (eds.) PKDD 2005. LNCS (LNAI), vol. 3721, pp. 445–452. Springer, Heidelberg (2005)
17. Sun, Y., Han, J., Zhao, P., Yin, Z., Cheng, H., Wu, T.: RankClus. In: EDBT, p. 565 (2009)
18. Celli, F., Di Lascio, F.M.L., Magnani, M., Pacelli, B., Rossi, L.: Social Network Data and Practices: The Case of Friendfeed. In: Chai, S.-K., Salerno, J.J., Mabry, P.L. (eds.) SBP 2010. LNCS, vol. 6007, pp. 346–353. Springer, Heidelberg (2010)
19. Magnani, M., Rossi, L.: Pareto Distance for Multi-layer Network Analysis. In: Greenberg, A.M., Kennedy, W.G., Bos, N.D. (eds.) SBP 2013. LNCS, vol. 7812, pp. 249–256. Springer, Heidelberg (2013)

A Flexible Framework for Probabilistic Models of Social Trust

Bert Huang, Angelika Kimmig*, Lise Getoor, and Jennifer Golbeck

University of Maryland, College Park, MD 20742

Abstract. In social networks, notions such as trust, fondness, or respect between users can be expressed by associating a strength with each tie. This provides a view of social interaction as a weighted graph. Sociological models for such weighted networks can differ significantly in their basic motivations and intuitions. In this paper, we present a flexible framework for probabilistic modeling of social networks that allows one to represent these different models and more. The framework, probabilistic soft logic (PSL), is particularly well-suited for this domain, as it combines a declarative, first-order logic-based syntax for describing relational models with a soft-logic representation, which maps naturally to the non-discrete strength of social trust. We demonstrate the flexibility and effectiveness of PSL for trust prediction using two different approaches: a structural balance model based on social triangles, and a social status model based on a consistent status hierarchy. We test these models on real social network data and find that PSL is an effective tool for trust prediction.

1 Introduction

Trust is a complex social phenomenon and a critical component of human social interaction. Modeling trust therefore plays an important role in social network analysis, with applications including viral marketing, collaborative filtering, and security. Computational modeling of trust provides added insight into the communication patterns, information flow, and behavior of social networks underlying these applications. In this paper, we present a computational framework for relational probabilistic modeling that is particularly well-suited for trust analysis in social networks. This framework is based on *probabilistic soft logic* (PSL) [1], an analysis engine that combines first-order rules with soft truth-values. PSL allows one to naturally capture structural ideas about the strength of trust, making it a natural, intuitive, and extensible framework for effective trust analysis.

The role of trust in social interactions has led to a vast body of work spanning many disciplines of science. Different types of factors influencing trust between two persons can be distinguished, relating to the trusting person (or truster), the trusted person (or trustee), the type of relationship between them, and the context in which trust occurs [2]. *Structural balance theory* in the context of trust suggests that social structures of trust can be stable or unstable. For example, social networks tend to exhibit *triadic closure*, which is loosely the concept that strong relationships are transitive [3]. Figure 1(a)

---

* Also at KU Leuven, Belgium.

A.M. Greenberg, W.G. Kennedy, and N.D. Bos (Eds.): SBP 2013, LNCS 7812, pp. 265–273, 2013.

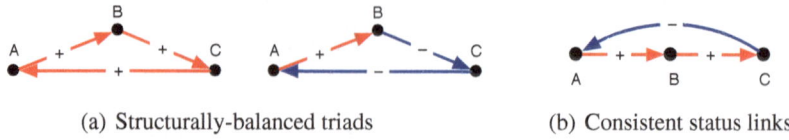

(a) Structurally-balanced triads   (b) Consistent status links

Fig. 1. Implied structures according to competing theories of *structural balance* and *status*. The positive trust relationships from A to B and B to C imply opposite relationships from C to A in the two models.

illustrates examples of such stable structures. If A strongly trusts B, and B strongly trusts C, then triadic closure implies that A will likely trust C (and vice versa). On the other hand, if A does not trust B, B does not trust C, and C does not trust A, this represents an unstable state that structural balance theory suggests should be less likely to occur, as the theory prefers triads with one or three strong trust links.

A competing idea is that these social systems are governed by *status* or *reputation*. This is related to ideas from social psychology on reputation [4], where individuals are trusted based on their expertise in a particular area. In a social status model, the notion of trust is that the trustee (i.e., the person being trusted) is of higher status than the truster (i.e., the person who is trusting). Thus, under a status model, individuals exist in a hierarchy from the most trustworthy to the least trustworthy, along which trust propagates in triangular structures. As for structural balance, if A strongly trusts B, and B strongly trusts C, then status also implies that A will likely trust C. However, as illustrated in Figure 1(b), in contrast to structural balance, status predicts that C will likely *not* trust A in this case. Similarly, if A does not trust B and B does not trust C, then status disagrees with structural balance and implies that A likely does not trust C.

1.1 Related Work

A large community of research focuses on computational modeling of social trust. Methods for analyzing trust include graph-based approaches [5,6,7], probabilistic models [8,9,10], as well as other logic-based approaches [11]. These contributions tend to be fixed computational models based on particular theories of trust, whereas in this paper, we propose PSL as a general tool that provides the flexibility to explore various models without the need to adapt and redesign inference algorithms.

The foundations for many of these computational approaches stem from the vast sociological and psychological literature on human behavior. Recent studies have analyzed some of these theories in the context of social media data, specifically comparing the structural balance- and status-based models we emulate in this work [12,13]. Trust is also an important topic in business analytics; for example, modeling of trust is a useful component for effective viral marketing and e-commerce [14].

2   Probabilistic Soft Logic

Probabilistic soft logic (PSL) [1] is a general purpose system for probabilistic modeling and reasoning in relational domains.[1] PSL uses a first-order language to specify features of graphical models over ground atoms with soft truth-values from the interval $[0, 1]$. In this setting, finding the most likely truth-value assignment (*most probable explanation* inference) can be done efficiently. We refer to Broecheler et al. [1] and Bach et al. [15] for the technical details of the formalism and latest advances in efficient inference, respectively, and restrict the discussion here to an illustration of the key ideas by example. We start from a social network given as ground facts of the form KNOWS(a, b), indicating that user a is acquainted with user b, where we are interested in the truth-values of ground facts of the form TRUSTS(a, b), indicating that user a trusts user b. The following two PSL rules model general constraints that we might expect to hold in this domain, namely, trust being mutual and transitive. They are part of the structural balance model discussed in full detail in Section 2.1.

$$\text{TRUSTS}(A, B) \overset{0.8}{\Rightarrow} \text{TRUSTS}(B, A)$$

$$\text{TRUSTS}(A, B) \wedge \text{TRUSTS}(B, C) \wedge \text{KNOWS}(A, C) \overset{0.5}{\Rightarrow} \text{TRUSTS}(A, C)$$

Rules have nonnegative weights (written above the implication operator $\Rightarrow$ in the previous example) indicating their relative importance. The probability of a truth-value assignment to all ground atoms is defined as a function of the weighted *distance to satisfaction* of each ground rule. Generalizing the notion of rule satisfaction from the Boolean case to continuous truth-values, a rule is satisfied if the truth-value of its head (i.e., consequence) is at least that of the body (i.e., antecedent). The distance to satisfaction of an unsatisfied rule is the difference between the truth-values of the body and head. In our example program above, we thus prefer trust networks where many links are mutual and respect transitivity if users know each other, where the (hypothetical) weights indicate that the mutuality is considered more important.

2.1   Modeling Trust in PSL

We now expand the sketch above into models of competing theories for social trust in PSL. As before, we reason about two predicates KNOWS and TRUSTS, representing an observed social network and trust relationships between individuals, respectively. For any two individuals A and B in the social network, we set KNOWS$(A, B) = 1.0$ if A is acquainted with B, and 0.0 otherwise. Soft truth-values for TRUSTS atoms represent degrees of trust. For instance, TRUSTS$(A, B) = 1.0$ indicates that A fully trusts B, while TRUSTS$(A, B) = 0.5$ indicates that A somewhat trusts B, and TRUSTS$(A, B) = 0.0$ indicates that A does not trust B. We assume that if A trusts B, A knows B, but not necessarily vice versa. We use these models to predict unobserved truth-values of TRUSTS(A, B) for pairs of individuals for whom KNOWS(A, B) is true. In all our models, we include a prior for the truth-value of an atom TRUSTS(A, B) centered around the global average of all observed trust scores.

---

[1] PSL is available as an open-source software package at http://psl.umiacs.umd.edu

In order to predict the degree of trust between two individuals, structural balance considers sixteen possible stable triangular structures involving the two individuals and a third individual. For example, an individual is likely to trust people his or her friends trust; this tendency is encoded as the first rule below. Simplified versions of the rules for each of these structures are:

$$
\begin{aligned}
\text{TR}(A,B) \wedge \text{TR}(B,C) &\Rightarrow \text{TR}(A,C), & \text{TR}(B,A) \wedge \text{TR}(B,C) &\Rightarrow \text{TR}(A,C), \\
\text{TR}(A,B) \wedge \neg\text{TR}(B,C) &\Rightarrow \neg\text{TR}(A,C), & \text{TR}(B,A) \wedge \neg\text{TR}(B,C) &\Rightarrow \neg\text{TR}(A,C), \\
\neg\text{TR}(A,B) \wedge \text{TR}(B,C) &\Rightarrow \neg\text{TR}(A,C), & \neg\text{TR}(B,A) \wedge \text{TR}(B,C) &\Rightarrow \neg\text{TR}(A,C), \\
\neg\text{TR}(A,B) \wedge \neg\text{TR}(B,C) &\Rightarrow \text{TR}(A,C), & \neg\text{TR}(B,A) \wedge \neg\text{TR}(B,C) &\Rightarrow \text{TR}(A,C), \\
\text{TR}(A,B) \wedge \text{TR}(C,B) &\Rightarrow \text{TR}(A,C), & \text{TR}(B,A) \wedge \text{TR}(C,B) &\Rightarrow \text{TR}(A,C), \\
\text{TR}(A,B) \wedge \neg\text{TR}(C,B) &\Rightarrow \neg\text{TR}(A,C), & \text{TR}(B,A) \wedge \neg\text{TR}(C,B) &\Rightarrow \neg\text{TR}(A,C), \\
\neg\text{TR}(A,B) \wedge \text{TR}(C,B) &\Rightarrow \neg\text{TR}(A,C), & \neg\text{TR}(B,A) \wedge \text{TR}(C,B) &\Rightarrow \neg\text{TR}(A,C), \\
\neg\text{TR}(A,B) \wedge \neg\text{TR}(C,B) &\Rightarrow \text{TR}(A,C), & \neg\text{TR}(B,A) \wedge \neg\text{TR}(C,B) &\Rightarrow \text{TR}(A,C),
\end{aligned}
\tag{1}
$$

where we write TR as shorthand for TRUSTS to save space, and the full version of each rule is of the form,

$$
\begin{aligned}
\text{KNOWS}(A,B) \wedge \text{KNOWS}(B,C) &\wedge \text{KNOWS}(A,C) \wedge \\
\text{TRUSTS}(A,B) \wedge \text{TRUSTS}(B,C) &\Rightarrow \text{TRUSTS}(A,C).
\end{aligned}
$$

In these full versions of the rules, a parallel, positive KNOWS atom is added for each TRUSTS atom, which ensures that the groundings for A, B, and C are relevant entities representing acquaintance triangles in the social network.

In addition to the triangle rules, a natural extension of the structural balance model may include reciprocation of trust, which is captured using the rules

$$
\begin{aligned}
\text{TRUSTS}(A,B) &\Rightarrow \text{TRUSTS}(B,A), \\
\neg\text{TRUSTS}(A,B) &\Rightarrow \neg\text{TRUSTS}(B,A).
\end{aligned}
\tag{2}
$$

The status model only makes predictions in the eight cases represented by the following simplified rules, where it agrees with structural balance on four triangular structures, but makes opposite predictions on the other four:

$$
\begin{aligned}
\text{TR}(X,Y) \wedge \text{TR}(Y,Z) &\Rightarrow \text{TR}(X,Z), & \text{TR}(Y,X) \wedge \neg\text{TR}(Y,Z) &\Rightarrow \neg\text{TR}(X,Z), \\
\neg\text{TR}(X,Y) \wedge \neg\text{TR}(Y,Z) &\Rightarrow \neg\text{TR}(X,Z), & \neg\text{TR}(Y,X) \wedge \text{TR}(Y,Z) &\Rightarrow \text{TR}(X,Z), \\
\text{TR}(X,Y) \wedge \neg\text{TR}(Z,Y) &\Rightarrow \text{TR}(X,Z), & \text{TR}(Y,X) \wedge \text{TR}(Z,Y) &\Rightarrow \neg\text{TR}(X,Z), \\
\neg\text{TR}(X,Y) \wedge \text{TR}(Z,Y) &\Rightarrow \neg\text{TR}(X,Z), & \neg\text{TR}(Y,X) \wedge \neg\text{TR}(Z,Y) &\Rightarrow \text{TR}(X,Z),
\end{aligned}
\tag{3}
$$

where again we use shorthand for space, and in our full implementation, we include positive KNOWS atoms mirroring each TRUSTS atom that appears in a rule.

In contrast to the structural balance model, a natural addition to enforce a consistent status hierarchy suggests the inversion of trust between pairs of individuals. We can represent this with the rules

$$\text{TRUSTS}(X, Y) \Rightarrow \neg\text{TRUSTS}(Y, X),$$
$$\neg\text{TRUSTS}(X, Y) \Rightarrow \text{TRUSTS}(Y, X). \tag{4}$$

3 Experiments

We now demonstrate the flexibility of trust modeling with probabilistic soft logic by evaluating different models on real social trust data.[2] We consider a structural balance model (referred to in our discussion below as PSL-Balance) comprised of the rules in (1), and a structural balance model with reciprocation (PSL-Balance-Recip), comprised of (1) and (2), as well as a status model (PSL-Status), comprised of (3), and a status model with inversion (PSL-Status-Inv), comprised of (3) and (4). We use the FilmTrust data set [17][3] as well as data from Epinions.com [7]. The FilmTrust data consists of a set of anonymized users, their trust values for other users, and their ratings for a set of movies (which we omit from this study). Users rate each other on a discrete scale of whole numbers from 1 to 10, which we normalize to $[0, 1]$, making each trust value interpretable as a soft truth-value. There are 1,754 users in the data set, among which there are 2,055 total user-to-user trust values. The trust values are directed and thus not symmetric. We sample via *snowball sampling* a network of 2,000 users from the Epinions data, which contains 8,675 discrete $\{-1, 1\}$ trust scores between users, which we treat as false and true TRUSTS predicate values.

The task we consider is collective prediction of trust values given the fully-observed social network. We generate eight folds where, in each fold, $1/8$ of the trust values are hidden at random. The prediction algorithm can use the remaining $7/8$ of the trust values and the full structure of the social network to learn parameters for a model and perform inference of the unknown trust values. For example, PSL learns weights for the rules in each given model from these observed trust values.

3.1 Baselines

We compare our PSL models to a range of baselines, including two popular approaches for computational trust modeling. As a simple baseline, we predict the average trust across all observed trust values for every prediction. EigenTrust [6] is a global metric that computes a trust value for each node by finding the left principle eigenvector of a normalized trust matrix. The trust matrix is normalized such that each row sums to 1.0, making the normalized trust matrix stochastic. EigenTrust's prediction is then the stationary distribution of the stochastic process described by the normalized trust matrix, or equivalently the limit on the probability of landing on each node as a random walk approaches infinity, where the probability of walking to a neighbor is proportional to how much the current node trusts the neighbor.

TidalTrust [5] is a graph-based algorithm that propagates trust values through neighbors by recursively using the weighted average of neighbor trust to decide a node's trust

---

[2] Early versions of these experiments appeared in [16].

[3] FilmTrust is a web service designed to leverage user-to-user trust values and user-to-movie ratings for movie recommendation. http://trust.mindswap.org/FilmTrust

Table 1. Average scores of FilmTrust trust predictions using mean average error (MAE), Kendall-tau statistic τ, and Spearman's rank correlation ρ for the full test set and the non-default predictions (MAE*, τ*, and ρ*). Each statistic is computed separately on each fold, and the average over all folds is listed here. Scores that are statistically equivalent to the best score according to a two-sample t-test with rejection threshold 0.05 in each metric are typed in bold.

Method	MAE	τ	ρ	MAE*	τ*	ρ*
Average	**0.210**	n/a	n/a	n/a	n/a	n/a
EigenTrust	0.339	−0.054	−0.074	0.339	−0.054	−0.074
TidalTrust	0.229	0.059	0.078	0.236	0.089	0.117
PSL-Balance	**0.207**	**0.136**	**0.176**	**0.193**	**0.235**	**0.314**
PSL-Balance-Recip	**0.207**	**0.139**	**0.188**	**0.193**	**0.241**	**0.318**
PSL-Status	0.224	**0.112**	**0.144**	0.230	**0.205**	**0.277**
PSL-Status-Inv	0.224	0.065	0.085	0.238	**0.143**	**0.189**

for another. TidalTrust predicts distinct trust values per link, rather than a single global trust value per node. To predict an unknown trust value from a source node to a sink node, the algorithm uses a breadth-first search to determine the set of minimum length paths from the source to the sink. TidalTrust then recursively computes the neighbor-weighted trust for the sink node along these paths, starting from the sink node until finally reaching the source, at which point it outputs the final weighted trust.

3.2   Results

On the FilmTrust data, since the ground truth is continuous-valued, we measure for each algorithm the average score over the eight folds for three metrics: mean average error (MAE), Kendall's τ statistic, and Spearman's rank correlation ρ. MAE measures the absolute error on the soft truth-values, while τ and ρ measure ranking performance. The average scores are listed in Table 1. Three PSL models, all but PSL-Status-Inv, are statistically tied for the best-performing method on all three metrics, according to a two-sample t-test with rejection threshold 0.05. This suggests that the inversion rules do not help in this setting. Both EigenTrust and TidalTrust do not do as well here as in their natural problem setup. Here, the prediction algorithms must do joint inference over many unknown trust values, where a significant fraction of the values are unknown. This can disrupt network-based methods that depend on the connectivity of observed information. For example, TidalTrust depends on the existence of alternate paths between nodes, and the removal of a full eighth of this already sparse network significantly increases the number of pairs for which a directed path does not exist. In these cases, we set TidalTrust to predict the global average of all trust values. Since EigenTrust returns a probability distribution over the nodes, its predictions are not on the same scale as the true values, thus making it difficult to directly compare the raw error. Nevertheless, the disconnected state of the network causes the spectral prediction to seemingly fail at recovering any signal from the data when measuring rank correlations. In contrast, PSL takes advantage of the edges with unobserved trust values to propagate information across the network during collective inference, and is thus more robust to the disconnections from the sampling process.

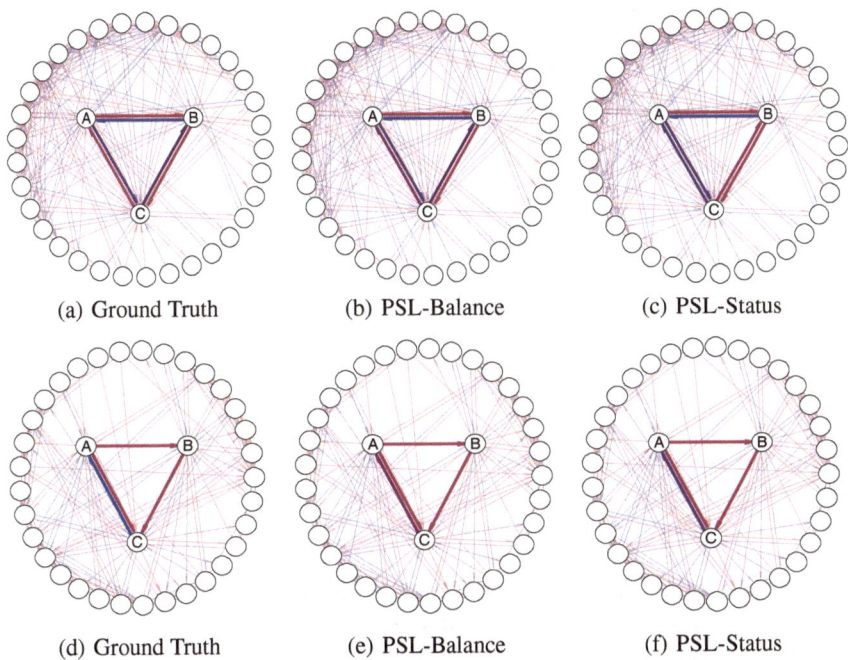

(a) Ground Truth   (b) PSL-Balance   (c) PSL-Status

(d) Ground Truth   (e) PSL-Balance   (f) PSL-Status

Fig. 2. Small example subgraphs of FilmTrust network. We plot the one-hop neighborhood around two triangles where the two primary PSL models disagree. The top row contains the network surrounding a triangle whose ground truth behaves consistently with the structural balance model, and the bottom row contains one surrounding a triangle that is more consistent with the status model. Edges are colored by trust scores, ranging from blue (no trust) to red (maximal trust). The left column (a,d) contains the ground truth trust network, the middle column (b,e) contains the predictions by the PSL-Balance model, and the right column (c,f) contains those by the PSL-Status model.

Because of the network's sparsity, especially after subsampling for testing, methods that propagate trust values suffer. In our experiments, we allow each method to predict the global average when no information is available to propagate, due to the query edge being disconnected from any observed edge. In PSL, the prior has a similar effect. To isolate performance on nontrivial predictions, we also measure accuracy statistics only on the edges for which the method predicts non-default values. We list these in Table 1 as MAE*, τ*, and ρ*. On the non-default predictions, the PSL models show a clear advantage over others, suggesting that their joint inference effectively propagates trust throughout the network.

Comparing the competing PSL models, of all 2055 edges, PSL-Balance and PSL-Status predict 514 trust scores that differ by at least 0.1. In Figure 2, we visualize a few cases where these predictions differ. Since both models produce similarly high accuracy, we suspect the trust behavior in the FilmTrust network follows a combination of the two models.

Table 2. Average area under precision-recall curves of various methods predicting Epinions trust relationships. We compute precision and recall with respect to finding the rarer non-trust links. Scores statistically equivalent to the best score are displayed in bold, where statistical significance is measured via a two-sample t-test with rejection threshold 0.05.

Method	AUC
Average	0.069762
PSL-Balance	**0.316843**
PSL-Balance-Recip	**0.343011**
PSL-Status	**0.296563**
PSL-Status-Inv	0.279580
EigenTrust	0.131159
TidalTrust	0.129785

On the Epinions data, since the ground truth is discrete-valued, we measure for each algorithm the precision and recall on the non-trust links. We compute these retrieval metrics on the non-trust links because the majority of links in the data set are trust links, with 7,974 positive links and only 701 negative links. Thus, we expect a better prediction to retrieve the rarer link type. Table 2 lists the average area under the precision-recall curve for each method. The PSL-Balance-Recip model again produces the best-scoring prediction, while PSL-Balance and PSL-Status produce statistically equivalent results, according to two-sample t-tests with rejection threshold 0.05.

4   Discussion

This paper proposes the use of probabilistic soft logic (PSL) as a natural framework for modeling trust in social networks. Such a generic framework allows for easy exploration of trust models based on different assumptions about social phenomena. To demonstrate the effectiveness of PSL for this task, we apply competing trust models based on structural balance and status to predict user trust data. Further exploration of the literature on trust within this framework is a promising direction for future work. For instance, one could model multiple relationship types and trust topics, capturing the intuition that a person may trust a sibling more than a co-worker about family issues, while trusting the co-worker more about career advice. Similarly, different people have varying degrees of expertise on particular topics, earning them different levels of trust dependent on the context. Finally, the structure of social trust is similar in form to various other phenomena in social networks, such as opinion, social influence, and complex contagion modeling. Each of these problems may benefit from the power and flexibility of a system such as PSL.

Acknowledgments. This work is supported by the Intelligence Advanced Research Projects Activity (IARPA) via Department of Interior National Business Center (DoI / NBC) contract number D12PC00337. The U.S. Government is authorized to reproduce and distribute reprints for governmental purposes notwithstanding any copyright annotation thereon. Disclaimer: The views and conclusions contained herein are those of the authors and should not be interpreted as necessarily representing the official policies or endorsements, either expressed or implied, of IARPA, DoI/NBC, or the U.S. Government. A. Kimmig is a postdoctoral fellow of the Research Foundation Flanders (FWO Vlaanderen).

References

1. Broecheler, M., Mihalkova, L., Getoor, L.: Probabilistic similarity logic. In: Conference on Uncertainty in Artificial Intelligence, UAI (2010)
2. Levin, D.: Trust. In: Clegg, S., Bailey, J. (eds.) International Encyclopedia of Organization Studies, pp. 1573–1579. Sage (2008)
3. Granovetter, M.: The Strength of Weak Ties. The American Journal of Sociology 78(6), 1360–1380 (1973)
4. Cosmides, L., Tooby, J.: Cognitive Adaptions for social exchange. The adapted mind: Evolutionary psychology and the generation of culture. Oxford University Press (1992)
5. Golbeck, J.: Computing and Applying Trust in Web-based Social Networks. PhD thesis, University of Maryland, College Park, College Park, MD, USA (2005)
6. Kamvar, S., Schlosser, M., Garcia-Molina, H.: The Eigentrust algorithm for reputation management in P2P networks. In: International Conference on World Wide Web, WWW (2003)
7. Richardson, M., Agrawal, R., Domingos, P.: Trust Management for the Semantic Web. In: Fensel, D., Sycara, K., Mylopoulos, J. (eds.) ISWC 2003. LNCS, vol. 2870, pp. 351–368. Springer, Heidelberg (2003)
8. Kuter, U., Golbeck, J.: Sunny: A new algorithm for trust inference in social networks using probabilistic confidence models. In: National Conf. on Artif. Intelligence, AAAI (2007)
9. Rettinger, A., Nickles, M., Tresp, V.: Statistical relational learning of trust. Machine Learning 82(2), 191–209 (2011)
10. Vydiswaran, V., Zhai, C., Roth, D.: Content-driven trust propagation framework. In: Knowledge Discovery and Data Mining, KDD (2011)
11. Jøsang, A., Hayward, R., Pope, S.: Trust network analysis with subjective logic. In: Australasian Computer Science Conference, ACSC (2006)
12. Leskovec, J., Huttenlocher, D., Kleinberg, J.: Signed networks in social media. In: Conference on Human Factors in Computing Systems, CHI (2010)
13. Guha, R., Kumar, R., Raghavan, P., Tomkins, A.: Propagation of trust and distrust. In: International Conference on World Wide Web, WWW (2004)
14. Salam, A., Iyer, L., Palvia, P., Singh, R.: Trust in e-commerce. Commun. ACM 48(2), 72–77 (2005)
15. Bach, S., Broecheler, M., Getoor, L., O'Leary, D.: Scaling constrained continuous Markov random fields with consensus optimization. In: Adv. in Neur. Information Proc. Sys. (NIPS) (2012)
16. Huang, B., Kimmig, A., Getoor, L., Golbeck, J.: Probabilistic soft logic for trust analysis in social networks. In: International Workshop on Stat. Relational Artif. Intelligence (2012)
17. Golbeck, J., Hendler, J.: FilmTrust: Movie recommendations using trust in web-based social networks. ACM Transactions on Internet Technology 6(4), 497–529 (2006)

Coauthor Prediction for Junior Researchers

Shuguang Han, Daqing He, Peter Brusilovsky, and Zhen Yue

School of Information Sciences, University of Pittsburgh
135 North Bellefield Ave. Pittsburgh 15213 United States
{shh69,dah44,peterb,zhy18}@pitt.edu

Abstract. Research collaboration can bring in different perspectives and generate more productive results. However, finding an appropriate collaborator can
be difficult due to the lacking of sufficient information. Link prediction is a related technique for collaborator discovery; but its focus has been mostly on the
core authors who have relatively more publications. We argue that junior researchers actually need more help in finding collaborators. Thus, in this paper,
we focus on coauthor prediction for junior researchers. Most of the previous
works on coauthor prediction considered global network feature and local network feature separately, or tried to combine local network feature and content
feature. But we found a significant improvement by simply combing local network feature and global network feature. We further developed a regularization
based approach to incorporate multiple features simultaneously. Experimental
results demonstrated that this approach outperformed the simple linear combination of multiple features. We further showed that content features, which
were proved to be useful in link prediction, can be easily integrated into our
regularization approach.

Keywords: Coauthor prediction, link prediction, social network, expert search.

1 Introduction

Identifying and maintaining appropriate collaboration relations are critical in a researcher's academic life [18] because collaboration can bring together diverse expertise to the same research problem and generate more influential results. The link prediction techniques developed in social network research community [12] can help
predict future collaboration and make researchers aware of the possible coauthors.
However, most of the research works considered only the core authors [12,20] who
have at least a certain number of publications both in the training dataset and the testing dataset (three in [12], and five in [20]). Considering the skewed distribution
between the number of authors and the number of publications [13], the selection
criteria will cut off a large proportion of authors. The conclusion from the core authors may not be useful for the rest authors, because predicting from sparse data is
more difficult [15]. Besides, the prediction in current works is in the global level, in
which top-k ranked pairs among the entire candidate pairs are selected as the predicted links (k is the number of links in the testing dataset). In the global level

A.M. Greenberg, W.G. Kennedy, and N.D. Bos (Eds.): SBP 2013, LNCS 7812, pp. 274–283, 2013.

prediction, there is no control of generating prediction for particular individuals; however, we believe that it is it is more useful if the prediction is for individuals, especially for junior authors. They usually don't have sufficient coauthors, and are more eager to form new connections.

Data sparseness is recognized as a major problem for the prediction of coauthors for junior researchers. To relieve data sparseness, content information was commonly used. Related techniques in expert search [1] utilize content information to find relevant experts. In the recommender system domain, a hybrid method combining both content information and social network feature was often used to solve the cold start problem [14,11,19]. In previous research that considered social network information, either the local network features (e.g. the direct connections) or global network features (e.g. shortest path) were used respectively. Our experiments showed that combing the local and global network features significantly improve the prediction performance, no matter content information was added or not.

Since multiple features are used in this task, a following question is how to combine them effectively. Linear combination is a simple solution, but it is difficult to scale different scores and tune the parameters. An alternative method is to treat the prediction as a binary classification problem [16,20,3] based on multiple features. However, to train a binary classifier, we need to use both positive subjects (real coauthor pairs) and negative subjects (real non-coauthor pairs). Negative subject sampling is difficult because not observing a coauthor link does not imply two authors not are real non-coauthor pair. It may because the coverage of the dataset is limited.

To sum up, the focus of this paper is to predict coauthors for junior researchers. Multiple features including local network features, global network features and content features are considered to improve the prediction performance. In the remainder sections of this paper, we first review related work in section 2. Then, in section 3, a new approach to combine multiple evidences using regularization framework is proposed. Then, we described the datasets and evaluation metrics in section 4. In addition, Empirical results analysis is discussed in section 5. In the final section, we summarize our findings and propose future directions.

2   Related Works

In the literature, coauthor prediction has been modeled as a similarity measuring problem, a recommendation problem, or a classification problem. When viewed as a similarity measuring problem, the similarities between any two authors are calculated, and then the author pairs are ranked and those in top positions are chosen as the predicted links [12]. The core of this approach is to define the vertex similarity [5]. Authors with high vertex similarity are assumed to have high probabilities of collaboration. Network topological features are usually used to measure the vertex similarity. Both local network measures such as common neighbor, Jaccard similarity, Adamic/Adar, preferential attachment, and global network measures such as the shortest path, simRank, and Katz index have been used before. All of these measures are mentioned and compared in [12].

Coauthor prediction can also be viewed as a personalized recommendation problem. The Collaborative Filtering (CF) method was extended in [19] for people-to-people recommendation; however, CF suffers from the cold-start problem when data is sparse. This problem is particularly important in our task because the junior researchers are usually lacking of coauthor information. A hybrid method that combines both social network information and content information can be adopted to relieve the data sparseness. The combination can be a simple linear combination [11,4], a regularization based combination [14], or a filtering based combination [17].

Other researchers [3,20,16] found that besides local and global network topological features, other features can also help improve the prediction performance. For example, the authors' keywords matching, the publication classification code matching [3,16,11] and the meta-path in heterogeneous information networks [20] were all found useful. In order to combine multiple features, the coauthor prediction was modeled as a binary classification problem.

The expert search in the information retrieval domain is also a related work. Related techniques of expert search were not well-studied until TREC's expert finding task [6], in which researchers are required to build an algorithm and rank candidates based on their relevance to the user issued queries. The widely adopted method for expert search is to construct expert profiles using the their previous publications or co-occurrence texts [1]. Expert search didn't model users' social context, which make it less useful than social network based method [11]. However, combining the expert profiles and social context information performs better than using them separately.

3   Methodology

3.1   Problem Definition

The prediction task is formalized as follows: we divide the dataset into the training dataset $\mathbb{D}$ and the testing dataset $\mathbb{D}'$. The division criteria are described in section 4. The test documents $\mathbf{D}' \subseteq \mathbb{D}'$ are defined as those documents with junior researchers as the first authors. Each document d in $\mathbf{D}'$ is further presented by a triple: $< u_1, \mathbf{u} - u_1, \mathbf{m} >$, which indicates the authors of d: u_1 is the first author (u_1 is a junior researcher), $\mathbf{u}$ represents all the authors of the document and $\mathbf{m}$ represents the metadata such as title and/or abstract. The junior researchers are defined as those people who published at least one first-author paper in $\mathbb{D}'$, and at least one but no more than five papers in $\mathbb{D}$. Our goal is to predict the collaborations between u_1 and the rest of the authors $\mathbf{u} - u_1$. However, if u_1 and any author in $\mathbf{u} - u_1$ are coauthors in $\mathbb{D}$, then that coauthor link is not included in our prediction because we are predicting the new coauthor links. $\mathbf{m}$ is used to simulate u_1's topic interest in document d, and we assume that u_1 has already known this information before he/she wants to build connections with authors $\mathbf{u} - u_1$.

3.2   Baseline Models

In terms of the baselines, we adopted two link prediction measures, i.e. the Adamic/Adar index, and the Katz index to represent the best practices using the local network topology features and the global network topology features. We also adopted the Balog Model 2, which is served as the best practice in content-based method. Besides, we considered the standard Collaborative Filtering algorithm which has been found as an effective method in recommendation systems. We adopt the similarity measuring approach for link prediction, the core of which is to rank candidate ca based on his/her similarity with author u_1.

The **Adamic/Adar** index [10] (**AA**) is a typical local network feature based method. In our task, we compute the similarity between candidate ca and u_1, i.e. $s(ca, u_1)$, using Formula (1). $\Gamma(z)$ denotes a set of neighbors of author z, and $|\Gamma(z)|$ denotes the size of $\Gamma(z)$.

$$S(ca, u_1) = \sum_{z \in \Gamma(ca) \cap \Gamma(u_1)} \frac{1}{\log|\Gamma(z)|} \tag{1}$$

The **Katz** [9] (**Katz**) index takes into account of the global network structure. It is defined as the summarization of all paths between candidate ca and u_1, which is computed using Formula (2). Path_{ca,u_1}^l is all the length l path between u_1 and ca. β is the damping factor that controls the weight of the path.

$$S(ca, u_1) = \sum_{l=1 \dots \infty} \beta^l \cdot \left|\text{Path}_{ca,u_1}^l\right| \tag{2}$$

In the content-based baseline model Balog Model 2 (ES) [1] the content similarity is calculated between the topic interest of ca and that of u_1 in paper d using Formula (3). The topic interest is represented by the bag-of-words in m and it is used to mimic user query in **ES**. $p(m|d)$ is estimated using the standard language modeling approach in information retrieval, and $p(ca|d)$ is the association between author ca and document d. In this paper, we used the uniform association for multi-authored papers, and each author receives the same weight of association regardless of author order.

$$S(ca, u_1) = p(m|ca, u_1) \propto \sum_d p(m|d)p(ca|d) \tag{3}$$

The fourth baseline is the user-based Collaborative Filtering (CF) algorithm [2]. The traditional scenario of CF consists of users, items and users' ratings on items. However, in the case of people-to-people recommendation, the user and item are both people and there are no explicit ratings on items. In order to apply the CF into the coauthor prediction, we treat people as both the user and the item, and the number of papers two people coauthored as the people's rating on each other, i.e. users' ratings on items. Using the simple average weighted aggregation, the similarity between u_1 and ca is calculated using Formula (4), in which C_k is k most nearest neighbors of u_1. $r_{u',ca}$ is u' 's ($u' \in C_k$) rating on ca, i.e. the number of coauthored papers of u' and ca. $w(u_1, u')$ measures the similarity of rating on items between user u_1 and u', which is calculated by the cosine similarity of their coauthors (see Formula (5).). κ is the normalized term.

$$S(ca, u_1) = \kappa \sum_{u' \in C_k} w(u_1, u') r_{u',ca} \tag{4}$$

$$w(u_1, u') = cosine(\Gamma(u_1), \Gamma(u')) \tag{5}$$

3.3   Multiple Objective Optimization Using Regularization

Each of the baseline models only considered one type of feature. Since the combination of multiple features has proven to be useful in many works [11,4,20,3,16] , a following problem is to combine multiple features more effectively. The simple linear combination works only when features in the combination are independent. As mentioned in [12], when β is small, Katz is very similar to the neighborhood based approach such as AA, which means these two features are not independent to each other. Therefore, here we propose to use a regularization based approach as suggested in paper [7].

Our first regularization based combination approach is named as **AAN**, in which local network feature based method AA is set as the base, and the objective is to combine features from global networks and/or content information. For each document d in $\mathbb{D}'$, we need to rank ca for u_1 based on their similarity score vector $\boldsymbol{S}$. $\boldsymbol{S}$ is initialized as a zero vector. $\boldsymbol{S}$ is updated according to an objective function Ω_1 defined in formula (6), in which $\boldsymbol{S}^*$ denotes the final score vector, $\boldsymbol{I} - \boldsymbol{M}$ (M is the adjacent matrix of coauthor networks) is the difference matrix, and $\|\cdot\|$ denotes the L2 norm of a vector. $\boldsymbol{S}^{*T}(\boldsymbol{I} - \boldsymbol{M})\boldsymbol{S}^*$ helps propagate local similarity scores through the global network while $\|\boldsymbol{S}^* - \boldsymbol{S}_{AA}\|^2$ ensures the final score $\boldsymbol{S}^*$ do not go far away from $\boldsymbol{S}_{AA}$, and μ_a is the importance parameter. To minimize the objective function, we set derivation of Ω_1 to $\boldsymbol{S}^*$ equals to 0, and the closed-form solution is shown in Formula (7). However, solving the inverse of a matrix is time consuming. An alternative method is to use the power iteration method as suggested in [7]. In each iteration, we can update the score $\boldsymbol{S}^*_{AAN}(t)$ using Formula (8) and the final solution for the iteration is $\boldsymbol{S}^*_{AAN}(t) = \boldsymbol{S}^*_{AAN}(\infty)$.

$$\Omega_1 = \boldsymbol{S}^{*T}(\boldsymbol{I} - \boldsymbol{M})\boldsymbol{S}^* + \mu_a \|\boldsymbol{S}^* - \boldsymbol{S}_{AA}\|^2, \mu_a > 0 \tag{6}$$

$$\boldsymbol{S}^*_{AAN} = (1 - \alpha)(1 - \alpha\boldsymbol{M})^{-1}\boldsymbol{S}_{AA}, \alpha = 1/(1 + \mu_a) \tag{7}$$

$$\boldsymbol{S}^*_{AAN}(t + 1) = \alpha\boldsymbol{M}\boldsymbol{S}^*_{AAN}(t) + (1 - \alpha)\boldsymbol{S}_{AA} \tag{8}$$

For the comparison purpose, we also proposed a linear combination model AANL that combines both local and global network feature. We computed two different similarity scores: the Adamic/Adar score $S_{AA}(ca, u_1)$ and the Katz score $S_{Katz}(ca, u_1)$. Then, the two scores are combined using Formula (9), in which λ indicates the importance of Katz score.

$$S_{AANL}(ca, u_1) = (1 - \lambda)S_{AA}(ca, u_1) + \lambda S_{Katz}(ca, u_1) \tag{9}$$

In order to introduce the second regularization based combination approach **AANE**, we first define a simple linear combination model AAE (shown in Formula 11) which incorporate content information with AA. **AANE** then incorporate both content and global network information with AA. The objective function Ω_2 in AANE is defined in Formula (11). The closed solution of Formula (11) is Formula (12). The power iteration method can also be used for **AANE** to optimize the objective function.

$$S^*{}_{AAE} = \gamma S_{ES} + (1 - \gamma)S_{AA}, \gamma = 1/(1 + \mu_b) \tag{10}$$

$$\Omega_2 = S^{*T}(I - M)S^* + \mu'_a\|S^* - S_{ES}\|^2 + \mu'_b\|S^* - S_{AA}\|^2, \mu'_a, \mu'_b > 0 \tag{11}$$

$$S^*{}_{AANE} = (I - \alpha'M)^{-1}(\gamma'S_{ES} + (1 - \alpha' - \gamma')S_{AA})$$

$$\text{Where, } \alpha' = 1/(1 + \mu'_a + \mu'_b), \gamma' = \mu'_a/(1 + \mu'_a + \mu'_b) \tag{12}$$

4   Dataset and Evaluation Design

The dataset used in this study contains 151,165 ACM hosted conference papers that were published between 2000 and 2011 in the ACM Digital Library. Each paper in the dataset includes a title and an abstract. The authors of these papers were disambiguated using the ACM author identifiers (In the ACM Digital Library, each author is assigned a unique identifier number). In total, there are 209,592 unique authors. Coauthor relations are extracted to create a coauthor network. A link between two authors is added if they co-published at least one paper.

The dataset is divided into three parts according to publishing time for evaluation: T1= [t_{2000}, t_{2003}], T2= [t_{2004}, t_{2007}] and T3= [t_{2008}, t_{2011}]. There are 3,760 papers in T2, and 5,914 papers in T3 that have junior researchers as the first author. These papers were selected for evaluation. T2 is the testing set when using T1 as the training set, while T2 is the training set when using T3 as the testing set. Therefore, as the two dataset used for evaluation are named asT1-T2 and T2-T3.

Two evaluation metrics were used. The first metric is the accuracy in top-10 positions (**WTP**), which examines whether the correct coauthor is ranked within the top-10 positions. However, the exact ranking position information is lost in this case. If two algorithms both can recommend results in top-10 positions, we cannot distinguish their performance using WTP. Therefore, another evaluation metric mean reciprocal rank (**MRR**) [22] was also used as it reflects the exact ranking position.

5   Result Analysis and Discussion

5.1   Parameter Selection

AA and ES were implemented directly as there are no explicit parameters in these two algorithms need to be tuned. For other algorithms, parameters were tuned and the one with best performance were selected. When the performances on WTP and MRR have conflictions, the parameter that has better performance on WTP was chosen.

For the user-based Collaborative Filtering (CF) algorithm, as shown in Formula (4), we tried different values of k (1, 3, 5, 7 and 9), and finally chose 5 because it has the best performance in terms of both MRR and WTP. This means that the 5 nearest neighbors were selected as the similar users. In the Katz index method, we follow the Gauss-Southwell algorithm [8]. A set of damping factors (i.e. the β) values are adopted and compared, including 0.1, 0.05, 0.005, 0.0005. Finally, β = 0.05 were selected as it is the one with best performance on both MRR and WTP. In AANL, λ is set to be 0.95 because it gives the best performance on both WTP and MRR.

For the rest three models AAN, AAE and AANE, both of the parameters α and γ are ranging from [0,1]. We set different values for the parameters from 0 to 1, with 0.1 as gradient step and chose the one with best performance ones: $\alpha = 0.5$ for T1-T2, and $\alpha = 0.1$ for T2-T3 in AAN; $\gamma = 0.1$ in T1-T2, $\gamma = 0.02$ is in T2-T3 for AAE, $\alpha' = 0.5$, $\gamma' = 0.01$ in T1-T2 and $\alpha' = 0.1$, $\gamma' = 0.002$ in T2-T3 for AANE. In AAE, the ES scores are usually small; therefore, we use a heuristic method to multiple them by 1000 in order to be able to combine with other scores.

5.2   Comparative Evaluation of Eight Models

The result analysis on each metric consists of two parts: a bar chart on how each model performed and a statistical test to reveal the significance of experimental results. Non-parametric test Wilcoxon Signed Ranks was used since the normality was not satisfied. The following results show the comparisons of eight models: AA (Formula 1), Katz (Formula 2), ES (Formula 3), CF (Formula 4), AAE (Formula 10), AANL (Formula 9), AAN (Formula 8) and AANE (Formula 12).

The evaluation results on WTP is shown in Figure 1 and results on MRR is shown in Figure 2. We found that the four proposed hybrid models (AAE, AANL, AAN and AANE) are all significantly better than the single feature based models (ES, CF, AA and Katz) on both WTP and MRR. It may suggest that different features actually reveal different aspects of data, and combing them can improve the performance. The

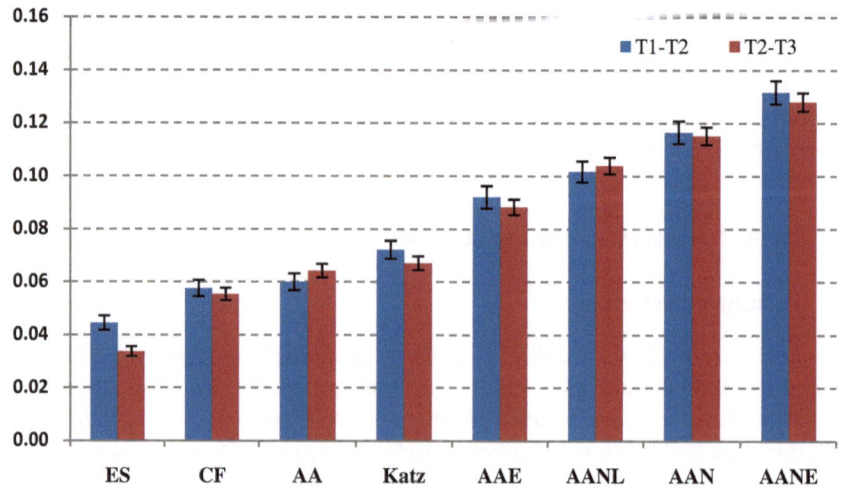

Fig. 1. WTP evaluation with stand errors

previous studies either only considered the local network feature or only the global network feature. The fact that AAN and AANL performs significantly better than both AA and Katz indicates that combining the local network features and the global network can improve the prediction accuracy. We also found that the regularization based model AAN is significantly better than linear combination models AANL. This indicates that the regularization based approach is a better approach for multiple feature combination compared to the simple linear combination. Among all the eight models, AANE performs the best. This indicates that incorporating all three features together using regularization based approach produce the best predication accuracy.

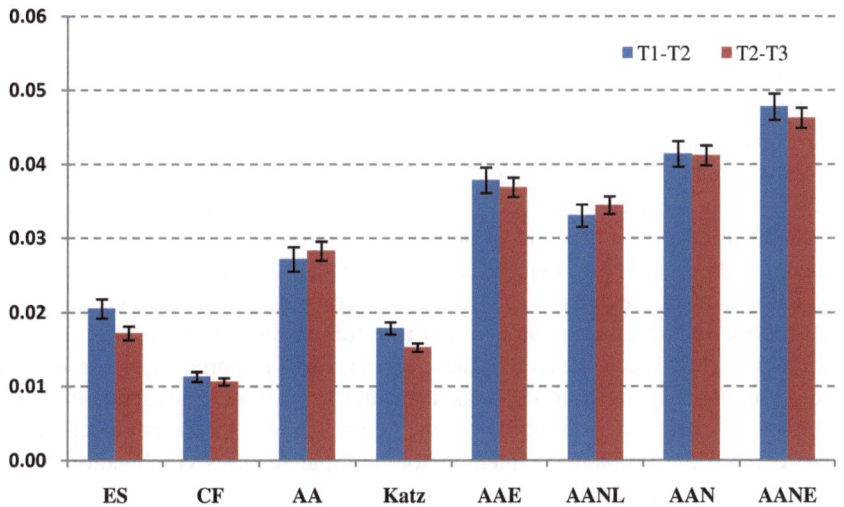

Fig. 2. MRR evaluation with stand errors

Some previous works found that combining content and network feature improves the predication. In our results, AAE is significantly better than content only model (ES) and network only model (CF, AA and Katz), which confirms the previous findings. In addition, we actually found that combing the local network feature and the global network feature helps more than combing the local network features with the content features. This is supported by the fact that AAN and AANL are significantly better than AAE on both WTP and MRR. The reason that AANL is better than AAE only on WTP is that MRR measures the exact rank positions while using content information can avoid ranking the right candidate in extreme low positions. We also found that pure network features based methods, i.e. CF, AA and Katz, are significantly better (p<0.001) than the pure content based method ES in terms of WTP evaluation. Content match focuses on finding potential coauthors using topic interest match, but people are unlikely to form coauthor relationships if they are not reachable to each other in the network even though they share similar topic interest. However, ES seems to be superior to CF on MRR, which may suggest that although it is unable to rank the right candidates to the top positions, it is also unlikely to rank the right candidates in extreme low positions.

AA and Katz have conflict performance on WTP and MRR. We found that in the T1-T2 dataset, Katz is significantly better than AA on WTP while it seems to be worse than AA on MRR evaluation in T2-T3. We think that AA suffers from data sparseness problem because it only considers local network feature. However, when only the global network feature is included in Katz, it introduces many noises.

6   Conclusion

In this paper, we look into the coauthor prediction problem for junior researchers who were actually ignored in previous works. The global network, the local network and the content-based feature were found to be useful in previous works for link prediction. We proposed two regularization based models to combine multiple features and optimize them simultaneously. Comparing to the four baseline models that each consider only a single feature, our proposed models performed significantly better on the predication accuracy. A particularly interesting finding is that propagating feature from the local network to the global network improves the performance significantly compared to the model that combine content and local network feature. This indicates that although many previous works focused on combing the content feature and the local network features, they actually didn't take full advantage of the network features by not taking global network feature into account. Most importantly, the results show that our proposed regularization approach is better than simple linear combination and can be easily expanded to multiple features combination. In the next step, we will further explore the propagation method for multiple features combination, such as random walk or belief propagation.

References

1. Balog, K., Azzopardi, L., Rijke, M.: Formal models for expert finding in enterprise corpora. Paper Presented at the Proceedings of the 29th Annual International ACM SIGIR Conference on Research and Development in Information Retrieval (2006)
2. Breese, J.S., Heckerman, D., Kadie, C.: Empirical analysis of predictive algorithms for collaborative filtering. Paper Presented at the Proceedings of the Fourteenth Conference on Uncertainty in Artificial Intelligence, Madison, Wisconsin (1998)
3. Chao Wang, V.S.: Srinivasan Parthasarathy Local Probabilistic Models for Link Prediction. In: Seventh IEEE International Conference on Data Mining (2008)
4. Chen, H.-H., Gou, L., Zhang, X., Giles, C.L.: CollabSeer: a search engine for collaboration discovery. Paper Presented at the Proceedings of the 11th Annual International ACM/IEEE Joint Conference on Digital Libraries, Ottawa, Ontario, Canada (2011)
5. Chen, H.-H., Gou, L., Zhang, X., Giles, C.L.: Discovering missing links in networks using vertex similarity measures. Paper Presented at the Proceedings of the 27th Annual ACM Symposium on Applied Computing, Trento, Italy (2012)

6. Deng, H., Han, J., Lyu, M.R., King, I.: Modeling and exploiting heterogeneous biblio-graphic networks for expertise ranking. Paper Presented at the Proceedings of the 12th ACM/IEEE-CS Joint Conference on Digital Libraries, Washington, DC, USA (2012)
7. Craswell, N., de Vries, A.P., Soboroff, I.: Overview of the trec-2005 enterprise track. In: Proceedings of the 14th Text Retrieval Conference (2005)
8. Bonchia, F., Esfandiar, P., Gleichc, D.F., Greifd, C., Lakshmanand, L.V.S.: Fast Matrix Computations for Pairwise and Columnwise Commute Times and Katz Scores. Internet Mathematics 8(1-2) (2011)
9. Katz, L.: A new status index derived from sociometric analysis. Psychometrika 18(1), 39–43 (1953)
10. Lada Adamic, E.A.: Friends and Neighbors on the Web. Social Networks 25, 211–230 (2002)
11. Lee, D., Brusilovsky, P., Schleyer, T.: Recommending Future Collaborators using Social Features and MeSH terms. Paper Presented at the Proceedings of the 74th Annual Meeting of the American Society for Information Science and Technology (2011)
12. Liben-Nowell, D., Kleinberg, J.: The link prediction problem for social networks. Paper Presented at the Proceedings of the Twelfth International Conference on Information and Knowledge Management, New Orleans, LA, USA (2003)
13. Lotka, A.J.: The frequency distribution of scientific productivity. Journal of the Washing-ton Academy of Sciences 16(12), 317–324 (1926)
14. Ma, H., Zhou, D., Liu, C., Lyu, M.R., King, I.: Recommender systems with social regula-rization. In: Paper Presented at the Proceedings of the Fourth ACM International Confe-rence on Web Search and Data Mining, Hong Kong, China (2011)
15. Shang, M.-S., Lü, L., Zeng, W., Zhang, Y.-C., Zhou, T.: Relevance is more significant than correlation: Information filtering on sparse data. EPL 88(6) (2009)
16. Hasan, M., Chaoji, V., Salem, S., Zaki, M.: Link Prediction Using Supervised Learning. In: SDM (2006)
17. Chaiwanarom, P., Ichise, R., Lursinsap, C.: Finding potential research collaborators in four degrees of separation. In: Cao, L., Zhong, J., Feng, Y. (eds.) ADMA 2010, Part II. LNCS, vol. 6441, pp. 399–410. Springer, Heidelberg (2010)
18. Kahn, R.L., Denis, J.P.: Interdisciplinary collaborations are a scientific and social impera-tive. The Scientist (1994)
19. Cai, X., Bain, M., Krzywicki, A., Wobcke, W., Kim, Y.S., Compton, P., Mahidadia, A.: Collaborative Filtering for People to People Recommendation in Social Networks. In: Li, J. (ed.) AI 2010. LNCS, vol. 6464, pp. 476–485. Springer, Heidelberg (2010)
20. Sun, Y., Barber, R., Gupta, M., Aggarwal, C.C., Han, J.: Co-Author Relationship Predic-tion in Heterogeneous Bibliographic Networks. In: International Conference on Advances in Social Networks Analysis and Mining, pp. 121–128 (2011)

Massive Media Event Data Analysis to Assess World-Wide Political Conflict and Instability

Jianbo Gao[1], Kalev H. Leetaru[2], Jing Hu[1],
Claudio Cioffi-Revilla[3], and Philip Schrodt[4]

[1] PMB Intelligence LLC, West Lafayette, IN 47996, USA
jbgao.pmb@gmail.com
http://www.gao.ece.ufl.edu
[2] Graduate School of Library and Information Science, University of Illinois,
Urbana-Champaign, IL 61820, USA
[3] Center for Social Complexity and Department of Computational Social Science,
George Mason University, Fairfax, VA 22030, USA
[4] Department of Political Science, Pennsylvania State University, PA 16802, USA

Abstract. Mining massive daily news media data to infer patterns of cultural trends, including political conflicts and instabilities, is an important goal of computational social science and the new interdisciplinary field called "culturnomics." While the sheer size of media data makes this task challenging, a greater hurdle is the nonstationarity of data, manifested in several ways, which invalidates surge in media coverage as a reliable indicator of political change. We demonstrate the use of advanced statistical, information-theoretic, and random fractal methods to analyze CAMEO-encoded political events data. In particular, we show that on the country level, event distributions obey a Zipf-Mandelbrot law, and interactions among countries follow an exponential law, indicating that local or prioritized events dominate the political environment of a country. Most importantly, we find that world-wide political instabilities, such as the Arab Spring, are associated with breakdown or enhancement of long-range correlations in political events.

1 Introduction: Motivation and Background

Archived news data are formidably large for any human individual or teams to attempt manual coding. Due to technological improvements, media data are now generated at an astronomical rate every minute. These data, old and new, contain valuable information on many aspects of culture. Analyses of these data to infer various kinds of sociopolitical trends, including political conflicts and instabilities, is a priority goal of computational social science and the new interdisciplinary field, "culturnomics" [1–4]. This is a very difficult task, since political processes can be extraordinarily complex. One of the major challenges is nonstationarity in news media coverage: A trivial event can attract huge media attention, while major political events may not be reported adequately. This is shown by daily event totals for USA in Fig. 1(a) (and other countries, not shown here). Such nonstationarity prevents researchers from judging the significance of

A.M. Greenberg, W.G. Kennedy, and N.D. Bos (Eds.): SBP 2013, LNCS 7812, pp. 284–292, 2013.
© Springer-Verlag Berlin Heidelberg 2013

an event based on media coverage. Here, we propose to use advanced statistical, information-theoretic, and fractal methods to analyze CAMEO-encoded political events data. Besides analyzing events time series, we also analyze their Goldstein-scale intensity [5], as shown in Fig. 1(b). By using the daily mean, the issue of whether the Goldstein scale accurately quantifies the significance of an event is mitigated. At a minimum, we show that such data provides more information than the tone or "mood" based on media coverage [2]. The local variance of the graph in Fig. 1(b) decreases with time, another form of nonstationarity. Below we demonstrate how fractal methods can eliminate such nonstationarity.

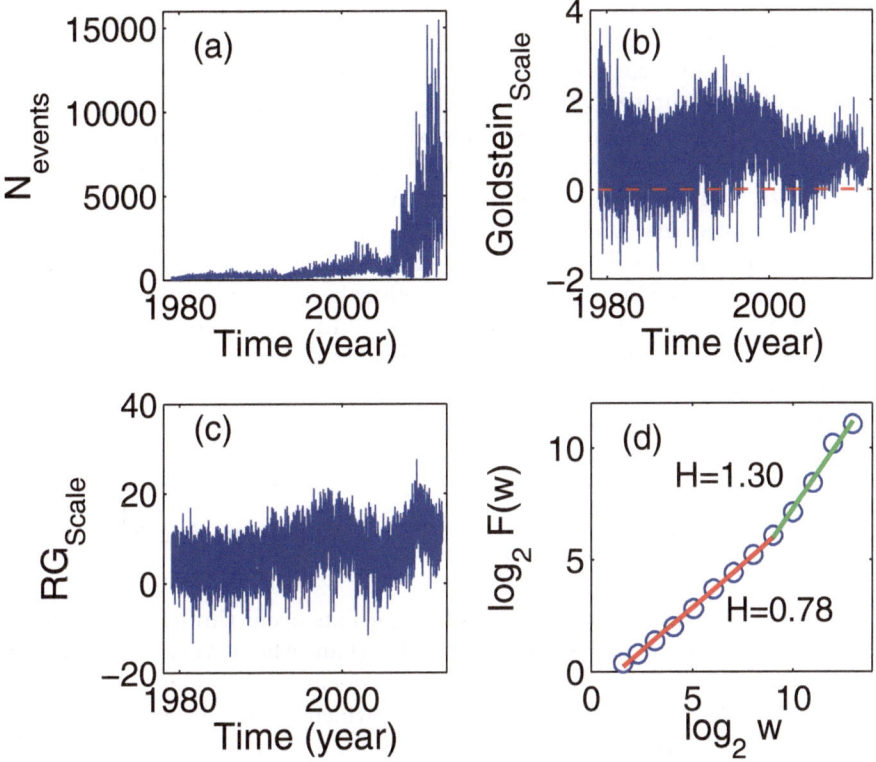

Fig. 1. Political events data in USA: (a) absolute daily number of events, (b) daily mean Goldstein Scale intensity, (c) rescaled daily mean Goldstein Scale, and (d) adaptive fractal analysis of the data shown in (c). For data source, see below for details.

2   Methodology

In this section we describe the new data set that we used in this study, followed by three analytical methods that are still relatively new to computational social science in general (automated information extraction) and culturnomics in particular.

2.1   Data

Political event data are nominal codes that record who did what to whom and when (i.e., codings based on units of analysis consisting of subject-verb-object (S-V-O) phrases), coded from news media[6]. The new event data set analyzed here is called Global Database of Events, Language, and Tone (GDELT). It includes more than 178 million unique events across all countries, during the period from 1979 to the present. GDELT is a new initiative based on tera-bytes of information to construct a catalog of all major human societal activity across all countries of the world. GDELT events are drawn from a wide array of news media, both in English and non-English, from across the world, ranging from international to local sources in nearly every country. These data were produced by the TABARI automated coding software (http://eventdata.psu.edu/software.dir/tabari.html) using the CAMEO event and actor coding system [7]. TABARI works with a very large set of verb-phrase (>15,000 phrases) and noun-phrase (>40,000 phrases) dictionaries in combination with shallow parsing of English-language sentences to identify grammatical structures such as subject-verb-object, compound subjects and objects, and compound sentences. CAMEO is an update of earlier (1960s) event coding systems, with changes introduced by automated coding and new behaviors, such as suicide bombings. CAMEO provides a detailed and systematic framework for coding contemporary political actors, including international, supranational, transnational, or internal actors. An earlier version of this system recently was successfully employed in the DARPA ICEWS project [8] to code 25 gigabytes of Asian news reports involving more than 6.7 million stories, which provided the key input for forecasting models, with accuracy, sensitivity, and specificity all exceeding DARPA's pre-set criteria.

2.2   Zipf-Mandelbrot Law for Ranked Events: Quasi-universal Scaling

Every day hundreds of thousands of political events occur throughout the world. However, some events occur more frequently than others. What is the general functional form of such distributions when events are ordered by occurrence frequency? The classic Zipfian rank-size law [9], which states that the frequency of an observed value is inversely proportional to its rank, is a theoretical hypothesis for the occurrence of political events. We focus on two classes of events: (1) those for the whole world in a short period of time, such as April 2012; and (2) events in a specific country during the past 30+ years, from 1979 to 2012. We show that the general shape for the probability density $p(k)$ in both cases is well characterized by a generalization of Zipf's law, the Zipf - Mandelbrot law [10], idefined as

$$p(k) = c/(k + q)^{\alpha}, \tag{1}$$

where α is the scaling exponent, k is the rank of the event ($k = 1$ for the most frequent event, $k = 2$ for the 2nd most frequent event, etc.), q is a constant parameter, and c a constant coefficient determined by the condition that total probability is 1. Three examples are shown in Fig. 2. The exponent α (slope of the red part of the graphs) varies across countries only slightly, and therefore, the scaling found here may be called quasi-universal.

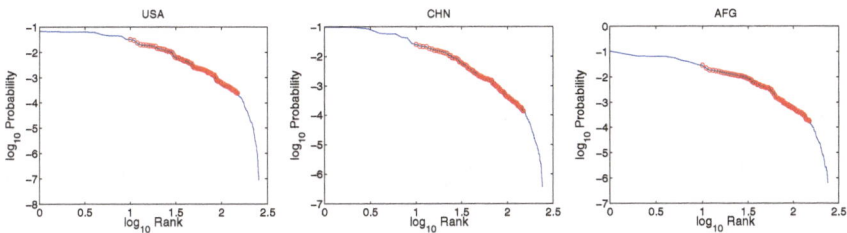

Fig. 2. The Zipf-Mandelbrot Distributions for the ranked events in USA, China, and Afghanistan, with the scaling exponents $\alpha = 1.95$, 2.12, and 2.05, respectively

2.3 Structure of Interaction Sub-networks

A country interacts with many other countries, as well as with groups within and outside the country. We quantify the interaction strength among the country and other actors by the total number of events occurring among them in a time period of 30+ years, from 1979 to 2012, normalized by the total number of interactions between the country and all other polities. Such interaction strength can be interpreted as a probability. Fig. 3 shows four examples. The distributions are close to exponential,

$$\text{Interaction strength}_i \sim e^{-\beta i}, \tag{2}$$

where i represents the i-th most significant interaction partner and $\beta > 0$ is a parameter, as shown by the nearly straight line when plotted in semi-log space. Exponential distributions in interaction sub-networks indicate that the political environments of a country are largely dominated by local or prioritized events.

2.4 Enhancement or Breakdown of Persistent Long-Range Correlations Arising from Political Instability

One of the most ubiquitous and puzzling features of complex systems is the so-called $1/f^\alpha$ noise, a form of temporal or spatial fluctuation characterized by a power-law decaying power spectral density. A sub-class of $1/f^\alpha$ noise is the stochastic process with long memory characterized by the Hurst parameter $0 < H < 1$. The process is said to have anti-persistent, short-range (or memory-less), or persistent long-range correlations, depending on whether $H < 1/2$, $H = 1/2$, or $H > 1/2$, respectively. The basic model for $1/f^\alpha$ noise is the fractional Brownian motion (fBm), which is a Gaussian random walk process with mean 0

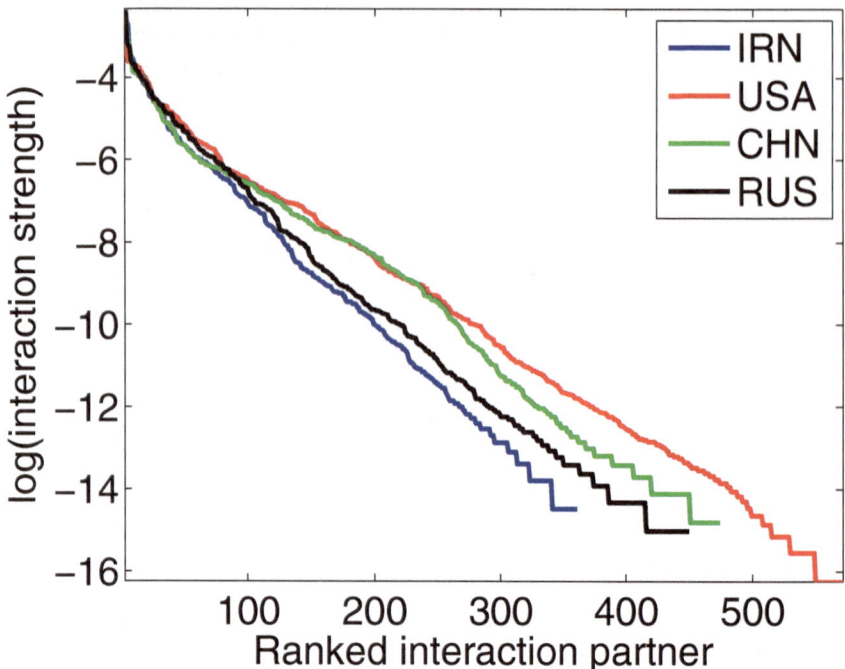

Fig. 3. Exponential distributions for the interaction sub-networks of USA, China, Russia, and Iran

and stationary increments. Its variance, covariance, and power spectral density (PSD) are given by $E[(B_H(t))^2] = t^{2H}$, $E[B_H(s)B_H(t)] = \frac{1}{2}\{s^{2H} + t^{2H} - |s - t|^{2H}\}$, and $E(f) \sim f^{-(2H+1)}$, respectively. Note that processes with $H > 1$, with long memory also exist, and may arise from a variety of situations including nonstationarity.

The Hurst parameter can be used to eliminate nonstationarity observed in daily mean Goldstein Scale data shown in Fig. 1(b). The idea is best understood by considering the effect of smoothing irregular data. Denote a stationary time series by $X = \{X_t : t = 0, 1, 2, \ldots\}$. We then construct a new time series

$$X^{(m)} = \{X_t^{(m)} : t = 1, 2, 3, \ldots\}, \ m = 1, 2, 3, \ldots,$$

obtained by simple nonoverlapping averaging,

$$X_t^{(m)} = (X_{tm-m+1} + \cdots + X_{tm})/m, \ \ t \geq 1 . \tag{3}$$

For ideal fractal processes, there is an interesting scaling law for the variance of $X_t^{(m)}$ on the aggregation level m [11, 12], given by the expression

$$var(X^{(m)}) = \sigma^2 m^{2H-2}, \tag{4}$$

where σ^2 is the variance of the original data. Eq. 4 means that if we multiply the daily mean Goldstein Scale data shown in Fig. 1(b) by $n^{1-H_{intraday}}$, where n and $H_{intraday}$ are the number of events and the Hurst parameter of events within a day, then the rescaled mean Goldstein Scale data will be largely stationary, in the sense that the local variance of the data will be almost constant. This is evident in Fig. 1(c). Note that $H_{intraday}$ may be different from H for daily mean Goldstein Scale data, which is for longer time scales.

There is a rich variety of methods to estimate H from a truly fractal process [11, 12]. However, care must be exercised when estimating H from real world data, especially when the data contain trends, nonstationarity, or signs of rhythmic activity [13]. One of the more reliable and convenient methods for dealing with these issues is the recently developed adaptive fractal analysis (AFA) [3, 14], which is a refinement of the celebrated detrended fluctuation analysis (DFA) method. Denote a time series of interest (e.g., the rescaled daily mean Goldstein Scale data) by $\{x_1, x_2, \cdots, x_N\}$. Its integral,

$$u(i) = \sum_{k=1}^{i}(x_k - \bar{x}), \quad i = 1, 2, \cdots, n, \tag{5}$$

is called a random walk process. For an arbitrary window size w, AFA involves first fitting a globally smooth trend signal $v(i)$ to the $u(i)$ time series. The residual, $u(i) - v(i)$, characterizes fluctuations around the global trend, and its variance yields the Hurst parameter H according to

$$F(w) = \left[\frac{1}{N}\sum_{i=1}^{N}(u(i) - v(i))^2\right]^{1/2} \sim w^H. \tag{6}$$

Fig. 1(d) shows an example of AFA using the rescaled daily mean Goldstein Scale data of USA. We observe two scaling regions. One is up to a time scale of 2^{10} days, with $H = 0.78$. The other is for longer time scales, with $H = 1.30$. The first scaling, having $1/2 < H < 1$, suggests that political events in USA have persistent correlations, at least for a time scale up to 3 years. This feature can be readily understood by noticing that in order for a movement or effort to shape a political landscape, it has to be persistently carried out for a while. History is full of such efforts and movements. Indeed, this feature is generic among a few dozens of countries that we have examined, albeit the exact value of H varies from one country to another. The second scaling, having $H > 1$, suggests that on time scales longer than 3 years, the political environment in USA is nonstationary.

Nonstationary political processes, and especially political instabilities worldwide from time to time suggest that a better strategy for examining the long-range-correlations in a country is to partition the rescaled daily mean Goldstein scale data into many short segments and then estimate H for each segment. The scaling break shown in Fig. 1(d) suggests that a good choice of the segment length is 3 years. As a compromise for getting good temporal resolution and computational speed, we have chosen to let successive segments overlap by 2 years. Five examples of the variation of the Hurst parameter with time are shown in Fig. 4. Since $H > 1/2$ in all those countries, we can conclude that the correlations there are always persistent.

Consistent with our earlier discussion, we argue that a stable political event process without much fundamental change in policies and general perception of the people in the country would yield an almost constant H. To understand how H may change with time, we may consider two political actors (polities), one old, holding onto its status quo policies (stable), and the other new, undergoing fundamental change in goals, personnel, or institutions (unstable). These two regimes are dynamically different. If the status quo prevails, current policies will be enhanced, manifested by a continual increase in the Hurst parameter value. On the other hand, if political change succeeds, then fundamental policy changes will follow and the Hurst parameter will decrease. Of course, if the new regime shares everything with the old power, except the "identity," then even if there is a switch in power, the variation of the Hurst parameter with time will follow the first pattern — to increase with time, indicating that the new regime, albeit with a different "name," is persisting with the old policies (status quo).

3   Results and Discussion

Fig. 4 shows the main results of the preceding methodological approach. Consider the graph for China. The first decrease in $H(t)$ occurred around 1982, coinciding with the leadership regime transition from Hua Guofeng to Deng Xiaoping. The second decrease, started shortly before 1990, was caused by the Tianmen Square event. The third (weak) decrease, which occurred around 2002, coincided with the more modest leadership regime transition from Jiang Zhemin and Hu Jintao. The broadly increasing segments of the graph, from about 1982 to 1989, and again from 1992 to 1996, signified regime persistence in terms of China's economic reform.

Next, consider the graph for Haiti. Here, we only observe two sharp decreases, around 1995 and 2004, respectively, coinciding with Aristide taking and being removed from the presidency — the time 1995 is slightly later than the actual time of his presidency, which was October 1994. This means it took a few months for him to change the political course of Haiti. Indeed, the old regime in Haiti was extremely powerful and consistent. Albeit the dictatorship of the Duvalier family ended in 1986, old policies were actually enhanced for years after 1986. In fact, although Aristide survived a coup attempt before his inauguration in 1991, he was soon overthrown. Only with the help of USA, he became a "true" President of Haiti in 1994. Apparently, the course he designed for Haiti was not good for the world, and again with the help of USA, he was removed from the presidency in 2004.

Finally, consider the $H(t)$ graphs for three key countries involved in the Arab Spring. Interestingly, the government of Yemen was soon overthrown during the movement. There is a fundamental reason for that: the increasing trend for Yemen's $H(t)$ starting before 2005. Although the process was perturbed by some events in Yemen around 2008, the basic pattern for the variation of $H(t)$ is still

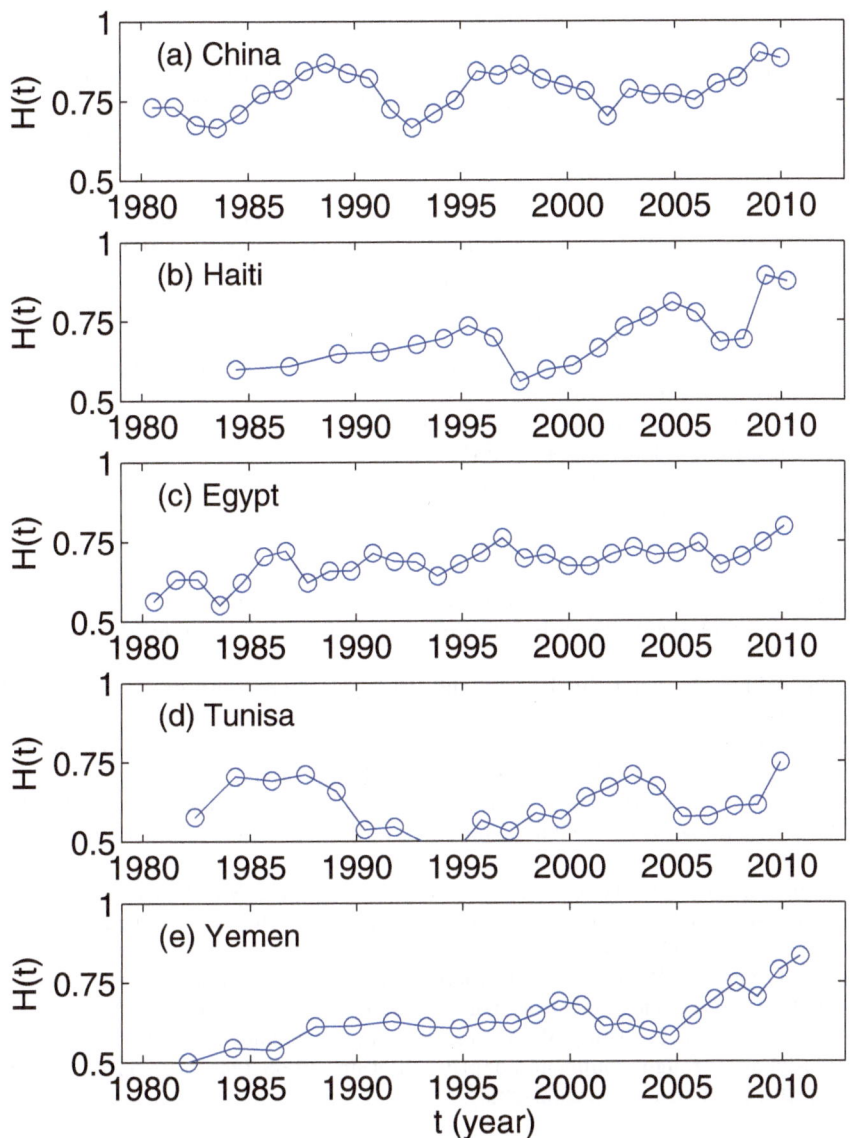

Fig. 4. Variation of the Hurst parameter with time for China, Haiti, Egypt, Tunisia, and Yemen

increasing. Therefore, even if the government in Yemen was overthrown, the basic political course in Yemen has remained similar. In some sense, the Arab Spring might not have been as revolutionary as freedom advocates would hope.

References

1. Michel, J.B., et al.: Quantitative analysis of culture using millions of digitized books. Science 331, 176–182 (2011), doi:10.1126/science.1199644
2. Leetaru, K.H.: Culturnomics 2.0: Forecasting Large-Scale Human Behavior Using Global News Media Tone in Time And Space. First Monday 16(9) (2011)
3. Gao, J.B., Hu, J., Mao, X., Perc, M.: Culturomics meets random fractal theory: Insights into long-range correlations of social and natural phenomena over the past two centuries. J. Royal Society Interface (2012), doi:10.1098/rsif.2011.0846
4. Petersen, A.M., Tenenbaum, J., Havlin, S., Stanley, H.E.: Statistical Laws Governing Fluctuations in Word Use from Word Birth to Word Death. Scientific Reports 2 (2012), doi:10.1038/srep00313.
5. Goldstein, J.S.: A Conflict-Cooperation Scale for WEIS Events Data. Journal of Conflict Resolution 36, 369–385 (1992)
6. Schrodt, P.A.: Precedents, Progress and Prospects in Political Event Data. International Interactions 38, 546–569 (2012)
7. Schrodt, P.A., Gerner, D.J., Ömür, G.: Conflict and Mediation Event Observations (CAMEO): An Event Data Framework for a Post Cold War World. In: Bercovitch, J., Gartner, S. (eds.) International Conflict Mediation: New Approaches and Findings (2009)
8. O'Brien, S.P.: Crisis Early Warning and Decision Support: Contemporary Approaches and Thoughts on Future Research. International Studies Review 12, 87–104 (2010)
9. Zipf, G.K.: Human Behavior and the Principle of Least Effort. Addison-Wesley (1949)
10. Mandelbrot, B.: Information Theory and Psycholinguistics. In: Wolman, B.B., Nagel, E. (eds.) Scientific psychology. Basic Books (1965)
11. Gao, J.B., Cao, Y.H., Tung, W.W., Hu, J.: Multiscale Analysis of Complex Time Series — Integration of Chaos and Random Fractal Theory, and Beyond. Wiley (August 2007)
12. Gao, J.B., Hu, J., Tung, W.W., Cao, Y.H., Sarshar, N., Roychowdhury, V.P.: Assessment of long range correlation in time series: How to avoid pitfalls. Phys. Rev. E 73, 016117 (2006)
13. Hu, J., Gao, J., Wang, X.: Multifractal analysis of sunspot time series: the effects of the 11-year cycle and fourier truncation. J. Stat. Mech. P02066 (2009)
14. Gao, J.B., Hu, J., Tung, W.W.: Facilitating joint chaos and fractal analysis of biosignals through nonlinear adaptive filtering. PLoS One 6(9), e24331 (2011), doi:10.1371/journal.pone.0024331

Sparsification and Sampling of Networks
for Collective Classification

Tanwistha Saha, Huzefa Rangwala, and Carlotta Domeniconi

Department of Computer Science
George Mason University
Fairfax, Virginia, USA
tsaha@gmu.edu, {rangwala,carlotta}@cs.gmu.edu

Abstract. Network analysis has been an active area of research for the
past few decades. Out of many open research questions that have been
extensively studied, relational classification, community detection, link
prediction are only to name a few. Collective classification is a well-known
relational classification method for classifying entities (nodes) within a
network which involves using both node based features and topological
features of each node. It involves *collective prediction* of the unknown
labels of *all* the test nodes in the network using label information of the
training nodes. Even though this has been a well researched topic for
years, very little has been done to address the following two challenges:
(1) how to actively select the labeled nodes from the network to be used
for training, and (2) how to efficiently obtain a sparse representation of
the original network without losing much information, so that learning
can scale to large networks. A lot of work has been done in theoretical
computer science which aims towards finding the best approximation of
large graphs. However, not much has been done from the perspective of
finding an approximate subgraph that will help in classification of network datasets. In this paper, our contribution is in proposing an efficient
graph sparsification method and a sampling technique which, along with
the state-of-the-art network classifiers, can give comparable runtime and
classification accuracies.

1 Introduction

Networks are usually represented as graphs where the vertices are entities and
the edges represent interaction between these entities. Of numerous aspects of
computational social science that are in practice, our research is primarily focused onto learning supervised models that can be used for making accurate
predictions on unseen data within the network. Although major work has been
done on finding efficient learning models, an important challenge arises while
trying to find the right samples (i.e, instances that represent the distribution
from which the original dataset was generated), using which we can learn our
models.

Learning in relational datasets is different from learning in non-relational
datasets, because of the presence of links between entities in networks. These

A.M. Greenberg, W.G. Kennedy, and N.D. Bos (Eds.): SBP 2013, LNCS 7812, pp. 293–302, 2013.
© Springer-Verlag Berlin Heidelberg 2013

links account for pairwise interactions between entities and hence, can be considered for extracting additional features about the network dataset. *Collective classification* considers link-based features along with the node specific features, while training a model. It deals with the simultaneous classification of neighboring instances in relational data, until a convergence criterion is reached. The rationale behind collective classification stems from the fact that an entity in a network (or relational data) is most likely influenced by the neighboring entities, and can be classified accordingly, based on the class assignment of the neighbors. Collective classification approaches jointly classify a set of related (linked) nodes by exploiting the underlying correlations between the labels and the attributes of the training nodes and that of the test nodes in a network [4].

Our contribution in this paper is two fold: (i) We propose an algorithm to *sample* subgraphs from the full network (a.k.a. *network sampling* [2]). (ii) We propose an algorithm to sparsify the network by removing noisy edges and find a suitable approximation that can facilitate collective classification, without compromising the accuracy. We explain the notion of sampling in detail in section 3.3. We use two benchmark network datasets: (i) Cora (directed graph representing citation network of scholarly papers)[1] and (ii) DBLP (undirected graph representing co-authorship network of computer science researchers)[2] to show that our sampling technique, along with a state-of-the-art collective classification method [8], can improve the classification accuracy. In addition, our graph sparsification method helps in reducing the number of edges in the network and reduces the execution time without compromising the classification accuracy.

2   Related Work

Given the aim of collective classification earlier in this paper, an important question that comes to mind is - how to train a collective classification model? In traditional learning theory, samples (set of instances) chosen randomly (with or without replacement) from the entire set of data points are assumed to represent the key characteristics of the original dataset [3]. In these cases, randomly chosen instances for training is sufficient for learning models which does not make use of link structure in the dataset. However, in network analysis, researchers have shown that randomly chosen nodes are not good representative samples of the network [2]. Many competitive approaches, e.g., forest fire sampling [5], node sampling, edge sampling and edge sampling with graph induction [1] have been proposed so far. Most of the proposed methods have either taken a generalized approach towards sampling or have considered networks that evolve over time. Recently, Ahmed et al. [2] have shown that, forest fire sampling has an overall good performance for classification in relational data. The goal of any sampling method is to preserve key features (e.g., sufficient statistics) of the underlying distribution. Which feature is a *key* depends on the problem we are trying to solve. Hence, instead of developing another generalized sampling algorithm for

---

[1] http://www.cs.umd.edu/~sen/lbc-proj/LBC.html
[2] http://www.informatik.uni-trier.de/~ley/db/

any learning method (supervised, semi-supervised or unsupervised), we propose
an adaptive version of forest fire sampling in order to facilitate collective classification in network datasets. This is different from the *active sampling* method
proposed recently [11], where the goal is to draw samples from the network that
have high probability of having a specific label. Our method can be categorized
as *active labeling*, where both the value of the label as well as the local structural
information of the labeled instance are acquired.

Our second contribution is a sparsification method for removing noisy edges
in the network. Attribute information and labels of neighboring nodes have been
useful for improving accuracy in collective classification problems. However, if
the collective classification algorithm fails to classify a node correctly, then the
misclassification error is propagated throughout the network. Certain edges in
a network do not carry much information and do not contribute towards better
accuracy for classification models. These edges, or *weak ties*, between entities can
be removed from the network to speed up the learning process without compromising much with the classification accuracy. There are many good sparsification
algorithms [15,16] which aim to find best approximation for large graphs in linear
time. Those algorithms are focused on reducing the time complexity for finding
the best approximation of large graphs. Our method is different from those approaches; our objective is to find a sparse network that will help in reducing the
runtime of collective classification algorithms. Our graph sparsification approach
is inspired by a local graph sparsification algorithm [13]. However, we emphasize
that this approach was meant for efficient clustering in graphs (or networks),
whereas, our approach is to facilitate classification in graphs.

3   Methodology

Let graph $G = (V, E)$ represent a relational dataset $D = (\mathcal{X}, \mathcal{Y})$, where $\mathcal{X} = \{x_1, x_2, \cdots, x_n\}$ is the set of node attributes and $\mathcal{Y} = \{y_1, y_2, \cdots, y_n\}$ is the
set of corresponding labels for the $|V| = n$ nodes in the network. Let K be the
number of classes, denoted by $k = 1, 2, \cdots, K$, to which the nodes can belong
($y_i = k$, where $i = 1, 2, \cdots, n$). Let $M_{|E| \times 2}$ be an edge-vertex incidence matrix
where row denotes an edge $e \in E$. The column of M indicates the two vertices
with which the edge e is associated. Let $A_{n \times n}$ be the weighted adjacency matrix
representing the interactions between any pair of nodes in the network.

3.1   Problem Definition

Given a sparsification ratio θ ($0 < \theta < 1$), our goal is to:

- Find a sparsified network $G_s = (V, E_s)$ (where $E_s \subset E$) using a sparsification
  algorithm *Adaptive Global Sparsifier* (Algorithm 1) where $|E_s| = \lceil \theta \times |E| \rceil$.
- Find a sample subgraph $G'_s = (V'_s, E'_s)$ from the network $G = (V, E)$ (where
  $V'_s \subset V$ and $E'_s \subset E$) using a sampling algorithm *Adaptive Forest Fire
  Sampling* (Algorithm 2), in order to train a classifier.

Algorithm 1. Adaptive Global Sparsifier (AGS)

Input: A graph $G = (V, E)$, sparsification ratio θ, edge-vertex incidence matrix $M_{|E| \times 2}$
Output: A sparsified graph $G_s = (V, E_s)$ where $E_s \subset E$
1: $E_s \leftarrow \emptyset$
2: $c_G = components(G)$
3: **for** each edge $e \in E$ **do**
4:     $(i, j) \leftarrow M_e$
5:     Compute $e.sim = Sim(i, j)$ as in equation (1)
6: **end for**
7: Sort and order the edges $e \in E$ of the graph according to decreasing order of $e.sim$
8: Add top $\lceil \theta \times |E| \rceil$ edges of G to the sparse graph G_s i.e. $|E_s| = \lceil \theta \times |E| \rceil$
9: **for** each edge $e_p \in E - E_s$ such that e_p is the edge with the lowest similarity score $(e.sim)$ in the sorted edge list **do**
10:     $E \leftarrow E - e_p$
11:     $c = components(G)$
12:     **if** $c > c_G$ **then**
13:         $E_s \leftarrow E_s \cup e_p$
14:     **end if**
15: **end for**

3.2   Adaptive Global Sparsifier

Given the node attributes for all the nodes in the graph G, we compute pairwise cosine similarity between any pair of nodes i and j, as follows,

$$Sim(i, j) = \frac{\mathbf{x}_i^T \mathbf{x}_j}{\|\mathbf{x}_i\|_2 \|\mathbf{x}_j\|_2} \tag{1}$$

We sort the edges according to descreasing order of similarity and keep the top $\lceil \theta \times |E| \rceil$ edges in the graph. For the rest of the edges, we start scanning from the bottom of the sorted list of edges. We remove an edge if this doesn't increase the number of connected components in the graph (function $components(G_s)$ in Algorithm 1, checks the number of connected components in the graph G_s), otherwise we put back that edge. A graph becomes more disjoint if the number of connected components increases. This situation is not encouraged in collective classification setting, since collective classification relies on the connectivity of the graph for propagating the labels from training nodes to test nodes. Our sparsification algorithm AGS aims to remove noisy edges from the graph such that removal of such an edge *does not increase* the number of connected components in the sparsified graph, as compared to the original graph. Checking for connected components within a graph has a time complexity of $O(|V| + |E|)$. Hence, the time complexity of Algorithm 1 is $O(|E| \times (|V| + |E|))$, since the outer *for-loop* (lines $9 - 15$) can run at most $|E|$ times. Our AGS algorithm is different from the global and local sparsification algorithms proposed by Satuluri et al. [13]. The global sparsification algorithm [13] adds only the top $\lceil \theta \times |E| \rceil$ edges in the graph based on the Jaccard similarity measures of the edges. The local sparsification algorithm [13] scans through each node in the graph, sorts the edges incident on the node according to the Jaccard similarity measure and keeps only d^e edges, where d is the degree of the node under consideration, and e is any predefined value between 0 and 1. For ease of reference, we call these two methods *Global Sparsifier (GS)* and *Local Sparsifier (LS)*, respectively. We consider these two sparsification algorithms as baseline methods for comparison.

3.3   Adaptive Forest Fire Sampling

Forest fire sampling as proposed by Leskovec et al. [6,5], starts by choosing a random seed node and burning a fraction of its outgoing links to neighboring nodes attached to it. This fraction is chosen from a geometric distribution with a predefined mean. It is similar to a random walk based approach, where the selection process is recursive and it halts when no new nodes are selected, or when the required sample size has been reached. Since collective classification propagates the label information from the training nodes to the test nodes in a network during the inference process, it would be helpful if a test node is connected to at least one training node in order to make use of the training node's characteristics for the test node during the training phase of the classifier. Our sampling algorithm is an adaptive version of forest fire sampling and is described in detail in Algorithm 2.

Algorithm 2. Adaptive Forest Fire Sampling

Input: A graph $G = (V, E)$, edge-vertex incidence matrix $M_{|E| \times 2}$, training ratio δ
Output: A sampled subgraph $G'_s = (V'_s, E'_s)$
1: $V'_s \leftarrow \emptyset$, $E'_s \leftarrow \emptyset$, $E_p \leftarrow \emptyset$
2: Sample size $s = \lceil |V| \times \delta \rceil$
3: Randomly select a seed node p from the graph G
4: **repeat**
5:     $V'_s \leftarrow V'_s \cup p$
6:     $d_p \leftarrow degree(p)$ /* $degree(p)$ returns degree of node p */
7:     Select $\lceil d_p \times \delta \rceil$ edges incident on node p and add to E_p
8:     **for each** edge $e \in E_p$ **do**
9:         $(p, v) \leftarrow M_e$
10:         $E'_s \leftarrow E'_s \cup e$
11:         **if** $v \notin V'_s$ **then**
12:             $V'_s \leftarrow V'_s \cup v$
13:             Add node v in the end of the queue Q
14:         **end if**
15:     **end for**
16:     **if** Queue Q is not empty **then**
17:         $p \leftarrow$ first node at the head of the queue Q
18:     **else**
19:         Randomly select a seed node p from the graph G such that $p \notin V'_s$
20:     **end if**
21:     $E_p \leftarrow \emptyset$
22: **until** $|V'_s| = s$

The complexity of the adaptive forest fire sampling algorithm mostly depends on the time for adding/removing elements (edges or nodes) from the queue Q which is run by the inner *for-loop* (lines $8 - 15$) in Algorithm 2 with $O(\lceil d_p \times \delta \rceil)$ time. Hence, the time complexity is a function of the degree of the seed nodes chosen randomly by the algorithm. Algorithm 2 aims to find at least one labeled node for each test node in the graph and also ensures that the subgraph is connected. To achieve this goal, the algorithm performs a check at each step by looking forward from the current node and selecting a certain percentage (equal to the percentage of training samples) of edges to be included in the sampled subgraph. It considers the nodes connected to those selected edges as *training nodes* for the sampled subgraph. This process continues until the required training sample size (of nodes) has been reached. We refer to this approach as *adaptive forest fire*, because at each step the algorithm *adapts* towards selecting the edges incident on that particular node residing at the head of the queue that keeps track of the processed nodes. In case there are multiple connected components in the graph, then this algorithm is repeated for each connected component, until the overall sample size for the entire graph has been reached. Selecting training nodes by traversing certain percentage of edges only, ensures that the nodes (that end up being test nodes) at the other end of the unselected edges are connected to at least one labeled node. However, readers should note that, by "sampling" from a graph we mean to extract a subset of nodes whose labels would be made available to the learning algorithm during the training phase such that it helps collective classification. To prove whether this sampled subgraph retains the essential properties of the original graph, is a part of future work.

3.4 Collective Classification

Collective Classification is a combinatorial optimization problem where we are given the set of nodes $V = \{v_1, v_2, v_3, \cdots, v_n\}$ over a graph $G = (V, E, \mathcal{X}, \mathcal{Y}, K)$, such that, E is the set of edges, each $\mathbf{x}_i \in \mathcal{X}$ is an attribute vector for node $v_i \in V$, each $y_i \in \mathcal{Y}$ is a label variable for node v_i, and K is the set of possible labels [14]. We are also given a set of known values Y_l for nodes $V_l \in V$ (the nodes for which we know the correct labels) such that $Y_l = \{y_i | v_i \in V_l\}$. The task is to infer Y_u (i.e., the value of y_i for the each node $v_i \in V - V_l$) for which we do not know the labels. Many sophisticated collective classification methods have been proposed [10,7,9]. Here we use the state-of-the-art weighted vote relational neighbor (wvRN) classifier [8] because it is simple, fast, and yet an efficient method. We also use another simple relational classifier RankNN [12](a multi-label classification algorithm, modified to support single label classification), and a multi-class Support Vector Machine classifier [17] to compare our methods against the baseline.

3.5 Connected Components

A graph $G = (V, E)$ is said to be connected if there is a path between any pair of vertices (u, v), such that $u \in V$ and $v \in V$. A *connected component*, $G_c = (V_c, E_c)$, is a *connected* subgraph of a graph G. A graph $G = (V, E)$ is said to have multiple connected components $G_1, G_2, \cdots, G_n$ (where $G_1 = (V_1, E_1), G_2 = (V_2, G_2), \cdots, G_n = (V_n, E_n)$, $V = V_1 \cup V_2 \cup V_3 \cdots \cup V_n$, and $E = E_1 \cup E_2 \cup E_3 \cdots \cup E_n$), if, for any pair of such components (say $G_i = (V_i, E_i)$ and $G_j = (V_j, E_j)$), there exists a pair of nodes (u, v) such that $u \in V_i$ and $v \in V_j$ and there is no path from u to v, or vice versa.

A graph becomes more disjoint if the number of connected components increases. This situation is not encouraged in collective classification setting since collective classification relies on the connectivity of the graph for propagating the labels from training nodes to test nodes. Our sparsification algorithm *AGS*

Table 1. Description of datasets

Dataset	n	K	$|E|$	$|E_s|$ with (AGS)	Time (sec) (AGS)	$|E_s|$ with (GS)	Time (sec) (GS)	$|E_s|$ with (LS)	Time (sec) (LS)
Cora	2708	7	5429	3850	5.4859	3800	2.5695	2429	2.3997
DBLP	5602	6	17265	12251	53.8278	12086	28.4722	6859	19.6182

aims to remove noisy edges from the graph such that removal of such an edge *does not increase* the number of connected components in the sparsified graph, as compared to the original graph.

4   Experimental Results

We processed two real-world datasets, Cora and DBLP, for testing our methods. The statistics of the datasets are given in Table 1, along with the time taken by different sparsifiers to obtain a sparsified graph from those networks (sparsification ratio $\delta = 0.7$). Cora is a directed citation network where each node represents a paper and each edge represents a citation link from one paper to another. Each paper can belong to one of seven research areas: Case-Based Reasoning, Genetic Algorithms, Neural Networks, Probabilistic Methods, Reinforcement Learning, Rule Learning and Theory. The DBLP network is an undirected co-authorship network between authors publishing research papers in computer science, which can belong to one of six research areas: Theory, NLP, Bioinformatics, Operating System, Distributed System and Networking. We have also reported the time taken by our proposed algorithm Adaptive Global Sparsifier (AGS) and by the baseline algorithms Global Sparsifier (GS)/Local Sparsifier (LS) in order to sparsify the original network datasets.

Figures 1(a) and 1(b) show the performance of our sparsification algorithm (AGS) with respect to the baseline methods (GS and LS) [13] on Cora and DBLP datasets respectively, using the wvRN classifier with iterative inference method and random sampling. On non-sparsified networks, the wvRN classifier performs best for both datasets (which is expected, since, due to sparsification we are losing some information from the network resulting into a loss of accuracy). But when networks are sparsified, our algorithm performs best compared to others (note that, GS and LS methods encourage removal of those edges that increase the number of connected components in the graph). Also, for DBLP, the classifier's accuracy is almost similar to that obtained using a non-sparsified graph.

Figures 1(c) and 1(d) show the performance of different classifiers with random node sampling and adaptive forest fire sampling on the Cora dataset. On non-sparsified networks, the wvRN classifier performs best with random sampling. Immediately following this, is the performance of the wvRN classifier used on networks sparsified by our adaptive global sparsification (AGS) algorithm. However, when adaptive forest fire sampling is used, all the three classifiers using

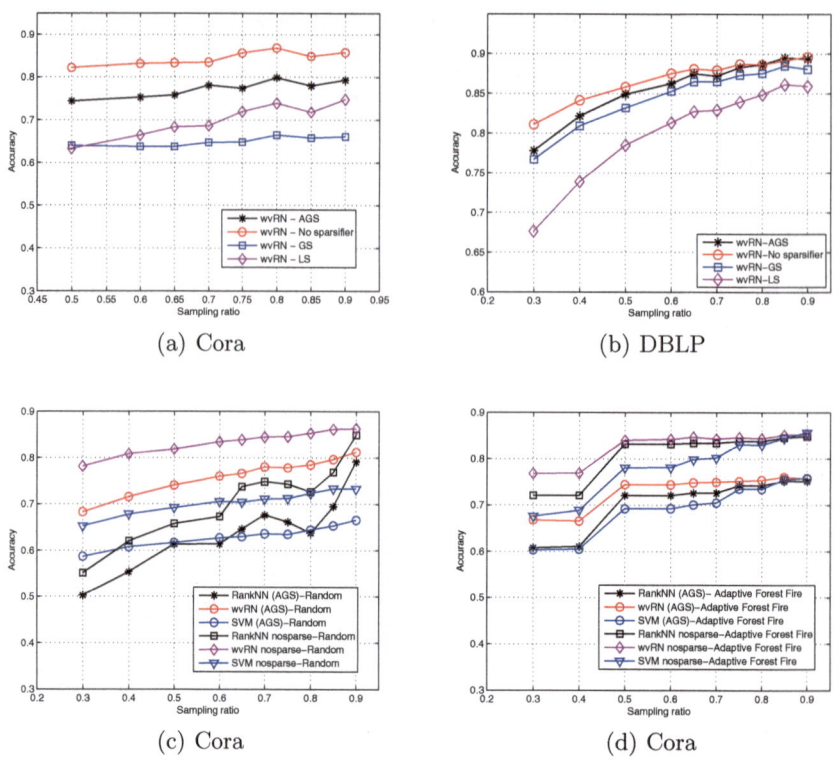

(a) Cora

(b) DBLP

(c) Cora

(d) Cora

Fig. 1. Collective Classification performance using graphs sparsified by AGS, GS and LS

non-sparsified networks, perform better in comparison with the classifiers using sparsified networks (which is natural, since sparsification of networks causes loss of information).

Figures 2(a) and 2(b) show the run time for the wvRN classifier using different graph sparsification algorithms on Cora and DBLP datasets respectively, while varying the training sample sizes. As expected, the wvRN classifier with sparsified networks, take less time for execution in both the datasets.

Figures 2(c) and 2(d) show the run time of the wvRN classifier with networks sparsified using our adaptive global sparsification algorithm (AGS) for different sparsification ratios. Time consumption is less for both the datasets (with different sampling algorithms) when AGS algorithm is used to sparsify the original input networks. We also observe that, the variation of time consumption for the wvRN classifier using the sparsified graphs becomes almost flat when sparsification ratio $\theta > 0.6$.

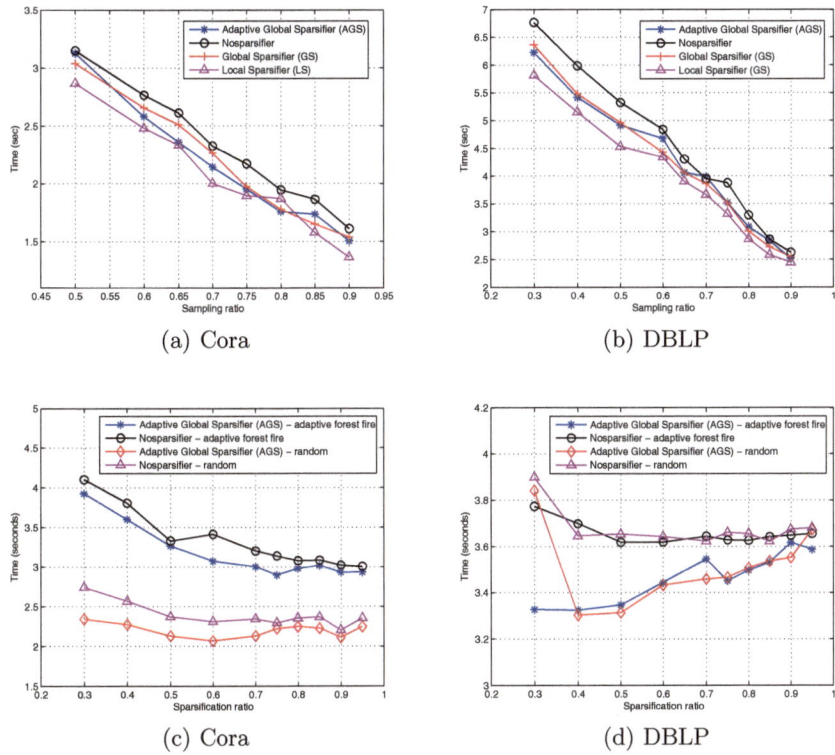

(a) Cora

(b) DBLP

(c) Cora

(d) DBLP

Fig. 2. Run Time of wvRN classifier using different sparsification methods

5   Conclusion

In this paper, we have proposed a useful and efficient technique to sparsify graphs (or networks of entities) that can help in improving the performance of a collective classification method, without compromising the accuracy. We also proposed a network sampling algorithm to extract samples from a network by actively labeling nodes in the original network. We have reported extensive experimental results on two real-world relational datasets, demonstrating that our proposed methods perform well, both with respect to accuracy and time complexity. However, these algorithms have been designed to treat single labeled network datasets only. In the future, we aim to design efficient graph sparsification algorithms for multi-labeled network datasets as well.

References

1. Ahmed, N., Berchmans, F., Neville, J., Kompella, R.: Time-based sampling of social network activity graphs. In: Proceedings of the Eighth Workshop on Mining and Learning with Graphs, pp. 1–9. ACM (2010)

2. Ahmed, N., Neville, J., Kompella, R.: Network sampling designs for relational classification. In: Sixth International AAAI Conference on Weblogs and Social Media (2012)
3. Friedman, J., Hastie, T., Tibshirani, R.: The elements of statistical learning. Springer Series in Statistics, vol. 1 (2001)
4. Jensen, D., Neville, J., Gallagher, B.: Why collective inference improves relational classification. In: Proceedings of the Tenth ACM SIGKDD International Conference on Knowledge Discovery and Data Mining, pp. 593–598. ACM (2004)
5. Leskovec, J., Faloutsos, C.: Sampling from large graphs. In: Proceedings of the 12th ACM SIGKDD, pp. 631–636. ACM (2006)
6. Leskovec, J., Kleinberg, J., Faloutsos, C.: Graphs over time: densification laws, shrinking diameters and possible explanations. In: Proceedings of the eleventh ACM SIGKDD, pp. 177–187. ACM (2005)
7. Lu, Q., Getoor, L.: Link-based classification. In: Proceedings of the 20th International Conference on Machine Learning (ICML 2003), vol. 20, pp. 496–503 (2003)
8. Macskassy, S., Provost, F.: Classification in networked data: A toolkit and a univariate case study. The Journal of Machine Learning Research 8, 935–983 (2007)
9. McDowell, L., Gupta, K., Aha, D.: Cautious inference in collective classification. In: Proceedings of the National Conference on Artificial Intelligence, vol. 22, p. 596. AAAI (2007)
10. Neville, J., Jensen, D.: Iterative classification in relational data. In: Proc. AAAI-2000 Workshop on Learning Statistical Models from Relational Data, pp. 13–20 (2000)
11. Pfeiffer III, J., Neville, J., Bennett, P.: Active sampling of networks. In: 10th International Workshop on Mining and Learning with Graphs (2012)
12. Saha, T., Rangwala, H., Domeniconi, C.: Multi-label collective classification using adaptive neighborhoods. In: 2012 11th International Conference on Machine Learning and Applications (ICMLA), vol. 1, pp. 427–432. IEEE (2012)
13. Satuluri, V., Parthasarathy, S., Ruan, Y.: Local graph sparsification for scalable clustering. In: Proceedings of the 2011 ACM SIGMOD, pp. 721–732. ACM (2011)
14. Sen, P., Namata, G., Bilgic, M., Getoor, L., Galligher, B., Eliassi-Rad, T.: Collective classification in network data. AI Magazine 29(3), 93 (2008)
15. Spielman, D., Srivastava, N.: Graph sparsification by effective resistances. SIAM Journal on Computing 40(6), 1913–1926 (2011)
16. Spielman, D., Teng, S.: Spectral sparsification of graphs. arXiv preprint arXiv:0808.4134 (2008)
17. Tsochantaridis, I., Hofmann, T., Joachims, T., Altun, Y.: Support vector machine learning for interdependent and structured output spaces. In: Proceedings of the 21st International Conference on Machine Learning, p. 104. ACM (2004)

A Comparative Study of Social Media and Traditional Polling in the Egyptian Uprising of 2011

Lora Weiss[1], Erica Briscoe[1], Heather Hayes[1], Olga Kemenova[1], Sim Harbert[1], Fuxin Li[1], Guy Lebanon[1], Chris Stewart[2], Darby Miller Steiger[2], and Dan Foy[2]

[1] The Georgia Institute of Technology, Atlanta, Georgia
{lora.weiss,erica.briscoe,heather.hayes,olga.kemenova,
sim.harbert}@gtri.gatech.edu,
{fli,lebanon}@cc.gatech.edu
[2] The Gallup Organization, Washington D.C.
{chris_stewart,darby_miller_steiger,dan_foy}@gallup.com

Abstract. Because social network sites such as Twitter are increasingly being used to express opinions and attitudes, the utility of using these sites as legitimate and immediate information sources is of growing interest. This research examines how well information derived from social media aligns with that from more traditional polling methods. Specifically, this research examines tweets from over 40,000 Egyptian users from both before and after the Egyptian uprising on January 25, 2011 and compares that information with polling data collected by The Gallup Organization during the same time period. This analysis ascertains trends in sentiment and identifies the extent to which these methodologies align over time. The results show that trends across the two sources are not consistent. Focusing solely on Twitter data, individuals expressed increasingly negative opinions after the uprising, whereas survey results indicated that individuals were increasingly positive post-uprising. We discuss the implications of these differences for the use of social media as a real-time information source.

Keywords: Social Media, Twitter, Sentiment analysis, Polling, Arab Spring.

1   Introduction

Increasingly, social networking sites have become an important tool for social interaction and information sharing. Use of social networks has increased exponentially in the past few years. Facebook went from 100 million users in 2008 to over 1 billion users in 2012 [13]. Twitter has also seen an exponential increase in users since its inception, where there are more than 500 million current users, up from 100 million in 2011 and 5.5 million in 2005 [7]. These users generate over 230 million tweets per day, which is an increase of 110% since 2011 [7].

Not only are social network sites used for self-expression, seeking and sharing of information, and social interaction, but these outlets also contribute to revolutionary movements [17]. Mainstream news and communication media are often restricted or

A.M. Greenberg, W.G. Kennedy, and N.D. Bos (Eds.): SBP 2013, LNCS 7812, pp. 303–310, 2013.
© Springer-Verlag Berlin Heidelberg 2013

blocked by governments, so that social network sites have become a prominent means of quickly disseminating information. Twitter is often associated with opinion, attitude, and rapid information sharing as compared to the more socially-interactive, group-oriented Facebook [12, 14]. Twitter is also used more often to track events that change quickly, such as disasters, accidents, and riots [8]. Twitter became increasingly popular in Egypt and other Arab countries during the Arab Spring. There was a significant increase in internet use for Egypt; namely, a 39.61% increase from 2010 to 2011 [2]. The greatest increase in users was during the first three months of 2011, where Egypt was in the top five Arab countries in number of Twitter users and tweets [3], with an increase of over one million tweets between the week of January 16, 2011 (122,319 tweets) to the week of January 24, 2011 (1.3 million tweets) [18], where the date associated with the Egyptian uprising is canonically January 25, 2011. Thus, Twitter effectively becomes a new paradigm for journalism, as evidenced by the increase of its use during the government-controlled blackout in Egypt after the 2011 uprising [3], [17].

The goal of the research reported in this paper is to ascertain how individuals' sentiment towards various topics in Egypt varied during the revolution and how this variance manifests in both Twitter and Gallup polling data. Polling data was obtained by The Gallup Organization by conducting major surveys in Egypt in 2010 and 2011 involving face-to-face interviews with approximately 1000 adults per survey [19, 20]. This paper describes research that examined a subset of many tweets from over 40,000 Egyptian users both before and after the Egyptian uprising on January 25, 2011 and compares that information with polling data collected by The Gallup Organization in Egypt during the same time period.

2   Methodology

A traditional method of assessing sentiment, opinions, and attitudes of a study population is through the use of administered surveys, which is one of the most common approaches to measuring a wide variety of constructs. Such methodologies are maximized through reliability and validity testing [15]. They can also serve as useful comparisons with newer, more open-ended, and less constricted analyses of data derived from open-source mediums such as social network sites. It would be ideal if the two methodologies yielded consistent results, thereby making them parallel measures of the same opinions or attitudes. Congruence would suggest that open-source data may serve as an augmentation or, when traditional methods are infeasible, a replacement of survey-collected information. In the current study, we consider the consistency that arises between the dynamic reflection of attitudes toward various topics, such as the economy or the government, as measured through Twitter data and Gallup polling.

Data was collected from two separate sample populations with possible, though unlikely, overlap. The first was Gallup poll data that surveyed Egyptians; the second was Egyptian Twitter users. Both populations are described below. Analysis of these populations included pre- and post-uprising attitudes, with the differentiation based on the date of the Egyptian uprising of January 25, 2011.

2.1   Survey Poll Data

Gallup conducted several nationally representative surveys of the Egyptian population during the target timeframe. Sampling involved multiple layers of stratification based on population size, geography, and randomized household and respondent selection procedures. This data was collected through face-to-face interviews in Arabic by native speakers and was weighted by key demographic variables including: household size, age, gender, education, and socioeconomic status to correct for disproportionalities across these variables. Surveys were conducted in October 2010 with 1,011 respondents ages 15+ and in April 2011 with 1,005 respondents. These dates correspond to periods of pre- and post-Egyptian uprising, centered about January 25, 2011.

From the results of a larger, comprehensive set of questions, respondents' opinions about four primary areas were obtained (see Table 1).

Table 1. Gallup poll questions categorized into indicator areas

Statement	Response Options
Positive Emotive Indicators	
You will invent or discover something that will change the world.	Agree, Disagree, Don't know
You never give up until you reach your goals, no matter what.	Agree, Disagree, Don't know
Even when things go wrong, you feel very optimistic.	Agree, Disagree, Don't know
Do you have relatives or friends who are living in another country whom you can count on to help you when you need them, or not?	Agree, Disagree, Don't know
Now, please think about yesterday, from the morning until the end of the day. Think about where you were, what you were doing, who you were with, and how you felt. Were you treated with respect all day yesterday?	Agree, Disagree, Don't know
Negative Emotive Indicators	
Did you experience the following feelings during A LOT OF THE DAY yesterday? How about Enjoyment?	Yes, No, Don't know
Did you experience the following feelings during A LOT OF THE DAY yesterday? How about Happiness?	Yes, No, Don't know
Economic Indicators	
Right now, do you feel your standard of living is getting better or getting worse?	Getting Better, The Same, Getting Worse, Don't know
Right now, do you think the economic conditions in Egypt, as a whole, are getting better or getting worse?	Getting Better, The Same, Getting Worse, Don't know
Institutional Indicators	
In Egypt, do you have confidence in each of the following, or not? How about Financial Institutions or banks?	Yes, No, Don't know
In Egypt, do you have confidence in each of the following, or not? How about Honesty of elections?	Yes, No, Don't know

2.2   Twitter Data

Twitter collection dates were September 9, 2010 to October 12, 2010 for the pre-uprising stage and March 8, 2011 to April 9, 2011 for the post-uprising stage. Initial analysis included tweets translated from Arabic, but validation testing of the machine translation of Arabic tweets by native Arabic speakers found these methods unreliable; therefore, we concentrated only on English tweets. Analysis of Twitter data required several steps to ensure samples were representative of Egyptian users. First, we obtained 6.5 million tweets based on relevant hash tags and keywords from the Twitter message archiving site TwapperKeeper.com. Tweets that included the hash tags "#jan25" or "#egypt" and any that included the word "Egypt" were kept. Then we filtered out those in Arabic, leaving 2.14 million English tweets. Next we filtered on those users who reported their location as somewhere in Egypt, which resulted in a pool of approximately 820,000 tweets from 48,077 Egyptian users. Next, categories of key words relevant to the focus areas of the Gallup surveys were created. For example, for job satisfaction, relevant key words included: chore, employment, occupation, put to work, busy, subcontract, farm out, job, etc.

This process of filtering and targeting relevant tweets resulted in a reduced number of tweets per topic: 509 tweets for job satisfaction, 661 tweets for confidence in military, 80 tweets for confidence in judicial system, 596 tweets for confidence in government, and 148 tweets for confidence in the honesty of elections. We then ran a sentiment analysis using a regressor trained on 146,291 LiveJournal blogs labeled with 32 different emoticons, where multi-dimensional scaling was performed on the mean document conditioned on each emoticon, in order to automatically retrieve a latent emotion manifold that accurately reflected human perceptions of the emotion space (see [21] for a more detailed description of this procedure). For this work, we concentrate on the mapping of the tweets to the first dimension of the manifold, which depicts the sentiment, ranging from depressed/sad to happy/excited. This analysis assigned a sentiment score on a scale from -1 (least positive) to 1 (most positive) to each tweet. A sample of the sentiment ratings, concentrating on those at the more extreme ends of the scale, was validated using human coders.

3   Analysis and Results

Before the analysis, poll question responses were selected for their appropriateness in representing each indicator area of sentiment (see Table 1). This step involved dichotomizing item responses as 1 for negative responses and 2 for positive responses (e.g., saying "yes" to a question in the "positive emotive indicator" section). Negative emotion scores were reverse scored such that higher scores represented lower negativity. The following two indices were investigated (i) the average (tetra-) correlations among items in each sentiment area (i.e., coefficient alpha) and (ii) each item's correlation with the total score for each dimension, as described above. In both cases, correlations were in the range of 0.60 - 0.80, which is acceptable for assuming that composite scores taken across individual item scores for each area of sentiment are appropriate for use in further analysis. A general sentiment score was also computed based on the sum of the specific composite scores.

For the Gallup analysis, there was inconsistency in the number of respondents retained in pre- versus post-test. Therefore, the data was "paired" by randomly sampling from the larger sample size (pre versus post). As a result, the sample sizes are lower than that originally attained. The mean of each composite score and the overall composite score were compared pre- and post-uprising, where the higher score meant a more positive sentiment. This analysis consisted of paired sample t-tests, in which the p-value was corrected for family-wise error ($p < 0.05/5 = 0.01$). Paired sample t-tests were used because the same set of individuals were used at two time points, resulting in the need to compare pre- and post- means via repeated measure analysis. Table 2 provides descriptive statistics. Table 3 provides t-test results. Table 3 shows an increase in positive sentiment for all areas as well as an overall increase. Based on the paired t-tests, all of these changes were statistically significant except for negativity (Table 3).

For the Twitter analysis, a similar event occurred wherein data was paired by random selection of larger sample size for pre versus post, resulting in a lower overall sample size. Sentiment for each area as well as overall sentiment was compared pre- and post-uprising similarly to that performed on the Gallup data. Based on the descriptive statistics provided in Table 4, it is apparent that respondents were generally more negative toward all topics except for job satisfaction. Moreover, sentimentality toward all areas, as well as overall sentiment, decreased from pre- to post-uprising. Based on the paired sample t-tests (Table 5), only one of these changes was significant: overall sentimentality. Thus, participants expressed more negatively in general after the uprising compared to before the uprising. Four main reasons are likely attributable to a lack in statistical significance in other changes: (i) the sample sizes were lower for other areas of sentiment and (ii) p-values were corrected for family-wise error such that statistical significance required a p of $0.05/7 = 0.007$; (iii) dimensions were flawed due to inadequate or inappropriate items used for a given dimension (e.g., too broad, doesn't necessarily mention Egypt or refer to current events); (iv) alternate sentiment and topic analysis methods could produce different results. Nevertheless, these findings demonstrate that survey and Twitter data did not provide consistent results and were therefore not well aligned.

Table 2. Descriptive Statistics for Gallup Data

Dimension	Sample Size	Mean Oct-10	Mean Apr-11	Standard Deviation Oct-10	Standard Deviation Apr 2011
Positive	1009	3.71	9.05	0.58	1.54
Economic Optimism	439	1.33	1.89	0.74	0.99
Institutional Confidence	678	1.61	2.6	0.49	0.99
Negative	1011	2.94	2.9	0.91	0.97
Overall Positivity	1011	8.89	15.3	1.67	2.49

Table 3. Paired Sample t-test Results for Gallup Data (* = p <0.01)

Dimension	Mean Difference	t- statistic	Degrees of Freedom	95% Confidence Interval Lower Bound	95% Confidence Interval Upper Bound
Positive	-5.34	-102.27*	1008	-5.44	-5.24
Economic Optimism	-0.56	-9.09*	438	0.44	0.68
Institutional Confidence	-0.99	-23.71*	677	-1.07	-0.09
Negative	0.03	0.85	1010	-0.05	0.11
Overall Positivity	-6.41	-65.79*	1010	-6.61	-6.22

Table 4. Descriptive Statistics for Twitter Data

Dimension	Sample Size	Mean Sept-Oct 2010	Mean Mar-Apr 2011	Standard Deviation Sept-Oct 2010	Standard Deviation Mar-Apr 2011
Work Satisfaction	156	0.09	0.046	0.176	0.167
Confidence in Army	43	-0.059	-0.063	0.115	0.144
Confidence in Judicial System	29	-0.018	-0.07	0.123	0.094
Confidence in Government	236	-0.014	-0.046	0.139	0.169
Confidence in Elections (honesty)	106	-0.03	-0.069	0.129	0,13
Overall Sentiment	1231	0.029	-0.04	0.169	0.139

Table 5. Paired Sample t-test Results for Twitter Data (* = p < 0.007)

Dimension	Sample Size	Mean Difference	t-statistic	Degrees of Freedom	95% Confidence Interval Lower Bound	95% Confidence Interval Upper Bound
Work Satisfaction	156	0.044	2.23	155	0.005	0.084
Confidence in Army	43	0.005	0.17	42	-0.053	0.062
Confidence in Judicial System	29	0.052	1.59	28	-0.015	0.12
Confidence in Government	236	0.031	2.24	235	0.004	0.059
Confidence in Elections (honesty)	106	0.039	2.18	105	0.004	0.075
Overall Sentiment	1231	0.068	12.47*	1230	0.058	0.079

4   Discussion

Analysis of sentiment in the Twitter data identified an increase of negativity experienced by individuals between pre- and post-uprising, whereas for the Gallup data, respondents demonstrated greater positivity post-uprising as compared to pre-uprising. Several possible explanations exist for this effect, such as the possibility that the unstructured nature of social media reflects a bias towards the expressing only the chaos and confusion that the population was experiencing following the uprising. Those who use Twitter are likely to be unrepresentative of the general population, where, for example, they may tend to be people who are inherently outwardly emotional. It may also be the case that, given the interview nature of the survey, respondents provided answers that were not as emotional as they would be in open-ended platforms, especially because polling often involves in-person interviews. Impression management and experimenter demands may explain an increase in positivity in the survey data.

The results of this analysis serve as a cautionary warning toward the use of social media as a 'pulse' of a population. While traditional survey methods may have limited reach, the nature of social media presents several different sources of error. First, there exists an obvious socio-economic bias in the sample of users of both the Internet and social media, where members of more disadvantaged populations are grossly under-represented (Facebook penetration in Egypt is only 5.49% as of December of 2010 [5]). Second, though increasingly intelligent methods of natural language processing of social media text are in development, the limitations afforded by processing data at such a scale necessitates the introduction of additional sources of error, such as the use of keywords that could be interpreted in multiple contexts (e.g., 'work' as in 'I hate going to work on Mondays' versus 'This election process is never going to work'). Third, in addition to the complexity of analyzing new language constructs constantly being created (e.g., the use of emoticons and abbreviations such as 'lol'), language translation proves difficult in informal and often ungrammatical usage such as in social media. Follow-up work to this study will include performing a comparison between sentiment analysis techniques, including human coding, and the traditional polling data. We also intend to perform an analysis of the Arabic tweets so as to determine if English tweets are significantly different in the attitudes conveyed.

5   Conclusion

This analysis sought to ascertain trends in sentiment relative to various topics for the Egyptian population using both traditional (polling) and new (social media) information sources and to identify the extent to which these two sources aligned during the 'Arab Spring'. The results showed that trends across the two sources are not exactly consistent. Focusing solely on Twitter data, individuals expressed increasingly negative opinions after the uprising, whereas polling indicated that individuals were increasingly positive post-uprising. These results suggest a greater need for caution when drawing conclusions based on the use of social media as a sole descriptor of a population or event.

References

1. Al Ahram Weekly,
   http://mohamed-salah.com/2010/11/17/Twitter-stats-in-egypt
2. ArabCrunch,   http://arabcrunch.com/2011/04/egypts-mcit-egypt-has-23-51-million-internet-users-71-46-million-mobile-subscribers-3972-ict-companies.html
3. Dubai School of Government: Arab Social Media Report Civil Movements: The Impact of Facebook and Twitter, 1 (2011)
4. Bailar, B., Bailey, L., Stevens, J.: Measures of interviewer bias and variance. Journal of Marketing Research 14, 337–343 (1977)
5. Mourtada, R., Salem, F.: Arab Social Media Report: Facebook Usage 1 (2011)
6. Crocker, L., Algina, J.: Introduction to Classical and Modern Test Theory. Harcourt Brace, New York (1986)
7. Dugan, L.: Unofficial Reports Suggest Twitter Surpassed 500M Registered Users in June,
   http://www.mediabistro.com/allTwitter/Twitter-500-million-registered-users_b26104
8. Farhi, P.: The Twitter explosion. American Journalism Review 31 (2009)
9. Flores-Macias, F., Lawson, C.: Effects of interviewer gender on survey responses: Findings from a household survey in Mexico. International Journal of Public Opinion Research 20, 100–110 (2008)
10. Howard, P.: The digital origins of dictatorship and democracy: Information technology and political Islam. Oxford University Press, London (2011)
11. Huberman, B.A., Romero, D.M., Wu, P.: Social networks that matter: Twitter under the micro-scope. First Monday 14, 1 (2009)
12. Hughes, D.J., Rowe, M., Batey, M., Lee, A.: A tale of two sites: Twitter vs. Facebook and the personality predictors of social media usage. Computers in Human Behavior 28(2), 561–569 (2012)
13. Number of active users at Facebook over the years,
    http://finance.yahoo.com/news/number-active-users-facebook-over-years-214600186-finance.html
14. Kwak, H., Lee, C., Park, H., Moon, S.: What is Twitter: A social network or news media. In: Proceedings of the 19th International Conference on the World Wide Web (2010)
15. Nunnally, J.C.: Psychometric Theory. McGraw Hill, New York (1967)
16. Papacharissi, Z.: A private sphere: Democracy in a digital age. Polity Press, Cambridge (2010)
17. Papacharissi, Z., de Fatima Oliveira, M.: Affective news and networked publics: The rhythm of news storytelling on Egypt. Journal of Communication 62(2), 266–282 (2012)
18. Sysomos,
    http://blog.sysomos.com/2011/01/31/egyptian-crisis-twitter
19. Gallup: Worldwide Research – Country Data Set Details (2010)
20. Abu Dhabi Gallup Center: Egypt: The Arithmetic of Revolution, An empirical analysis of social and economic conditions in the months before the January 25 uprising (2011)
21. Kim, S., Li, F., Lebanon, G., Essa, I.: Beyond Sentiment: The Manifold of Human Emotions, arXiv:1202.1568 (2011)

Hashtag Lifespan and Social Networks during the London Riots

Kimberly Glasgow[1] and Clayton Fink[2]

[1] Johns Hopkins University Applied Physics Laboratory, College of Information
Studies, University of Maryland
[2] Johns Hopkins University Applied Physics Laboratory
{kimberly.glsgow,clayton.fink}@jhuapl.edu

Abstract. Social media is a powerful medium for rapidly sharing information and organizing response in times of crisis or extreme events. Twitter users have adopted a convention of hashtags to support this and other uses of microblogging services. Using Twitter data from the 2011 London riots, we analyze emergent social networks directly relating to response to crisis. We examine networks of riot response oriented around cleanup or prayer activities. These networks differ in size, structure, general membership, and prominent actors. We explore whether temporal patterns observed in social media, such as hashtag "lifespan," may relate to observed social processes and behaviors.

Keywords: social media, Twitter, social networks, hashtag, crisis informatics.

1 Introduction

Social networks play an important role in helping individuals cope with, navigate, and mitigate challenges in their environments, including natural disasters and other crises. The growth of social media such as microblogs enables individuals to swiftly share time-critical information both with members of their social networks or the broader society. Social media also provides a window of visibility into social interaction and response that may influence expression of personal views, either to encourage or to dampen these phenomena. Thus examining the London riots from a perspective incorporating social network and temporal phenomena may be informative.

Hashtags can be a powerful mechanism in social media for sharing information, indicating salience, organizing response, and enabling formation of networks and communities in the wake of an extreme event. Yet little is known about the temporal and social aspects of these phenomena. Building on previous work on social networks observed in social media in times of crisis [1], we explore how temporal phenomena, such as hashtag lifespan (days until final occurrence) and hashtag half-life (time from initial to median occurrence), may relate to social behaviors and social networks coupled to crisis response. Could a hashtag that describes some non-ephemeral activity have a component of its lifespan affected by observation of social and network behavior relating to that activity? Might certain network behaviors, which

A.M. Greenberg, W.G. Kennedy, and N.D. Bos (Eds.): SBP 2013, LNCS 7812, pp. 311–320, 2013.

produce network structures, help support lifespan (longevity) or use of a tag, or produce the opposite effect?

We examine the use of the microblogging service Twitter during one of the most tumultuous and violent events in the recent history of London to explore these questions. London, a metropolis with a population of 8 million, was shaken by a series of riots in August 2011. A few days after the police shooting of an unarmed man in Tottenham, an ethnically diverse working class London neighborhood, peaceful protests were followed by an outburst of violence, arson, and mass rioting. Hundreds of people were injured and five were killed. Arrests numbered in the thousands, and over one hundred million pounds of damage was done to the city. Other cities in England also experienced outbreaks of rioting. During these events, Twitter was used extensively in London.

We leverage the Twitter convention of hashtags to identify relevant communications from within the larger stream of tweets. Hashtags are preceded by the symbol "#", such as *#prayforlondon* or *#riotcleanup*. They are words, abbreviations, acronyms, or phrases that signal association with the topic or meaning of a term. Use of a hashtag makes it simple to discover tweets on that topic. Hashtags enable following and participating in conversations around a topic, in contrast to following specific Twitter users. Hashtags often appear in Twitter's trending topics.

After categorizing the hashtags appearing in this data, we select two categories of hashtags relating to the riots. Tweets containing hashtags from these categories were used to construct social networks, and to provide data on hashtag usage over time. The two categories reflect different responses to the damage, destruction, and suffering resulting from the events. The first category is riot cleanup, and the second is prayer. Both categories relate to circumstances and responses that would be appropriate across the timeframe of data collection. Rebuilding after the riots continued for many months after the events. And physical destruction can be often dealt with far more easily than emotional or psychological loss. However, riot cleanup is an active, often communal, public response. Debris is swept from the streets. Damaged objects are removed or replaced, and public spaces are restored their previous state. Prayer is a very different behavior. It may be performed in private, unobserved by others. It may seek to provide a sense of comfort or control through invocation of a higher power. (Interestingly, reference to a prayer-related response that bridges into the public sphere, the vigil, does not appear in any hashtags with any frequency.) Given these distinctions, networks of riot response oriented around cleanup or prayer activities might be expected to differ in size, structure, and membership. The visibility offered by social media such as Twitter into social behavior and response may have an impact on the temporal behavior of associated hashtags. We find significant differences in the networks of cleanup and response behavior, and observe distinct temporal patterns for their related hashtags.

Social network analysis (SNA) provides insights into individual, group, and societal phenomena, from a framework in which patterns of interaction are represented as networks [2]. The structure of these networks, the positions of individuals within them, their dynamics, and any underlying social processes or mechanisms can inform our comprehension of an event and our expectations of related ongoing social activity.

2   Related Work

2.1   Twitter

Hashtags in Twitter have been examined in terms of "stickiness" and contagion [3], trending topics [4], political polarization [5], clustering and time-series [6], and numerous other topics. Both the scale of social media data such as Twitter and its temporal aspects provide unique opportunities for research into social phenomena [7] [8].

2.2   Disaster Response and Social Media

Response to emergencies and disaster has been studied from perspectives ranging from social psychology and organization theory to public policy and emergency management [9]. Self organization of individuals into informal, emergent groups to fill gaps and respond flexibly during crisis periods is well known [10], but may take new forms in a social media-enabled world [11], [12]. Local populations may assist with preparations for an impending event, help prevent or minimize damage, or support recovery efforts [13], [14]. For residents coping with a crisis, information that is both current and local is critical, and social media may be preferred over mass media in this regard [15].

2.3   Social Networks

Social networks are collections of actors linked together by socially meaningful connections. They are central for providing social support [16], acquiring important resources and information [17], and for mitigating the effects of psychological trauma [18] or stress, to include consequences of a natural disaster [19].

2.4   Role of Prayer

The potential for prayer as a potential positive mechanism for coping with and responding to crisis has been well documented [20]. Religious coping strategies, including prayer and attendance at worship services have been associated with decreased stress and improved mental health outcomes. Prayer was among the most common responses to the September 11 terrorist attacks in the United States according to a national survey of stress responses to those attacks [21]. Ninety percent of respondents reported turning to prayer, religion, or spirituality to help cope with the disaster.

3   Methods and Approach

3.1   Data Collection

Twitter data used in the research was collected from Twitter after the riots started on August 6, 2011 using the Twitter search Application Program Interface (API), which supports location-based queries. We used the center coordinates of the Greater

London administrative area of 51.502 latitude and -0.127 longitude, with a radius of 20 miles. To improve on the sample returned from the API, we also used the Search API to query against the users whose tweets were returned for the London query. We continued this process through September 30, 2011, covering the riots as and the following weeks of remedy and recovery. Location metadata is returned for each tweet when making location-based queries. This data may be in the form of a location name (*London, England*) or a pair of coordinates. We used the Geonames gazetteer[1] to match this data to an actual place name or, if coordinates were available, the populated place closest to the coordinate location. Tweets obtained via queries against user names usually do not have location data. In these cases the last known tweet location for the user from the location-based query was used. 14.3 million tweets were collected and 12 million were grounded to the Greater London area. Only the tweets grounded to London were used for analysis.

3.2 Hashtags in the London Riots Data

The 12 million tweets contained over 352,000 unique hashtags. Most hashtags were neither frequently used nor long-lived. Fewer than 32% of hashtags appeared more than once, and frequency of usage drops dramatically in the data (which could be fit to a power law distribution with $\alpha=2.96$). During the initial week of data collection, the most frequent hashtag was *#londonriots*, appearing 25,000 times. Many other hashtags relating to the London riots emerged spontaneously in the early days of the event. The *#prayforlondon* hashtag was a Twitter trending topic. However, even during these extreme events, hashtags pertaining to everyday activities, such as entertainment or shopping, continue to appear. Riot-related hashtags described locations of rioting and destruction, expressed concern about the events, or recommended responses to rioters or rioting activity.

3.3 Social Networks in the London Riots Data

We compare Twitter communications relating to one aspect of recovery, cleaning up the damage from the riots with a second aspect, prayer. A review of the most frequently used hashtags in the first week of the disaster identified a seed set of hashtags for riot cleanup. That set was expanded a final set of 65 unique hashtags relevant to riot cleanup to include *#riotcleanup, #riotwombles,* and *#londoncleanup*. These hashtags were used to identify relevant communications for riot cleanup social networks, derived from the distinct tweet-based social behaviors of quoting (retweeting) others, talking to (directly addressing) others, or talking about (mentioning) others, as described in [1]. The same process was followed for prayer hashtags.

A sender of a tweet can choose any of these three behaviors when tweeting. Connecting to others in each of these ways is sociologically and semantically distinct, and may provide insights into roles played by members in these networks, how other

---

[1] http://www.geonaes.org

network members perceive them, and even into the robustness or brittleness of the network itself [22].

3.4   Temporal Information for the London Riots Data

All tweets include a timestamp. The timestamp was captured for all instances of tweets containing a cleanup or prayer related hashtag. A "lifespan" was calculated for each hashtag, indicating how long in days it continued to appear, from initial to final occurrence. A "half-life" was also determined, reflecting time until half of all usages of that hashtag appeared (based on the timespan from first to median tweet).

4   Findings and Discussion

4.1   Network Findings

Analysis of social networks for prayer and riot cleanup found substantial differences in network size (both number of actors and number of ties) and in network structure and centralization. Different actors were prominent across the cleanup and prayer networks (based on social network metrics of in-degree, out-degree, in-2step reach, and out-2step reach computed using UCINET 6 [23]). Prominent actors differentially engage in *talking to*, *talking about* or *quoting* others, and are subject to the same phenomena from network members. See [24] for more detail. Further, while prominent actors in riot cleanup include grassroots activists and organizers of neighborhood cleanup events, no spiritual leaders or clergy are prominent in the prayer networks. This may be an indication of underlying social forces or processes influencing behavior.

Figure 1 shows the riot cleanup network on the left, and the prayer network on the right, for *talking about* behavior. Differences in network structure and size are apparent. The main component of the cleanup network comprises over 70% of network actors, while the proportion for prayer is just 10%. The cleanup network's actors are both more numerous, and more active in mentioning others (1.24 to .76 mentions per actor, p<.01).

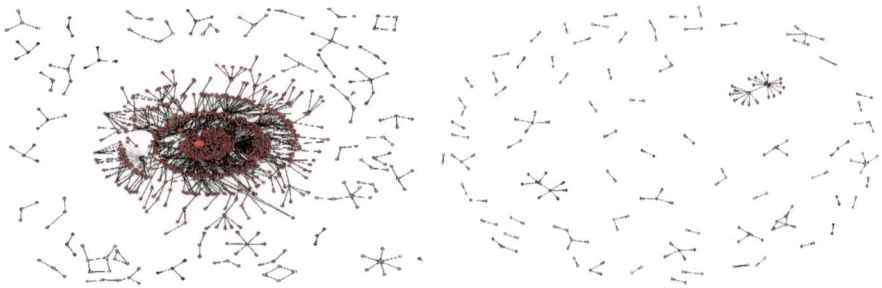

Fig. 1. Riot Cleanup and Prayer Networks: Talking About (mentioning)

Table 1. Hashtag Frequency, Lifespan, and Half-life

Hashtag	Count	Lifespan (days)	Half-life	Hashtag	Count	Lifespan (days)	Half-life
#riotcleanup	4874	52	19hr, 44min	#prayforlondon	2778	41	1 day, 2hr, 28min
#riotwombles	234	28	5hr, 20min	#prayforuk	398	15	18hr, 59min
#manchester-cleanup	108	2	3hr, 12min	#prayforengland	316	7	12hr, 5min
#salfordri-otcleanup	106	4	10hr, 18min	#prayfortheuk	133	3	22hr, 8min
#londoncleanup	58	7	13hr, 11 min	#pray-forbirmingham	74	2	1hr, 51min
#riotscleanup	57	14	13hr, 12min	#prayforman-chester	49	1	18hr, 28min
#riotcleanups	44	2	1 day, 14 hr, 59min				
Total cleanup	5481	52	20 hr, 13min	Total prayer	3748	41	1 day, 5hr, 3 min

4.2   Temporal Findings

For this collection of Twitter data, there was a 53-day range in which these hashtags could appear. (Counts of hashtags for the first and final days are suspect, and may reflect an artifact of data collection starting and stopping.) From the sets of cleanup and prayer hashtags, we examine the lifespan and half-life (adapted from the approach described by bit.ly[2]) for the more frequently occurring hashtags (n>40, top 1%), and for aggregations of both prayer and cleanup hashtags. Most of the less frequent hashtags occurred once, and many appear have been typographic errors.

Both cleanup and prayer activity peaks in the first few days. Then both decrease, with cleanup doing so more slowly than prayer. (Fitting a Poisson regression finds regression coefficients of -.267 and -.422 respectively, though the fit is quite imperfect,

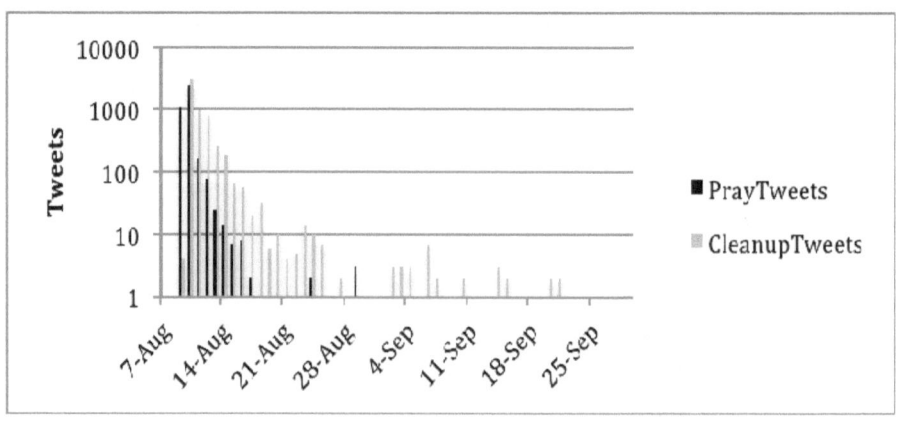

Fig. 2. Riot Cleanup and Prayer hashtags

[2] http://blog.bitly.com/post/35341087592/halflife-by-topic

with large residuals.) From another angle, prayer has its first day with 0 events at day 11 of its lifespan, and its first 2-day run of no events at day 12. Cleanup last 17 days until its first day with 0 events, and 34 days before a 2-day run of no events.

4.3   Observations

Both types of response at first glance are equally reasonable and normal in the aftermath of disaster, yet cleanup is the more successful response from both a network and temporal perspective. A large majority of Londoners are religious and typically Christian [25]. Prayer expends less energy and effort than cleanup. Since hashtags from both the categories of *cleanup* and *prayer* were among the most common in the data (in the top 10 during the first week of the data collection), and were both trending topics, lack of awareness of the #pray topic seems implausible. Other factors may be contributing.

We consider whether social processes may be a factor. By definition, social media is intended to be a social, not an individual, activity. Thus social influences and forces are likely to have played a role in the actions of individuals using social media. Burt has discussed how actors moderate their information sharing and statement of opinions based upon perception of network members' positions. They may either withhold information, or slant what they do express in order to "echo" other's predispositions [26].

A large percentage of riot cleanup communications were positive about the activity, and about those organizing or participating in it. Few, if any, were critical:

> *students' unions and students -if you're near somewhere affected, do your bit and get out and help with #riotcleanup #loveyourneighbourhood*

> *Getting the clean up together -Meet outside Tackle Shop, Roman Road, hackney 9am in the morning to help local shops clean up. #riotcleanup*

> *My faith in humanity is restored. Thanks to the good & kind people who get together to clean-up their communities #riotcleanup*

Prayer communications present a more complex picture. In some instances, individuals evoked prayer in the context of describing rioting and rioters, and the emotional response they were experiencing:

> *scared stiff. Omg the looters are here. My hometown:(#prayforuk*

> *A girl died in a burning house while saving her brother and sis-ter.She sacrificed her life for them.I'm crying. #prayforlondon.*

Some tweets were clearly sympathetic and encouraging of prayer as an appropriate response to events:

> *#prayforUK ... my heart with all those affected by the riots -please keep my family in your thoughts <3*

> *Let's all #prayforlondon and continue to heal the world through love, creativity and positivity. Don't let the dark forces win!!!*

However other tweets expressed judgmental, negative attitudes towards praying for London as an appropriate response to the crisis. Some found the riots did not merit invocation of a greater power, while others found prayer itself an inadequate response compared to direct action:

> *#controversialmood #londonriots its actually really really boring+ #prayforlondon -blimey how bout actually helpin sort out some root issues*

> *Vomit! #PrayForLondon is trending! If you're going to hassle God, maybe #PrayForSomalia, #PrayForTalibanWomen, #PrayForChildSoldiers.*

> *Don't #PrayForLondon we aren't that lame. We have free education and healthcare. We get paid even if we don't have a job*

London Twitter users could easily have been exposed to negative responses to prayer, seeing other criticized. Those who held neutral or mildly positive views, or no views would find themselves in a dissonant social media environment. Both selective exposure to prayer-related views, based in the homophily of social network members, and selective disclosure, motivated by a desire to appear congruent with group norms, bias the information environment and influence individual's behavior in a larger social context [27]. This could discourage adoption and use of prayer hashtags in an environment with hostile reactions. Conversely, these same users would encounter universally positive responses to cleanup.

To be sure, tweeting and use of hashtags to affiliate oneself with a topic may to some degree reflect pre-existing views, including personal beliefs about appropriateness. Yet knowledge that one's actions will be broadly socially visible, and subject to judgment and potential social reproval by others, is inescapable. This would decrease the likelihood of tweeting about prayer as a response as well as the likelihood of *talking to*, *quoting*, or *talking about* others in that context. Riot cleanup is not subject to these inhibitors.

5   Future Directions

The research examines Twitter communications originating in London beginning with the riots and continuing through the initial weeks of recovery from a social network and temporal perspective. As Londoners experienced these traumatic events, their emotions, behavior, interpretations, and communications would likely evolve. Finergrained analysis might reveal new insights. Deeper inspection of the sentiment of tweets containing hashtags of prayer or cleanup could strengthen understanding of the role of negative and positive expression, related social forces, and network and temporal consequences. Models of hashtag lifespan could be improved.

6   Conclusion

This work explores the microblogging landscape of London during the worst period of rioting and public disorder in many decades. Social networks are constructed from subsets of riot communications relating to specific riot responses, *cleanup* and *prayer*. Computing social network measures on network actors allows us to identify prominent actors and assess their similarity. We find numerous differences between networks of *cleanup* and *prayer* in terms of size, structure and network membership. Patterns differentiating the networks and actors may indicate social processes and forces at work. The cleanup network is more active, and more robust. Similarly, temporal analysis of prayer and cleanup hashtags shows differences in lifespan. Cleanup as a topic of conversation lives longer, and decays more slowly from its peak.

References

1. Glasgow, K., Ebaugh, A., Fink, C.: #Londonsburning: Integrating Geographic, Topical and Social Information during Crisis. In: Proceedings of the 6th International AAAI Conference on Weblogs and Social Media, ICWSM 2012 (April 2012)
2. Wasserman, S., Faust, K.: Social network analysis: methods and applications. Cambridge University Press (1994)
3. Romero, D.M., Meeder, B., Kleinberg, J.: Differences in the mechanics of information diffusion across topics: idioms, political hashtags, and complex contagion on twitter. In: Proceedings of the 20th International Conference on World Wide Web [Internet], pp. 695–704. ACM, New York (2011),
   http://doi.acm.org/10.1145/1963405.1963503
4. Kwak, H., Lee, C., Park, H., Moon, S.: What is Twitter, a social network or a news media? In: Proceedings of the 19th International Conference on World Wide Web [Internet], pp. 591–600 (2010)
5. Conover, M.D., Ratkiewicz, J., Francisco, M., Gonccalves, B., Flammini, A., Menczer, F.: Political polarization on twitter. In: Proc. 5th Intl. Conference on Weblogs and Social Media [Internet] (2011),
   http://www.aaai.org/ocs/index.php/ICWSM/ICWSM11/
   paper/download/2847/3275 (November 14, 2012)
6. Yang, J., Leskovec, J.: Patterns of temporal variation in online media. In: Proceedings of the Fourth ACM International Conference on Web Search and Data Mining, pp. 177–186 (2011)
7. Savage, M., Burrows, R.: The Coming Crisis of Empirical Sociology. Sociology 41(5), 885–899 (2007)
8. Watts, D.: A twenty-first century science. Nature 445(7127), 489 (2007)
9. Quarantelli, E.L., Dynes, R.R.: Response to Social Crisis and Disaster. Annual Review of Sociology 3, 23–49 (1977)
10. Stallings, R.A., Quarantelli, E.L.: Emergent citizen groups and emergency management. Public Administration Review 45, 93–100 (1985)
11. Liu, S., Palen, L., Sutton, J., Hughes, A., Vieweg, S.: In search of the bigger picture: The emergent role of on-line photo sharing in times of disaster. In: Proceedings of ISCRAM 2008 (2008)
12. Starbird, K., Palen, L.: Voluntweeters: Self-organizing by digital volunteers in times of crisis. In: Proceedings of the 2011 Annual Conference on Human Factors in Computing Systems, pp. 1071–1080 (2011)

13. Starbird, K., Palen, L., Hughes, A.L., Vieweg, S.: Chatter on the red: what hazards threat reveals about the social life of microblogged information. In: Proceedings of the 2010 ACM Conference on Computer Supported Cooperative Work, pp. 241–250 (2010)

14. Vieweg, S., Hughes, A.L., Starbird, K., Palen, L.: Microblogging during two natural hazards events: what twitter contribute to situational awareness. In: Proceedings of the 28th International Conference on Human Factors in Computing Systems, pp. 1079–1088 (2010)

15. Shklovski, I., Palen, L., Sutton, J.: Finding community through information and communication technology in disaster response. In: Proceedings of the 2008 ACM Conference on Computer Supported Cooperative Work [Internet], pp. 127–136. ACM, New York (2008), http://doi.acm.org/10.1145/1460563.1460584

16. Gottlieb, B.: Social Networks and Social Support: An Overview of Research, Practice, and Policy Implications. Health Education & Behavior 12(1), 5–22 (1985)

17. Granovetter, M.S.: The Strength of Weak Ties. The American Journal of Sociology 78(6), 1360–1380 (1973)

18. Flannery, R.B.: Social support and psychological trauma: A methodological review. Journal of Traumatic Stress 3(4), 593–611 (1990)

19. Bland, S.H., O'Leary, E.S., Farinaro, E., Jossa, F., Krogh, V., Violanti, J.M., et al.: Social network disturbances and psychological distress following earthquake evacuation. The Journal of Nervous and Mental Disease 185(3), 188 (1997)

20. Meisenhelder, J.B.: Terrorism, Posttraumatic Stress, and Religious Coping. Issues in Mental Health Nursing 23(8), 771–782 (2002)

21. Schuster, M.A., Stein, B.D., Jaycox, L.H., Collins, R.L., Marshall, G.N., Elliott, M.N., et al.: A national survey of stress reactions after the September 11, 2001, terrorist attacks. New England Journal of Medicine 345(20), 1507–1512 (2001)

22. Borgatti, S.P., Mehra, A., Brass, D.J., Labianca, G.: Network Analysis in the Social Sciences. Science 323(5916), 892–895 (2009)

23. Borgatti, S.P., Everett, M.G., Freeman, L.: Ucinet for Windows: Software for social network analysis. Harvard Analytic Technologies (2002-2006)

24. Glasgow, K., Fink, C.: From push brooms to prayer books: Social media and social networks during the London riots. In: iConference 2013 (accepted, 2013)

25. Office for National Statistics. Table View [Internet] (2007),
    http://www.neighbourhood.statistics.gov.uk/dissemination/
    LeadTableView.do;jsessionid=d9W9PnnNcyZ11Yw22J1bjYWDPHgJ2wNL
    pZz6mNLnKM0JsZpxqySJ!1149657076!1336862605420?a=7&b=6059498&
    c=london&d=81&e=13&g=336453&i=1001x1003x1006&k=religion&o=25
    4&m=0&r=1&s=1336862605420&enc=1&domainId=16&dsFamilyId=95&ns
    js=true&nsck=true&nssvg=false&nswid=1600 (May 12, 2012)

26. Burt, D.R.S.: Bandwidth and Echo: Trust, Information, and Gossip in Social Networks. Social Networks 60637, 30–74 (2001)

27. Kitts, J.A.: Egocentric bias or information management? Selective disclosure and the social roots of norm misperception. Social Psychology Quarterly, 222–237 (2003)

A Text Cube Approach to Human, Social and Cultural Behavior in the Twitter Stream

Xiong Liu[1], Kaizhi Tang[1], Jeffrey Hancock[2], Jiawei Han[3], Mitchell Song[1],
Roger Xu[1], and Bob Pokorny[1]

[1] Intelligent Automation, Inc.
[2] Cornell University
[3] University of Illinois at Urbana-Champaign
xliu09@gmail.com, {ktang,msong,hgxu,bpokorny}@i-a-i.com,
jeff.hancock@cornell.edu, hanj@illinois.edu

Abstract. Twitter is a microblogging website that has been useful as a source for human social behavioral analysis, such as political sentiment analysis, user influence, and spread of news. In this paper, we discuss a text cube approach to studying different kinds of human, social and cultural behavior (HSCB) embedded in the Twitter stream. Text cube is a new way to organize data (e.g., Twitter text) in multiple dimensions and multiple hierarchies for efficient information query and visualization. With the HSCB measures defined in a cube, users are able to view statistical reports and perform online analytical processing. Along with viewing and analyzing Twitter text using cubes and charts, we have also added the capability to display the contents of the cube on a heat map. The degree of opacity is directly proportional to the value of the behavioral, social or cultural measure. This kind of map allows the analyst to focus attention on hotspots of concern in a region of interest. In addition, the text cube architecture supports the development of data mining models using the data taken from cubes. We provide several case studies to illustrate the text cube approach, including public sentiment in a U.S. city and political sentiment in the Arab Spring.

1   Introduction

Human Social Cultural Behavior (HSCB) analysis and modeling is an emerging research area that focuses on understanding, predicting, and shaping human behaviors cross-culturally [1]. As social media becomes more prevalent, HSCB analysis can take advantage of the availability of in-situ data for real-world applications. Nowadays, many social media sites provide Application Programming Interface (API) invocation via web services. For example, Twitter is a microblogging website that has been useful as a source for HSCB analysis (e.g., political sentiment analysis [2], user influence [3], and spread of news [4]). The Twitter streaming API allows applications to have real-time access to tweet objects. Using this API, we can design code to automatically extract live tweets for a topic, transform and load them into the textual database for subsequent analysis.

A.M. Greenberg, W.G. Kennedy, and N.D. Bos (Eds.): SBP 2013, LNCS 7812, pp. 321–330, 2013.
© Springer-Verlag Berlin Heidelberg 2013

In this paper, we introduce a text cube approach to social media analysis, especially sentiment analysis. In data warehousing, data cube is a way to organize data in multiple dimensions and multiple hierarchies for information query and visualization from multiple perspectives [5]. A data cube allows data to be aggregated and viewed from multiple perspectives, and it is defined by measures and dimensions. The measures (or facts) are numeric values that are usually additive (e.g., sales of a product). Analysts need to look at measures using some "by" conditions. The "by" conditions are dimensions. For example, in order to analyze sales volume, analysts often want to see its measure by day and by location. In this sense, dimensions are the perspectives with respect to which an analyst wants to aggregate or view measures. The advantage of using the data cube approach over using the relational database approach to organize multidimensional data is that data cube performs very well with complex queries and analysis of very large datasets [5].

A data cube can be extended to summarize and navigate structured data together with unstructured text. Such a cube is called text cube [6,7]. Unlike a traditional data cube where measures are directly retrieved from the original databases, text cube provides an advanced text analytics capability for extracting HSCB measures from unstructured text streams. With the HSCB measures defined we are now able to view the text cube and perform analyses. This includes slicing, dicing and drilling through cube cells. Text cube also supports complex analysis methods, such as topic modeling [8], online analysis [9], and keyword-based exploration [10].

We have added two additional functionalities beyond the standard text cube methodology. First, we have added the capability to display the contents of the cube on a heat map. Basically, the heat map shows each geographic region with a shade of red. The degree of opacity is directly proportional to the value of the measure; the larger the measure is the more opaque the color. This kind of map allows the analyst to focus attention on hotspots of concern in the region of interest. Second, we have added support for the development of prediction models using a data mining approach [11,12]. Therefore, text cube is a practically useful approach for organizing and analyzing text-based communications to assess HSCB dimensions (e.g., sentiment or affect, deception, group identity) for a given group and to predict current belief states and likely intended actions.

This paper is organized as follows. We first describe the text cube approach to HSCB analysis. Then we present several case studies to illustrate the text cube approach, including public sentiment in a given region and political sentiment in the Egyptian Revolt. Finally, we conclude the paper and discuss future research.

2   A Text Cube Approach

We have developed a dynamic data cubing and mining framework, called *SocialCube* [6], for large amounts of HSCB data. SocialCube is an advanced data cube architecture that allows analysts to summarize and navigate structured data together with unstructured text for efficient query and analysis. In SocialCube, linguistic feature analysis is a preliminary step for developing text-based data cubes or text cubes.

2.1   HSCB Linguistic Analysis for Sentiment

We have designed a comprehensive HSCB linguistic feature analysis framework that allows for an extensible set of HSCB dimensions, such as affect/sentiment [13], deception [14], sense of fatalism vs. mastery, and power structure. Here we focus on the linguistic feature analysis of sentiment.

Perhaps one of the most important social and cultural dynamics for humans is their sense of emotion [15]. Emotion reflects not only how an individual is reacting to ongoing events, but can also reflect to how an individual generally views the world and his/her place in it. While emotion was long ignored by cognitive psychologists, a wide preponderance of data suggests that understanding an individual or group's emotional state can provide important insight and prediction into their decision-making, cognitive responses, and future behavior [16].

Although emotion is often assumed to be only communicated nonverbally [17], a number of recent studies suggest that humans convey their emotions in text-based communication, such as emails, blogs, instant messaging, and other forms of textual communication through linguistic cues. In one study [18], for example, individuals were asked to communicate only by text, and one partner was induced to feel sad before the interaction. Under these conditions, their partner was able to detect the negative emotion in the emotionally induced participant, indicating that emotion can be detected in text-based communication. Importantly for the present research, these data suggest that emotions can be detected from text-based communication.

There are specific linguistic patterns of emotional expression in verbal content, and there are a number of established tools that can extract relevant emotional content, including the Linguistic Inquiry and Word Count program and the Dictionary of Affect in Language program. Using these tools, Hancock and colleagues [19] have found that when people are sad they tend to use fewer words, disagree more, use more negative-affect words, and respond more slowly.

These kinds of verbal patterns are extractable not only at the dyadic or group level, but also at the organization and even national level. Consider Kramer's work [20] on Facebook status updates and his assessment of the Gross National Happiness index. The Gross National Happiness index assesses the emotional context of the United States by extracting positive and negative emotional indicators from 100 million Facebook users. This analysis, based on the textual content from status updates, correlates highly with self-reported satisfaction as well as culturally and emotionally significant calendar events (e.g., Christmas, death of a politician, etc.). A more recent study provides even more powerful evidence that affect words from tweets reflect actual emotional states. Golder and Macy [21] analyzed positive and negative affect words from 550 million tweets collected from cultures around the world. Their data revealed that both positive and negative affect in tweets tracked precisely with daily circadian rhythms, with positive affect peaking in the mid-morning and mid-evening, and negative affect peaking mid-afternoon and very early morning.

Taken together, these data suggest that emotional indicators or an individual or group can be extracted from verbal content present in text-based communication, and that these features, dynamically tracked over time, can predict emotionality of an individual or even a group.

In our current research, we use the Linguistic Inquiry and Word Count (LIWC) tool [22] to extract emotional features. LIWC counts word frequency along the approximately 65 dimensions of language in the default LIWC dictionary. These dimensions include function word categories, such as pronouns, articles and auxiliary verbs, as well as psychological categories, such as affect and cognition-related words, and social dimensions, such as family words. The output for each word category from the LIWC default dictionary represents a feature in our analysis. Note that some LIWC measures have hierarchical relationships. For example, "affective processes" can be divided to "positive emotion" and "negative emotion"; and "negative emotion" can be further divided into "anger", "anxiety", and "sadness".

2.2   Text Cube Construction for Sentiment

Our goal is to organize linguistic features/indicators of sentiment in a data cube model, a new way to organize data in multiple dimensions and multiple hierarchies for efficient information query and visualization from multiple perspectives [7]. A data cube allows data to be aggregated and viewed in multiple dimensions. It is defined by dimensions and facts (or measures). In general terms, dimensions are the perspectives with respect to which an organization wants to keep records (e.g., by time, by location, etc.). Each dimension may have a table associated with it called a dimension table. Facts are numerical measures that are quantities by which we want to analyze relationships between dimensions.

The star schema is a multidimensional data model to design the data cube. In a star schema, there are one or more fact tables referencing any number of dimension tables. The fact table contains the names of the facts (measures), as well as keys to each of the related dimension tables. We have designed a star schema to store the extracted linguistic features for different HSCB dimensions including sentiment. They are stored as measures in the Fact table. Based on star schema, we have designed a data cube architecture to allow users to conveniently view aggregated statistics of sentiment relevant measures along different dimensions, such as time and location.

3   Case Studies on Cubing and Visualization

Several data cubes have been designed and implemented for demonstration purposes, including one cube looking at tweets originating from the Washington, D.C. region, and one cube looking at tweets on the topic of the Egypt revolt.

3.1   Affective Processes Cube in Washington, D.C.

We collected ~0.5 million tweets in the Washington D.C. region for the period of May to July 2011. **Fig. 1** shows the database screen shot with sample tweets. Then we performed HSCB linguistic analysis of the tweets to extract sentiment measures, such as "positive emotion" and "negative emotion". The extracted measures and other structured information are stored in a star schema. The dimensions for this schema are Date, Zone, User, Location, Event, and Tag, and are represented by database tables date_dim, zone_dim, user_dim, location_dim, event, and tags, respectively. The facts (i.e. measures) are stored in the fact table pycholinguistic_facts.

date	time	tweet_text
2011-05-21	17:22:45	My Mom said, "if anything weird happens for the 'end of the world', come to the house..." "uhhh...oh...kayyyyy"...
2011-05-21	11:46:38	An Igbo man fell into a well one day & started screaming for help. His wife rushed off to buy a rope to sav... (cont)...
2011-05-21	19:12:50	@NickyyBamess naw a clothin joint
2011-05-21	08:52:45	@OzzieNeutron awwwe that's nice
2011-05-21	18:11:01	Did Medieval Times for dinner. Was a great show and big fun.
2011-05-21	14:12:15	I'm at Union Street Public House (121 S Union St, at Prince St, Alexandria) http://4sq.com/lTNObs
2011-05-21	11:00:27	#endoftheworldconfessions i keep a roll of toilet paper in my truck in case i have to go in a public bathroom and ...
2011-05-21	09:11:18	So who went to bed last nite thinking the world was gonna end today?
2011-05-21	08:43:32	I have heard good things about Xoom too @RdLessTkn Until I use them I am not the best to decide
2011-05-21	15:04:53	BRAVA Heyyy! @ JussCallMeOzz
2011-05-21	13:07:12	I'm at Fiesta Asia Street Fair (Pennsylvania Avenue NW, btw 3rd & 6th, Washington) w/ 6 others http://4sq.com...
2011-05-21	10:22:08	When I get my abs I'm going running shirtless.
2011-05-21	12:09:02	Chilling in da crib @ heathaplexVISION http://gowal.la/c/4gHzZ
2011-05-21	15:57:05	We get it by now

Fig. 1. Sample tweets

In order to view and perform analysis on the data, a cube must be defined and the dimensions and measures mapped to corresponding database table columns. This is done in an XML file referred to as a cube schema. A cube schema contains a logical model, consisting of cubes, hierarchies, and members, and a mapping of this model onto a physical model. With the schema defined we are now able to view the tweets cube and perform analysis using an Online Analytical Processing (OLAP) tool. This includes slicing, dicing and drilling through cells.

Fig. 2 is a cube interface of the Affective Processes Cube for the Washington, D.C. Region with Date as the horizontal dimension and Zone as the vertical dimension. Zone, as we mentioned, represents a rectangular area on the Earth. Each measure is associated with a zone as determined by the tweet location (i.e. latitude and longitude the tweet originated from). Measures are shown along the horizontal axis. Fig. 2 shows the average negative emotion for each day for each zone. The cube interface can display a different view of the same cube by showing the average measure for all days combined for each of the zones. The cube interface also supports many charting options (e.g., 3D vertical bar chart).

⧩Date	2011-05-21	2011-05-22	2011-05-23	2011-05-24	2011-05-25	2011-05-26
	Measures	Measures	Measures	Measures	Measures	Measures
⧩Zone	⬧ Negative Emotion	⬧ Negative Emotion	⬧ Negative Emotion	⬧ Negative Emotion	⬧ Negative Emotion	⬧ Negative Emotion
⧩All Zones	2.469	1.526	0.985	1.197	1.079	1.233
⧩unknown	3.05	1.78	0.9	1.161	1.091	1.322
⧩Zone_1	2.684	1.486	0.819	1.289	1.667	2.145
⧩Zone_1_3	0	5		0		0
⧩Zone_1_4	2.765	1.216	0.819	1.428	1.667	2.264
⧩Zone_2	1.308	1.316	1.714	0	0.94	0.275
⧩Zone_2_2		0				
⧩Zone_2_3	1.379	1.513	1.818	0	1.019	0.275
⧩Zone_2_4	0	0	0	0	0	
⧩Zone_3	3.72	0	0.59	0.909	0.312	2.077
⧩Zone_3_1					12.5	
⧩Zone_3_2	3.469	0	0.616	1.186	0	2.439
⧩Zone_3_3					0	0
⧩Zone_3_4	5.557	0	0	0	0	0
⧩Zone_4	2.036	1.374	1.038	1.314	1.124	1.113
⧩Zone_4_1	2.098	1.363	1.065	1.329	1.154	1.093
⧩Zone_4_2	0.392	1.838	0	1	0	0.769
⧩Zone_4_3	4.763			4.76		
⧩Zone_4_4	0		0	0	1.111	1.704

Fig. 2. Affective Cube for individual dates

3.2   Heat Map Visualization

Along with viewing and analyzing tweets data using cubes and charts we have also added the capability to display the contents of the cube on a heat map. **Fig. 3** is a representative heat map. It shows each zone with a shade of red. The degree of opacity is directly proportional to the value of the measure; the larger the measure is the more opaque the color. The blue area on the right lists the map type options as well as the overlays that may be turned on or off by either checking or unchecking the respective boxes. Each overlay represents a column of the cube currently being displayed. For example, the first overlay (after the Show Grid overlay) is the first column from **Fig. 2** for the May 21, 2011 date.

If a measure is in the lower 25 percent range, its corresponding zone is given an opacity level of 0.2 (light pink). If a measure is within the 25 percent and 50 percent range, its corresponding zone is given an opacity level of 0.4. If a measure within the 50 percent and 75 percent range, its corresponding zone is given an opacity level of 0.6. Finally, zones of measures above the 75 percent are given an opacity level of 0.8 (red). In this example, the heat map allows the analyst to quickly visualize the areas that have the highest negative emotions on a given day or set of days.

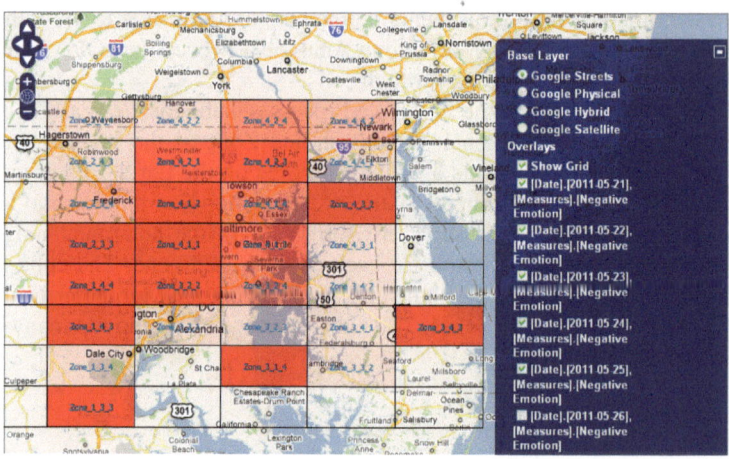

Fig. 3. Heat map interface

In addition, we have studied the affect (e.g., negative emotion) change in a given region. **Fig. 4** shows the heat map plot of negative emotion in the Washington DC area. We note that Zone 4-3-2 has relative high negative emotion on both May 21 and May 24, suggesting that these regions may be of interest for further investigation, such as correlating the emotion with known events in the area, crime statistics, or even weather conditions. Schwarz and Clore (1983) [23] showed that peoples' emotional state is driven by the weather in their location, but that emotion reports would return to "normal" once the weather's influence was pointed out. In other words, the weather may be an unidentified cause of emotion at the aggregate level that to date has not been fully investigated.

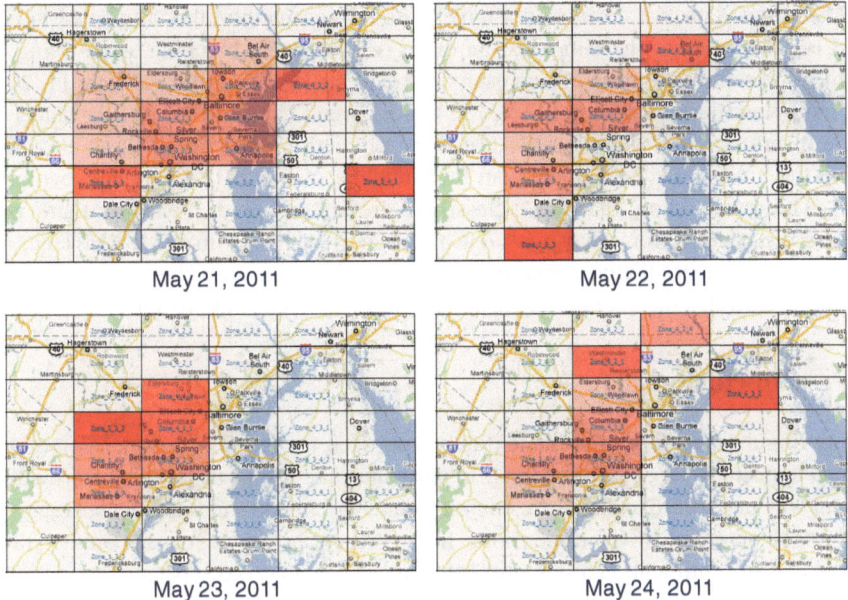

Fig. 4. An example of negative emotion change

3.3 Affective Process Cube for the Egypt Revolt

As an example of political sentiment, we built an Affective Processes cube for the Egypt revolt (see **Fig. 5**). The horizontal dimension is the cities from which the tweets originated from and the vertical dimension is the time period. Its measures include Affective Processes, Positive Processes, Negative Processes, Anger, Anxiety, Sadness, Religion, and Social. We can expand the time dimension to drill down to specific days. We can also aggregate the sentiment measures for all locations by "shrinking" the location dimension. This will generate the dataset for trend analysis of sentiment regardless of locations.

Time								
+All Periods								
Measures								
Location	Affective Processes	Positive Emotion	Negative Emotion	Anger	Anxiety	Sadness	Religion	Social
=All Locations	3.491	1.96	1.521	0.775	0.202	0.308	0.647	5.201
	3.634	2.018	1.491	0.825	0.188	0.131	0.547	4.187
"Arlington, VA"	0.893	0	0.893	0.893	0	0	0	7.011
"Baltimore, MD"	4.35	0	4.35	0	4.35	0	0	0
"Boston, MA"	4.559	3.633	0.926	0	0	0	0	4.456
"Cairo, Egypt / Dubai"	5.904	3.96	1.908	1.096	0.195	0.268	0.355	8.083
"Cairo, Egypt"	5.156	2.799	2.358	1.896	0	0.463	2.381	8.779
"Cambridge, UK"	5.968	2.38	3.588	1.138	1.138	0	0	4.732
"Doha, Qatar"	1.779	1.118	0.727	0.447	0	0.28	0.29	5.394
"Edinburgh, UK"	2.6	1.15	1.45	0	0	1.45	0	5.903
"Edmonton, Alberta, Canada"	0	0	0	0	0	0	0	9.09
"Exeter, UK"	0	0	0	0	0	0	0	0
"Flemington, New Jersey"	0	0	0	0	0	0	0	6.695
"Fucknutsville, DC"	5.88	0	5.88	5.88	0	0	0	5.88
"Goes, The Netherlands"	3.992	2.056	1.936	0.877	0	0.458	0.196	5.981
"Hell's Breakfast Nook, NYC"	4.331	2.345	1.394	0.558	0.309	0.263	0.702	5.884

Fig. 5. Affective Processes Cube for the Egypt Revolt

Using this data, we developed a novel data mining method that attempted to connect tweet sentiment measures and an event (e.g., protest, bombing, etc). Using the aggregated sentiment measures on each day between Jan. 25, 2011 and Feb. 11, 2011, we built classifiers, including libSVM [24], REPTree [25], and IBK [26], to detect the scale of protests for each day. We tested the prediction accuracy of these classifiers using the ground truth obtained from the Timeline of the 2011 Egyptian Revolution [27]. The results show that these classifiers are able to predict whether there were large-scale protests for a day with reasonable performance: libSVM, REPTree, and IBK achieved a prediction accuracy of 83.33%, 83.33%, and 73.33%, respectively. This demonstrates the data mining capability of the text cube approach.

4   Conclusions

Using linguistic features and the text cube approach for the HSCB dimension of sentiment appears to be quite promising. In particular, the use of the text cube provides a useful way to explore and analyze complex data, and in particular to connect language patterns to potential HSCB dimensions. In the present paper we focused on the HSCB dimension of sentiment. While sentiment analysis has received substantial attention in the literature, we believe that it is novel to apply the text cube approach on this dimension.

We have added two new capabilities to the text cube. The first is the heatmap functionality that provides a geographical overlay of the relevant HSCB derived from linguistic data extracted from the Twitter stream. In the present case, we showed how sentiment varied within a region of interest in a US city, highlighting areas that were producing more negative affect than surrounding areas. We believe this kind of visualization tool can provide meaningful data to analysts that are often overwhelmed by the sheer volume of data produced by Twitter feeds. We also believe that this approach will be useful in other geographic regions, as we suggested in our research on the Arab spring.

The second capability is the data mining functionality. The text cube architecture supports the development of prediction models using the data taken from cubes. For example, models that detect events, such as large-scale protests in the Egyptian Revolution, can be built using the sentiment features stored in an event data cube.

Our future research will expand the text cube approach to other HSCB dimensions. We are currently working on group dynamics, leadership, fatalism, and deception, but believe that the analytic community will find value in the text cube approach for many other HSCB dimensions.

Acknowledgement. We thank the reviewers for the valuable comments. Part of this research was funded by Navy Grant # N00014-11-M-0102 awarded to IAI.

References

1. Numrich, S.K., Tolk, A.: Challenges for Human, Social, Cultural, and Behavioral Modeling. SCS M&S Magazine 1(1) (January 2010)

2. Tumasjan, A., Sprenger, T.O., Sandner, P.G., Welpe, I.M.: Predicting elections with Twitter: What 140 characters reveal about political sentiment. In: International AAAI Conference on Weblogs and Social Media (2010)
3. Cha, M., Haddadi, H., Benevenuto, F., Gummadi, K.P.: Measuring User Influence in Twitter: The Million Follower Fallacy. In: Fourth International AAAI Conference on Weblogs and Social Media (2010)
4. Lerman, K., Ghosh, R.: Information Contagion: An Empirical Study of the Spread of News on Digg and Twitter Social Networks. In: Fourth International AAAI Conference on Weblogs and Social Media, Washington, DC, May 23-26 (2010)
5. Gray, J., Chaudhuri, S., Bosworth, A., Layman, A., Reichart, D., Venkatrao, M., Pellow, F., Pirahesh, H.: Data Cube: A Relational Aggregation Operator Generalizing Group-by, Cross-Tab, and Sub Totals. Data Mining and Knowledge Discovery 1(1), 29–53 (1997)
6. Liu, X., Tang, K., Hancock, J., Han, J., Song, M., Xu, R., Manikonda, V., Pokorny, B.: SocialCube: A Text Cube Framework for Analyzing Social Media Data. In: Proceedings of ASE International Conference on Social Informatics, Washington, DC (December 2012)
7. Lin, C., Ding, B., Han, J., Zhu, F., Zhao, B.: Text Cube: Computing IR Measures for Multidimensional Text Database Analysis. In: Proc. 2008 Int. Conf. on Data Mining, Pisa, Italy (December 2008)
8. Zhang, D., Zhai, C., Han, J.: Topic Cube: Topic Modeling for OLAP on Multidimensional Text Databases. In: Proc. 2009 SIAM Int. Conf. on Data Mining, Sparks, NV (April 2009)
9. Zhang, D., Zhai, C., Han, J.: MiTexCube: MicroTextCluster Cube for Online Analysis of Text Cells. In: Proc. 2011 NASA Conf. on Intelligent Data Understanding, Mountain View, CA (October 2011)
10. Zhao, B., Lin, C.X., Ding, B., Han, J.: TEXplorer: Keyword based object ranking and exploration in multidimensional text databases. In: Int. Conf. on Information and Knowledge Management (October 2011)
11. Liu, X., Tang, K., Buhrman, J.R., Cheng, H.: An agent-based framework for collaborative data mining optimization. In: IEEE International Symposium on Collaborative Technologies and Systems (2010)
12. Tang, K., Liu, X., Tang, Y., Manikonda, V., Buhrman, J.R., Cheng, H.: ABMiner: A scalable data mining framework to support human performance analysis. In: International Conference on Applied Human Factors and Ergonomics (July 2010)
13. Brown, C., Frazee, J., Beaver, D., Liu, X., Hoyt, F., Hancock, J.: Evolution of Sentiment in the Libyan Revolution (2011), White Paper at
https://webspace.utexas.edu/dib97/libya-report-10-30-11.pdf
14. Liu, X., Hancock, J., Zhang, G., Xu, R., Bazarova, N.: Exploring linguistic features for deception detection in unstructured text. In: Hawaii International Conference on System Sciences, January 4-7 (2012)
15. Ekman, P.: Telling Lies: Clues to Deceit in the Marketplace, Politics, and Marriage. Norton & Company Inc., New York (2001)
16. Russell, J.A.: A circumplex model of affect. Journal of Personality and Social Psychology 39, 1161–1178 (1980)
17. Mehrabian, A.: Nonverbal communication. Aldine-Atherton, Chicago (1972)
18. Hancock, J.T., Landrigan, C., Silver, C.: Expressing emotion in text. In: Proceedings of the ACM Conference on Human Factors in Computing Systems (CHI 2007), pp. 929–932 (2007)
19. Hancock, J.T., Gee, K., Ciaciaco, K., Mae, J.: I'm sad you're sad: Emotional contagion in CMC. In: Proceedings of the ACM Conference on Computer-Supported Cooperative Work (2008)

20. Kramer, A.D.I.: An unobtrusive behavioral model of "Gross National Happiness". In: Proceedings of the ACM Conference on Human Factors in Computing Systems (2010)
21. Golder, S., Macy, M.: Diurnal and Seasonal Mood Vary with Work, Sleep and Daylength across Diverse Cultures. Science 333, 1878–1881 (2011)
22. Pennebaker, J.W., Booth, R.J., Francis, M.E.: Linguistic Inquiry and Word Count: LIWC. LIWC, Austin, http://www.liwc.net
23. Schwarz, N., Clore, G.L.: Mood, Misattribution, and Judgments of Well-Being: Informative and Directive Functions of Affective States. JPSP 45, 513–523 (1983)
24. Fan, R.-E., Chen, P.-H., Lin, C.-J.: Working set selection using the second order information for training SVM. Journal of Machine Learning Research 6, 1889–1918 (2005)
25. Witten, I.H., Frank, E.: Data Mining: Practical Machine Learning Tools and Techniques, 2nd edn. Morgan Kaufmann, San Francisco (2005)
26. Aha, D., Kibler, D., Albert, M.: Instance-based learning algorithms. Machine Learning 6, 37–66 (1991)
27. http://en.wikipedia.org/wiki/Timeline_of_the_2011-2012_Egyptian_revolution

Mapping Cyber-Collective Action among Female Muslim Bloggers for the *Women to Drive* Movement

Serpil Yuce, Nitin Agarwal, and Rolf T. Wigand

Department of Information Science, University of Arkansas at Little Rock, USA
{sxtokdemir,nxagarwal,rtwigand}@ualr.edu

Abstract. Social media platforms have been lauded for their democratizing potential. They serve as facilitating platforms for activists seeking to replace or alter authoritarian regimes and to promote freedom and democracy. However, regardless of the prominent role of Twitter, Facebook, and other social media platforms in various recently observed social movements, there is a scarcity of rigorous studies that go beyond mere descriptive tendencies and suggest theoretical underpinnings for the manifestations of cyber-collective actions. In this study, we propose a methodology to gain deeper insights into online collective action by analyzing how decentralized online individual actions transform into cyber-collective actions. The proposed model is experimentally analyzed on the data collected for the Saudi women campaigns on driving prohibition. The data consists of female Muslim bloggers' postings from 23 different countries during 2007 and 2012, including various events organized through the Internet (primarily via social media), such as the *Saudi Arabian Women* campaign of September 2007, *International Women's Day* of March 2008, and *Women to Drive* campaign of June 2011. As conceptualized, utilized and illustrated in the study, our novel methodological approach highlights several key contributions to the fundamental research on online collective action as well as computational studies on social media. The tools and methodologies proposed here enable the study of collective actions in broader settings, such as digital/hashtag activism for equitable human rights and citizen engagement for better governance.

Keywords: Online collective action, *Women to Drive*, female Muslim bloggers, social movements.

1   Introduction

Social media provides an easy-to-use and almost ubiquitous platform for Internet users to voice opinions, share thoughts, and participate in discussions. Using various forms of social media, individuals can report first-hand accounts of various events and even organize mass protests and other types of collective actions that eventually may transform into social movements. The emergence of cyber-collective movements has driven much attention and frequently made headlines in the news. The *Saudi Arabian Women* campaign for the right to drive (September 2007 – January 2008), *Wajeha al-Huwaider*'s campaign (2008), and *Manal al-Sharif*'s Facebook campaign named *Teach me how to drive so I can protect myself* (2011), also known as *Women2Drive*

A.M. Greenberg, W.G. Kennedy, and N.D. Bos (Eds.): SBP 2013, LNCS 7812, pp. 331–340, 2013.
© Springer-Verlag Berlin Heidelberg 2013

campaign, are a few examples. However, very little research is devoted to deepen our understanding of this phenomenon, especially those happening in the female Muslim population in Saudi Arabia. Using the global female Muslim blogosphere as an epitome, the proposed research aims to understand the complexity of cyber-collective movements and methodologically track their formations.

Overall these studies support that the Internet can empower political movements in these countries since it provides a suitable infrastructure for expressing alternative views and mobilizing voices for bottom-up actions. We study the female Muslim blogosphere because: First, while research shows that three of four females online are active social media users [1], there is very little research attempting to understand social, cultural and political roles of female bloggers and collectivity among female social groups. Second, the domain epitomizes an important contrast deserving attention between socio-political systems where women are frequently denied freedom of expression and active political uses of social media by female Internet users. Female Muslim bloggers find the blogosphere as a digital recourse to exercise their freedom of speech if compared to their physical and repressively controlled spaces.

2   *Women to Drive* Movement

Saudi women face some of the most inequitable laws and practices when compared to international standards, including the prohibition of driving motorized vehicles. On November 6, 1990 47 Riyadh women staged a remarkable protest against this prohibition. Protesters were imprisoned for a day, had their passports confiscated and some of them even lost their jobs. After more than a decade, in September 2007, Wajeha al-Huwaider and Fawzia al-Uyyouni submitted a 1,100-signature petition to King Abdullah asking for women's freedom to drive. On *International Women's Day* in 2008, Wajeha al-Huwaider filmed her driving and posted the video on YouTube, which garnered international media attention. As a follow up to those actions, in 2011, a group of women, including Manal al-Sharif, started the Facebook campaign supporting women's driving rights in Saudi Arabia. The following months of the campaign, al-Huwaidar filmed al-Sharif driving a car and posted the video on YouTube and Facebook. The consequences were inevitable; she was arrested the following day. Although she was released on bail, there were intolerant conditions, including a ban on driving or talking to media. During the following days, several Saudi women protesters posted their videos while they were driving in reaction to al-Sharif's arrest. In June 2012, to celebrate the anniversary of the June 2011 driving campaign, a member of the *My Right to Dignity* women's right campaign drove her car in Riyadh. Figure 1 illustrates the timeline of the *Women to Drive* campaign depicting various events during the movement.

The campaigns discussed above demonstrate the important role of social media in facilitating cyber-collective actions. They further afford studying how individual sentiment diffuses within the social media network, shapes into collective sentiment, and transforms into collective action. The overarching question is: How are decentralized online individual actions transformed into cyber-collective actions? To follow up on these questions we will provide a theoretical background in Section 3.

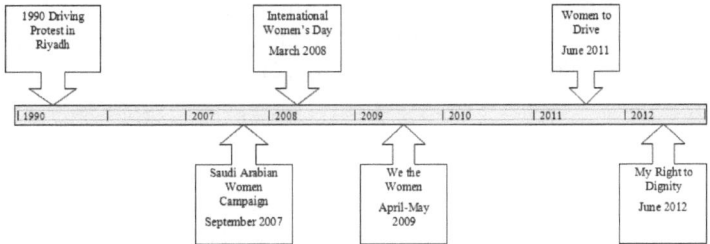

Fig. 1. Timeline for *Women to Drive* campaign ([2-5])

3   Literature Review

Considering the novelty of social media, only few studies have attempted to map collective social dynamics in cyberspace. The first one is the map of the American political blogosphere of the 2004 U.S. elections [6]. The authors studied linking patterns and discussion topics of political bloggers to measure the degree of interaction between liberal and conservative blogs and to uncover differences in the structure of the two communities. The American blogosphere reflected the polarization theory [7]. Similarly the Iranian blogosphere was observed to be clustered along ideological lines [8]. Both studies show that most bloggers tend to read, write about and link to similar things, usually sources that reinforce their own views. This supports the view that homophily [9] has a strong influence on the organization of social networks. However, the homophily principle alone, while useful, does not help in fully explaining how and why a cyber-collective action comes into being. In order to build a strong theoretical framework for the proposed research, in the following section we will revisit two relevant theoretical domains – collective action and social network analysis.

3.1   Collective Action in the Age of Internet

Collective action can be defined as all activity involving two or more individuals contributing to a collective effort on the basis of mutual interests and the possibility of benefits from coordinated action [10]. Theories of collective action are integral to explanations of human behavior. Perspectives on collective action have been useful in explaining diverse phenomena, including social movements [11], membership in interest groups [12, 13], the operation of the international alliance [14], establishment of electronic communities [15], formation of inter-organizational relationships [16], formation of standards-setting organizations [17, 18], and even bidding behaviors [19]. This range of actions accounted by collective action perspectives illustrates the centrality of this body of theory to social science. Traditional collective action theory dates back to 1937, when Ronald Coase sought to explain how some groups mobilize to address free market failures. Yet even when Mancur Olson began updating the theory in 1965 to explain "free-riding" the high-speed, low-cost communications now enjoyed were not imaginable [20]. New information and communication technologies (ICTs), especially the Internet, have completely transformed the landscape of collective action. Lupia and Sin (2003) explain that the burden of internal communication is no longer a hindrance to collective actions, so larger groups are no longer more successful than smaller ones (at least not by virtue of their size). E-mail, chat rooms, blogs, and bulletin boards enable efficient communication, organization, and even deliberation within collective actions of any size [21].

3.2   Social Network Analysis

With the rise of collective action facilitated by online social network media, it is natural for social scientists to embrace the concept of social network in collective action analysis. In analyzing collective actions, social networks can be presented as networks of individuals. Social networks thus contribute extensively and substantially to individual participation. Here, prior social ties operate as a basis for recruitment and established social settings are the locus of movements' emergence. Social network analysis (SNA) has emerged as a set of methods geared towards an analysis of social structures and investigation of their relational aspects [22, 23]. SNA studies social relations among a set of actors assuming a varying degree of importance of relationships among interacting nodes representing individuals, groups, organizations, etc.. Growing interest and increased use of SNA has formed a consensus about the central principles underlying the network perspective. In addition to the use of relational concepts, we note the following as being important [24]: (a) actors and their actions are viewed as interdependent rather than independent autonomous units; (b) relational ties (linkages) between actors are channels for transfer or flow of resources (either material or non-material); (c) network models focusing on individuals and view network structural environments as providing opportunities or constraints on individual actions, and (d) network models conceptualize structure (social, economic, cultural, political) as lasting patterns among actors. Computational SNA (CSNA) helps in the utilization of SNA concepts by providing a rich set of methodologies to examine and summarize large information networks to observe and explain characteristic patterns including: community extraction, expert identification, information diffusion, preferential attachment, and the small-world phenomenon.

4   Methodology

The web, including blogs, could be mined to track information and data about emerging trends and behaviors in almost any area (e.g., political trends and opinions, drug use, racial tension, new films, new products, etc.). Moreover, such data may also demonstrate and reveal information about precisely how ideas diffuse and how trends develop and take hold. We will delve into evolving individual opinions and their development into cyber collective movements, and in so doing delineate the challenges and propose research methodologies. The objective of this paper is to analyze the female Muslim blogosphere data and identify individual blogger sentiments for a specific event, i.e. *Women to Drive*; observe the polarity of sentiments and analyze conflicting views; and study and model the propagation of sentiment in a socially and culturally diverse setting. Specifically, by following the research methodology in Figure 2, we try to answer – do the sentiments of individual bloggers converge to a collective sentiment as time progresses?

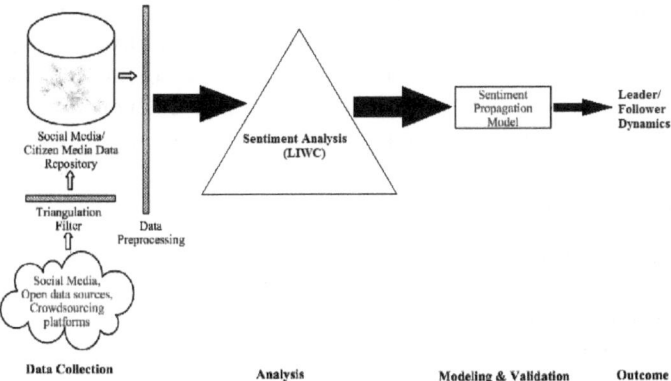

Fig. 2. Research methodology to study collective action for *Women to Drive* campaign

Data Collection for the *Women to Drive* Movement
The content from around 300 blog sites from 23 different countries were collected. Bloggers are included based on three shared characteristics: they are women over the age of 18, they are Muslim, and they primarily blog in English. Other available demographic information, such as nationality, current residence, and name is also included. Since these blogs are updated with frequencies varying between two to three blog posts per day to one blog post per month, a crawler (viz., Web Content Extractor, www.newprosoft.com/) was configured with the above mentioned nuances running constantly to automatically collect, parse, and index the data.

Blog Crawling with Web Content Extractor
The crawler allows us to store the extracted data in a variety of formats, including CSV, TXT, HTML, XML, or directly to an ODBC data source. Collected data includes the title of the blog post, blog post content, the timestamp when the blog post was created, followers' reactions in the form of comments, and the category/tags of the blog post, which can be system-defined or user-defined. We preferred a relational database to store the data due to reliability, scalability, platform independence, and most importantly fast indexing to handle millions of records. From the crawled blog sites, 45 blog sites consisting of 300 blog posts were talking about *Women to Drive* movement at different time periods.

Sentiment Extraction with Linguistic Inquiry and Word Count (LIWC)
After we had collected the blog posts, sentiments were extracted using LIWC (www.liwc.net) software, in order to study the transformation of individual opinions to collective sentiments. LIWC provides an efficient and effective method for studying various emotional cognitive and structural components present in individuals' verbal and written speech samples. LIWC outputs approximately 80 variables. The variables include 4 general descriptor categories, 22 standard linguistic dimensions (e.g., percentage of words in the text that are pronouns, articles, auxiliary verbs, etc.), 32 word categories tapping psychological constructs (e.g., affect, cognition, biological processes), 7 personal concern categories (e.g., work, home, leisure activities), 3

paralinguistic dimensions (assents, fillers, nonfluencies), and 12 punctuation categories (periods, commas, etc.). We have mainly focused on affective processes embedded within psychological processes. The affective processes include 406 positive
emotion words (e.g., love, nice, sweet) and 499 negative emotion words (e.g., hurt,
ugly, nasty). Negative emotions are further categorized into anxiety, anger, and sadness feelings. Scores for positive and negative emotions were obtained from LIWC
for the blog posts and the associated comments.

Towards Online Collective Movements

Existing knowledge on online collective actions mostly relies on quick, short-term,
often journalistic observations; indicating a lack of in-depth studies in this area. The
proposed methodology suggests a computational model advancing our understanding
of the transformation of individual opinions to collective sentiment, a precursor to the
manifestation of collective action as collective movement. Our methodology continues to embrace conventional collective action theories and helps reshape them further
to better understand the implication of new forms of communication.

5   Experiments and Results

For our experiments, we focused on the events occurred during 2007 and 2011. Figure
3 shows the distribution of blog entries and follower comments related to the *Women
to Drive* movement events that occurred between 2007 and 2011. We apply the LIWC
on both blog entries and follower comments to observe how bloggers' sentiments
diffuse into the community, transform into collective sentiment and, eventually, into
collective action.

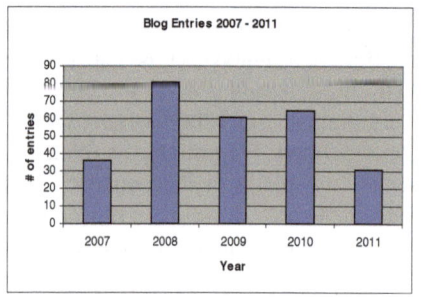
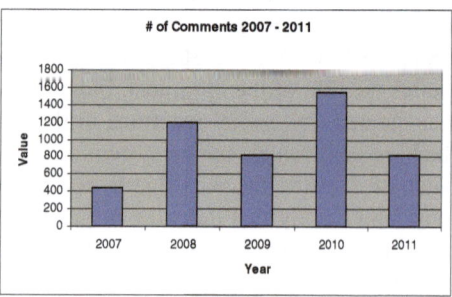

Fig. 3. Volume of blog posts (left) and comments (right) for *Women to Drive* campaign

5.1   Al-Huwaider's Protest in 2008

Figure 3 indicates that the highest traffic of the blog entries was observed in 2008. In
2008, women's driving ban in Saudi Arabia was protested by al-Huwaider on the
International Women's Day by driving her car, although the protest was in March
2008, al-Huwaider posted her video to YouTube in September 2008. Many female
Muslim bloggers expressed their thoughts about the *Women to Drive* movement in
2008. As shown in Figure 4, several blog posts were submitted to create awareness of

the protest in January and February 2008. The reactions of the community are depicted by the increasing number of comments in the following months. We ran LIWC on the 2008 data to identify individual blogger sentiments for this specific event. The results are shown in Figure 5.

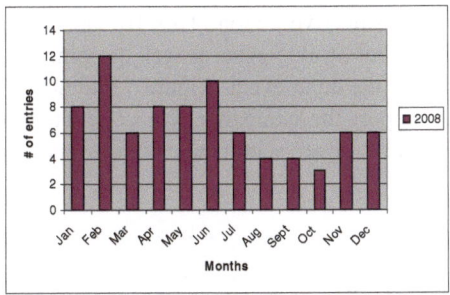 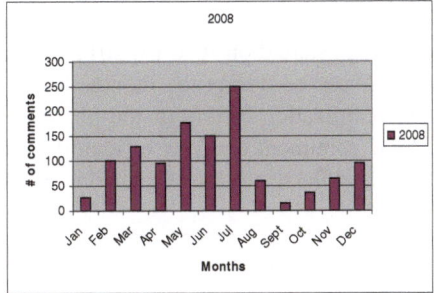

Fig. 4. Data distribution of the blog posts (left) and follower comments (right) in 2008

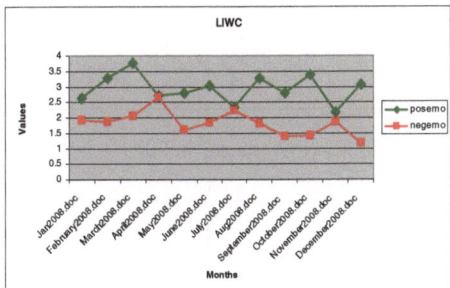 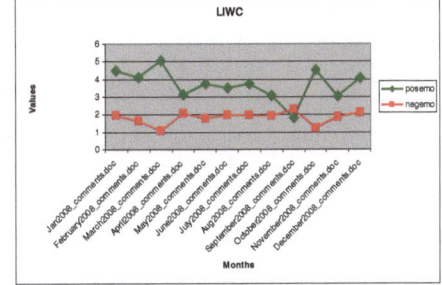

Fig. 5. LIWC results for 2008 data (blog posts on left and follower comments on right)

As seen in Figure 5, bloggers' sentiments about al-Huwaider's protest in March 2008 was largely positive, indicating support for the campaign. A similar sentiment distribution was observed in the comments, indicating sentiment diffusion among the followers. This indicates a sense of solidarity among female Muslim bloggers for al-Huwaider's protest. Again, when al-Huwaider uploaded her video to YouTube in September 2008, a similar reaction was triggered among the female Muslim bloggers as observed by an increase in positive emotions among the blogger and followers.

5.2 Anniversary of the March 2008 al-Huwaider's Protest and *We The Women* Campaign

March 2009 marked the one-year anniversary of the al-Huwaider's 2008 protest. In March 2009 there was a rise in positive emotions regarding the *Women to Drive* movement. Looking at the statistics in Figure 6, we see that there is a very high rise in the blog entries as well as the users' positive reactions regarding the *Women to Drive* movement. After March 2009, between April 2009 and May 2009, the *We the Women* campaign occurred. Again we observe an increase in the positive emotions among the bloggers as well as the followers. These observations indicate and clearly demonstrate the transformation of individual opinions of blog leaders to collective sentiment among the followers via blog interactions.

5.3   The 2011 Facebook *Women2Drive* Campaign

A similar analysis was conducted for the 2011 Facebook campaign to see if the *Women to Drive* movement again led to an online collective action. This *Women2Drive* campaign started on June 17, 2011. A large number of blog posts as well as comments were observed during the months leading up to the June event (Figure 7), indicating the mobilization efforts of the bloggers to support the campaign. We conducted an LIWC sentiment analysis to identify individual blogger sentiments for this specific event and the results are depicted in Figure 8.

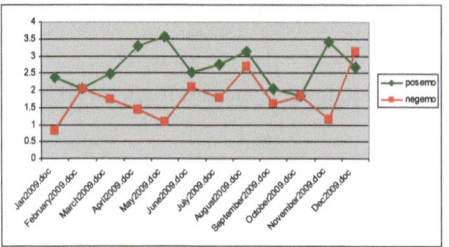 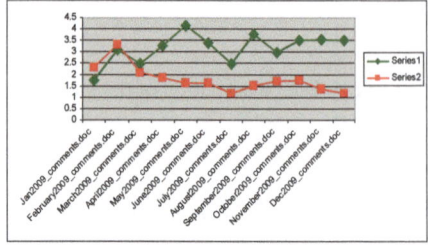

Fig. 6. Data distribution of blog posts and follower comments in 2009

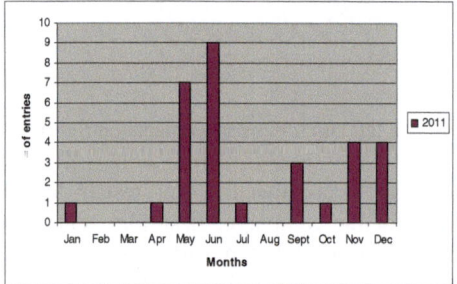 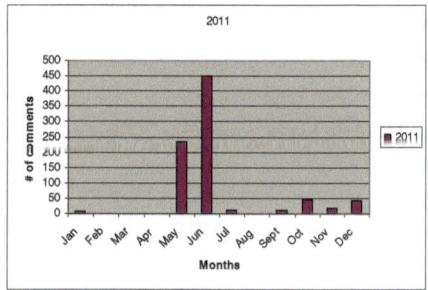

Fig. 7. Data distribution of the blog posts and follower comments in 2011

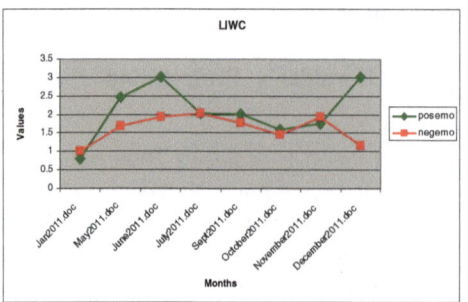 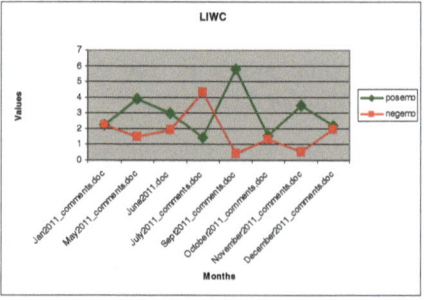

Fig. 8. LIWC results for the *Women2Drive* 2011 campaign

As we see from the charts in Figure 8, bloggers' thoughts about the *Women2Drive* campaign was increasingly positive, supporting and creating awareness towards the campaign. This demonstrates our ability to observe, capture, and model online collective action among female Muslim bloggers. However, we observe a decrease in followers' positive emotions regarding this movement. Upon investigating, we found some anti *Women2Drive* followers' reactions coming primarily from males resulting in the decrease in the positive emotions and increase in negative emotions. Although we carefully handpicked the bloggers as female Muslim users, our data collection does not filter out the male commenters. This suggests a need for considering demographic based filtering of the followers for a more accurate analysis. This remains to be a future direction for our research. Based on our results from the LIWC analysis, we conclude that there was indeed an online collective action observable among Muslim female bloggers for the *Women to Drive* movement.

6   Conclusion

We proposed a novel methodological approach highlighting several key contributions to the fundamental research of online collective action as well as computational studies of social media. Through a rigorous study of various organized events in the *Women to Drive* campaign, we offer: (1) a new framework to understand the evolution and the diffusion of sentiments in online blogger networks; (2) a new approach focusing on the transformation of individual opinions to collective sentiments providing a powerful explanatory model; and, ultimately, (3) a new understanding of the relationship between online collective actions and the rapidly changing online environment. Moreover, our findings highlight a need to discover further pathways of knowledge to fully understand people's cognitive and social behavior, individually and collectively, in online environments with diverse social, cultural, and political backgrounds. Our future research thus will attempt to perform cross-cultural analysis, including data from non-English speaking communities.

We encourage the reader to envision our case study and analysis in the broader context, as our research also lends insights into the relationship between social media and governance. The al-Huwaider case study presented in this contribution shows that collective action is a form of citizen engagement, acting as a corrective mechanism and it is in itself a part of a governance system. In addition, such actions often also enable new organizational forms as well as refreshingly new forms of citizen and government engagement. Social media lend themselves to give citizens a new voice to be heard and, conversely, encourage citizens to engage and participate. Ongoing citizen participatory actions through social media, such as online citizen journalism, can provide a mechanism to pursue a better governance through public monitoring for better decision-making, transparency, and accountability. Consequently, social media can potentially be a bridge to connect the government and its citizenry, have dialogues, and together pursue democratic forms of governance.

Acknowledgments. This research was funded in part by the National Science Foundation's Social Computational Systems (SoCS) research program (Award Numbers: IIS-1110868 and IIS-1110649) and the US Office of Naval Research (Grant number: N000141010091).

References

1. THE BLOGHER-IVILLAGE, Social Media Matters Study, co-sponsored by Ketchum and the Nielsen Company (2010),
   `http://www.blogher.com/files/Social_Media_Matters_2010.pdf` (retrieved)
2. Shmuluvitz, S.: The Saudi Women2Drive Campaign: Just Another Protest in the Arab Spring? Tel Aviv Notes 5(14) (2011)
3. Saudi Arabian Women Campaign for the Right to Drive (2007-2008),
   `http://nvdatabase.swarthmore.edu/content/saudi-arabian-women-campaign-right-drive-2007-2008` (retrieved)
4. `http://en.wikipedia.org/wiki/Women_to_drive_movement`
5. `http://www.care2.com/causes/we-the-women-the-campaign-to-drive-in-saudi-arabia.html`
6. Adamic, L., Glance, N.: The political blogosphere and the 2004 U.S. election: divided they blog. In: LinkKDD 2005: Proceedings of the 3rd International Workshop on Link discovery, Chicago, pp. 36–43 (2005)
7. Wilhelm, A.: Democracy in the Digital Age: Challenges to Political Life in Cyberspace. Routledge (2000)
8. Kelly, J., Etling, B.: Mapping Iran's Online Public: Politics and culture in the Persian blogosphere. Technical report, Berkman Research Center, Harvard Law School (2008)
9. Mcpherson, M., Smith-Lovin, L., Cook, J.: Birds of a Feather: Homophily in Social Networks. Annual Review of Sociology 27, 415–444 (2001)
10. Marwell, G., Oliver, P.: The Critical Mass in Collective Action. Cambridge University Press (1993)
11. Tarrow, S.: Power in Movement: Social Movements and Contentious Politics. Cambridge University Press (1998)
12. Berry, J.: The Interest Group Society. Little Brown, Boston (1984)
13. Olson, M.: The Logic of Collective Action. Harvard University Press, Cambridge (1965)
14. Olson, M., Zeckhauser, R.: An economic theory of alliances. Review of Economics and Statistics 48, 266–279 (1966)
15. Rafaeli, S., Larose, R.: Electronic Bulletin Boards and 'Public Goods' Explanations of Collaborative Mass Media. Communication Research 20(2), 277–297 (1993)
16. Flanagin, A., Monge, P., Fulk, J.: The value of formative investment in organizational federations. Human Communication Research 27, 69–93 (2001)
17. Wigand, R., Steinfield, C., Markus, M.: IT Standards Choices and Industry Structure Outcomes: The Case of the United States Home Mortgage Industry. Journal of Management Information Systems 22(2), 165–191 (Fall 2005)
18. Markus, M., Steinfield, C., Wigand, R., Minton, G.: Industry-wide IS Standardization as Collective Action: The Case of the US Residential Mortgage Industry. MIS Quarterly 30, 439–465 (2006); (Special Issue on Standard Making)
19. Kollock, P.: The Economies of Online Cooperation: Gifts and Public Goods in Cyberspace. In: Smith, M., Kollock, P. (eds.) Communities in Cyberspace, pp. 220–239. Routledge, London (1999)
20. Lupia, A., Sin, G.: Which Public Goods Are Endangered? How Evolving Technologies Affect The Logic of Collective Action. Public Choice 117, 315–331 (2003)
21. Bimber, B., Flanagin, A., Stohl, C.: Reconceptualizing collective action in the contemporary media environment. Communication Theory 15(4), 365–388 (2005)
22. Wigand, R.: Communication Network Analysis: A History and Overview. In: Goldhaber, G., Barnett, G. (eds.) Handbook of Organizational Communication, pp. 319–358. Ablex, Norwood (1988)
23. Scott, J.: Social Network Analysis. Sage, Newbury Park (1992)
24. Wasserman, S., Faust, K.: Social Network Analysis. Cambridge University Press, Cambridge (1994)

Discovering Patterns in Social Networks with Graph Matching Algorithms

Kirk Ogaard[1], Heather Roy[1], Sue Kase[1], Rakesh Nagi[2], Kedar Sambhoos[2],
and Moises Sudit[2]

[1] Tactical Information Fusion Branch, Computational and Information Sciences
Directorate, U.S. Army Research Laboratory, Aberdeen Proving Ground, MD
{kirk.a.ogaard.ctr,heather.e.roy2.ctr,sue.e.kase.civ}@mail.mil
[2] Department of Industrial and Systems Engineering, Center for Multisource
Information Fusion, University at Buffalo (SUNY), Buffalo, NY
{nagi,kps6,sudit}@buffalo.edu

Abstract. Social media data are amenable to representation by directed graphs. A node represents an entity in the social network such as a person, organization, location, or event. A link between two nodes represents a relationship such as communication, participation, or financial support. When stored in a database, these graphs can be searched and analyzed for occurrences of various subgraph patterns of nodes and links. This paper describes an interactive visual interface for constructing subgraph patterns called the Graph Matching Toolkit (GMT). GMT searches for subgraph patterns using the Truncated Search Tree (TruST) graph matching algorithm. GMT enables an analyst to draw a subgraph pattern and assign labels to nodes and links using a mouse and drop-down menus. GMT then executes the TruST algorithm to find subgraph pattern occurrences within the directed graph. Preliminary results using GMT to analyze a simulated collection of text communications containing a terrorist plot are reported.

Keywords: social network analysis, visualization software, graph matching.

1 Introduction

The modeling and analysis of social networks is one of the most powerful tools available to intelligence analysts attempting to understand the social structure present in the battlespace. Analysis of the relationships between people, organizations, locations, and other entities and concepts can reveal the organization of enemy forces and the relationships between key elements of the local population. These types of relationships can be modeled using attributed directed graphs.

Graphical network models can be applied in a wide range of applications such as Cyber Security, Asymmetric Warfare, Disease Surveillance, and Intelligence and Knowledge Discovery. Walsh [1] suggested the creation of a "global graph" as a representation of the world (or area of interest) from which information and relationships can be discovered to support Intelligence Preparation of the Battlespace and Command and Control applications. The global graph is structured as a classical

A.M. Greenberg, W.G. Kennedy, and N.D. Bos (Eds.): SBP 2013, LNCS 7812, pp. 341–349, 2013.

graph format with objects (also called nodes or entities) interrelated by relationships (also called edges or links). Nodes can represent people, organizations, locations, individuals, or facilities. Links represent relationships such as communication, participation, or association. In addition to nodes and links, attributes (or labels) can store details about objects and their relationships.

A complex situation or hypothesis of interest to an intelligence analyst can be formulated as a template graph made up of nodes and links. In real-time an analyst wants to determine the occurrences of the situation represented by the template graph within a database of information in the form of a graph (also called a graph database or data graph). Sometimes searching for a one-to-one correspondence between the template graph and data graph is too strong, necessitating inexact (or fuzzy) graph matching where two graphs are compared even though they are semantically or topologically different. Inexact subgraph matching between the template graph and the data graph allows analysts to focus on a set of most likely situations.

Section 2 introduces the Graph Matching Toolkit (GMT) —a visual interface for drawing template graphs. GMT serves as a front end to the stochastic Truncated Search Tree (TruST) algorithm [2] detailed in Section 3. Using the template graph, TruST performs fuzzy graph matching on the data graph. In Section 4, we present a use case in which a user, an analyst, hypothesizes relationships between individuals of a social network thought to be members of a terrorist organization. In the use case the analyst uses GMT to draw template graphs (i.e., graphical representations of queries) to search the data graph for evidence that supports or refutes the hypotheses.

2   GMT

GMT was implemented in the C++ language using the Qt cross-platform application framework [3] for its Graphical User Interface (GUI). The front end for GMT has been successfully tested on multiple operating system platforms. GMT interfaces with the graph matching algorithm via flat files. GMT stores the subgraph query (i.e., the template graph) in a GraphML file prior to executing the TruST algorithm (see Fig. 1). The TruST algorithm loads the GraphML file created by GMT, executes the specified subgraph query, and stores the results in an XML file. Finally, GMT loads the XML file containing the results from the TruST algorithm, and displays the top k matching subgraphs with the highest overall scores in the results subwindow.

The GUI for GMT is divided into four subwindows (see Fig. 2): 1) the toolbar, 2) the canvas, 3) the link ontology legend, and 4) the results subwindow. The toolbar provides four tools (represented by appropriate icons) for adding nodes, adding links between nodes, removing nodes and associated links, and undoing a previous action. The canvas allows users to draw the subgraph patterns for their search queries. The link ontology legend shows the color coding scheme for the links based on the categories in the currently loaded ontology. The results subwindow displays the search results obtained by executing the subgraph query on the graph database (i.e., the data graph) with the graph matching algorithm (i.e., the TruST algorithm).

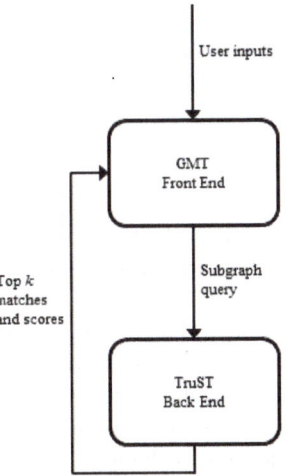

Fig. 1. The flowchart for user interactions with GMT

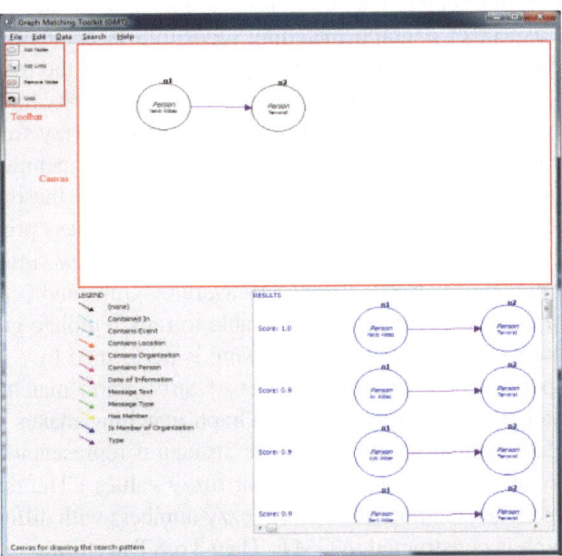

Fig. 2. GMT with the canvas for drawing a template graph (large top window outlined in red) and the search results window (bottom window in blue)

To add nodes or links to the subgraph pattern, the user clicks on the appropriate tool in the toolbar, then, clicks the mouse on the canvas to create nodes or links at those positions. The user annotates the subgraph pattern with categories from the ontology by right-clicking on the nodes and selecting categories from the context menus. After the user has drawn and annotated the subgraph, the user initiates the search for matching subgraph patterns in the graph database. Both the subgraph patterns and graph database are stored in flat files with the GraphML file format. The back end of GMT then executes the graph matching algorithm selected by the user. The primary graph matching algorithm currently utilized by GMT is TruST which is detailed in the next section.

3   Stochastic TruST Algorithm

TruST [4, 5] is a heuristic search algorithm optimization of best-first search. One disadvantage of best-first search is an optimal solution is not guaranteed. However, the majority of the time, the reached sub-optimal solution is a very good one and through the control of the truncation parameters various efficiency tradeoffs can be obtained. Zhang [6] has shown that truncated branch-and-bound as a heuristic gives better results than most heuristics methods. The search tree is developed dynamically during the search and initially consists of only the root. At each iteration of the algorithm, a sub-problem is selected for exploration from the pool of live sub-problems using scores of the current match. TruST uses a strategy similar to the breadth first search strategy found in the literature [11]. The basic principle is to process all the nodes at one level of the search tree before processing any node at a deeper level.

An important aspect of a graph matching algorithm is calculating the degree of similarity between a situation of interest (template graph) and the real world (cumulative data graph). There are two main categories of similarity metrics, those that provide a point estimate similarity and those that provide a fuzzy similarity. The motivation for incorporating uncertainty into situation awareness techniques is the ability to quantify the uncertainty of any final conclusions made by these techniques. This desire to preserve uncertainty throughout the situation awareness process necessitates the use of a fuzzy similarity measure. Currently, the best fuzzy similarity measure is based on the complementary fuzzy distance measure by Guha and Chakraborty [7].

The TruST graph matching heuristic [2] is able to rank template graph – data graph matches. In the case of crisp scoring, this ranking is performed by a simple numerical ordering. To properly exploit the fuzzy scores of dirty graph matching, a method of ranking fuzzy numbers must be used. (Dirty Graph matching makes use of inaccurate observations or data estimates, and inaccurate structural representations of a state of interest, thus accounting for the uncertainties or fuzzy values.) The method to perform this is the Chen and Chen method for ranking fuzzy numbers with different spreads [8].

A template graph is constructed in GMT. Then TruST performs subgraph matching with ranking and fuzzy scores passed back to GMT to display to the user. The application of this process to intelligence analysis is demonstrated in a use case scenario described in the next section.

4   Use Case

GMT is designed to enable a user to: create subgraph patterns of nodes and links in the form of a visual query; execute a subgraph matching algorithm to search for matching and nearly matching subgraph patterns in a large graph database; and evaluate the ranked subgraph matching results. These capabilities can assist analysts to formulate hypothesized relationships of interest—typically derived from information requirements, expert judgment, or intuition. Analysts can then apply these capabilities to test their hypotheses to determine if these relationships are supported by the data.

In this use case, an analyst is presented a simulated dataset of over 600 text communications containing a hidden terrorist plot taking place in England [9]. The analyst will use GMT and the TruST graph matching algorithm to iteratively search the social network constructed from the text communications for pertinent information.

To begin this scenario, the analyst receives a Police Report that "Yakib Abbaz" is a suspected terrorist. The police are requesting additional information about Yakib Abbaz and his relationships. The analyst wants to search the graph database of text communications for other Tactical Reports (TacReps) referencing Yakib Abbaz. To perform the search the analyst draws a two-node template graph in GMT's canvas subwindow. The template graph has node 1 (n1) as a "Person" labeled "Yakib Abbaz" and node 2 (n2) as a "TacRep" with no label (Fig. 3, left side). The nodes are connected by a "Contained In" link (Fig. 3, light blue arrow) indicating Yakib Abbaz's name may appear in other reports.

Once the template graph is drawn, the analyst initiates the graph matching search by selecting the TruST algorithm from the menu. The results of the search return the top five scoring matches (Fig. 3, right side). These matches indicate specific TacReps containing information about Yakib Abbaz. The score of 0.9 for each match means a nearly perfect match was found to the template graph. The analyst then examines the specific TacReps corresponding to those top five matches in greater detail.

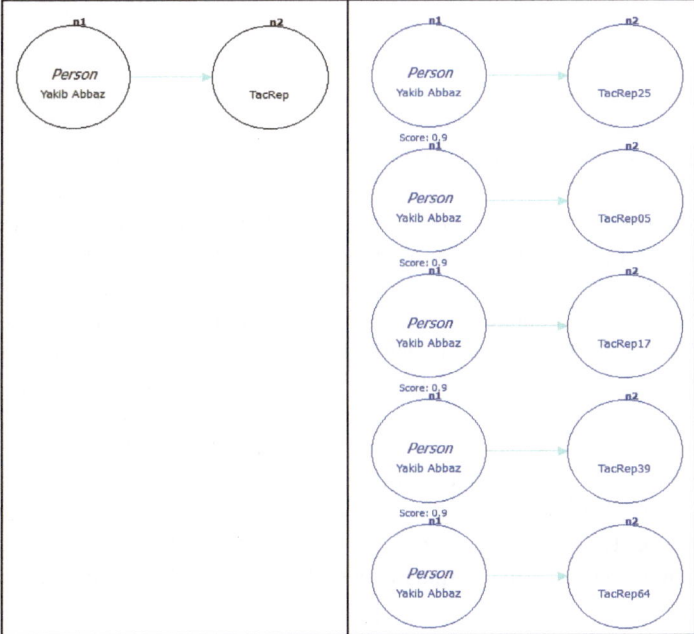

Fig. 3. Two-node template graph for a search (left side) and search results (right side)

From the text contained in the TacReps listed in the search results, more information is gleaned about Yakib Abbaz. Yakib communicates with Salam Seeweed who appears to be a financier for an operation referred to as "baking a cake" (Fig. 4, TacRep 25). Also, Yakib may be recruiting "apprentice bakers" at the East Side Mosque (Fig. 4, TacReps 5 and 39). In TacRep 5, Yakib is reporting his success to Imad Abdul. And in TacRep 64, Yakib is being "ordered" by Abdul. Thus, Imad Abdul appears to be acting in a leadership capacity at a level above Yakib.

TACREP 25 *(14 July 2003)*
In London, Salam Seeweed informed Yakib Abbaz that money would be available for all who would assist in baking a big, delicious cake. Money would also be available for the newcomers to the party. Yakib expressed his gratitude.

TACREP 5 *(06 June 2003)*
In London, Yakib Abbaz reported to Imad Abdul that he (Yakib) was able to identify several high quality prospects at the East Side Mosque. Abdul praised Yakib for his good work.

TACREP 17 *(28 June 2003)*
In London, Yakib informed Salam about the misfortunes of two of his friends and asked if financial aid could be forthcoming. Salam promised to bring up the issue with Abdul and Sheikh.

TACREP 39 *(19 Aug 2003)*
In London, Raed told Tarik that Yakib Abbaz had done a very good job in bringing apprentice bakers on the mission.

TACREP 64 *(14 October 2003)*
In the town of Henley, Abdul told ordered Yakib to cease his activity in London and to take a vacation in the northern part of England.

Fig. 4. The messages from the dataset corresponding to the search results

With the new information gathered from the TacReps, the analyst can characterize Yakib Abbaz's relationships using a concept map. The TacReps can be imported into a social network analysis tool such as AXIS Pro for further analysis. AXIS Pro is the preferred intelligence analysis tool used by the U.S. Army [10]. Fig. 5 shows an AXIS Pro concept map of Yakib Abbaz's relationships derived from GMT search results.

Because Imad Abdul appears to be high in the chain of command and the analyst is interested in Yakib Abbaz's social network, the next step is to use GMT to search the database for communications involving both Imad Abdul and Yakib Abbaz. A three-node template graph could be constructed with node 1 (n1) as a "Person" labeled "Imad Abdul," and node 2 (n2) as a "Person" labeled "Yakib Abbaz," connected with "Contains Person" links to a node 3 (n3) as a "TacRep" (Fig. 6). When the search is executed, GMT displays the top five scoring matches in the form of TacRep numbers. These specific TacReps could then be pulled from the database for more in depth analysis, and the concept map expanded with the additional information.

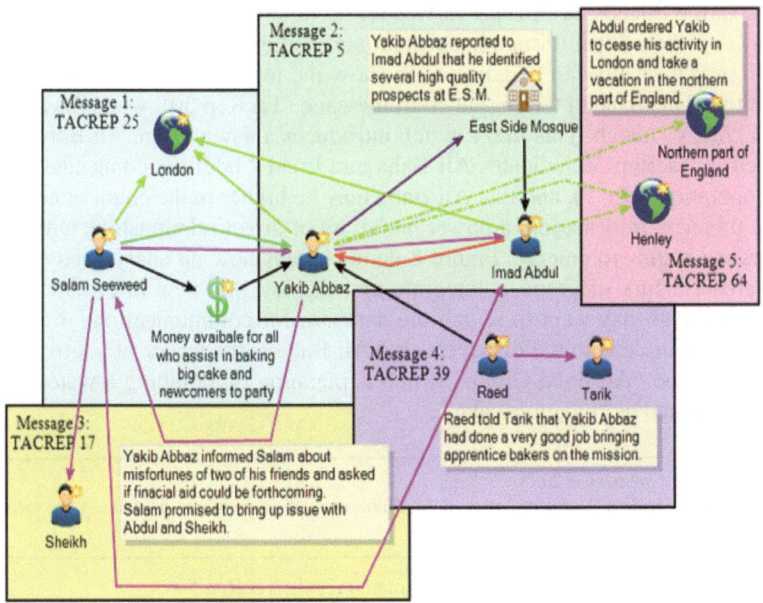

Fig. 5. AXIS Pro concept map of GMT search results outlined by TacRep number

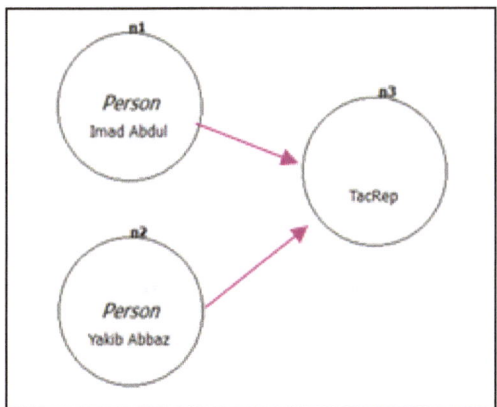

Fig. 6. GMT three node search template

Results from the three-node template graph search reveal five matches with a score of 0.9. Two of the five subgraph matches are new TacReps, the other three were contained in the previous search results. The information gained from the new TacRep messages can be integrated into the analyst's concept map. The first new message, TacRep1, provides further evidence that Abdul is higher in the chain of command than Yakib by stating "Imad tasked Yakib". In this message, Yakib is tasked to visit several mosques in eastern London to identify persons sympathetic with their cause. In the second new message, TacRep 19, Yakib reports back to Abdul that the response at the Sheepside Mosque in western London was less receptive than eastern London. Together, this information substantiates Yakib's role as a recruiter.

GMT search results can be adjusted to show the top five to ten scored matches. If the result set is expanded to ten, the sixth message (TacRep 50), which has a score of 0.8, is a "fuzzy" match. This fuzzy match introduces a new person, Ali Baba. The text contained in TacRep 50 indicates Ali Baba and Imad Abdul are connected by a cake baking operation (Fig. 7), and that Ali Baba may be higher in the chain of command.

With this new information, a power hierarchy of direct relationships within the social network begins to emerge. Figure 8 demonstrates how an analyst may use AXIS Pro to create a link diagram to integrate and display this new information. At this point, the analyst may want to search the database for communications involving Ali Baba. This would lead to the discovery that Ali Baba is the leader of a terrorist organization called the "Ali Baba Group" which is planning on bombing a water treatment facility referenced as "baking a cake".

TACREP 50 *(29 September 2003)*
In the town of Henley, Abdul reported to Ali Baba that all preparations for baking the cake were proceeding on schedule.

Fig. 7. GMT fuzzy match result: TacRep 50

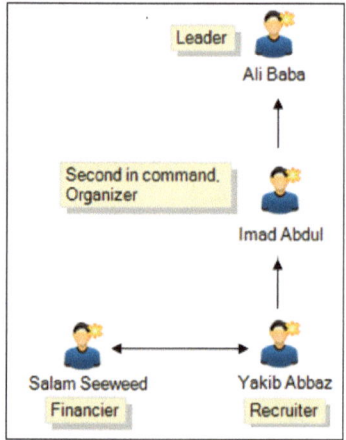

Fig. 8. AXIS Pro link diagram of the power hierarchy of the Ali Baba Group

5   Conclusion

This paper introduces the Graph Matching Toolkit (GMT). GMT was tested with a use case involving the analysis of a set of simulated text communications containing a hidden terrorist plot. GMT provides analysts the capability to build template graphs to iteratively search a graphical database of communications. GMT executes the TruST graph matching algorithm to find matching subgraph patterns in the data graph. Matches are scored and displayed in a ranked order which offers a set of alternatives to explore with additional template graph queries. In summary, GMT enables an analyst to quickly hone in on high value targets within a social network containing a large number of nodes that might otherwise require an exhaustive manual text search.

GMT is designed to serve as a general GUI for graph matching algorithms. Although GMT can find matching subgraph pattern occurrences in a data graph with the TruST algorithm, it does not support any constraints on those searches. Search constraints, such as restricting the search to data from a specific time frame, would facilitate the efficient analysis of data graphs derived from massive datasets. Future work will focus on incorporating such a search constraint capability into GMT.

Acknowledgments. This project was supported in part by an appointment to the Internship/Research Participation Program for the U.S. Army Research Laboratory administered by the Oak Ridge Institute for Science and Education through an agreement between the U.S. Department of Energy and the USARL.

References

1. Walsh, D.: Relooking the JDL Model for Fusion on a Global Graph. In: National Symposium for Sensor Data Fusion (2010)
2. Sambhoos, K., Nagi, R., Sudit, M., Stotz, A.: Enhancements to High Level Data Fusion using Graph Matching and State Space Search. Information Fusion (2009) (in press, corrected proof)
3. Qt SDK, http://qt.digia.com/Product/Qt-SDK
4. Sambhoos, K.: Graph Matching Applications in High Level Information Fusion [Dissertation]. State University of New York at Buffalo, Buffalo (2007)
5. Sudit, M., Nagi, R., Stotz, A., Sambhoos, K.: A Graph-Based Framework for Fusion: From Hypothesis Generation to Forensics. In: 9th International Conference on Information Fusion. IEEE Press, New York (2006)
6. Zhang, W.: Depth-First Branch-and-Bound versus Local Search: A Case Study. In: 17th National Conference on Artificial Intelligence, pp. 930–935. AAAI Press, Palo Alto (2000)
7. Guha, D., Chakraborty, D.: A New Approach to Fuzzy Distance Measure and Similarity Measure between Two Generalized Fuzzy Numbers. Applied Soft Computing 10(1), 90–99 (2010)
8. Chen, S., Chen, J.: Fuzzy Risk Analysis Based on Ranking Generalized Fuzzy Numbers with Different Heights and Different Spreads. Expert Systems with Applications 36(3), 6833–6842 (2009)
9. Mittrick, M., Roy, H., Kase, S., Bowman, E.: Refinement of the Ali Baba Data Set. U.S. Army Research Laboratory, ARL-TN-0476 (2012)
10. AXIS Pro, http://www.overwatch.com/products/axis_pro.php
11. Knuth, D.: The Art of Computer Programming, 3rd edn. Fundamental Algorithms, vol. 1. Addison Wesley Longman Publishing Co. Inc., Redwood City (1997)

Critiquing Text Analysis in Social Modeling: Best Practices, Limitations, and New Frontiers

Peter A. Chew

Galisteo Consulting Group, Inc., 4004 Carlisle Blvd NE, Suite H,
Albuquerque, NM 87107, USA
PeterAChew@gmail.com

Abstract. Natural language processing (NLP) is an important contributor to the field of social modeling. Language is a social artifact; it is how people express opinions, persuade, or convey what they believe is important. It is thus rightly recognized that computational tools can automate at least some of the analytical work of reading an ever-increasing volume of textual data, reducing time and costs. Language is also, however, a complex and variegated system, creating a challenge for social modelers. In this paper, we contend NLP's full potential is commonly not being exploited, leading to unnecessary work and lower-quality results, and that social modelers using NLP should understand at a high level what NLP problems are, and are not, solved. Our findings have implications for both the practice and validation of NLP in the social modeling community.

Keywords: Validation, text analysis, methodological innovation.

1 Introduction

Managing data is an increasing challenge for virtually all organizations. A recent Gartner Group report [1] estimates that enterprise data capacity is growing at a rate of 40% to 60% a year – a phenomenal rate – owing to factors including an explosion in unstructured data such as e-mails and documents. Almost half the respondents to the survey documented in the Gartner report identified data growth as one of their top three challenges. These issues simply reflect a trend which is observable within the wider world – that data is being generated faster and is becoming more widely available: for example, the internet has grown from around 25 million webpages in 1996 to over a trillion today. Furthermore, the internet is linguistically diverse (see Figure 1 overleaf), with increasing adoption by non-English speakers and particularly outside Western countries, with the Arab Spring being a recent and pertinent example.

While a challenge, the growth of data also creates many opportunities. Here, we list cases where the 'state of the art' has been considerably improved thanks to, rather than despite, larger datasets:

- Search engines [3] (whose purpose, it should be noted, is to organize unstructured information) generally tend to find relevant results more reliably than their counterparts of 15 years ago.

A.M. Greenberg, W.G. Kennedy, and N.D. Bos (Eds.): SBP 2013, LNCS 7812, pp. 350–358, 2013.

- A new data-driven approach to machine translation has emerged [4] and has achieved initial successes which eluded the rule-based machine translation approaches of the Cold War.
- And, in the field of social modeling, data-driven approaches have enabled social networks to be studied on a scale previously unimaginable (see e.g. [5]).

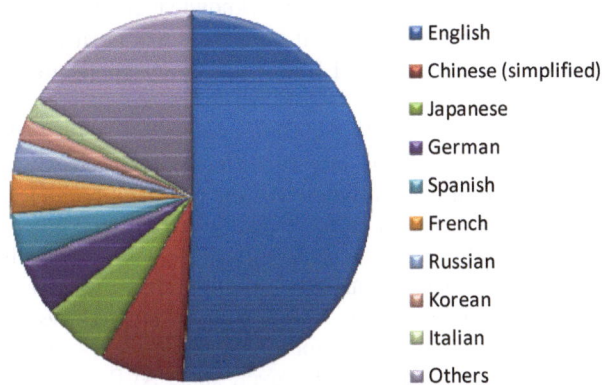

Fig. 1. The World Wide Web by Language (from [2])

In all cases, the mere existence of larger datasets has acted as a catalyst in the development of new and better approaches. For one thing, these datasets allow algorithms to be trained and tested on more representative datasets. For another, many data mining algorithms are highly sensitive to subtle patterns in data and therefore tend to work better on larger datasets than on smaller ones.

In the area of social modeling specifically, where unstructured data abounds (for example, in the form of social media posts), much hope rests on harnessing similar algorithms, for example, to provide a 'social radar' [6]. It is this area that we focus on in this paper. And since a significant portion of unstructured data is text [1], we believe it is useful in particular to focus on how text is analyzed in social modeling, and how current approaches sometimes fail to use, or misuse, the current state of the art in natural language processing.

To set the stage, in section 2 we provide a brief overview of the field of natural language processing (NLP) and some applications of NLP to social modeling. We then look at the relationship of NLP to data analysis more generally in section 3. Specific social-modeling uses of NLP are then critiqued in section 4, focusing on cases where social modeling fails to make full use of the existing state-of-the-art. We conclude in section 5 with specific recommendations for better use of NLP generally within the field of social modeling.

2   An Overview of NLP and Its Relationship to Social Modeling

A basic goal of NLP is to 'get computers to perform useful tasks involving human language, tasks like enabling human-machine communication, improving human-human communication, or simply doing useful processing of text or speech' [7]. Applications of NLP in the field of social modeling are widespread and diverse. In surveying just the most recent proceedings of the International Conference on Social Computing, Behavioral-Cultural Modeling, and Prediction (SBP), we find analysis of social media [8], group membership detection from online text [10], crime prediction from Twitter posts [11], use of social media for disaster relief [12], and modeling social networks from academic acknowledgements using named-entity recognition (NER) [13]. In addition to all this, NLP has been used for example in sentiment analysis [14]. Of course, NLP also has many applications which are not directly related to social modeling.

It is instructive to consider how NLP differs in outlook from social modeling. The key difference is that NLP is highly empirical; most published work contains some type of empirical evaluation of a proposed method against a baseline, for example in terms of accuracy; [14] is a case in point. Social models, on the other hand, are often intrinsically difficult to validate [15] (this is not a criticism of, but rather a fact of life in, social modeling); if NLP's standards of empiricism were applied in social modeling, much useful social modeling work would not be published.

There are good reasons, therefore, for the different roles played by empirical methods in NLP and social modeling. However, since NLP is so widely applied in, and is a foundation for so much of, social modeling, it behooves the social modeler to keep the empirical basis of NLP clearly in view when using techniques from the field. The following example will illustrate why. Suppose a modeler wishes to build a model of a social network using references to named entities (people) in online documents, as in [16]. Here, the modelers decided they needed to pre-process the documents using various NLP component workflows including de-duplication of named entities, correction of typographical errors, expansion of contractions and abbreviations, pronoun resolution, and identification of 'concept-changing n-grams' (explained below in section 4.2). Code to achieve these goals was incorporated into the social model. The model produces output believed to be useful, but which cannot be validated in the classical sense because there is no 'ground truth' against which to compare. It might appear in this case that empirical testing is irrelevant. Yet just because the social model cannot be validated, this does not mean its individual components cannot be. For example, it is standard practice in NLP to test a named-entity recognition (NER) program or technique for precision and recall against some 'ground truth' dataset. Let us further suppose that this program is found to have 60% recall (meaning that of the actual named entities in some test dataset, it finds 60%) and 65% precision (meaning that of the items it identified as named entities, 65% were actually named entities). Derivatively, one can reasonably infer that these empirical measures are also characteristics of the social model as a whole, as far as the model deals with named entities. Further, given a choice between a NER model that has 60% recall and 65% precision, and another suitable model that has 75% recall and 70% precision, we would prefer to incorporate the latter into the social model. In other words, empirical accuracy can be a 'hidden' characteristic of a social model, even if it is not directly observable. But too often, this may not be given due consideration in the development of a social model.

Choices between NLP techniques also have important consequences beyond the realm of the empirical. While considerations such as accuracy should be a concern, equally as important should be the efficiency and generality of methods used. The more general a method, the more different types of situation it can be deployed to. This is key for organizations concerned about containing costs, because the costs of developing a reusable method can be amortized over many different projects. The concept is very similar to that of software reuse, which has long been recognized as a best practice in computer programming [17]; and it is also related to the principle of Occam's razor in science.

One might think that simplicity and generality in NLP, or other areas, would tend to militate against empirical measures like accuracy. We will show, through several NLP-related examples which have direct relevance to social modeling (in that similar techniques have actually been used in social models), that this is often not the case. In fact, greater simplicity and generality often *increase* accuracy. Social modelers who ignore this fact run twofold risks – one, of developing suboptimal models, and two, of spending unnecessary resources reinventing already-invented 'wheels'.

3   The Place of NLP in Data Analytics

In the last couple of decades, a key thrust in the field of data analytics (also referred to as data mining) has been reflected in increasing interest in, and use of, *unsupervised learning* methods. Broadly speaking, the goal of unsupervised learning is to find hidden structure in unlabeled data (i.e., data where a human has not already added the structure that is to be found, for example in annotations or metadata). Unsupervised learning contrasts with two other ways in which data analysis is still commonly approached: (1) a heuristic, rule-based approach, and (2) supervised learning.

Machine translation offers a good illustration of the contrast between rule-based and unsupervised (or at least minimally supervised) approaches to data analysis. The earliest attempts at machine translation (the 1954 IBM-Georgetown experiment) [18] were essentially rule-based: the IBM machine translation system incorporated 250 vocabulary items and 6 grammar rules. While perceived at the time as a success, approaches of this type did not deliver on promises that machine translation would be a solved problem within 3-5 years. The basic problem is that rule-based systems quickly become unmanageably complex, with rules interacting in often unforeseeable ways. However, a new approach, statistical machine translation (SMT) [4], took root in the mid-1990s and has reignited interest in, and (tellingly) funding for machine translation. The basic premise of SMT is that parallel corpora – collections of text aligned across multiple languages, such as the European Parliament proceedings [19], where speeches are currently translated into 21 languages – exhibit various types of latent but algorithmically-learnable structure. This can then be used as the basis for predicting the most likely rendition in a target language of text in (another) source language. Because SMT relies on the intrinsic structure of data, rather than rules, it is inherently more general than the early rule-based attempts at machine translation. Furthermore, it is worth noting that SMT is possible only because of the growth of electronically available text, for example on the World Wide Web; SMT, in common with other unsupervised learning methods, tends to perform better with more input data. Finally, a measure of SMT's increasingly widespread use is, for example, its presence in Google Translate (translate.google.com).

Supervised learning, on the other hand, requires a labeled dataset as input; predictions about unlabeled data are then made on the basis of the input. For an example of supervised learning, we can turn to [14]. In this case, the input was movie reviews (text) labeled with star ratings (one star denoting an unfavorable rating, five stars denoting a favorable rating), and the task was to predict whether an unseen movie review was positive or negative based only on the text of the review. To accomplish this, various different supervised learning algorithms were presented with examples of positive and negative reviews, along with the star-rating labels. With these examples, the algorithms 'learn' associations between certain words and positive or negative sentiment (without a programmer having to encode these associations explicitly). The disadvantage of supervised learning is that labeling data is usually expensive and time-consuming, if it is even feasible at all.

Many applications which use NLP have been highly successful of late; one measure of this is commercial adoption, and a prime example is Google. It is noteworthy that of the prominent examples of commercially successful NLP, it is hard to find many built extensively upon supervised or heuristic/rule-based algorithms. In Google's case, the PageRank algorithm [3] (which is not an NLP algorithm, but is unsupervised nonetheless) and SMT are both unsupervised. Recall that unsupervised learning excels at finding hidden structure in *unlabeled* data, and the World Wide Web is noisy, unlabeled data par excellence. In Google's case, it would simply be impossible, given the size of the internet (estimated in 2010 at a trillion web pages [20]), to maintain heuristic-based rules for indexing web pages in hundreds of languages and on innumerable topics. Social modelers who deal with data from the World Wide Web would be well-advised to keep the prevalence of unsupervised techniques at the forefront of their minds, and at the same time to remember that NLP, in common with many other types of data analysis, is currently most successful when it is used in its unsupervised incarnations. And by 'success', we do not mean just commercial success; it has been shown that the replacement of heuristic-based approaches with unsupervised and more general approaches often results in superior accuracy [21].

In the remaining sections, we shall explore some specific ways in which this has implications for social modeling.

4   Failing to Take Full Advantage of NLP

In this section, we shall examine ways in which NLP has actually been approached in social modeling, and show how each of these fails to take full advantage of the 'unsupervised' paradigm, resulting in unnecessary work and suboptimal accuracy. It is not our intention to single out particular instances of prior work here; however, [16] offers convenient cases in point: [16] makes widespread use of text-processing, and is representative of, and influential in, much social modeling. Our intention is simply to be helpful in showing how the state of the art in the field can be improved, and we feel that specific examples are most conducive to this.

4.1   Identifying Key Words in Text

It has long been recognized in NLP that not every word is equally important in text. In the previous sentence, for example, 'word' is intuitively more important to the overall meaning than 'has'. A subtler point is that the importance of a word is not necessarily correlated to its frequency. Again, from the first sentence, it should be intuitively apparent that 'in' is less important than 'word' to the overall meaning, even though the frequency of 'in' is double that of 'word'. There are various strategies commonly used to discriminate between words with differing levels of importance. We shall review two here (stoplists and term-weighting), with a particular focus on stoplists.

A stoplist is essentially a list of words considered to be unimportant to the meaning of text: examples of such words in English might include 'and', 'the', 'in'. After a corpus is tokenized (split into its constituent words), words in the stoplist are discarded and become irrelevant to further processing. The use of stoplists is exemplified in [16] (who do not use the term 'stoplist' but instead refer to a '"noise" list for concepts to be deleted when cleaning the text').

Using stoplists has drawbacks, however. The main disadvantage is that most stoplists are compiled on a heuristic basis (in other words, someone reviews words and determines somewhat arbitrarily which are stopwords and which are not). The availability of pre-compiled lists[1], while lending stoplists a 'scientific' aura, does not diminish the heuristic/arbitrary nature of these lists. In fact, one should question whether a single list can really be applicable to the wide variety of text that exists, even when dealing just in a single language. A clue to this is in [16]'s conclusion:

> Several challenges exist, whose resolution will further speed and enhance this process. These include..., need to handle embedded concepts that are not English, ... improved automated removal of meta-data from texts, utilization of auto-topic identification to supplement and direct user defined topic identification...

A further hint to the problematic nature of stoplists is in the phrase 'need to handle embedded concepts that are not English', quoted above. What happens when we need to process non-English text? Using the stoplist approach, we are forced to develop or find a stoplist for each new language. If no stoplist exists, we then need the expertise of a linguist to determine which words are unimportant, and moreover we have to rely on the linguist's judgment. Clearly, this is the problem the authors of [16] ran into: creating stoplists for multiple languages quickly becomes an expensive, time-consuming proposition.

The key to avoiding this time and expense, yet still obtaining satisfactory results in determining which words are most 'important', can in our view be found in term-weighting schemes (see [21], [22]). These aim to quantify the importance of words in text, often using principles from information theory. For example, under the log-entropy weighting scheme [21] (which we do not wholeheartedly endorse, as will become apparent), the weighting of a term is the product of its local and global weights; the local weight is the log of the term's frequency in a particular document, and the global weight is a function of the ratio of the term's entropy in a corpus to the maximum possible entropy, as follows:

---

[1] See e.g.
`http://nlp.cs.nyu.edu/GMA_files/resources/english.stoplist`

$$w_{i,j} = \log(f_{i,j} + 1) \times \sum_{x=1}^{n} \frac{\frac{f_{i,j}}{f_i} \times \log\frac{f_{i,j}}{f_i}}{\log n}$$

where:

$w_{i,j}$ is the frequency of term i in document j weighted according to log-entropy;

$f_{i,j}$ is the frequency of term i in document j;

n is the number of documents in the corpus being analyzed; and

f_i is the frequency of term i in the entire corpus.

Note that entropy is an information-theoretic measure which can be computed from term frequencies as shown above; a term like 'the' tends to have a high entropy, leading to a low global weight which tends to counteract high local weighting.

In actual fact, many approaches (e.g. [23]) use both stoplists and term-weighting (in our view, this combination fails to capitalize on the ability of term-weighting both to save labor and simplify the system overall – which, all else being equal, should be preferred under the principle of Occam's razor).

We argued above that social modelers may fail to consider empirical ramifications in choices between NLP techniques. With regard to the identification of key words in text, this line of argument extends beyond the choice between stoplists and term-weighting, to choices *between* term-weighting schemes. For example, [21] shows, using a multilingual document-clustering task (which could be relevant, for example, in the reconstruction of a social network from multilingual Twitter data), that the choice between log-entropy weighting (which is in widespread use) and the lesser-known Pointwise Mutual Information (PMI) weighting can lead to sometimes dramatic differences in accuracy – at the extreme, 23.37% accuracy for the former and 88.18% for the latter. As stated before, even if accuracy cannot directly be measured in a social model, the above results would be reasonable a priori evidence that a social model using PMI could be vastly better than one using log-entropy. Of course, this is with the caveat that it may take the assessment of an NLP expert to verify that a particular NLP approach is suitable given a particular problem.

4.2   Extraction of Significant n-Grams

One of the steps in [16] is to extract 'concept-changing n-grams', a phrase used in [16] to mean phrases of $n > 1$ words, where the meaning of the phrase is not simply the aggregate of the meaning of the constituent parts. (An example would be 'black market', which is not simply a market which is black.) This can be an important step in recognizing named-entities, many of which may be referred to by a sequence of terms, such as 'United States'. The authors of [16] imply they achieve this by comparison to lists (gazetteers and a 'meta-network ontology' are specifically mentioned).

Again, the disadvantage of this approach is that it is non-robust, because a given list may work for one problem but not another; and the clearest illustration of non-robustness is the fact that a list will generally be limited in applicability to a single language at a time. Finally, construction of lists can be labor-intensive if the lists do not already exist. This may help social modelers fund graduate students, but it does not help funders achieve results most expediently, nor does it really advance science.

The use of gazetteers and ontologies just described is, in fact, another example of the heuristic/rule-based approaches described in section 3. It turns out that the extraction of significant n-grams in an unsupervised fashion is to at least some extent a solved problem [24]. The approach described in [24] can easily be applied to text from any domain or language, since it relies on the identification of patterns intrinsic to the data. As described in section 3, this should be much less costly and labor-intensive in the long run, and with the potential to produce more reliable results.

5   Conclusion

In this paper we have argued that natural language processing (NLP) is a key contributor to social modeling, because so much social data is found in text. We have argued that social modelers who use NLP would be well-advised to understand, at a high level, what problems in NLP are (and are not) solved. By placing NLP within the general context of data analytics, with specific reference to the distinction between heuristic-based, supervised, and unsupervised methods, we aimed to give social modelers insight into the fact that NLP is most powerful when used to solve problems in an 'unsupervised' way – which means finding patterns (including groupings, similarities, or anomalies) intrinsic to data. By contrast, NLP commonly consumes far more resources, and may at the same time be less reliable, when solving problems in a heuristic-based or supervised fashion, as is the case in sentiment analysis [14].

With this understanding, and with the appropriate input from NLP experts, social modelers will be much better placed to utilize NLP effectively, saving time, resources, and producing better results.

References

1. Gartner Group: User Survey Analysis: Key Trends Shaping the Future of Data Center Infrastructure Through 2011. Gartner Report ID G00208112 (2011)
2. Pimienta, D., Prado, D., Blanco, Á.: Twelve Years of Measuring Linguistic Diversity in the Internet: Balance and Perspectives. UNESCO (2009), http://unesdoc.unesco.org/images/0018/001870/187016e.pdf (accessed on January 27, 2013)
3. Page, L., Brin, S., Motwani, R., Winograd, T.: The PageRank Citation Ranking: Bringing Order to the Web. Technical Report, Stanford InfoLab (1999)
4. Brown, P., Della Pietra, V., Della Pietra, S., Mercer, R.: The Mathematics of Statistical Machine Translation: Parameter Estimation. Computational Linguistics 19, 263–311 (1993)
5. Tang, L., Liu, H.: Community Detection and Mining in Social Media. Morgan & Claypool (2010)
6. Costa, B., Boiney, J.: Social Radar. MITRE Technical Report #120088 (2012)

7. Jurafsky, D., Martin, J.: Speech and Language Processing, 2nd edn. Pearson Education Inc., Upper Saddle River (2009)
8. Fink, C., Kopecky, J., Bos, N., Thomas, M.: Mapping the Twitterverse in the Developing World: An Analysis of Social Media Use in Nigeria. In: [9], pp. 164–171 (2011)
9. Yang, S., Greenberg, A., Endsley, M. (eds.): Social Computing, Behavioral-Cultural Modeling & Prediction. Springer, Berlin (2011)
10. Ellen, J., Kaina, J., Parameswaran, S.: Implicit Group Membership Detection in Online Text: Analysis and Applications. In: [9], pp. 222–230 (2011)
11. Wang, X., Gerber, M., Brown, D.: Automatic Crime Prediction Using Events Extracted from Twitter Posts. In: [9], pp. 231–238 (2011)
12. Abbasi, M.-A., Kumar, S., Filho, J., Liu, H.: Lessons Learned in Using Social Media for Disaster Relief – ASU Crisis Response Game. In: [9], pp. 282–289 (2011)
13. Khabsa, M., Koppman, S., Giles, C.: Towards Building and Analyzing a Social Network of Acknowledgements in Scientific and Academic Documents. In: [9], pp. 357–364 (2011)
14. Pang, B., Lee, L.: Thumbs Up? Sentiment Classification Using Machine Learning Techniques. In: Proceedings of the ACL 2002 Conference on Empirical Methods in Natural Language Processing, pp. 79–86 (2002)
15. Turnley, J., Chew, P., Perls, A.: Beyond Validation: Alternative Uses and Associated Assessments of Goodness for Computational Social Models. In: [9], pp. 147–155 (2011)
16. Carley, K., Bigrigg, M., Papageorgiou, C., Johnson, J., Kunkel, F., Lanham, M., Martin, M., Morgan, G., Schmerl, B., van Holt, T.: Rapid Ethnographic Assessment: Data-to-Model. In: Proceedings of HSCB Focus 2011: Integrating Social Science Theory and Analytic Methods for Operational Use, Chantilly, VA, February 8-10 (2011), http://www.casos.cs.cmu.edu/publications/papers/2011RapidEthnographicAssessment.pdf (accessed on October 25, 2012)
17. McIlroy, M.: Mass Produced Software Components. In: Software Engineering: Report of a Conference Sponsored by the NATO Science Committee, Garmisch, Germany, vol. 79 (1969)
18. Hutchins, J.: The First Public Demonstration of Machine Translation: the Georgetown-IBM System, 7th January 1954. In: AMTA Conference (1954)
19. Koehn, P.: Europarl: a Parallel Corpus for Statistical Machine Translation. MT Summit 5 (2005)
20. Kelly, K.: What Technology Wants. Viking Penguin, New York (2010)
21. Chew, P., Bader, B., Helmreich, S., Abdelali, A., Verzi, S.: An Information-Theoretic, Vector-Space Model Approach to Cross-Language Information Retrieval. Journal of Natural Language Engineering 17(1), 37–70 (2011)
22. Dumais, S.: Improving the Retrieval of Information from External Sources. Behavior Research Methods, Instruments, and Computers 23, 229–236 (1991)
23. Dunlavy, D., Shead, T., Stanton, E.: ParaText: Scalable Text Modeling and Analysis. In: Proceedings of HPDC 2010, pp. 344–347 (2010)
24. Lin, D.: Automatic Identification of Non-Compositional Phrases. In: Proceedings of the 37th Annual Meeting of the Association for Computational Linguistics, pp. 317–324 (1999)

Which Targets to Contact First to Maximize Influence over Social Network

Kazumi Saito[1], Masahiro Kimura[2], Kouzou Ohara[3], and Hiroshi Motoda[4]

[1] School of Administration and Informatics, University of Shizuoka
52-1 Yada, Suruga-ku, Shizuoka 422-8526, Japan
`k-saito@u-shizuoka-ken.ac.jp`
[2] Department of Electronics and Informatics, Ryukoku University
Otsu 520-2194, Japan
`kimura@rins.ryukoku.ac.jp`
[3] Department of Integrated Information Technology, Aoyama Gakuin University
Kanagawa 229-8558, Japan
`ohara@it.aoyama.ac.jp`
[4] Institute of Scientific and Industrial Research, Osaka University
8-1 Mihogaoka, Ibaraki, Osaka 567-0047, Japan
`motoda@ar.sanken.osaka-u.ac.jp`

Abstract. We address a new type of influence maximization problem which we call "target selection problem". This is different from the traditionally thought influence maximization problem, which can be called "source selection problem", where the problem is to find a set of K nodes that together maximizes their influence over a social network. The very basic assumption there is that all these K nodes can be the source nodes, i.e. can be activated. In "target selection problem" we maximize the influence of a new user as a source node by selecting K nodes in the network and adding a link to each of them. We show that this is the generalization of "source selection problem" and also satisfies the submodularity. The selected nodes are substantially different from those of "source selection problem" and use of the solution of "source selection problem" results in a very poor performance.

Keywords: Information diffusion, influence degree, target node selection.

1 Introduction

The emergence of Social Media such as Facebook, Digg and Twitter has provided us with the opportunity to create large social networks, which plays a fundamental role in the spread of information, ideas, and influence. Such effects have been observed in real life, when an idea or an action gains sudden widespread popularity through gword-of-mouthh or gviral marketingh effects. This phenomenon has attracted the interest of many researchers from diverse fields [11], such as sociology, psychology, economy, computer science, etc.

A substantial amount of work has been devoted to the task of analyzing and mining information diffusion processes in large social networks [15,13,1]. The

A.M. Greenberg, W.G. Kennedy, and N.D. Bos (Eds.): SBP 2013, LNCS 7812, pp. 359–367, 2013.
© Springer-Verlag Berlin Heidelberg 2013

main focus over the past decade has been on optimization problems in which the goal is to maximize the spread of information through a given network, either by selecting a good subset of nodes to initiate the cascade [7] or by applying a broader set of intervention strategies such as node and link additions [18,21]. Widely used information diffusion models in these studies are *independent cascade (IC)* [7], *linear threshold (LT)* [22] and their variants [8,19,6,20]. These two models focus on different aspects of information diffusion. IC model is sender-centered (information push) and each active node *independently* influences its inactive neighbors with given diffusion probabilities. LT model is receiver-centered (information pull) and a node is influenced by its active neighbors if their total weight exceeds the threshold for the node. Basically the former models diffusion process of how a disease spreads and the latter models diffusion process of how an opinion or innovation spreads.

In this paper we deal with a new type of influence maximization problem. Traditionally this problem is defined to be finding a subset of nodes of size K that maximizes the influence degree with K as a parameter under a given information diffusion model and a given social network. It is unconditionally assumed that the information is guaranteed to start spreading from the selected K nodes. We call this problem as "Source selection problem" to distinguish it from our problem. We rather select K nodes and send information to these nodes. There is no guarantee that these nodes become the information source nodes. Suppose we want to spread our idea or opinion using a twitter, you must acquire reliable followers in the first place. To do this you have to carefully select the target users. Those users who have many followers already may not necessarily be good targets if they have many followees. Our problem is defined to be creating new links to a subset of nodes of size K from a new user such that the influence degree of this user is maximized. We call this problem as "Target selection problem" and analyze it for both LT and IC models.

"Target selection problem" also carries the same problem of 1) computational complexity of estimating influence degree of a given user which is defined to be the expected number of influenced nodes at the end of diffusion process that started from this user and 2) combinatorial explosion of search space in finding the optimal K target nodes. Fortunately, the influence degree is submodular, i.e. its marginal gain diminishes as the size K becomes larger in "Source selection problem", and the greedy solution has a lower bound which is 63% of the true optimal solution [7]. We prove that this submodularity also holds to "Target selection problem", and use a greedy algorithm at the expense of optimality. Various techniques have been devised to reduce the computational cost of solving "Source selection problem". These include bond percolation [9], pruning [8], lazy evaluation [14,5], burnout [19], heuristics [2,3], belief propagataion [16] and linear sytem approximation [23]. In this paper we use our own previous work, i.e. bond percolation [9], pruning [8] and burnout [19].

We compare the influence degree of "Target selection problem" with three other methods using four different real social networks. One is to use the solution of "Source selection problem" as target nodes. The other two are to use nodes selected from the largest out-degree and nodes randomly selected. In this paper we show only the results of LT model due to the page limitation. The results clearly show that the solution of "Target selection problem" is different from that of "Source selection problem" and the influence degree using the solution of "Source selection problem" is only half of the influence degree of "Target selection problem".

2   Information Diffusion Models

We consider a network represented by a directed graph $G = (V, E)$, where V and E ($\subset V \times V$) are the sets of all the nodes and links, respectively. Below we revisit the definition of IC and LT models according to the literatures [7,10]. In both models the diffusion process proceeds from an initial active node in discrete time-step $t \geq 0$, and it is assumed that nodes can switch their states only from inactive to active (*i.e.*, the SIR setting).

IC model has a *diffusion probability* $p_{u,v}$ with $0 < p_{u,v} < 1$ for each link (u, v) as a parameter. Suppose that a node u first becomes active at time-step t, it is given a single chance to activate each currently inactive child node v, and succeeds with probability $p_{u,v}$. If u succeeds, then v will become active at time-step $t+1$. If multiple parent nodes of v first become active at time-step t, then their activation trials are sequenced in an arbitrary order, but all performed at time-step t. Whether u succeeds or not, it cannot make any further trials to activate v in subsequent rounds. The process terminates if no more activations are possible.

LT model has a *weight* $q_{u,v}$ (> 0) with $\sum_{u \in B(v)} q_{u,v} \leq 1$ for each link (u, v) as a parameter, where $B(v) = \{u \in V; (u, v) \in E\}$ is the set of parent nodes of node v. First, for any node $v \in V$, a *threshold* θ_v is chosen uniformly at random from the interval $[0, 1]$. An inactive node v is influenced by its active parent nodes. If the total weight from the active parent nodes of v at time-step t is at least the threshold θ_v, *i.e.*, $\sum_{u \in B_t(v)} q_{u,v} \geq \theta_v$, then v will become active at time-step $t+1$. Here, $B_t(v)$ stands for the set of all the parent nodes of v that are active at time-step t. The process terminates if no more activations are possible.

For a set of initial active nodes $W(\subset V)$, let $\varphi(W; G)$ denote the number of active nodes at the end of the random process. It is noted that $\varphi(W; G)$ is a random variable. We denote the expected value of $\varphi(W; G)$ by $\sigma(W; G)$, and call it the *influence degree of W*.

3   Target Selection Problem

We first give the formal definition of the source selection problem, or the traditional influence maximization problem [7,14,10,3,2]. Given a network $G = (V, E)$ and a constant K, the problem is to find a set of K nodes $W_K(\subset V)$ that maximizes the influence degree $\sigma(W_K; G)$, which is formally defined as follows:

$$\underset{W_K \subset V}{\operatorname{argmax}} \, \sigma(W_K; G). \tag{1}$$

On the other hand, in the target selection problem tackled in this paper, we are given not only a network G and a constant K, but also an external information source node $x \notin V$ and values $\{r_{x,v} \mid v \in V\}$, each associated with link (x, v), where $r_{x,v} \in [0, 1]$ corresponds to a diffusion probability $p_{x,v}$ in case of IC model and a weight $q_{x,v}$ in case of LT model. Then, we seek a set of K nodes $W_K \subset V$ that maximizes the influence degree of x in an extended network $G'(W_K)$ resulted from adding K links from u to each node $w \in W_K$ into G, which is formally defined as follows:

$$\underset{W_K \subset V}{\operatorname{argmax}} f(W_K), \tag{2}$$

where $f(W_K) = \sigma(\{x\}; G'(W_K))$ and $G'(W_K) = (V \cup \{x\}, E \cup \{(x, w) | w \in W_K\})$. In case of LT model, we assume that each of the original weights to the target nodes, expressed as $q_{v,w}$ where $w \in W_K$ and $v \in B(w)$, is weakened to $(1 - r_{x,w})q_{v,w}$ due to the constraints on weights for LT model. Here we should emphasize that the target selection problem is a natural extension to the source selection problem because we obtain $\operatorname{argmax}_{W_K \subset V} \sigma(W_K; G) = \operatorname{argmax}_{W_K \subset V} f(W_K)$ by setting $r_{x,w} = 1$ for each $w \in W_K$. This is because all of the nodes selected in the target selection problem are definitely activated.

As mentioned in Section 1, since the function σ is submodular, i.e., $\sigma(W' \cup \{v\}; G) - \sigma(W'; G) \geq \sigma(W \cup \{v\}; G) - \sigma(W; G)$ if $W' \subseteq W$, we can approximately solve the source selection problem with a greedy method that recursively finds out W_k based on W_{k-1} by adding node v that maximizes $\sigma(W_{k-1} \cup \{v\}; G)$ to W_{k-1} starting from $W_0 = \emptyset$. Fortunately, in the target selection problem, the function f can be proven to be submodular from the following relation:

$$f(W_K) = \sum_{A \in 2^{W_K}} \sigma(A; G) \prod_{w \in A} r_{x,w} \prod_{w \in (W_K \setminus A)} (1 - r_{x,w}), \tag{3}$$

where 2^{W_K} denotes the power set of W_K. Recall that $r_{x,w}$ corresponds to a diffusion probability $p_{x,w}$ in case of IC model and a weight $q_{x,w}$ in case of LT model. Thus we can easily see that Equation (3) deals with each possible activation pattern A for the target set W_K with the probability that the pattern A happens. Here we should note that in case of LT model, each of the original weights to the target nodes $q_{v,w}$ is weakened to $(1 - r_{x,w})q_{v,w}$. Namely, under the condition that the external source node x fails to activate the target node w, the probability that the node v succeeds to activate the target node w is equivalent to $q_{v,w}$.

From Equation (3), since $f(W_K)$ is a non-negative linear combination of submodular functions $\sigma(\cdot)$, it is also submodular. Thanks to this property, we can solve the target selection problem in the same fashion as the source selection problem with a greedy method. As mentioned earlier, we can efficiently calculate such greedy solutions by using the techniques such as bond percolation [9], pruning [8] and burnout [19].

4   Experiments

Using large real-world networks, we experimentally evaluated the performance of the proposed method for solving the target selection problem on network $G = (V, E)$. We show only the results of LT model due to the page limitations. We chose to show LT model because this model is better suited to opinion spread where we came up with the notion of "target selection".

4.1   Datasets and Settings

In our experiments, we employed four datasets of real networks, where all the networks are represented as directed graphs. The first one is the Ameblo network, which is a reader network of Japanese blog service site "Ameaba" [1] (see [4] for more details). The Ameblo network has $56,604$ nodes and $734,737$ links. The second one is the Blog network, which is a trackback network of Japanese blogs used in [10]. The Blog network has $12,047$ nodes and $53,315$ links. The third one is the Cosme network, which is a fan-link network of "@cosme", [2] a Japanese word-of-mouth communication site for cosmetics (see [17] for more details). The Cosme network has $45,024$ nodes and $351,299$ links. The last one is the Enron network, which is derived from the Enron Email Dataset [12] (see [17] for more details). The Enron network has $19,603$ nodes and $210,950$ links.

We compared the proposed method with three other heuristic methods as mentioned in Section 1. The first one is to use the solution of the source selection problem for the original network $G = (V, E)$ and add links to the selected K nodes from an external source node. Here, we employed the combined methods of our previous work (bond percolation [9], pruning [8] and burnout [19]). We refer to this method as the *InflMaxSrc* method. The second one is to select nodes in order of decreasing out-degrees, where the out-degree of a node means the number of outgoing links from the node. This is a method often used in the field of complex networks science. We refer to this method as the *Out-degree* method. The third one, which serves as the crude baseline, is to simply select nodes uniformly at random. We refer to this method as the *Random* method.

We evaluated the performance, $f(W_K) = \sigma(\{x\}; G'(W_K))$, where W_K is the selected K nodes by each method. The influence degree $\sigma(\{x\}; G'(W_K))$ was estimated by the empirical mean of the number of active nodes obtained from $10,000$ independent runs of information diffusion, with each run based on the bond percolation [9], pruning [8] and burnout [19]. For the parameters of LT model, we set $q_{u,v} = 1/B(v)$ $(\forall u, v \in V)$.

4.2   Experimental Results

Figures 1a, 1b, 1c and 1d show the results for the Ameblo, Blog, Cosme and Enron networks, respectively. Here, we plot the value of the objective function f (influence degree) as a function of the number k of target nodes, where the circles, crosses, squares and triangles indicate the results for the proposed, InflMaxSrc, Out-degree and Random methods, respectively. First, we see that the proposed method significantly outperformed the InflMaxSrc, Out-degree and Random methods for all four networks. The Random method is by far the worst. We can say that the proposed method can spread the information twice as much as the best of the other two methods can do We also note that the performance of the Out-degree method strongly depends on the characteristics of the network

---

[1] http://www.ameba.jp/
[2] http://www.cosme.net/

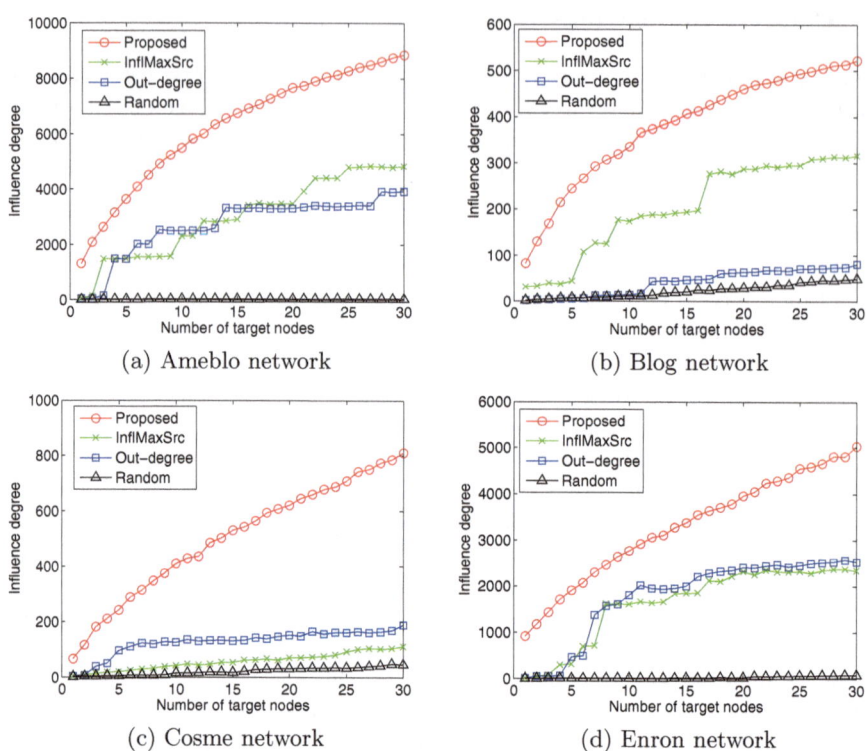

(a) Ameblo network   (b) Blog network

(c) Cosme network   (d) Enron network

Fig. 1. Performance comparison for the target selection problem

structure, and in some cases it is better than the InflMaxSrc method. We know
that the InflMaxSrc method always outperforms the Out-degree and Random
methods for the source selection problem (see [10]), but for the target selection
problem it is not always the case that the InflMaxSrc method outperforms the
Out-degree method. This is attributed to the fact that the information source
node x is not necessarily able to activate all of the target nodes W_*. For instance,
the influence degree f of the proposed method is 1.7 to 7.2 times as much as
that of the InflMaxSrc method for $k = 30$.

This means that the selected nodes must be substantially different from each
other for the four methods. To verify this we measured the solution similarity by
F-measure $\mathcal{F}(k) = |W_k^* \cap W_k|/k$, where k stands for the number of target nodes
for the target selection problem, and W_k^* and W_k are the solutions extracted
by the proposed and one of the other three methods, respectively. The largest
F-measure is 0.33 for Amebro network with W_3^* of InflMaxSrc. For the other
networks, F-measure is much smaller, e.g., nearly 0 for Cosme network with all
k and all other three methods. We confirmed that the proposed method found
a solution dramatically different from that by the other three methods.

We next show the in- and out-degrees of the selected nodes in Fig. 2 to
investigate why the influence degree achieved by the proposed method is much
better than the influence degree by the other methods, *i.e.*, why the selected

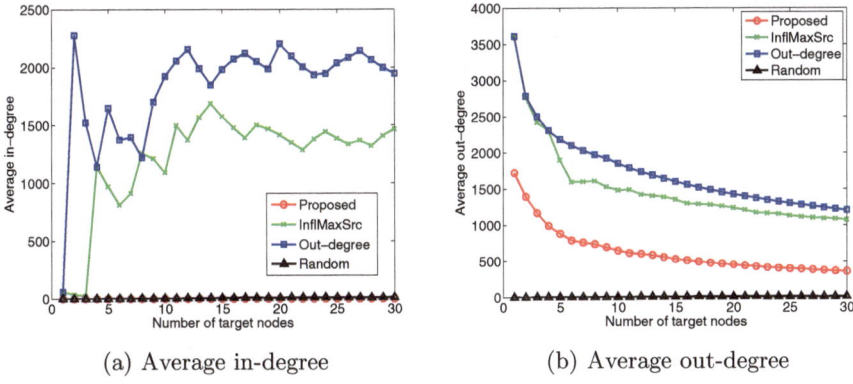

(a) Average in-degree   (b) Average out-degree

Fig. 2. Average degrees of target nodes for the Ameblo network

nodes are different. Here we only show the result of the Ameblo network due to a space limitation, but quite similar results have been obtained from the other networks. From this figure, it is found that, both the in- and out-degrees of the nodes selected by the InflMaxSrc and Out-degree methods tend to be high, while the out-degree of the target nodes selected by the proposed method is not so high, but their in-degree is always low. The InflMaxSrc and Out-degree methods select the target nodes independently of their in-degree. This is self-evident for the Out-degree method by definition. In case of the InflMaxSrc method the target nodes are always active at the beginning of the information diffusion process by definition and the in-degree of the target nodes never affects their influence degree. Thus, it tends to select nodes that have many children as the target nodes. It is noted that in the LT model, nodes that have fewer parents have better chance to get activated than those that have many parents. This is because the weights from the parent nodes are larger in the former case and even a small number of active parents can activate the child nodes. Thus, the target node selected by the proposed method are more likely to get activated by the information source node than those selected by the InflMaxSrc and Out-degree methods. This is the main reason of the large difference in the selected target nodes and thus in the resulting influence degree.

5   Conclusion

In this paper we proposed a new type of influence maximization problem, which we call "target selection problem". Traditionally influence maximization problem assumed unconditionally that the selected nodes can be the source nodes, e.g., can be activated, thus can be called "source selection problem", and was the simplest model for viral marketing, e.g. which 1000 persons to send direct mails to promote a new product. We thought it more natural and realistic to view this problem from a slightly different angle. We maximized the influence of a new user (source node) who is outside of a community by selection a fixed number of target nodes in the existing community (social network) and adding a link to each of the target nodes. Acceptance of the information of the target nodes from the source node follows a probabilistic information diffusion model as well as the spread of information from the target nodes to the other nodes in the network does so. This "target selection problem" is a generalization of "source selection problem" and carries similar properties, e.g. submodularity and high computational complexity of estimating influence degree which is the expected number of activated nodes at the end of information diffusion. We estimated the influence degree by the bond percolation and selected target nodes by a greedy algorithm. We solved "target selection problem" in four real world networks, each with slightly different characteristics. Our findings are 1) The solution of "target selection problem" is substantially different from the solution of "source selection problem, 2) Use of the selected nodes of "source selection problem" results in very poor performance (information spread is only half), 3) there is basically no or very few overlap of the nodes selected. This implies that care should be taken in selecting whom to contact first to maximize influence over a social network. We conjecture that such target nodes can be notable mediators, who play an important rolefor widely spreading information. Our immediate future work is to validate this claim using available real information propagation data.

Acknowledgments. This work was partly supported by Asian Office of Aerospace Research and Development, Air Force Office of Scientific Research under Grant No. AOARD-13-4042, and JSPS Grant-in-Aid for Scientific Research (C) (No. 23500194).

References

1. Bakshy, E., Hofman, J., Mason, W., Watts, D.: Everyone's an influencer: Quantifying influences on twittor. In: Proceedings of the 4th International Conference on Web Search and Data Mining (WSDM 2011), pp. 65–74 (2011)
2. Chen, W., Wang, C., Wang, Y.: Scalable influence maximization for prevalent viral marketing in large-scale social networks. In: Proceedings of the 16th ACM SIGKDD International Conference on Knowledge Discovery and Data Mining (KDD 2010), pp. 1029–1038 (2010)
3. Chen, W., Yuan, Y., Zhang, L.: Scalable influence maximization in social networks under the linear threshold model. In: Proceedings of the 10th IEEE International Conference on Data Mining (ICDM 2010), pp. 88–97 (2010)
4. Fushimi, T., Saito, K., Kimura, M., Motoda, H., Ohara, K.: Finding relation between pageRank and voter model. In: Kang, B.-H., Richards, D. (eds.) PKAW 2010. LNCS, vol. 6232, pp. 208–222. Springer, Heidelberg (2010)
5. Goyal, A., Lu, W., Lakshmanan, L.: Influence spread in large-scale social networks - a belief propagation approach. In: Proceedings of the 20th International World Wide Web Conference (WWW 2011), pp. 47–48 (2011)
6. Gruhl, D., Guha, R., Liben-Nowell, D., Tomkins, A.: Information diffusion through blogspace. SIGKDD Explorations 6, 43–52 (2004)
7. Kempe, D., Kleinberg, J., Tardos, E.: Maximizing the spread of influence through a social network. In: Proceedings of the 9th ACM SIGKDD International Conference on Knowledge Discovery and Data Mining (KDD 2003), pp. 137–146 (2003)

8. Kimura, M., Saito, K., Motoda, H.: Efficient estimation of influence functions fot sis model on social networks. In: Proceedings of the 21st International Joint Conference on Artificial Intelligence (IJCAI-2009) (2009)

9. Kimura, M., Saito, K., Nakano, R.: Extracting influential nodes for information diffusion on a social network. In: Proceedings of the 22nd AAAI Conference on Artificial Intelligence (AAAI-2007), pp. 1371–1376 (2007)

10. Kimura, M., Saito, K., Nakano, R., Motoda, H.: Extracting influential nodes on a social network for information diffusion. Data Mining and Knowledge Discovery 20, 70–97 (2010)

11. Kleinberg, J.: The convergence of social and technological networks. Communications of ACM 51(11), 66–72 (2008)

12. Klimt, B., Yang, Y.: The enron corpus: A new dataset for email classification research. In: Boulicaut, J.-F., Esposito, F., Giannotti, F., Pedreschi, D. (eds.) ECML 2004. LNCS (LNAI), vol. 3201, pp. 217–226. Springer, Heidelberg (2004)

13. Leskovec, J., Adamic, L.A., Huberman, B.A.: The dynamics of viral marketing. In: Proceedings of the 7th ACM Conference on Electronic Commerce (EC 2006), pp. 228–237 (2006)

14. Leskovec, J., Krause, A., Guestrin, C., Faloutsos, C., VanBriesen, J., Glance, N.: Cost-effective outbreak detection in networks. In: Proceedings of the 13th ACM SIGKDD International Conference on Knowledge Discovery and Data Mining (KDD 2007), pp. 420–429 (2007)

15. Newman, M.E.J., Forrest, S., Balthrop, J.: Email networks and the spread of computer viruses. Physical Review E 66, 035101 (2002)

16. Nguyen, H., Zheng, R.: Influence spread in large-scale social networks – A belief propagation approach. In: Flach, P.A., De Bie, T., Cristianini, N. (eds.) ECML PKDD 2012, Part II. LNCS (LNAI), vol. 7524, pp. 515–530. Springer, Heidelberg (2012)

17. Ohara, K., Saito, K., Kimura, M., Motoda, H.: Effect of in/Out-degree correlation on influence degree of two contrasting information diffusion models. In: Yang, S.J., Greenberg, A.M., Endsley, M. (eds.) SBP 2012. LNCS, vol. 7227, pp. 131–138. Springer, Heidelberg (2012)

18. Richardson, M., Domingos, P.: Mining knowledge-sharing sites for viral marketing. In: Proceedings of the 8th ACM SIGKDD International Conference on Knowledge Discovery and Data Mining (KDD 2002), pp. 61–70 (2002)

19. Saito, K., Kimura, M., Motoda, H.: Discovering influential nodes for sis models in social networks. In: Gama, J., Costa, V.S., Jorge, A.M., Brazdil, P.B. (eds.) DS 2009. LNCS (LNAI), vol. 5808, pp. 302–316. Springer, Heidelberg (2009)

20. Saito, K., Kimura, M., Ohara, K., Motoda, H.: Learning continuous-time information diffusion model for social behavioral data analysis. In: Zhou, Z.-H., Washio, T. (eds.) ACML 2009. LNCS (LNAI), vol. 5828, pp. 322–337. Springer, Heidelberg (2009)

21. Sheldon, D., Dilkina, B., Elmachtoub, A., Finseth, R., Sabharwal, A., Conrad, J., Gomes, C., Shmoys, D., Allen, W., Amundsen, O., Vaughan, W.: Maximizing the spread of cascades using network design. In: Proceedings of the Twenty-Sixth Conference Annual Conference on Uncertainty in Artificial Intelligence (UAI 2010), pp. 517–526. AUAI Press, Corvallis (2010)

22. Watts, D.J., Dodds, P.S.: Influence, networks, and public opinion formation. Journal of Consumer Research 34, 441–458 (2007)

23. Yang, Y., Chen, E., Liu, Q., Xiang, B., Xu, T., Shad, S.A.: On approximation of real-world influence spread. In: Flach, P.A., De Bie, T., Cristianini, N. (eds.) ECML PKDD 2012, Part II. LNCS (LNAI), vol. 7524, pp. 548–564. Springer, Heidelberg (2012)

Intruder or Welcome Friend: Inferring Group Membership in Online Social Networks

Ofrit Lesser[1], Lena Tenenboim-Chekina[2], Lior Rokach[1], and Yuval Elovici[1,2]

[1] Department of Information Systems Engineering, Ben Gurion University of the Negev, Israel
[2] Telekom Innovation Laboratories at Ben-Gurion University of the Negev
lessero@post.bgu.ac.il,{lenat,liorrk,elovici}@bgu.ac.il

Abstract. Inferring Online Social Networks (OSN) group members may help to evaluate the authenticity of an applicant asking to join a certain group, and secure vulnerable populations online, such as children. We propose machine learning based methods, which associate OSN members' affiliation with virtual groups based on personal, topological, and group affiliation features. The study applies and evaluates the methods empirically, on two social networks (Ning and TheMarker). The experimental results demonstrate that one can accurately determine the group genuine members. Our study compares personal, topological and group based classification models. The results show that topological and group affiliation attributes contribute the most to group inference accuracy. Additionally, we examine the relations among the groups and identify group clustering tendencies where some groups are more tightly connected than others.

Keywords: social networks, group prediction, machine learning.

1 Introduction

Online Social Network (OSN), characteristics differ from each other, however the majority of these networks contain two main capabilities: connecting two members via a friendship connection and a group creation. These mechanisms simulate real life scenarios. While a friendship connection signifies or correlates with a tie between two individuals (e.g. relatives, friends, acquaintances, etc.), the group formation correlates with a community of multiple individuals based on similarity (i.e. residence, workplace, interests, etc.). Due to the OSN's rise and the opportunity to obtain large scale datasets, many recent studies have focused on various different aspects of the social network, such as: structure, evolution, security aspects, complex network common characteristics, and more [1] [2] [3].

The problem we contemplate is community membership inference in social networks: inferring group membership of a user based on friendship connections (i.e. topological structures), personal attributes, and affiliations with other groups. Our methodology may be used for the automatic screening of applications to join groups, protecting vulnerable internet populations (e.g. children), group recommendations and more. We believe our work is the first of its kind to study this group inference

A.M. Greenberg, W.G. Kennedy, and N.D. Bos (Eds.): SBP 2013, LNCS 7812, pp. 368–376, 2013.

problem variation, as defined here, map it to a classification problem, examine four classification models, and compare these models.

Here, we propose new group affiliation inference methods using machine learning techniques based on three sets of features: topological, groups, and personal. These methods are examined and validated on several real world datasets. Four prediction models for the OSN affiliation analysis problem were developed: (a) structural/topological features set based model. (b) group affiliation features set based model, (c) personal features set based model, and (d) a model that combines all features in the previous models (i.e., topological, group affiliation, and personal features). We evaluate our proposed models using sample datasets of two social media websites: TheMarker[1] (an online social network site in Hebrew) and Ning [2] (an online social network for Ning creators). Specifically, our novel contributions are:

- Proposal of four novel group affiliation predicting methods using machine learning techniques based on three types of features: topological, membership in other groups, and personal features.
- Introduction of several new topological measures that encompass information from the two graph structures of OSN: the social ties graph, and the group membership bipartite graph.
- We demonstrate that group clusters can be identified easily by calculating the information gain among the groups.

We describe related studies in the next section. Section 3 outlines the problem formal definition and our methods. Then, we describe the evaluation in section 4, and experiment results in section 5. We conclude and discuss future works in section 6.

2   Related Work

Identifying graph communities has been a prevalent topic in recent years. In 2002, Girvan and Newman published an innovative algorithm, which detects such communities by isolating them as separate graph components [1]. Since then, additional methods for community detection have been presented. These can be found in Fortunato's comprehensive survey on community detection [2].

While recent studies have focused on community structure in social networks [5] [3] [2], these studies concentrated mostly on the communities' detection on complex graphs, and examined their structure and dynamics. Traud et Al. [6][7], applied network analysis tools to study the role of university organizations and affiliations in structuring the social networks of students by examining a snapshot of the Facebook "friendships" graph at five American universities. They also compared the relative contributions of different personal characteristics to the community structure of universities.

---

[1] http://cafe.themarker.com
[2] http://creators.ning.com

The classical community detection problem concentrates on detecting communities within a social network based on the friendship connections between friends [1] [2]. Unlike studies where each member is associated with a unique community, our study concentrates on OSN groups where a member may belong to multiple groups or communities. Detection of the overlapping communities' problem has been addressed by Friggeri et al. [8], who introduced cohesion metrics based on network to topological features and triangles counting. Similarly, we focus on OSNs, where a member may join overlapping interest groups. OSNs include a rich set of features and information. Members' information includes social ties, group membership, and personal data. The personal and group information does not require extensive computation; therefore, our model may use this additional information in order to perform group prediction efficiently.

3   Problem Definition and Methods

We represent a social network as a graph $G = (V,E,H)$, where V is a set of n nodes (OSN members), of the same type, E is a set of edges (the friendship links), H is a set of groups that nodes can belong to, and A is a set of node attributes. The graph edge $e_{i,j} \in E$ represents an undirected link between node v_i and node v_j. We describe a group as a hyper-edge $h \in H$ among all the nodes that belong to that group; $h.V$ denotes the set of users who are connected through hyper-edge h. A user profile has a unique ID with which the user forms links and participates in groups. The goal is to predict for a user $v_i \in V$, whether $v_i \in h.V$, while we do not know whether v_i is a member of $h.V$ but all other group members of $h.V$ (i.e. other users who are members of the group h) are given. An alternate goal is to predict for a certain group $h \in H$, which users should be included as members in $h.V$. For both cases the same method may be used, but the evaluation process is different

We chose to use machine learning methods and develop group inference classifiers, which aim to estimate the probability that a specific user is a member in the target group. Therefore, we presented our problem as a binary classification problem where each user (OSN member), is represented as an instance that is characterized by multiple attributes (also known as features). The target class is a binary attribute indicating whether a user is a member of the group or not. For each group, a specific dataset is generated. The dataset feature attributes contain a users' information about their personal characteristic (age, gender, etc.), social ties structure (aka topological features), and affiliation with other groups. Thus, the features can be divided into three categories:

Personal Characteristics Features (PRS) - The personal information refers to the information in users' profile and usually includes demographic details such as: gender, age, residence, etc. We included each one of these information categories as a feature for our machine learning model. We assume that these personal characteristics may indicate users' groups due to homophily [9]. For example in a fan group of a kids' TV show we would expect that most of the members will be children. Therefore an applicant with an older man profile would be suspicious.

Group Affiliation Features (GRP) - This set of attributes denotes the user affiliation with all the social network groups, except for the target class group. Every instance includes a Boolean vector, where each dimension corresponds to a unique group and includes group membership information. If user v is a member in group $h_i \in H$, meaning $v \in h_i . V$, then the corresponding attribute is TRUE. The motivation for using other groups' affiliations is derived from the fact that similar users tend to register to the same set of groups. For example, many users that are registered to the "Data Mining" group are also registered to the "Big Data" group.

Network Topological Features (TPL) - These features are extracted from the topological structure of the graph. For each group, we extracted a set of topological features. These features assist in estimating the chances that a given user is a member in the group. For each member $v \in V$ and a group $h \in H$ we calculated a set of 8 topological features as displayed below in Table 1.

Table 1. Topological Features

Attribute Name	Indication
$degree(v)$	Degree, number of immediate friends
$GRP\_F(v, h)$	Number of v's friends in group h
$GRP\_FN(v, h)$	$GRP\_F(v, h)$ normalized by total number of friends
$GRP\_CF(v, h)$	Number of v's friends connected with at least one other of v's friend in group h
$GRP\_CFN(v, h)$	$GRP\_CF(v, h)$ normalized by v's degree
$GRP\_CC(v, h)$	Number of connections v's friends in group h have among themselves
$GRP\_CCN(v, h)$	$GRP\_CC(v, h)$ normalized by number of all possible such connections
$GRP\_F\_L2(v, h)$	Number v's friends of friends in the group h sub-graph (exactly 2 hops from v)

The following definitions are the formal definitions of the topological features:
The neighbourhood $\Gamma(v)$ of v is defined as the set of v's friends, namely, vertices that are adjacent to v. The following is the formal definition of neighbourhood:

$$\Gamma(v) := \{u | (u, v) \in E\} \tag{1}$$

The group-neighbourhood $\Gamma(v, h)$, of $v \in V$ and $h \in H$, is the set of v's friends who are also   members of group h. The following is the formal definition of group-neighbourhood:

$$\Gamma(v, h) := \{u \mid u \in h \ \& \ (u, v) \in E\} \tag{2}$$

Based on the group-neighbourhood definition, we define *ingroup-common-friends* of user v to be the set of v's friends who are members of group $h \in H$ and have at least another friend in this set. We denote this set of nodes as *ICF*:

$$ICF(v, h) := \{u \mid u \in \Gamma(v, h) \ \& \ (\exists \, u' \in \Gamma(v, h) \ \& \ (u, u') \in E)\} \tag{3}$$

Based on the group-neighbourhood definition, we define *ingroup-common-connections* of user v to be all the pairs of v's friends who are members of group $h \in H$ and are also friends with each other. We denote this set of nodes as ICC:

$$ICC(v,h) := \{(u_i, u_j) \,|\, u_i, u_j \in \Gamma(v,h) \,\&\, (u_i, u_j) \in E\} \qquad (4)$$

Using the above definitions, we can create the following features for vertex v:

Degree: We defined the vertex v degree as the number of vertices user v has a friendship connection with. We formally define it as:

$$degree(v) := |\Gamma(v)| \qquad (5)$$

Ingroup-friends (GRP_F): We define the number of v's friends who are members in group h, as:

$$GRP\_F(v,h) := |\,\Gamma(v,h)| \qquad (6)$$

Ingroup-friends-l2 (GRP_F_L2): We define the number of v's friends of friends who are all members in group h (i.e., at two hops distance from v within group h subgraph) as:

$$GRP\_F\_L2(v,h) := |\,\{u_j \,|\, (u_i, u_j) \in E \,\&\, u_j \in h, \, u_i \in \Gamma(v,h) \,\&\, u_j \notin \Gamma(v,h)\}| \qquad (7)$$

Grp-common-friends (GRP_CF): We define the number of v's friends in group h who are connected   with at least one other v's friend in group h as:

$$GRP\_CF(v,h) := |ICF(v,h)| \qquad (8)$$

Grp-friends-connections (GRP_CC): We define the number of connections, which v's friends, who belong to   group h, have with other v's friends in group h as:

$$GRP\_CC(v,h) := |ICC(v,h)| \qquad (9)$$

Ingroup-friends-ratio (GRP_FN): We normalize **GRP_F** by the number of v's friends who are members in group h with with v's degree and define it as:

$$GRP\_FN(v,h) := \frac{|\Gamma(v,h)|}{|\Gamma(v)|} \qquad (10)$$

GRP_CFN is the normalized value of **GRP_CF**, obtained by dividing it by v's degree.

GRP_CCN is the normalized value of **GRP_CC,** by dividing it with the number of all possible such connections between v's friends. i.e. $\left(\frac{degree(v)(degree(v)-1)}{2}\right)$.

4 Evaluation

We performed an evaluation of the proposed methodology on two OSN datasets TheMarker and Ning, and compared the four suggested group prediction models. The social networks datasets were collected using a dedicated Web crawling code. The properties of the datasets are presented in Table 2.

Table 2. Properties Of The Datasets

Property	TheMarker	Ning
Number of users	87,905	11,011
Number of links	1,644,848	76,263
Number of groups	85	81
Average degree	37.4	7.4
Number of groups per user: Average (Range)	2.4(0-84)	0.4(0-53)
Group size: Average/ (Range)	2,465/ (92-8,360)	59/(1-698)
Number of personal features	28	3

Note that group affiliation inference is a highly imbalanced problem. For most of the groups there are many more non-members than members among all of the OSN users. Formally, there are many more negative links than positive links. Imbalanced datasets pose difficulties for induction algorithms as standard machine learning techniques may be "overwhelmed" by the majority class and in result ignore the minority class. For overcoming this problem, we followed the under-sampling approach in which a balanced training set is generated and used to train a classifier, which is then tested on an imbalanced test set. Two non-overlapping subsets of data, train and test, were selected from each original group data set. For the train set, half of the total of positive and the equal number of negative examples were selected, thus creating a balanced set. The rest of the positive and negative examples were used for test set (imbalanced).

We used the area under the ROC curve (AUC) measure, which is not influenced by the imbalance distribution of the classes [10], for evaluation of different classification models. Additionally, we used the Precision and Recall measures in order to verify the ranking performance of our algorithm. We ran the experiments with WEKA [11], a popular machine learning software suite, and used the Bagging algorithm due to its high performance and relatively low run time [4]. The Bagging algorithm was setup with its default configuration parameters and J48 (Weka's implementation of the well-known C4.5 decision tree algorithm), with the minimal number of instances per leaf set to 10 as the base learning method.

5 Experimental Results

The AUC results of the Bagging algorithm, on TheMarker and Ning networks, using various sets of attributes for the 13 selected groups are presented in Table 3. The groups were selected randomly for TheMarker network, and the largest groups were selected for the Ning. The ALL column presents the results achieved using the combination of all three subsets. For each group, the best result among all the evaluated four models is marked in bold font. The best result among the three attribute subsets (GRP, PRS and TPL) is underlined.

Table 3. AUC Results on the TheMarker (a) and Ning (b) Datasets. Baseline AUC = 0.5

Group	Attributes subset				Group	Attributes subset			
Size	ALL	GRP	PRS	TPL	Size	ALL	GRP	PRS	TPL
339	**0.839**	0.822	0.694	0.747	102	**0.909**	0.813	0.596	0.891
353	**0.773**	0.759	0.574	0.733	103	0.854	0.854	**0.895**	0.818
362	0.859	**0.854**	0.570	0.778	122	**0.831**	0.665	0.611	0.778
563	0.913	**0.918**	0.622	0.807	127	**0.901**	0.550	0.842	0.856
949	**0.888**	0.871	0.627	0.695	141	**0.879**	0.840	0.537	0.797
1366	**0.882**	0.852	0.735	0.728	150	**0.866**	0.805	0.475	0.849
1671	**0.875**	0.763	0.767	0.736	152	**0.851**	0.775	0.464	0.793
1751	**0.838**	0.800	0.675	0.771	204	**0.934**	0.527	0.919	0.899
1930	**0.867**	0.838	0.710	0.753	239	**0.872**	0.759	0.585	0.849
2210	**0.823**	0.768	0.585	0.727	239	**0.910**	0.804	0.528	0.881
2248	**0.770**	0.696	0.656	0.712	378	**0.883**	0.673	0.689	0.873
3788	**0.840**	0.764	0.725	0.772	582	0.788	0.695	0.502	**0.789**
7600	**0.840**	0.767	0.656	0.689	698	0.876	0.568	**0.886**	0.875

(a)   (b)

It can be seen that the best AUC is achieved by using all attributes for most of the groups in both networks. Among the evaluated attribute subsets, the group's affiliation provides the best prediction results for most of the groups in TheMarker network, and topological attributes perform the best for most groups of the Ning network. This suggests that the optimal model is OSN related, and depends on the social network properties. As shown in Table 2, TheMarker groups are larger, and the number of groups per user is higher compared to Ning (2.4 vs. 0.4 accordingly), which may explain the strength of the GRP attribute set in the case of TheMarker.

Interestingly, the prediction accuracy achieved using the ALL attribute set versus the GRP attribute set is very similar in many cases. This suggests that the GRP model (including group's affiliation attributes only), may be used in certain cases, such as in large social networks with many members and various interest groups. In these cases it allows for a better computational performance (as these attributes are easy and quick to compute), with the lowest (if any), loss in prediction accuracy. It can also be noted that in the TheMarker network, personal attributes provide AUC values only slightly above the baseline AUC (equal to 0.5). Contrarily, in the Ning network, personal attributes provide relatively high AUC values, especially in the groups which are country or language related. This strengthens our conclusion that the type of most predictive attributes depends on the investigated network itself.

The Precision and Recall results at various K sizes for one of the groups from the TheMarker network are presented in Fig. 1(a), and compared to the optimal and baseline values of these measures. The vertical dashed line specifies the K equal to the actual number of positive examples in the test set (i.e. number of group members). While the absolute Precision and Recall values improvement is desired, they are much better than baseline (computed as percentage of group members out of the total amount of users in the network). These results suggest that the developed models can already be used for reducing the load and cost of group inference related tasks. These set of tasks include identifying the most suitable individuals to the group or the most suitable groups for an individual. The results on other groups in both networks follow a very similar pattern, and are thus discarded from this publication.

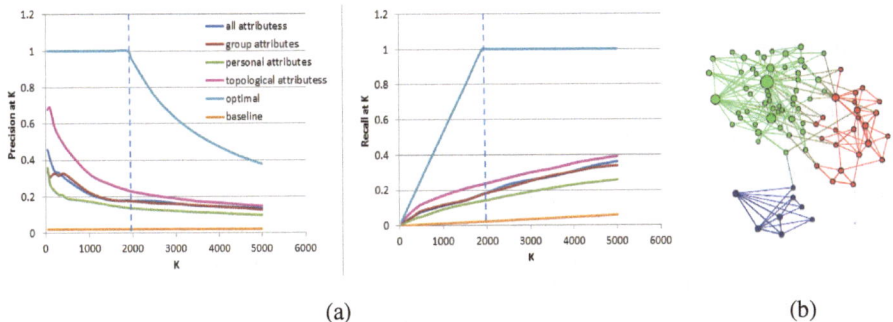

(a)   (b)

Fig. 1. (a) Precision, Recall for different K's on the "The design works" group, TheMarker dataset; (b) TheMarker groups' clusters using information gain evaluation

Additionally, to obtain an indication of the usefulness of the various individual features, we analysed their importance using Weka's information gain attribute selection algorithm. Generally, the results of this analysis correspond with AUC and Precision/Recall results. We also calculated the information gain between all possible group pairs in the TheMarker network and represented this information using a graph (see Fig. 1(b)). The groups in this graph are represented by nodes; directed edges (u, v) represent information gain value for group u when attempting to predict membership in target group v. For each group we chose the three edges with the highest information gain values to be included in the graph. The size of the nodes in the graph is proportional to the information gain value the group has for inferring other groups. We applied the Louvain method for community detection [12], which divided the graph into three communities. The clustered nature on this graph indicates that some groups are more tightly connected than others. Further evaluation of this clustering in OSN group graphs is on our future task plans.

6   Conclusions and Future Work

This study presents the group inference problem in OSN and proposes machine learning based methods to address it. The classification models are based on personal, topological and group affiliation features. Generally, we can see that a relatively high predictive accuracy (an AUC of about 0.8 on average, while baseline is 0.5), can be achieved using all the attributes along with the simple and quick bagging classification algorithm. The model yielding the highest accuracy varies across OSNs, and may depend on OSN properties. The precision and recall measurements also demonstrated significantly higher results compared to baseline values. Additionally, our study demonstrates that information gain values between different groups can be used for analyzing their relations and detecting group clusters. These clusters can then be used as a target class instead of individual groups.

We believe that predictive performance can be further improved using more sophisticated classification methods and by devising additional topological features. An evaluation of our approach with such methods on additional OSN datasets is one of our nearest future tasks.

References

1. Girvan, M., Newman, M.E.J.: Community structure in social and biological networks. Proc. Natl. Acad. Sci. (2002)
2. Fortunato, S.: Community detection in graphs. Physics Reports 486(3-5) (2010)
3. Boccalettia, S., Latora, V., Moreno, Y., Chavezf, M., Hwang, D.-U.: Complex networks: Structure and dynamics. Physics Reports 424(4-5), 175–308 (2006)
4. Fire, M., et al.: Link Prediction in Social Networks Using Computationally Efficient Topological Features. IEEE SOCIALCOM (2011)
5. Girvan, M., Newman, M.E.J.: Community structure in social and biological networks. Proceedings of the National Academy of Sciences 99 (2002)
6. Traud, A.L., Kelsic, E.D., Mucha, P.J., Porter, M.A.: Community structure in online collegiate social networks. eprint arXiv:0809.0690 (2008)
7. Traud, A.L., Mucha, P.J., Porter, M.A., Kelsic, E.D.: Comparing community structure to characteristics in online collegiate social networks. SIAM Review (2011)
8. Friggeri, A., Chelius, G., Fleury, E.: Triangles to Capture Social Cohesion. IEEE Passat 2011 (2011)
9. McPherson, M., Smith-Lovin, L., Cook, J.M.: Birds of a feather: Homophily in social networks. Annu. Rev. Sociol. 27, 415–444 (2001)
10. Menon, A.K., Elkan, C.: Link prediction via matrix factorization. In: Gunopulos, D., Hofmann, T., Malerba, D., Vazirgiannis, M. (eds.) ECML PKDD 2011, Part II. LNCS, vol. 6912, pp. 437–452. Springer, Heidelberg (2011)
11. Hall, M., Frank, E., Holmes, G., Pfahringer, B., Reutemann, P., Witten, I.H.: The weka data mining software: an update. SIGKDD Explor. Newsl. 11, 10–18 (2009)
12. Blondel, V.D., Guillaume, J.-L., Lambiotte, R., Lefebvre, E.: Fast unfolding of communities in large networks. J. Stat. Mech.: Theory and Experiment 2008 (2008)

Identifying Influential Twitter Users in the 2011 Egyptian Revolution*

Lucas A. Overbey[1], Christopher Paribello[2], and Terresa Jackson[1]

[1] SPAWAR Systems Center Atlantic, P.O. Box 190022, North Charleston,
SC 29419-9022
[2] MathWorks, 3 Apple Drive, Natick, MA 01760

Abstract. Recent international events surrounding contentious political environments have uncovered a new utility for social media. Communities now use resources such as Facebook and Twitter to quickly spread information and project influence amongst potentially geographically disparate people. In this work, we investigate Twitter activity in Egypt during the 2011 protests and revolution, and introduce a model to automatically ascertain key individuals within these networks. The model takes advantage of a more sparse network on Twitter than the traditional follower/following network by leveraging direct communications. Furthermore, we employ a measure of alpha centrality, which incorporates both directionality of network connections and a measure of external importance. The model is applied to topic-based communities within Twitter rather than previously introduced measures of influence that focus on the cascading spread of single messages or broad, topic-invariant measures. Results indicate a model successful at automatically identifying users that are active and influential within a given community, agreeing well with heuristics and comparable to other influence models but with particular advantages such as tunability and robustness to incomplete data.

Keywords: social media, influence, social network analysis, contentious politics.

1 Introduction

The study of a select number of key opinion leaders or "influentials" [1] has become central to the study of the diffusion of information and influence [2,3]. There are many different communication media from which influential entities may emerge, e.g. physical person-to-person, mass media, telecommunications, writings, etc. However, it is difficult to accurately trace and attribute the spread of information or influence in most such networks [4]. The expanding popularity of social networking applications have provided a tool for which word-of-mouth-like communications can be more easily analyzed.

---

* The rights of this work are transferred to the extent transferable according to title 17 U.S.C. 105.

A.M. Greenberg, W.G. Kennedy, and N.D. Bos (Eds.): SBP 2013, LNCS 7812, pp. 377–385, 2013.

In particular, the microblogging site Twitter has become a source for observing the interactions between persons in contentious politics [5]. Twitter is a free social networking service that allows users to post and read messages of up to 140 characters, displayed on the author's page and delivered to the author's subscribers (followers). Twitter is the eighth most popular website in the world based on web traffic [6], with 76.8% of that traffic emanating from outside the United States. Twitter provides a standardized means of information sharing that results in communications between users in roughly the same form. In addition, all Twitter users do not exert the same influence, yet it is at least possible to measure and compare communications in a standard way.

While some claims have disputed the role of social media as the leading cause of recent events related to social contention [7], at a minimum, social networking tools such as Twitter provide an enabling means for speedy communication across physical boundaries to interested and potentially amenable parties. In this work, we investigate a data-based approach to identifying whom these important users are in a Twitter network centered around contentiuous politics. Specifically, we look at communications on Twitter centered around the political unrest in Egypt during 2011 [8].

2   Identifying Key Users

Influence on Twitter can be measured in a variety of ways. For example, one may wish to identify who is, in general, the most popular. This interpretation can lead to the straightforward connection between number of subscribers (followers) or amount of activity (number of messages) and general popularity. While this measure gives a sense of how many eyes are on a user's posted message, it does not take into account whether that information is meaningful regarding a specific topic, whether any real or pertinent information is being shared, or whether the user's messages will actually result in a change or strengthening of others' opinions. Furthermore, a person could be overall quite popular and/or influential with regard to a specific subject matter, but very unqualified to comment on a completely different subject matter. As a result, the weight assigned by readers to a message outside of the individual's domain of expertise might be very low. Recent research has shown that these measures of popularity correlate only moderately with influence, if at all, especially within a specific domain [9].

An alternative line of research focuses on influence as a means to identify the relative success of an advertising campaign through the word-of-mouth recommendations of a product. Here, influence is based on the likelihood that an individual message or meme is passed from one individual to another [10,11,12,13]. For example, a user on Twitter may post a message with an uniform resource locator (URL) link to an external source of information. A measure of influence could therefore be the likelihood that this particular link is read and/or shared with others connected to that user. This line of research has led to studies and models focusing on chains of "information cascades" and how they propagate through a medium such as Twitter. Watts and Dodds [10] assert that the structure of the network itself may have a larger impact on whether information cascades occur than the presence of specific influentials. Agarwal et al. [11] attempt to identify influential bloggers in a community using a recursive model primarily based on inlinks and outlinks between posts. Bakshy et al. [12] look at followers, followed by, tweets, account creation date, and historical influence to measure influence. They find that high influencers can be identified, but only on average, for a marketing application involving link sharing.

This link-tracking approach tends to result in metrics that describe the spread of individual pieces of information, which can be aggregrately correlated to the originator of this message. The described problem of domain-specific expertise can be addressed with this more focused measure of influence. On the other hand, the most influential users in a domain may not necessarily be the *originator* of the message; further, a meme such as a URL link can actually have multiple originating points. In addition, an influential individual may not even exert their influence through messages with external links to other sources of information. While we have observed messages of this URL-sharing type, we wish to identify a measure of influence that incorporates the above ideas but represents a broader definition of influence. For this work, we wish to measure influence based on both the availability of information from a user (e.g. how many eyes are on a message) and the perceived relevance of a message in terms of both the topic area and the effect on the population reading them.

2.1   The Alpha Centrality

As with previous works [10,11,12,13], we apply a social network approach to model this description of influence. We adopt the use of the alpha centrality [14], which is related to the eigenvector centrality [15], but can incorporate directionality and an external measure of importance outside of the network structure. Like the eigenvector centrality, the alpha centrality is a measure of a certain aspect of status based on those with whom a user is in contact. Thus, an individual that is connected to other central individuals is also central. Defining a graph $G := (V, E)$, with n the number of vertices, the alpha centralities, x, are expressed as:

$$x = \alpha A^\top x + e, \tag{1}$$

where $A = (a_{i,j})$ is the adjacency matrix of G and x is a vector of centrality scores for each vertex i. e is a vector of exogenous status or importance that does not depend on their connection to others. $\alpha < \frac{1}{\lambda_1}$, where λ_1 is the first eigenvalue of the eigenvector centrality, is a parameter that reflects the trade-off between the importance of that prior importance e and the endogenous importance determined by network connections. Eq. (1) has a matrix solution:

$$x = (I - \alpha A^\top)^{-1} e. \tag{2}$$

Along with the aforementioned advantage of solving the eigenvector centrality problem for directed networks, the alpha centrality has other benefits as well. For one, the alpha centrality provides the opportunity to incorporate factors exogenous to the network using e, whose relative importance can be tuned based on qualitative factors for a given network and process being modeled. Secondly, unlike the eigenvector centrality or PageRank [16], the alpha centrality is a steady state solution of linear non-conservative diffusion, making it a non-conservative metric for representing the process of information sharing or influence over time [13]. Network diffusion can be generally modeled as a dynamic stochastic process that distributes some quantity (weight) across a graph, analogous to a random walk with birth; at each point in time, a random walker can give birth to new random walkers. In other words, non-conservative diffusion are processes where the quantity across the network does not remain constant. Formally, given a diffusion function $\mathcal{F} : G_t \rightarrow G_{t+1}$, where t is some instance in time, a non-conservative diffusion process will result in $||x_{t+1}|| \neq ||x_t||$. Here, $x_{i,t}$ is the weight of vertex i at time t. This process can be represented using a diffusion matrix A and scaling factor α, by

$$x_{t+1} = \alpha x_t A. \tag{3}$$

As $t \rightarrow \infty$, the steady-state solution becomes

$$x_{t \rightarrow \infty} = x_0 \sum_{k=0}^{t \rightarrow \infty} (\alpha A)^k = x_0 (I - \alpha A)^{-1}, \tag{4}$$

for $\alpha < \frac{1}{\lambda_1}$. Notice that this equation is identical to Eq. (2) for $x_0 = e$.

2.2   What Network?

A key consideration for the measure of importance described in Eq. 2 is the network, represented by the adjacency matrix A. A straightforward solution for defining the network is to build a digraph based on Twitter followers. However, recent research [9,12,17] has shown this network representation is largely uncorrelated to actual Twitter influence. People often subscribe to a large number of other users but only directly interact with an extremely reduced subset of these users. Diffusion of influence is believed to be much more correlated to these direct interactions within this more sparse underlying network, congruous with the word-of-mouth focused influentials theory mentioned previously [1]. Within Twitter, many of these direct messages (DMs) are predominately identified through the presence of retweets (e.g. "RT @username" or some derivative substring). In fact, Starbird and Palen [18] assert that retweets actually represent a social information filtering mechanism, in which only messages deemed important get passed on to a user's respective followers. Boyd et al. [19] similarly suggest that retweet mechanisms are associated with direct conversations or to validate someone's thoughts. Consequently, a form of user influence can be based on a user's decision to retweet a user that posted a message they deem worthy of passing on. The total number of retweets has actually been employed as a measure of influence in itself [9]. However, unlike the information cascade-based score [12], this mechanism does not allow us to examine how *far* a message propagates through a network from its origination. For our measure of influence, we wish to take advantage of the network described by the retweet mechanism while also accounting for the spread of information beyond this first link. Therefore, we construct the retweet network by drawing links from the user making the retweet to the user that was retweeted, where the link is in the direction of influence (and exactly opposite to the direction of information passing). We then use this newtork as the adjacency matrix A in Eq. (2). Retweets were identified using common Twitter syntax similar to Starbird and Palen [5].

2.3   Measuring External Influence

Most previous analyses incorporating the alpha centrality focuses solely on the effects of network structure by defining e as a vector of ones [13,14]. Alternatively, we would like a way to reflect the amount (quantity) of information a user shares as a part of our definition of influence, e.g. a reflection of the number of "eyes" on a message. Thus, an active user that has sent messages to a large number of subscribers starts out with a larger exogenous importance than a non-active user with a low number of followers. Then, the underlying network structure will exhibit how the diffusion of these messages propagates across the network as messages are chosen to be shared (or not shared) and passed on (or not passed on). In order to avoid biasing the network based on extreme outliers, and to bound the values of e, we define e_i based on cumulative distribution functions (CDFs) of the pertinent network properties:

$$e_i = P_i(m, \alpha(m)) \cdot P_i(f, \alpha(f)), \tag{5}$$

where, $P_i(\cdot, \alpha(\cdot))$ is the percentile the user falls under for the distribution of messages (m) and followers (f) compared to other users on the network, and $0 \le e \le 1$. We will evaluate this measure of exogenous importance by comparing the use of Eq. (5) with a vector of ones.

3   Results

We collected Twitter data from December 10, 2010 to October 11, 2011 using a hashtag-based search on the four most common tags from the Egyptian revolution (#Egypt, #Jan25, #mubarak, and #tahrir). A hashtag-based search does not obtain every message that has been posted regarding the Egyptian revolution. Nevertheless, hashtags are employed in Twitter to broadcast to a specific community of interest. While our approach does not assure that messages about a specific topic are unmissed, it inherently bounds the network to a reasonable size, with few false inclusions. We further filter the data to include only those individuals stating they are located within Egypt. We will inherently miss individuals for whom their location is indeterminate, but we wish to capture those individuals representing "on the ground" [5], actively involved participants and witnesses to the events surrounding revolution.

Table 1 displays the top 10 users using our model from Eq. (2) for the four most common hashtags related to the Egyptian revolution. Not surprisingly, there are

Table 1. Top users by influence (x), for hashtags related to the Egyptian revolution

Rank	#Egypt	#Jan25	#tahrir	#mubarak
1	dostornews	dostornews	lilianwagdy	shorouk_news
2	earhim	ghonim	gsquare86	monaeltahawy
3	ghonim	sandmonkey	carloslatuff	adamakary
4	almasryalyoum_a	alaa	shorouk_news	lilianwagdy
5	ontveg	3arabawy	leilzahra	raafatology
6	3arabawy	almasryalyoum_a	salmasaid	almasryalyoum_a
7	egyfeeds	monasosh	3arabawy	egyptocracy
8	alaa	dima_khatib	pakinamamer	carloslatuff
9	carloslatuff	wael	monasosh	almasryalyoum_e
10	zeinobia	egyptocracy	sandmonkey	daliaziada
Total	**28,730**	**20,422**	**16,901**	**7,600**

several common users across multiple hashtags, as these are all interrelated topics. However, there are also users that appear prominently in one hashtag but have a much lower influence in another. For example, @monaeltahawy has the second highest influence score under the hashtag #mubarak but ranked 32nd, 20th, and 166th in #Egypt, #Jan25, and #tahrir, respectively. She had a relatively large number of tweets that specifically addressed the regime and President Mubarak. Alternatively, @lilianwagdy ranked much higher for #tahrir, as most of her messages focused on events unfolding on the ground, at the location of the protests. Many of the top influencers are news sources rather than individual people (e.g. @dostornews, @earhim, @shorouk_news, @almasryalyoum_a).

Validation of this type of model is difficult, as there is no actual quantitative "ground truth." Thus we are limited to comparisons of influence using other approaches. Here we compare our results to (1) a heuristic study and (2) other measures of influence introduced in previous work. Table 2 shows a set of heuristically selected users and their relative rankings based on our model. The heuristic study involved qualitiative research of Twitter and other activities surrounding the revolution [8], conducted prior to achieving the results from our model and specifically singling out *individuals* rather than aggregate media sources. Expected influential users rank very high in our influence model, as desired. Of particular note is @ghonim, who, despite having a very low e, still ranks in the > 98th percentile of all users. This user was actually captured and detained for a large portion of the early Egyptian protests, but was considered by other Egyptian Twitter revolutionaries very influential both before, during, and after detainment. Therefore, his connectivity in the network led to a relatively high importance despite not even having the chance to post messages for a significant time, and posting zero messages using our four hashtags. Those messages that he did post (before and after detainment) were mentioned and retweeted frequently by a large number of other users. One could obviously perceive how an adjustment of e or α could raise or lower this type of influencer in the model. The last column of Table 3 shows how these scores change when e is set to a constant value of one for all users. While in this case, @ghonim and others with low e scores do not move up drastically, other users with high e scores noticeably move down in rankings. @ghonim and @alaa do not move up in rankings primarily because they are already second and fourth, respectively, and

Table 2. Influence (x) rankings and ranking percentiles for a selection of heuristically-determined important users, for four hashtags related to the Egyptian revolution

Name	#Egypt Rank	Percentile	#Jan25 Rank	Percentile	#tahrir Rank	Percentile	#mubarak Rank	Percentile
@3arabawy	18	99.94%	5	99.98%	7	99.96%	38	99.50%
@gsquare86	25	99.91%	15	99.93%	2	99.99%	20	99.74%
@ghonim	4	99.99%	2	99.99%	141	99.17%	126	98.34%
@sandmonkey	54	99.81%	3	99.99%	10	99.94%	15	99.80%
@alaa	21	99.93%	4	99.98%	24	99.86%	764	89.95%
@waelabbas	79	99.73%	56	99.73%	107	99.37%	52	99.32%
@monasosh	1535	94.66%	7	99.97%	9	99.95%	1470	80.66%
@wael	151	99.47%	9	99.96%	235	98.61%	61	99.20%
@pakinamamer	23	99.92%	11	99.95%	8	99.95%	18	99.76%
@lilianwagdy	20	99.93%	39	99.81%	1	99.99%	4	99.95%

the number one ranked Twitter account (@dostornews) has a significant margin ahead of the others.

Table 3 also displays this same selection of users for several influence measures, including number of tweets, followers, following, and exogenous importance (e). We further include the url-link-cascade based influence measure described previously [12] and the general retweet network outdegree. Note that not only @ghonim, but also @alaa and @monasosh did not have a recorded message using the four hashtags collected on, so do not have scores for any influence measures that require this characteristic. @alaa and @monasosh wrote almost exclusively in Arabic, rarely using English hashtags and did not have a single message in our dataset. Yet, because other users passed on their messages and appended these hashtags, they became incorporated into the network model and can actually achieve quite high influence measures. This result shows how we can effectively expand the dataset to include relevant users and message-passing exchanges that were not a part of the original dataset simply through the inference created by the retweet mechanism that leads to a link formed in the network. Thus, a particular advantage of using a retweet-based network model is that missing and incomplete data can still be incorporated indirectly, and lead to users with very high influence measures if they are still central to the network.

Table 3. Comparison between influence (x) rankings and several other measures of influence, for a selection of heuristically-determined important users and the #Jan25 hashtag

Name	Followers	Tweets	Cascade Score	e	Retweet Outdegree	x	$x(e = 1))$
@3arabawy	28	21	3	4	6	5	7
@gsquare86	45	100	17	16	21	15	18
@ghonim	NA	NA	NA	NA	2	2	2
@sandmonkey	24	23	5	3	4	3	3
@alaa	NA	NA	NA	NA	5	4	5
@waelabbas	23	487	1556	68	73	56	73
@monasosh	NA	NA	NA	NA	7	7	6
@wael	43	112	29	18	13	9	11
@lilianwagdy	51	111	59	22	34	39	41
@pakinamamer	91	59	26	15	14	11	15
R vs. x	0.467	0.395	0.418	0.879	0.624	1.00	0.660
R vs. $x(e = 1)$	0.535	0.606	0.802	0.229	0.993	0.660	1.00

The last two rows of Table 3 reveal the Pearson correlations between our measure of influence (with and without constant e) and the other measures of influence over the complete dataset. We find a moderate correlation between our measure and followers or tweets, analogous to previous findings [12]. We find a strong correlation to either the information cascade-based measure of influence or our exogenous importance score depending on whether that e was used as a part of the alpha centrality. This result is expected, as both the cascade score and the alpha centrality are based on network properties associated with filtered recommendation mechanisms employed by Twitter users, and will be more correlated if less weight is given to pre-existing measures of exogenous importance in our measure. An advantage of our measure of influence over these alternatives that it is *tunable* by varying α or by making e variable or constant in Eq. (2). Using the retweet outdegree achieves scores even more closely aligned to our measure with a constant e, as the alpha centrality with a low path length is related to retweet network outdegree. Therefore, using the retweet outdegree by itself is not unreasonable, but again does not allow for tunability based on how much we wish to account for this exogenous importance.

For these results, we used an α very close to its maximum ($\alpha < \frac{1}{\lambda_1}$). Setting α instead close to zero enables us to probe only the local structure of the network. As α increases, longer paths become important, and the measure becomes more global, with a path length of diffusion equal to $\frac{1}{1-\alpha\lambda_1}$ [13]. Hence, using a low α will result in a measure of influence much closer to each user's exogenous importance e, and the variability of x with α ranges from a number somewhat close to e and the x shown.

4   Discussion

In this work, we introduce a model for automatically identifying opinion leaders, e.g. key users with a higher likelihood of influencing others, for contentious politics in internet social media. Specifically, we have examined Twitter activity related to the revolution that occured beginning in early 2011 in Egypt. These protests represented one of the first documented cases where internet social media was intrinsically involved and facilitated knowledge sharing, planning, and situational awareness for communities of interest in contentious political actions.

The model allows for an exogenous influence based on the initial spread of a user's content to other users (e.g. the number of eyes on the content) and a network-based measure using the alpha centrality that incorporates directionality and can be viewed in terms of a non-conservative diffusion process. The measure takes advantage of the underlying, direct communication network to model the diffusion of ideas or information that are actually adopted or, at the least, filtered based on relevance of content. Our measure agrees strongly with heuristic expectations and compares well with other influence measures. Advantages of this measure are its tunability based on the application of interest and its ability to infer influence despite missing or incomplete data. The measure is also a broad, but topic-specific measure of influence. Straightforward influence measures like Twitter followers are constant across any topic area and therefore cannot identify influence for a particular domain or for a specific community of interest. On the other hand, this measure is community wide, and so provides a broader scope than measures that are limited to those users posting messages with external URL links [12].

Acknowledgments. The authors acknowledge support and funding through the SPAWARSYSCEN Atlantic Innovation Program. The second author would also like to acknowledge funding through the Naval Research Enterprise Internship Program (NREIP).

References

1. Merton, R.: Patterns of influence: local and cosmopolitan influentials. In: Merton, R. (ed.) Social Theory and Social Structure, pp. 441–474. Free Press, New York (1968)
2. Roch, C.: The dual roots of opinion leadership. Journal of Politics 67 (2005)
3. Rogers, E.: Diffusion of Innovations. Free Press, New York (1995)
4. Berk, R.: An introduction to sample selection bias in sociological data. American Sociological Review 48(3), 386–398 (1983)
5. Starbird, K., Palen, L. (How) will the revolution be retweeted? information diffusion and the 2011 egyptian uprising. In: CSCW 2012 (February 2012)
6. Alexa (2012), http://www.alexa.com/topsites/global
7. Gladwell, M.: Why the revolution will not be tweeted. The New Yorker (2010)
8. Nunns, A., Idle, N.: Tweets from Tahrir: Egypt's Revolution as it Unfolded, in the Words of the People Who Made it. OR Books (2011)
9. Cha, M., Haddadi, H., Benevenuto, F., Gummadi, K.: Twitter frees iran: An evaluation of twitter's role in public diplomacy and information operations. In: Communications and Policy Forum. Network Insight Institute (2009)
10. Watts, D., Dodds, P.: Influentials, networks, and public opinion formation. Journal of Consumer Research 34(4), 441–458 (2007)
11. Agarwal, N., Liu, H., Tang, L., Yu, P.: Identifying the influential bloggers in a community. In: WSDM 2008. ACM (February 2008)
12. Bakshy, E., Hofman, J., Mason, W., Watts, D.: Everyone's an influencer: quantifying influence on twitter. In: WSDM 2011. ACM (February 2011)
13. Ghosh, R., Lerman, K., Surachawala, T., Voevodski, T., Teng, S.: Non-conservative diffusion and its application to social network analysis (2011), http://arxiv.org/pdf/1102.4639.pdf
14. Bonacich, P., Lloyd, P.: Eigenvector-like measures of centrality for asymmetric relations. Social Networks 23, 191–201 (2001)
15. Wasserman, S., Faust, K.: Social Network Analysis: Methods and Applications. Cambridge University Press (1994)
16. Brin, S., Page, L.: The anatomy of a large-scale hypertextual web search engine. Computer Networks and ISDN Systems 30, 107–117 (1998)
17. Huberman, B., Romero, D., Wu, F.: Social networks that matter: Twitter under the microscope. First Monday 14, 1–5 (2009)
18. Starbird, K., Palen, L.: Pass it on? retweeting in mass emergency. In: ISCRAM 2010 (May 2010)
19. Boyd, D., Golder, S., Lotan, G.: Tweet, tweet, retweet: Conversational aspects of retweeting on twitter. In: HCSS, pp. 1–10. The University of Hawaii at Manoa (2010)

Analytical Methods to Investigate the Effects of External Influence on Socio-Cultural Opinion Evolution

Subhadeep Chakraborty

Department of Mechanical, Aerospace, and Biomedical Engineering,
University of Tennessee, Knoxville, Tennessee

Abstract. This paper extends the framework for merging sociologically inspired rational cognitive models of decision making with social media inspired feedback mechanisms. This model, with certain simplifying assumptions, is used to analyze the effects of external influence on the dynamics of opinion evolution in a fully connected society. The master equation is shown to have the form of the Fokker-Planck equation, and the necessary and sufficient conditions for a polynomial solution are investigated. It is proved that the parameters of the model guarantees the existence of a polynomial solution.

Keywords: Socio Cultural Decision Modeling, Effective Influence, Behavioral Dynamics.

1 Introduction

Recent events have amply demonstrated that cyber warfare is an extremely powerful weapon that, in the wrong hands, can potentially cause rapid destabilization and meltdown in an apparently stable society. On the other hand, proper tools can help us understand, predict and influence the dynamics of opinion formation to steer a population towards stability during political turmoil and protect critical markets and infrastructure.

These objectives can only be achieved with deep and concurrent understanding of both complex microscopic human logic mechanisms, as well as the underlying network of social interactions. To date, cognitive architectures have largely been applied to individual behavior. The concept of unified theories of cognition [1] implemented computationally as cognitive architectures [2] have produced an impressive collection of scientific contributions. In contrast, network science focuses on the interaction between individuals and the social phenomena that emerge from these interactions, while making oversimplifying assumptions about individual cognition. These models are based on a macroscopic approach utilizing a master equation or Boltzmann-like equations for global variables [3]. Graph-theoretic studies of opinion dynamics have been pioneered by Bikhchandani S. et. al. [5], where the concept of *information cascades*, based on *observational learning theory* was formally introduced. Watts D.J. [6] has studied

A.M. Greenberg, W.G. Kennedy, and N.D. Bos (Eds.): SBP 2013, LNCS 7812, pp. 386–393, 2013.

the origin of rare cascades in terms of a sparse, random network of interacting agents using *generating functions*. The more recent Sznajd model [7], based on the "united we stand" principle is capable of providing insight into the dynamics of opinion formation and is of particular relevance to this paper. Analytical results for the Sznajd model has been given by Slanina [8].

In an earlier work [9] we have put forward a Socio-Cultural Opinion Evolution (SCOPE) modeling paradigm that utilizes a new language-measure based reward maximization framework for stochastic simulation of a person's decision making process and studies them in the context of social influences and global (political) human induced perturbations. This paper provides some analytical results for a simplified version of the model described in [9]. It is shown that under certain assumptions the language theoretic model converges to the Sznajd model. In addition, the effect of including external influencing agents has been addressed and analytical solutions have been provided.

For completeness, the main features of the language theoretic model is very succinctly described in the next section. Section 3 enumerates the assumptions needed for the analysis and sets up the model equations to account for external influencing agents. The results are derived in Section 4 and the paper is concluded in Section 5.

2 Discrete Choice Modeling (DCM)

The assumptions inherent to the SCOPE model are as follows:

- Each decision scenario can be expressed as a finite set of discrete choices $Q \equiv \{q_1, q_2, ..., q_n\}$, where n is finite,
- There exists an identifiable equivalent normative (rational) perspective,
- The mechanism of decision making, while abiding by the rules of the common sociologically defined normative perspective, is still probabilistic, varying from person to person as well as from one time to another,
- Individuals can make extempore probabilistic decisions (internal events) or decisions influenced by induced perturbations.

The assumption of normative perspective allows rational behavior, to be encoded as a Probabilistic Finite State Automaton (PFSA). The PFSA is conceived as an abstract mathematical construct that consists of a finite number of states (or discrete choices, in the case of cognitive modeling). The individual cognitive mechanism modeled by the PFSA, can probabilistically move from one state to another when initiated by a triggering event or condition, called a transition. The probabilistic automaton generalizes the concept of a Markov chain. As before, let the discrete choice behavior be modeled as a PFSA:

$$G_i \equiv (Q, \Sigma, \delta, q_i, Q_m) \tag{1}$$

where $Q = \{q_1, q_2, ..., q_n\}$ is the finite set of choices and the initial state $q_i \in Q$. $\Sigma = \{\sigma_1, \sigma_2, ...\sigma_m\}$ is the (finite) alphabet of events; the Kleene closure of Σ is denoted as Σ^*; the (possibly partial) function $\delta : Q \times \Sigma \times Q \to [0, 1]$ represents probabilities of state transitions, $\delta^* : Q \times \Sigma^* \times Q \to [0, 1]$ is an extension of δ; and $Q_m \subseteq Q$ is the set of marked (i.e., accepted) states.

Definition 1. *The reward from each state $\chi : Q \to [-1, 1]$ is defined as a characteristic function that assigns a real weight to each state q_i, such that*

$$\chi(q_j) \in \begin{cases} [-1, 0) & \text{if } q_j \in Q_m^-, \\ \{0\} & \text{if } q_j \notin Q_m, \\ (0, 1] & \text{if } q_j \in Q_m^+. \end{cases} \quad where \quad Q_m = Q_m^+ \cup Q_m^- \quad (2)$$

Q_M^+ and Q_m^- are considered 'good' and 'bad' marked states respectively. It may be noted that such notions of good and bad are transient and changes with the cultural climate.

Definition 2. *The event transition cost, conditioned on a PFSA state at which the event is generated, is defined as $\tilde{\pi} : \Sigma^* \times Q \to [0, 1]$ such that $\forall q_j \in Q, \forall \sigma_k \in \Sigma, \forall s \in \Sigma^*$,*

(1) $\tilde{\pi}[\sigma_k, q_j] \equiv \tilde{\pi}_{jk} \in [0, 1);$
(2) $\tilde{\pi}[\sigma, q_j] = 0$ *if* $\delta(q_j, \sigma, q_k) = 0 \, \forall k; \quad \tilde{\pi}[\epsilon, q_j] = 1;$
(3) $\tilde{\pi}[\sigma_k s, q_j] = \tilde{\pi}[\sigma_k, q_j] \, \tilde{\pi}[s, \delta(q_j, \sigma_k)].$
The event cost matrix, ($\tilde{\Pi}$-matrix), is defined as: $\tilde{\mathbf{\Pi}} = \begin{bmatrix} \tilde{\pi}_{11} & \tilde{\pi}_{12} & \dots & \tilde{\pi}_{1m} \\ \tilde{\pi}_{21} & \tilde{\pi}_{22} & \dots & \tilde{\pi}_{2n} \\ \vdots & \vdots & \ddots & \vdots \\ \tilde{\pi}_{n1} & \tilde{\pi}_{n2} & \dots & \tilde{\pi}_{nm} \end{bmatrix}$

The characteristic vector $\bar{\chi}$ is chosen based on the individual state's impact. For example, if the states represent various job choices, the remuneration from these jobs, normalized between $[-1, 1]$ can serve as the characteristic vector. The event cost is an intrinsic property of the nominal perspective.

Definition 3. *The state transition function of the PFSA is defined as a function $\pi : Q \times Q \to [0, 1)$ such that $\forall q_j$, and $q_k \in Q$,*

(1) $\pi(q_j, q_k) = \sum\limits_{\sigma \in \Sigma: \, \delta(q_j, \sigma, q_k) \neq 0} \tilde{\pi}(\sigma, q_j) \equiv \pi_{jk}$
(2) and $\pi_{jk} = 0$ *if* $\{\sigma \in \Sigma : \delta(q_j, \sigma) = q_k\} = \emptyset,$
The state transition matrix, (Π-matrix), is defined as: $\mathbf{\Pi} = \begin{bmatrix} \pi_{11} & \pi_{12} & \dots & \pi_{1n} \\ \pi_{21} & \pi_{22} & \dots & \pi_{2n} \\ \vdots & \vdots & \ddots & \vdots \\ \pi_{n1} & \pi_{n2} & \dots & \pi_{nn} \end{bmatrix}$

Definition 4. *A real measure $\mu(q_i)$ over the discrete state space may be recursively defined as [10]*

$$\mu(q_i) = \sum_j \pi_{ij} \mu(q_j) + \chi(q_i) \quad (3)$$

The reader is referred to [10] for the complete derivation of Eqn. 3, but conceptually, the measure of state q_i can be perceived as the weighted expected value of χ over all time-steps in the future for the Markov process that begins in state q_i. The measure in vector form yields

$$\bar{\mu} = (\mathbb{I} - \Pi)^{-1} \bar{\chi} \quad (4)$$

This expression is a simplified version of the measure derived in [9] without using any discount for more immediate transitions. Higher measure for a state implies that the expected reward from that state is higher; consequently the incentive to transition to that state is proportionately higher.

3   Simplifications for a 2 State System with External Influence Groups

The previous section provides a stochastic framework for modeling cognitive processes and strategic decision making for maximizing future rewards. Let us now make a few more constrictive assumptions in order to derive some interesting dynamical characteristics of this system.

1. Only 2 states, i.e. $|Q| = 2$. Let $Q = \{q_1, q_2\}$ as shown in Fig. 1,
2. The complex social network is approximated by a fully-connected network (the complete graph) of N nodes. Here, any two nodes are neighbors.

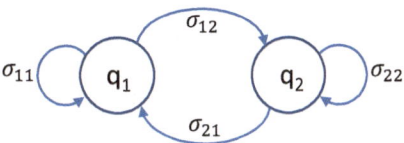

Fig. 1. PFSA modeling a 2 state decision process

For the 2 state system, conventionally represented by $\{+, -\}$ or $\{\uparrow, \downarrow\}$ to correspond to Ising spins, the transition matrix is simply $\mathbf{\Pi} = \begin{bmatrix} \pi_{11} & \pi_{12} \\ \pi_{21} & \pi_{22} \end{bmatrix}$, and Eqn. 3 becomes,

$$\mu(q_1) = \pi_{11}\mu(q_1) + \pi_{12}\mu(q_2) + \chi_1$$
$$\mu(q_2) = \pi_{21}\mu(q_1) + \pi_{22}\mu(q_2) + \chi_2 \tag{5}$$

We can use $\mu(q_i)$ as probability mass function ($P_i = \mu(q_i)$) defined over the discrete decision states only if $P_1 + P_2 = 1$. Using Eqn. 5, and $P_1 + P_2 = 1$, it can be easily shown that

$$P_1 = 0.5 + \frac{\chi_1}{2\pi_{12}}$$
$$P_2 = 0.5 + \frac{\chi_2}{2\pi_{21}} \tag{6}$$

Compact support conditions for the probability measure also dictates that

$$\frac{\chi_1}{\pi_{12}} + \frac{\chi_2}{\pi_{21}} = 0$$
$$-\pi_{12} \leq \chi_1 \leq \pi_{12}$$
$$-\pi_{21} \leq \chi_2 \leq \pi_{21} \tag{7}$$

The reward function χ gets propagated through a network and evolves constantly due to social feedback, media propaganda, external perturbations etc. In one specific scenario, let us adopt the following rule for an individual's reward estimate update mechanism:

- Pick a node (j) at random
- Pick a neighbor of node k at random
- If at that instant

  - k is in state q_1, node j sets $\chi_1 = 0.5$, $\chi_2 = -0.5$ and $\pi_{12} = \pi_{21} = 0.5$.
    $\Rightarrow P^j(q_1) = 1$ and $P^j(q_2) = 0$.
  - k is in state q_2, node j sets $\chi_1 = -0.5$, $\chi_2 = 0.5$ and $\pi_{12} = \pi_{21} = 0.5$.
    $\Rightarrow P^j(q_1) = 0$ and $P^j(q_2) = 1$.

That means with this reward update mechanism, the probability of node j moving to a particular state is the same as the probability of his randomly chosen neighbor being in that state. This is the exact same framework suggested by Ochrombel's simplification [11] to the Sznajd [7] model.

Let us now consider the scenario depicted in Fig. 2. At a certain instant, there are N_1 nodes in state q_1, N_2 nodes in state q_2. $N = N_1 + N_2$ nodes make up the vertices of a complete graph. In addition, there are $I = I_1 + I_2$ external agents with respective pre-conceived allegiance to states q_1 and q_2. In an election scenario, N can be thought of as undecided voters, while members of the influence groups have already made up their minds one way or the other.

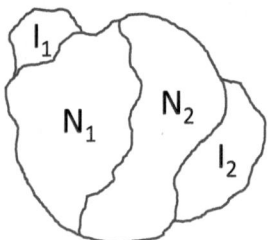

Fig. 2. Population composition and external influence groups

Let us define the control variable $u = \frac{I_1 - I_2}{I}$ and the magnetization parameter $m = \frac{N_1 - N_2}{N}$. We can then proceed to compose the master equation as

$$\dot{P}_m = r_{m + \frac{2}{N}} P_{m + \frac{2}{N}} + g_{m - \frac{2}{N}} P_{m - \frac{2}{N}} - (r_m + g_m) P_m \tag{8}$$

where,
$$r_{m + \frac{2}{N}} = P(m + \frac{2}{N} \to m) = \left(\frac{N_1 + 1}{N}\right)\left(\frac{N_2 - 1 + I_2}{N + I - 1}\right)$$

$$g_{m - \frac{2}{N}} = P(m - \frac{2}{N} \to m) = \left(\frac{N_2 + 1}{N}\right)\left(\frac{N_2 - 1 + I_1}{N + I - 1}\right)$$

$$r_m = P(m \to m - \frac{2}{N}) = \left(\frac{N_1}{N}\right)\left(\frac{N_2 + I_2}{N + I - 1}\right)$$

$$g_m = P(m \to m - \frac{2}{N}) = \left(\frac{N_2}{N}\right)\left(\frac{N_2 + I_1}{N + I - 1}\right) \tag{9}$$

Using Eqn. 9 in Eqn. 8, we get,

$$
\begin{aligned}
\dot{P}_m =\ & \left(\frac{N_1+1}{N}\right)\left(\frac{N_2-1+I_2}{N+I-1}\right) P_{m+\frac{2}{N}} \\
& + \left(\frac{N_2+1}{N}\right)\left(\frac{N_2-1+I_1}{N+I-1}\right) P_{m-\frac{2}{N}} \\
& - \left[\left(\frac{N_1}{N}\right)\left(\frac{N_2+I_2}{N+I-1}\right) + \left(\frac{N_2}{N}\right)\left(\frac{N_2+I_1}{N+I-1}\right)\right] P_m
\end{aligned}
\tag{10}
$$

After some simple manipulation, the terms can be separated to yield,

$$
\begin{aligned}
\dot{P}_m =\ & \frac{1}{N(N+I-1)}\left[(1-m^2)\frac{\partial^2 P_m}{\partial m^2} - 4m\frac{\partial P_m}{\partial m} - 2P_m\right] + \\
& \frac{1}{N(N+I-1)}\left[\frac{2}{N}\frac{\partial P_m}{\partial m}(N_1 I_2 - N_2 I_1) + I_2 P_{m+\frac{2}{N}} + I_1 P_{m-\frac{2}{N}}\right]
\end{aligned}
\tag{11}
$$

In the limit $N \to \infty$, assuming that $I \ll N$, $I_2 P_{m+2/N} + I_1 P_{m-2/N} \approx I P_m$. Also with proper scaling of time as $\tau = t/N^2$ and noting that $N_1 I_2 - N_2 I_1 = \frac{NI}{2}(m-i)$, the master equation can be cast in its final form as,

$$
\frac{\partial P_m}{\partial \tau} = \frac{1}{2}\frac{\partial^2}{\partial m^2}[B(m)P_m] - \frac{\partial}{\partial m}[A(m)P_m]
\tag{12}
$$

$$
\text{where,} \quad B(m) = 2\left(1-m^2\right)
\tag{13}
$$

$$
A(m) = I\left(u-m\right)
\tag{14}
$$

4 Conditions for Solution of the Master Equation

At this stage it is obvious that Eqn. 12 describing the time evolution of the probability density function of the magnetization parameter m is the Fokker-Planck equation and can be treated with generic methods developed for such stochastic differential equations. Wong [12] has reported certain general conditions under which the problem reduces to an eigenvalue problem of the Sturm-Liouville type and gives rise to polynomial solutions. If it is assumed that an equilibrium density function exists, and

$$
\lim_{\tau \to \infty} \frac{\partial P_m}{\partial \tau} = 0
\tag{15}
$$

then it is simple to show that the equilibrium density $p_e(m)$ satisfies

$$
\frac{d}{dm}\left((1-m^2)p_e(m)\right) - I(u-m)p_e(m) = 0
\tag{16}
$$

if the constants of integration are assumed to be 0. Substituting $P_m(\tau) = f(\tau)p_e(m)\varphi(m)$, in Eqn. 12 and using separation of variables, we get

$$
\frac{df(\tau)}{dt} = -\lambda f(t)
\tag{17}
$$

$$
\begin{aligned}
\frac{d^2}{dm^2}\left((1-m^2)p_e(m)\varphi(m)\right) &- \frac{d}{dm}\left(I(u-m)p_e(m)\varphi(m)\right) \\
&= -\lambda p_e(m)\varphi(m)
\end{aligned}
\tag{18}
$$

Assuming discrete eigenvalues, Eqn. 17 can be easily solved to yield,

$$f_n(\tau) = k_n e^{-\lambda_n \tau} \tag{19}$$

while using Eqn. 16 in Eqn. 18 gives the Sturm-Liouville form,

$$\frac{d}{dm}\left((1-m^2)p_e(m)\frac{d\varphi(m)}{dm}\right) + \lambda p_e(m)\varphi(m) = 0 \tag{20}$$

Necessary and sufficient conditions for Eqn. 20 to yield a complete orthonormal set of polynomials as eigenfunctions have been studied by Wong et. al. [12]. They can be summarized as follows:

$$B(m_1)p_e(m_1) = B(m_2)p_e(m_2) = 0, \quad \text{where} \quad m_1 \leq m \leq m_2 \tag{21}$$
$$A(m) = am + b \tag{22}$$
$$B(m) = cm^2 + dm + e \tag{23}$$

and

$$\int_{m_1}^{m_2} m^n p_e(m)dm < \infty, \quad n = 0,1,...,n < \infty \tag{24}$$

From Eqn. 13,14 and noting that $-1 \leq m \leq 1$, it is easy to see that the necessary and sufficient conditions are satisfied. The above conditions restrict the density function $p_e(m)$ to be of the form [12],

$$p_e(m) = \frac{1}{2^{\alpha+\beta+1}}\frac{\Gamma(\alpha+\beta+2)}{\Gamma(\alpha+1)\Gamma(\beta+1)}(1-m)^\alpha(1+m)^\beta, \quad \alpha, \beta > -1 \tag{25}$$

while the polynomial eigenfunctions $\varphi_n(m)$ orthonormalized with respect to the equilibrium density function $p_e(m)$ are the Jacobi polynomials,

$$\varphi_n(m) = \frac{(-1)^n}{2^n}$$

$$\times \sqrt{\frac{(2n+\alpha+\beta+1)\Gamma(n+\alpha+\beta+1)\Gamma(\alpha+1)\Gamma(\beta+1)}{\Gamma(\alpha+\beta+2)\Gamma(n+\alpha+1)\Gamma(n+\beta+1)n!}} \tag{20}$$

$$\times (1-m)^{-\alpha}(1+m)^{-\beta}\frac{d^n}{dm^n}\left[(1-m)^{n+\alpha}(1+m)^{n+\beta}\right]$$

For $p_e(m)$ defined as in Eqn. 25, the functions

$$A(m) = \gamma(\beta-\alpha) - \gamma(\alpha+\beta+2)m = Iu - Im \quad \text{from Eqn.14}$$
$$B(m) = 2\gamma(1-m^2) = 2(1-m^2) \quad \text{from Eqn.13} \tag{27}$$
$$\text{and} \quad \lambda_n = \gamma n(n+\alpha+\beta+1)$$

Solving 27 yields $\gamma = 1$, $\lambda_n = n(n+I-1)$, $\alpha = I_2 - 1$ and $\beta = I_1 - 1$. This restricts $I_1, I_2 \geq 1$.

The joint probability density function $p(m_0, m; \tau)$ have the form,

$$p(m_0, m; \tau) = p_e(m_0)p_e(m)\sum_{n=0}^{\infty} e^{-\lambda_n \tau}\varphi_n(m_0)\varphi_n(m) \tag{28}$$

where p_e, φ_n and λ_n are given by respectively Eqns. 25, 26 and 27, and $m_0 = m(\tau_0)$. This completely specifies the progression of the joint probability density function.

5 Conclusion and Future Work

The main contributions of this paper are twofold.

This paper provides the groundwork for merging sociologically inspired rational cognitive models of decision making with the diffusion and feedback mechanism of opinions in a fully connected society. Under certain assumptions, the cognitive model can replicate the Sznajd model, but in its full generality, this model is more realistic and flexible since it is inspired by strategic thinking and policy optimization procedures.

Secondly, using this model, this paper provides a thorough analysis of the opinion dynamics in a society which is influenced by external agents. Future studies along the same lines would be to adopt the solution for more complex decision space ($|Q| > 2$), more complex diffusion characteristics of the reward function χ and control algorithms to drive the probability distribution $P(m_0, m; \tau)$ to a desired equilibrium distribution $p_e^d(m)$, using the control input u. These problems are currently under investigation.

References

1. Newell, A.: Unified Theories of Cognition. Harvard University Press, Cambridge (1990)
2. Keltner, D., Gruenfeld, D.H., Anderson, C.: Power, approach, and inhibition. Psychological Review 110(2), 265–284 (2003)
3. Helbing, D.: Quantitative sociodynamics: Stochastic methods and models of social interaction processes. Springer (2010)
4. Galam, S.: Rational group decision making: A random field Ising model at $T = 0$. Physica A: Statistical Mechanics and its Applications 238(14,15), 66–80 (1997)
5. Bikhchandani, S., Hirshleifer, D., Welch, I.: A Theory of Fads, Fashion, Custom, and Cultural Change as Informational Cascades. The Journal of Political Economy 100(5), 992–1026 (1992)
6. Watts, D.J.: A simple model of global cascades on random networks. PNAS 99(9), 5766–5771 (2002)
7. Sznajd-Weron, K., Sznajd, J.: Opinion evolution in closed community. International Journal of Modern Physics C 11(6), 1157–1165 (2000)
8. Slanina, F., Lavicka, H.: Analytical results for the Sznajd model of opinion formation. The European Physical Journal B-Condensed Matter and Complex Systems 35(2), 279–288 (2003)
9. Chakraborty, S., Mench, M.M.: Socio-cultural Evolution of Opinion Dynamics in Networked Societies. In: Yang, S.J., Greenberg, A.M., Endsley, M. (eds.) SBP 2012. LNCS, vol. 7227, pp. 78–86. Springer, Heidelberg (2012)
10. Ray, A.: Signed real measure of regular languages for discrete event supervisory control. International Journal of Control 78(12), 949–967 (2005)
11. Ochrombel, R.: Simulation of Sznajd sociophysics model with convincing single opinions. International Journal of Modern Physics C 12(7), 1091 (2001)
12. Wong, E., Thomas, J.B.: On polynomial expansions of second-order distributions. Journal of the Society for Industrial and Applied Mathematics 10(3), 507–516 (1962)

Who Shall We Follow in Twitter for Cyber Vulnerability?

Biru Cui, Stephen Moskal, Haitao Du, and Shanchieh Jay Yang

Department of Computer Engineering
Rochester Institute of Technology, Rochester, New York 14623
{bxc2868,sfm5015,hxd1011,jay.yang}@rit.edu

Abstract. Twitter has become a key social media for sharing information, not only for casual conversations but also for business and technologies. As the Twitter community continues to grow, an intriguing question is to determine how to obtain most valuable information the earliest by following fewest Tweeters or Tweets. This multi-criteria optimization problem exhibits similar features as in the information cascade problem for blogs. This work revises an information cascade outbreak detection algorithm to find critical Twitter accounts that disseminate the most cyber vulnerabilities the earliest. Three award functions are defined to evaluate every account's contribution per topic from three aspects: timeliness, originality and influence. Critical users are selected according to their total contribution on a specific security category. Experiments were conducted using Tweets containing CVE information over a five-week period, to compare the proposed algorithm with account selections based on the number of followers and based on the PageRank algorithm. The results show that with the same number of users and tweets, our algorithm outperforms in both information coverage and timeliness.

1 Introduction

The volume of tweets has been continuously increasing significantly, from 340 million a day in March 2012 as released by Twitter to 400 million this June [1]. Comparing to other social media or blogs, Twitter aims at providing timely information in a concise format. Besides sharing people's daily life or personal opinions, Twitter can also play an role in broadcasting emergency information to public when facing nature disasters. Sakaki *et al.* [2] first use Twitter as social sensors to detect earthquakes in Japan. Every person who has a connection to Internet can be a potential sensor which collects information and publishes on Twitter. Similar uses can be applied to cyber security when new vulnerabilities or exploits are discovered. Unfortunately, as the volume of tweets and the number of Twitter accounts arise, research has shown that only a small portion of tweets are worth reading [3]. The overwhelming number of retweets, duplicated data, and trivial information makes this powerful social media less useful than it should be. This paper addresses the question on how to obtain most valuable information from least accounts and least tweets in the most timely fashion for the cyber vulnerability related information in Twitter.

A.M. Greenberg, W.G. Kennedy, and N.D. Bos (Eds.): SBP 2013, LNCS 7812, pp. 394–402, 2013.
© Springer-Verlag Berlin Heidelberg 2013

Research in social network has been addressing similar questions, including the classical HITS [4] algorithm that defines links between web pages in hubs and authorities. PageRank [5] evaluates the rank of web pages according to the probability a user surfs a page. These analysis rank web pages according to the authority or the origin of the information. Another set of research investigates the influence models in social network, such as [6]. Both ranks and influence models help find network centers, which allows to reach most nodes in a network.

Twitter has some distinctive characteristics, making direct applications of these algorithms not effective. First, the Twitter follower-network gives the structure of popularity but not necessarily the effectiveness and timeliness of information dissemination. Experts and celebrities may not the first ones posting the breaking news. Second, the nature of retweets suggests the information dissemination flow, but only forms loosely connected information cascades. Third, for a given specific topic area, it is unclear how to define the coverage of information. Recognizing these limitations, this work develops a Twitter Critical Account Discovery (TCAD) algorithm to find the optimal accounts that covers most topics in the timeliest fashion for a given set of categories. The algorithm is an adaptation of the multi criteria information cascade outbreak detection algorithm developed for blogs [7]. This work will focus primarily on cyber vulnerabilities to demonstrate the professional uses of critical information in Twitter.

2   Problem Definition and Related Work

The problem may be formally described as follows. For a specific security vulnerability category (C), find the set of critical accounts $S \subset V$ so that the total reward function $R(S)$ is maximized subject to the cost function $C(S) \leq L$. Later on, we will integrate the cost constraint into our total reward function and solve the multi criteria problem heuristically.

$$\max_{S \subseteq V} R(S) \text{ subject to } cost(S) \leqslant L \tag{1}$$

The key in solving the problem is to properly define the reward and cost functions. For this purpose, this work looked into related works that analyze the importance of information flow in social media, particularly Twitter. Java $et\ al.$ [8] studied the topological and geographical properties of the large-scale Twitter network and analyzed the community with users that share the same intention. Haewoon $et\ al.$ [9] compared the accounts' influence based on the PageRank, the number of followers, and the number of retweets. Cha $et\ al.$ [10] analyzed accounts influence in Twitter by their indegree, retweets, and number of mentions. Cha found influential accounts hold significant influence over a variety of topics. Other related works studied the information diffusion problem using a threshold model [11] and a cascade model [12]; both applied a similar concept as epidemic disease propagation model. In the threshold model, a node adapts when the weights connect to it crosses a threshold; while in the cascade model, a node is infected by its neighbor with a probability. Gruhl $et\ al.$ [13] characterized how topics propagate from individual to individual, and used the theory of infectious diseases to model the flow. Gomez-Rodriguez *et al.* [14] discovered the influence relations beneath the information diffusion network by using an approximation algorithm to best explain the time sequences the nodes adopt information. Leskovec *et al.* [7] proposed a way to detect the information outbreak among blogs.

The aforementioned set of work motivated the use of three reward functions: timeliness, originality and influence, as well as model the cost as the number of tweets needed to cover an information topic. We further acknowledge that the algorithm developed in this work has been significantly influenced by [7], with an adaptation to fit specifically to Twitter and Cyber Vulnerability topics.

3   TCAD (Twitter Critical Account Discovery)

The first challenge to identify critical accounts in a specific topic area, *i.e.*, Cyber Vulnerabilities for this paper, is to determine a filtering mechanism of tweets related to the topic. In general, this requires an intelligent semantic processing, allowing to determine not only the set of relevant tweets but also the context of the tweets. In the context of Cyber Vulnerability, a well defined and relatively unambiguous set of keywords is the Common Vulnerabilities and Exposures index (CVE xxxx xxxx), an 8 digit number following 'CVE' - a dictionary of publicly known information security vulnerabilities [15]. Using CVE tags as keywords finds tweets that are explicitly addressing the cyber vulnerabilities, and allows this work to concentrate on developing the algorithm that finds the critical accounts from the relevant tweets. The semantic meanings of each tweets are not treated in this work and will be addressed in future work.

Every CVE tag is assigned a CWE (Common Weakness Enumeration) type by a third party organization [16], which defines a software security weakness. Several CWEs make up a security vulnerability category. For example, CWE 119 (Failure to Constrain Operations within the Bounds of a Memory Buffer) and CWE 20 (Improper Input Validation) belong to the same category that causes Denial-of-Service (DoS). Fig. 1 illustrates the structure of CVE, CWE and security vulnerability categories along with the Twitter accounts and retweet relationship using directed edges.

In this information hierarchy, each CVE tag is treated as a unique information topic within a security vulnerability category, and tweets containing this CVE tag are attached to this topic. Each node in Fig. 1 represents a tweet, an original or a retweet. The directed edges reflects the retweet(s) cite the information from the original tweet. Clearly, the original tweets is more timely than the retweets and the time stamps further differentiate quantitatively the timeliness among all tweets. Each Twitter account may post tweets on multiple topics. An account is considered covering a topic if it publishes a tweet in that CVE tag. The more topic an account covers using fewest and timely tweets, the more critical the account is. Three award functions are defined as follows to address this multi-criteria problem.

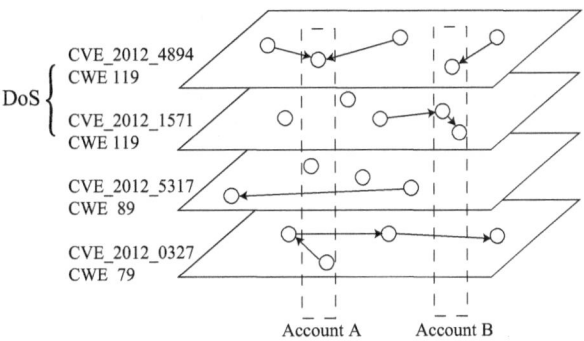

Fig. 1. Each layer stands for a CVE topic; each CVE has a CWE type which belongs to a category. Circles in layers are tweets. Each user may publish tweets among topics.

3.1   Reward Functions

The reward functions are defined according to what we expect to gain from selected accounts. In this case, the reward comes from three aspects:

Timeliness (R_T)**.** Given multiple tweets about the same news, the first one reveal the information is highly valued, with descending value for tweets thereafter. The reward function R_T for a tweet W published by account A in a CVE topic T is defined as:

$$R_T(A, W, T) = 1 - \frac{D(W, W_0(T))}{\max_{T \in C} D(W_L(T), W_0(T))} \tag{2}$$

where $W_0(T)$ and $W_L(T)$ are the first and last tweet in topic T; $D(W, W_0(T))$ defines the time difference between the tweet W and W_0; $\max_{T \in C} D(W_L(T), W_0(T))$ is the longest time period of CVE topics in category C.

Originality (R_O)**.** Though retweets help propagate information, it is still the original tweets that are referenced. People also prefer the original tweets but not the retweets. The intuition of reward R_O is that a tweet's originality is shared by all of its retweets. This is similar to the way PageRank defines the relations between web pages.

$$R_O(A, \tilde{W}, T) = \frac{R_O(P, W, T)}{\tilde{N}(W) + 1} \tag{3}$$

where $\tilde{W}$ is a retweet of tweet W; $\tilde{N}(W)$ is the total number of retweets for W; P is the account for the original tweet W. This is a recursive definition. $R_O(A, \tilde{W}, T)$ defines the originality of A's retweet $\tilde{W}$ that splits the originality of its parent tweet $R_O(P, W, T)$.

Influence (R_F)**.** If two tweets from accounts A and B contain the same information, same time stamp and are all original tweets. A should be rewarded more if its tweet has been retweeted by more accounts, implying more people are influenced by the tweet from A. Thus, besides the timeliness and originality, we define the reward of influence in (4).

$$R_F(A, W, T) = \frac{\tilde{N}(W)}{N(T)} \tag{4}$$

where $N(T)$ is the total number of posts in the topic T.

3.2 Total Reward

With the three reward functions R_T, R_O, R_F, the problem becomes finding optimal set S that maximizes the reward with multiple reward functions. Note that it is possible to have two candidates sets S and S', where $R_T(S) > R_T(S')$ but $R_O(S) < R_O(S')$. A common approach in solving such problem is to find the Pareto-optimal with respect to the following objective function:

$$\max_{S \subseteq V} \sum \lambda_1 R_T(S) + \lambda_2 R_O(S) + \lambda_3 R_F(S) \tag{5}$$

where λ_i are weights to trade off the importance of the reward functions. As mentioned in Sec. 3.1, it is ambiguous to decide the reward of an account by selecting a subset of these three reward functions. There is no reason to strengthen or weaken one aspect reward. In this paper, we use the same weights for the three rewards while normalizing the rewards in [0 1].

Recall that the optimization should be subject to the cost of the selected accounts. This work defines the cost to be the total tweets $N(A, T)$ published by each selected account A on a topic T. This is equivalent to find the reward an account has on a topic per tweet. The final total reward function an account A gain on topic T per post can then be written as:

$$R(A, T) = \sum_{W \in T} \frac{R_T(A, W, T) + R_O(A, W, T) + R_F(A, W, T)}{N(A, T)} \tag{6}$$

For a set S, the total reward of this set on a topic T is:

$$R(S, T) = \frac{\sum_{A \in S} R(A, T) \times N(A, T)}{\sum_{A \in S} N(A, T)} \tag{7}$$

The total reward obtained by selecting a set of accounts in a category is the summation of the reward on all topics belong to this category. Then the reward by selecting an account set S in a category C is:

$$R(S, C) = \sum_{T \in C} R(S, T) \tag{8}$$

3.3 Greedy Algorithm

Based on $R(S, C)$ to select K from N Twitter accounts for the optimal total reward could be computationally complex using a brute force approach which takes time exponential in K, especially when N is huge. In fact N can be significant since Twitter has reached 500 million active accounts in July 2012 [17]. $R(S, C)$ is not monotonic since by selecting an account who always posts outdated retweets may decrease the total $R(S, C)$. It is also not submodular because it does not guarantee $\forall S \subseteq S', A \notin S', R(S \cup A, C) - R(S, C) \geq R(S' \cup A, C) - R(S', C)$. This work developed a simple greedy algorithm, which exhibits outstanding performance. The algorithm, independently executed for each category C, begins by selecting one account A into S to maximize R(S, C). Then, during each iteration, it adds an account from the remaining set into S to maximize R(S,C). This greedy algorithm has a complexity of $O(KN)$ that each round iterates left accounts and takes K rounds.

4   Experiment Design and Result

We collected about 5,000 CVE related tweets between Sep 25 and Nov 2 of 2012. These tweets come from about 1,600 accounts and cover about 900 CVE topics. Nine common security categories are selected according to [17]: GI (Gain Information), RV (Reserved), EC (Execute Code), US (Unspecified), MC (Memory Corruption), DoS (Denial of Service), DT (Directory Traversal), GP (Gain Privilege), CSS (Cross-site Script). The collected data is divided into two sets, one for selecting critical accounts, another for testing how selected accounts perform on unknown data.

We compare TCAD with two other approaches: PageRank and number of retweeters. Account A is B's retweeter if A has one or more retweets from B. The same definition is used to define the directed links between accounts for PageRank. Each approach selects top 10 critical accounts per category. The performance is evaluated in terms of information coverage and timeliness. The information coverage is defined as the topics mentioned by critical accounts over the total topics in a category; the timeliness is defined as the normalized $R_T(A, W, T)$ for all $W \in C$. Both of them are normalized in [0 1]. A higher score stands for a better topic coverage or obtaining information earlier.

Figures 2 and 3 show the information coverage and timeliness achieved by the three algorithms for the two sets of data. It is clear that, by following the selected accounts, TCAD significantly outperforms the other two popularity (retweeter) based algorithms. The only category the two popularity-based algorithms not perform terribly is DT (Directory Traversal). The reasons is that there are only 21 accounts in the collected data for DT, while there are typically hundreds of accounts in other categories. Selecting 10 accounts out of 21 guarantees a reasonable information coverage and timeliness. Coincidentally, the two popularity-based approaches choose the same 10 accounts, which is why they have the same performance in DT.

We ran additional experiments to test the scenario where the two popularity-based approaches are executed in a category agnostic manner, *i.e.*, accounts are selected by mixing all tweets from all categories. The results were worse or the same in all categories. This suggests a different conclusion from Cha *et al.* [12],

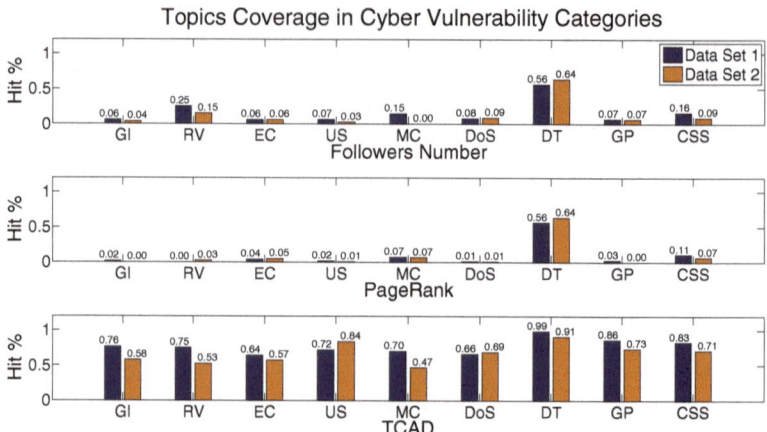

Fig. 2. Topics coverage of selected accounts in security vulnerability categories

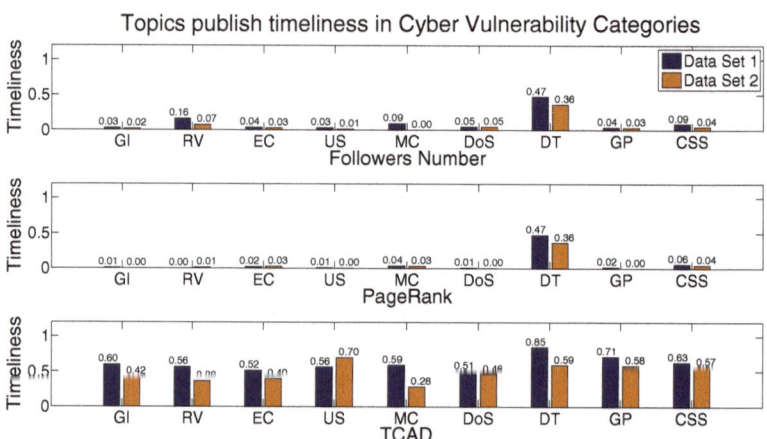

Fig. 3. Topic publish timeliness of selected accounts in security vulnerability categories

which found that "influential accounts hold significant influence over a variety of topics." Another interesting observation is that some critical accounts selected are Twitter machines such as scripts which publish posts when there is a security alert. These accounts may not be the preferred human expert accounts but do provide the most valuable information the earliest.

5   Conclusion

In this paper, we develop a Twitter Critical Account Discovery (TCAD) algorithm to find critical accounts for Cyber Vulnerabilities based on rewards from three aspects: timeliness, originality and influence. Compared to other two approaches: PageRank and Number of Retweeters, TCAD algorithm outperforms both in information coverage and timeliness. The results support our hypothesis that ranking or influence model focusing on node popularity may not effectively capture information needed. Also, the critical accounts selected from one security category can be very different to another, thus suggesting that popular accounts across many categories may not be the best account to follow for specific needs.

References

1. Bennett, S.: Twitter now seeing 400 million tweets per day (June 2012),
   http://www.mediabistro.com/alltwitter/
   twitter-400-million-tweets2_b23744
2. Sakaki, T., Okazaki, M., Matsuo, Y.: Earthquake shakes twitter users: real-time event detection by social sensors. In: Proceedings of the 19th International Conference on World Wide Web, WWW 2010, pp. 851–860. ACM, New York (2010)
3. Spice, B.: Press release: Quarter of tweets not worth reading, twitter users tell researchers (February 2012),
   http://www.cmu.edu/news/stories/archives/2012/
   february/feb1_twitterresearch.html
4. Gibson, D., Kleinberg, J., Raghavan, P.: Inferring web communities from link topology. In: Proceedings of the Ninth ACM Conference on Hypertext and Hypermedia: Links, Objects, Time and Space—Structure in Hypermedia Systems: Links, Objects, Time and Space—Structure in Hypermedia Systems, HYPERTEXT 1998, pp. 225–234. ACM, New York (1998)
5. Page, L., Brin, S., Motwani, R., Winograd, T.: The pagerank citation ranking: bringing order to the web (1999)
6. Anagnostopoulos, A., Kumar, R., Mahdian, M.: Influence and correlation in social networks. In: Proceedings of the 14th ACM SIGKDD International Conference on Knowledge Discovery and Data Mining, KDD 2008, pp. 7–15. ACM, New York (2008)
7. Leskovec, J., Krause, A., Guestrin, C., Faloutsos, C., Van Briesen, J., Glance, N.: Cost-effective outbreak detection in networks. In: Proceedings of the 13th ACM SIGKDD International Conference on Knowledge Discovery and Data Mining, KDD 2007, pp. 420–429. ACM, New York (2007)
8. Java, A., Song, X., Finin, T., Tseng, B.: Why we twitter: understanding microblogging usage and communities. In: Proceedings of the 9th WebKDD and 1st SNA-KDD 2007 Workshop on Web Mining and Social Network Analysis, WebKDD/SNA-KDD 2007, pp. 56–65. ACM, New York (2007)
9. Kwak, H., Lee, C., Park, H., Moon, S.: What is twitter, a social network or a news media? In: Proceedings of the 19th International Conference on World Wide Web, WWW 2010, pp. 591–600. ACM, New York (2010)
10. Cha, M., Haddadi, H., Benevenuto, F., Gummadi, K.P.: Measuring user influence in twitter: The million follower fallacy. In: 4th International AAAI Conference on Weblogs and Social Media, ICWSM, vol. 14(1), p. 8 (2010)
11. Granovetter, M.: Threshold models of collective behavior. American Journal of Sociology, 1420–1443 (1978)
12. Goldenberg, J., Libai, B., Muller, E.: Talk of the network: A complex systems look at the underlying process of word-of-mouth. Marketing letters 12(3), 211–223 (2001)

13. Gruhl, D., Liben-Nowell, D., Guha, R., Tomkins, A.: Information diffusion through blogspace. SIGKDD Explor. Newsl. 6(2), 43–52 (2004)
14. Gomez-Rodriguez, M., Leskovec, J., Krause, A.: Inferring networks of diffusion and influence. ACM Trans. Knowl. Discov. Data 5(4), 21:1–21:37 (2012)
15. Common Vulnerabilities and Exposures (September 2012), http://cve.mitre.org/
16. Common Weakness Enumeration (October 2012), http://cwe.mitre.org/
17. Security Vulneribility Datasource (October 2012), http://www.cvedetails.com/

Respondent-Driven Sampling in Online Social Networks[*]

Christopher M. Homan[1], Vincent Silenzio[2], and Randall Sell[3]

[1] Rochester Institute of Technology, Rochester, NY
cmh@cs.rit.edu
[2] University of Rochester Medical Center, Rochester, NY
vincent_silenzio@URMC.Rochester.edu
[3] Drexel University, Philadelphia, PA
rls82@drexel.edu

Abstract. Respondent-driven sampling (RDS) is a commonly used method for acquiring data on hidden communities, i.e., those that lack unbiased sampling frames or face social stigmas that make their members unwilling to identify themselves. Obtaining accurate statistical data about such communities is important because, for instance, they often have different health burdens from the greater population, and without good statistics it is hard and expensive to effectively reach them for prevention or treatment interventions. Online social networks (OSN) have the potential to transform RDS for the better. We present a new RDS recruitment protocol for (OSNs) and show via simulation that it outperforms the standard RDS protocol in terms of sampling accuracy and approaches the accuracy of Markov chain Monte Carlo random walks.

1 Introduction

Respondent-driven sampling (RDS) [Hec97, SH04, Hec07, VH08, WH08] is a commonly used method to survey such communities as IV drug users, men who have sex with men, and sex workers [MJK+08]; jazz musicians [HJ01]; unregulated workers [BSPar]; native American subcommunities [WS02]; and other hidden communities. RDS is a variant of snowball sampling [Tho92] that uses a clever recruitment protocol that: (1) helps ensure the confidentiality of respondents and the anonymity of the target community and (2) generates a relatively large number of recruitment waves, which hypothetically leads to unbiased sampling estimators.

Unfortunately, in terms of sampling accuracy there is still a large gap between theory and practice [Wej09, GH10, TG11, GS10]. A small body of work [SH04, VH08, SH04, Hec07, VH08, WH08], most of which focuses on improving the estimators on which RDS depends, deals with closing that gap.

This paper describes a new approach: leveraging the features of online social networks (OSNs) to improve the sampling design. We believe OSNs have the potential to dramatically transform RDS by enabling better neighborhood recall,

[*] This material is based upon work supported by the National Science Foundation under Grant No. 1111016, and through Grant No. K23MH079215 from the National Institute of Mental Health, National Institutes of Health.

A.M. Greenberg, W.G. Kennedy, and N.D. Bos (Eds.): SBP 2013, LNCS 7812, pp. 403–411, 2013.

randomized and confidential recruitment, and other improvements that allow it to better meet the assumptions on which the estimators rest. Here we focus on one particular modification, which is based on the network that a recruitment protocol generates, i.e., the network consisting of all respondents as actors and having directed ties between each respondent and those whom the respondent recruits. The estimators for RDS typically assume that these so-called *recruitment networks* are arbitrary, although in practice they are essentially trees. Gile and Handcock show [GH10] in simulation that this discrepancy is a major source of the poor performance they observe in established RDS estimators.

Our main contribution is a new protocol where the recruitment networks are directed acyclic graphs (DAGs). This protocol, while likely infeasible in many other settings, seems well suited for RDS over OSNs. Using the same simulation-based experimental framework that Gile and Handcock [GH10] and Tomas and Handcock [TG11] developed in their rather comprehensive assessments of RDS, we show that this new protocol dramatically outperforms the standard RDS protocol and approaches the sampling accuracy of a Markov chain Monte Carlo (MCMC) random walk (a process that typically satisfies standard RDS sampling assumptions). It even outperforms a recruitment protocol that, superficially at least, more closely resembles MCMC walks than does ours.

Our work is related to that of Gjoka et al. [GKBM11], who use the established RDS estimators to compare the performance of several different methods for passively—without the active participation of its users—crawling Facebook, including MCMC random walks and breadth-first search. By contrast, we are concerned primarily with methods that, due to confidentially concerns, require the active participation of those sampled, and this leads different sampling dynamics.

In another closely related study, Wejnert and Heckathorn develop a tool for conducting RDS over the World-Wide Web they call WebRDS [WH08]. Their system explicitly fixes the recruitment graph to be a tree. We, on the other hand, study what happens precisely when we relax this constraint.

2   A Brief Overview of Respondent-Driven Sampling

Heckathorn introduced RDS as a sampling protocol paired with an estimator [Hec97]. The protocol begins with a small number of seed respondents from the target community, who may be recruited in any fashion. Each respondent takes a survey, and is then given a small number of recruitment coupons (e.g., three) to distribute among other members of the target community, each of which allows whomever redeems it to take the survey (assuming that he or she meets the inclusion criteria). Each respondent is paid for taking the survey and for each of the redeemed coupons he or she distributed. The process continues until a target number of either recruitment waves or samples is reached. Thus, RDS uses the social network of the hidden population itself to do the work of subject identification, and in this regard it has been very successful in finding hidden communities. Couponing ensures the confidentiality of all those surveyed, which is often a crucial concern for the communities RDS is designed to reach.

Though the recruitment protocol has remained stable, the estimators have evolved significantly over time as questions are raised about each successive generation of estimators. We present here what is known as the Volz-Heckathorn (VH) estimator [VH08]. Although probably not as widely used as an earlier estimator due to Salganik and Heckathorn [SH04], it is newer and has been the subject of recent papers [GH10, TG11, GKBM11] that experimentally test its performance. In particular, Handcock and Gile show that the VH estimator frequently outperforms the Salganik-Heckathorn estimator [GH10]. The assumptions underlying the VH estimator are:

1. The network is connected and aperiodic.
2. Each respondent recruits exactly one person into the survey.
3. Each respondent chooses whom to recruit uniformly at random from all network relationships.
4. All relationships are reciprocal.
5. Respondents are sampled with replacement (i.e., may be rerecruited into the survey).
6. Respondents can accurately recall the number of people in the target community that they know.

It is fairly clear that in practice these assumptions, except possibly the first one, never hold. In this paper, we are particularly interested in assumption 5. In typical RDS settings most people lack the time to respond more than once, since doing so often involves travel, so this assumption fails. Consequently, recruitment networks tend to look like trees.

It is worth noting that prior estimators rested on even stronger assumptions [Hec97, SH04]. More recently, Handcock and Gile [HG10] proposed newer estimators that depend on fewer assumptions and that seem in their experiments to outperform earlier estimators [GH11] (see also [Gil11, TG11, GJS12]). Though their approach seems very promising, it is model based, and such approaches themselves depend on assumptions that can be difficult or impossible to validate.

Let $\{y_1, \ldots y_n\}$ be samples of some scalar property of a networked population. Let each $d_i \in \{d_1, \ldots, d_n\}$ be the degree (number of network ties) of the person associated with each sample. When the VH assumptions do hold, Markov chain Monte Carlo (MCMC) theory suggests $\hat{y} = (\sum_{i=1}^{n} y_i/d_i)/(\sum_{i=1}^{n} 1/d_i)$ as an asymptotically unbiased estimator for the mean of $\{y_1, \ldots, y_n\}$.

3   Simulation-Based Experiments for Assessing RDS

Gile and Handcock [GH10] and Tomas and Gile [TG11] provide a pair of thorough critiques of the VH estimator. We adopted their methods to test our new recruitment protocol, so we present them here in detail.

They simulate RDS over graphs drawn randomly from an exponential random graph model (ERGM). In each experiment, 20% of the network nodes are labeled "infected" and the remaining are "uninfected." The goal in these experiments is to estimate the proportion of infected nodes in the population. Each experiment fixes the ERGM and recruitment parameters, then repeats the following steps 1000 times:

1. Generate a test graph from the ERGM.
2. Run an RDS simulation on the test graph; stop when 500 samples are made.
3. Estimate the proportion of infected nodes using VH.

The ERGM parameters Gile and Handcock use are based on a CDC study [AQHM+06]. Network size ranges from 525 to 1000. They fix the expected degree at seven. Expected *activity ratio* is the mean degree of the infected nodes divided by the mean degree of the uninfected nodes. This ranges from one to three. Expected homophily is defined here as the expected number of relationship between infected actors divided the expected number of relationships between infected and uninfected actors. This ranges from two to thirteen.

Seed nodes are drawn at random in proportion to their neighborhood size, either from all nodes, just the infected nodes, or just the non-infected nodes. The number of seeds ranges from 4 to 10.

For the recruitment parameters, each chosen node recruits exactly two new nodes uniformly at random from its "eligible" network neighbors, where "eligible" is either all neighbors (for sampling with replacement) or all neighbors who have not yet been sampled (for sampling without replacement). We call the without-replacement protocol "RDS" and the with-replacement one "REP." Note that RDS produces trees as recruitment networks and REP produces arbitrary graphs.

4   A New DAG-Based Recruitment Protocol

As Gile and Hancock show (see also Fig. 1–5, which reproduce in part their results), the RDS protocol, even with perfect randomness and response in the recruitment process, results in significantly degraded performance under the VH estimator. But what if sampling with replacement were feasible? It seems plausible do to so in an online setting, i.e., where the survey is administered via the Web: if a respondent is recruited a second time, all the respondent needs to do is log in to the website where the survey is administered and the system can automatically count the respondent's survey a second time (and send the respondent additional electronic recruiting coupons) without requiring the respondent to return to a physical polling site.

The trickier part is in the recruitment dynamics. If we let respondents rerecruit freely, as in the REP protocol, then, in order to gain more money from survey incentives, they could collude to rerecruit each other many more times than chance would predict, thus skewing the results. We propose to discourage this behavior by allowing respondents to be rerecruited only if doing so does not result in the recruitment graph containing a directed cycle. The resulting recruitment graph is thus a directed acyclic graph. We call this protocol "DAG."

5 Experiments and Results

We use the same methods as Gile and Handcock, as we described in section 3. The major difference is that we consider two additional variants of the RDS protocol: "MCMC," in which each respondent recruits only one person (with replacement), chosen from that person's friend list uniformly at random, i.e., it is a Markov chain Monte Carlo random walk and serves as a control case; and "DAG," as described in Sect. 4.

Figures 1–5 show some of our results. Here we run a series of tests, analogous to those Gile and Handcock [GH10]. All tests shown used a seed size of six. The first three figures show the effects of drawing seeds from the entire population, just the infected population, and just the uninfected population, respectively. Together, they show the effects of recruitment bias on the performance of the estimators.

Additionally, we consider *burn-in*, a feature of most MCMC-based sampling in which a fraction of the earliest samples are dropped, because they more heavily depend on the seeds—and are thus more biased—than the later samples, which are ideally independent of the seeds. The last two figures show the effects of recruitment bias after a burn-in of the first 100 samples.

The parameters considered within each figure are the network sizes 1000, 715, and 525 and the activity ratios (labeled "w") 1.1 and 3.

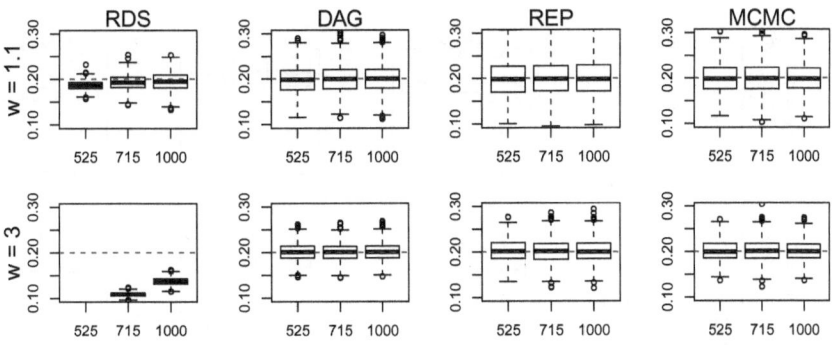

Fig. 1. Estimated size of infected population where seeds are drawn from the entire population with no burn-in

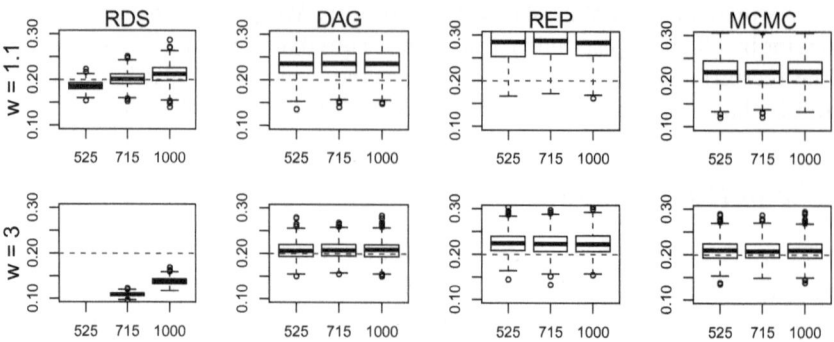

Fig. 2. Estimated size of infected population where seeds are drawn from the infected population only with no burn-in

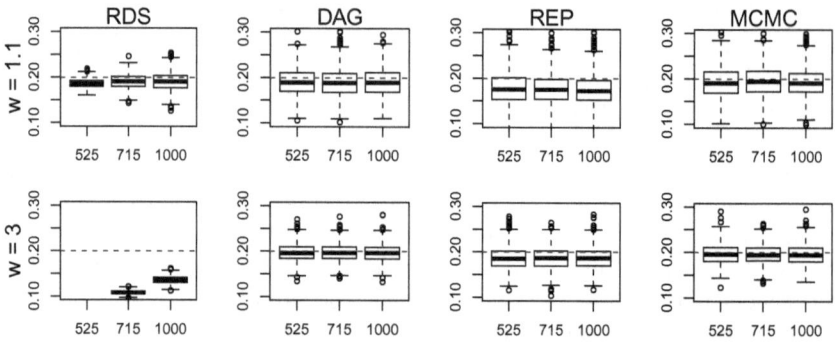

Fig. 3. Estimated size of infected population where seeds are drawn from the noninfected population only with no burn-in

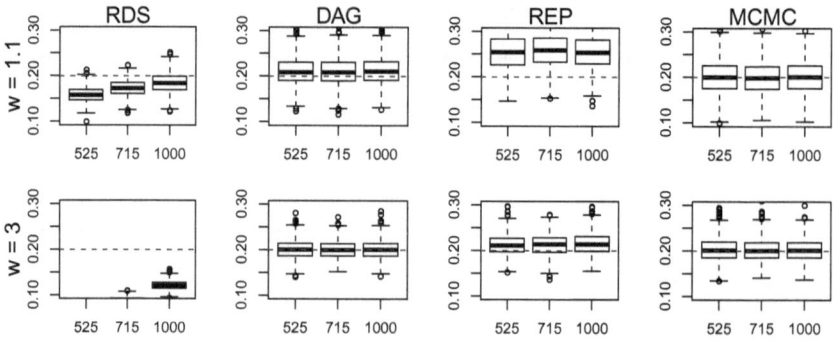

Fig. 4. Estimated size of infected population where seeds are drawn from the infected population only with the first 100 nodes of each sample are discarded as "burn-in"

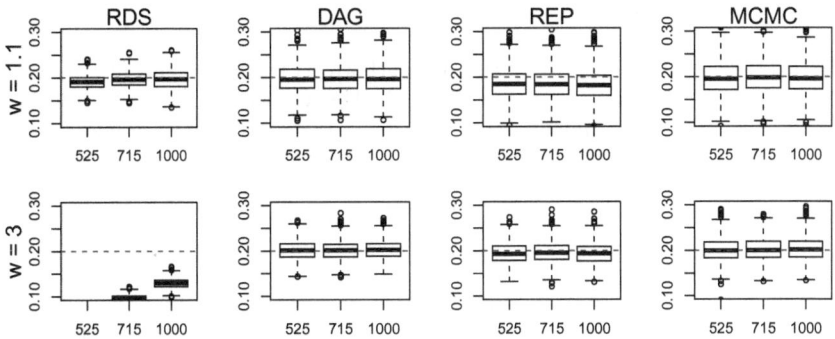

Fig. 5. Estimated size of infected population where seeds are drawn from the non-infected population only with the first 100 nodes of each sample are discarded as "burn-in"

6   Discussion and Conclusion

The results for RDS and REP essentially replicate for comparison purposes those of Gile and Handcock. One reason RDS performance degrades so dramatically as network size decreases is that the probability that any node is sampled approaches one as the network size decreases, but the VH estimator still weighs each sample as if it had been chosen in proportion to its network neighborhood.

Of all the protocols we test, MCMC performs best, which is what we would expect as it represents RDS in the impractical case when all the VH assumptions hold. Surprisingly to us, DAG was clearly second best, outperforming even REP, the protocol which seemed to us to be the most like MCMC (note that both REP and MCMC produce arbitrary recruitment networks). The only test in which DAG did not perform at a level comparable to MCMC was when all seed nodes were drawn from the infected population and the activity ratio was low, though a 100 node burn-in almost corrects this. We are investigating why DAG performs as well as it has. Space prevents us from giving details, but we have seen that the recruitment graphs created by DAG have clustering coefficients and average path lengths that are closer than the other protocols to MCMC.

We hope that this study shows that creative thinking about how RDS is implemented in OSNs may lead to significant improvements in its sampling accuracy. We have ideas about how human-computer inferace methods on OSNs can improve neighborhood size recall and the randomness of the recruitment process, neither of which we have space to discuss here. Additional open issues remain, such as the inherent biases of OSNs and the degree of realism that the ERGM models used here and in related work provide. In future work we plan to conduct field studies of these issues and others, using an actual implementation of RDS over Facebook.

References

[AQHM+06]   Abdul-Quader, A.S., Heckathorn, D.D., McKnight, C., Bramson, H., Nemeth, C., Sabin, K., Gallagher, K., Des Jarlais, D.C.: Effectiveness of respondent-driven sampling for recruiting drug users in New York City: findings from a pilot study. Journal of Urban Health 83(3), 459–476 (2006)

[BSPar]   Bernhardt, A., Spiller, M., Polson, D.: All work and no pay: Violations of employment and labor laws in Chicago, Los Angeles and New York City. Social Forces (to appear)

[GH10]   Gile, K., Handcock, M.: Respondent-driven sampling: an assessment of current methodology. Sociological Methodology 40(1), 285–327 (2010)

[GH11]   Gile, K.J., Handcock, M.S.: Network model-assisted inference from respondent-driven sampling data. arXiv preprint arXiv:1108.0298 (2011)

[Gil11]   Gile, K.J.: Improved inference for respondent-driven sampling data with application to HIV prevalence estimation. Journal of the American Statistical Association 106(493), 135–146 (2011)

[GJS12]   Gile, K.J., Johnston, L.G., Salganik, M.J.: Diagnostics for respondent-driven sampling. arXiv preprint arXiv:1209.6254 (2012)

[GKBM11]   Gjoka, M., Kurant, M., Butts, C., Markopoulou, A.: A walk in facebook: Uniform sampling of users in online social networks. Technical Report 0906.0060v4, arXiv (February 2011)

[GS10]   Goel, S., Salganik, M.J.: Assessing respondent-driven sampling. Proceedings of the National Academy of Sciences of the United States of America 107(1515), 6743–6747 (2010)

[Hec97]   Heckathorn, D.: Respondent-driven sampling: A new approach to the study of hidden populations. Social Problems 44(2), 174–199 (1997)

[Hec07]   Heckathorn, D.: Extensions of respondent-driven sampling: Analyzing continuous variables and controlling for differential recruitment. Sociological Methodology 37(1), 151–207 (2007) (in press)

[HG10]   Handcock, M.S., Gile, K.J.: Modeling social networks from sampled data. The Annals of Applied Statistics 4(1), 5–25 (2010)

[HJ01]   Heckathorn, D., Jeffri, J.: Finding the beat: Using respondent-driven sampling to study jazz musicians. Poetics 28(4), 307–329 (2001)

[MJK+08]   Malekinejad, M., Johnston, L.G., Kendall, C., Kerr, L.R.F.S., Rifkin, M.R., Rutherford, G.W.: Using respondent-driven sampling methodology for HIV biological and behavioral surveillance in international settings: a systematic review. AIDS and Behavior 12, 105–130 (2008)

[SH04]   Salganik, M.J., Heckathorn, D.D.: Sampling and estimation in hidden populations using respondent-driven sampling. Sociological methodology 34(1), 193–240 (2004)

[TG11]   Tomas, A., Gile, K.J.: The effect of differential recruitment, non-response and non-recruitment on estimators for respondent-driven sampling. Electronic Journal of Statistics 5, 899–934 (2011)

[Tho92]   Thompson, S.: Sampling. John Wiley & Sons, New York (1992)

[VH08]   Volz, E., Heckathorn, D.: Probability based estimation theory for respondent driven sampling. Journal of Official Statistics 24(1), 79–97 (2008)

[Wej09]   Wejnert, C.: An empirical test of respondent-driven sampling: Point estimates, variance, degree measures, and out-of-equilibrium data. Sociological Methodology 39(1), 73–116 (2009)

[WH08]   Wejnert, C., Heckathorn, D.: Web-based network sampling: Efficiency and efficacy of respondent-driven sampling for online research. Sociological Methods & Research 37(1), 105–134 (2008)

[WS02]   Walters, K., Simoni, J.: Health survey of two-spirited native americans (2002)

Trade-Offs in Social and Behavioral Modeling in Mobile Networks

Yaniv Altshuler[1], Michael Fire[2], Nadav Aharony[1], Zeev Volkovich[3], Yuval Elovici[2],
and Alex (Sandy) Pentland[1]

[1] MIT Media Lab
{yanival,shmueli,sandy}@media.mit.edu
[2] Deutsche Telekom Lab,
Department of Information Systems Engineering, Ben-Gurion University
{mickyfi,elovici}@bgu.ac.il
[3] Department of Software Engineering, Ort Braude College
vlvolkov@braude.ac.il

Abstract. Mobile phones are quickly becoming the primary source for social, behavioral, and environmental sensing and data collection. Today's smartphones are equipped with increasingly more sensors and accessible data types that enable the collection of literally dozens of signals related to the phone, its user, and its environment. A great deal of research effort in academia and industry is put into mining this raw data for higher level sense-making, such as understanding user context, inferring social networks, learning individual features, and behavior prediction. In this work we investigate the properties of learning and inferences of real world data collected via mobile phones. In particular, we look at the dynamic learning process over time with various sizes of sampling groups and examine the interplay between these two parameters. We validate our model using extensive simulations carried out using the "*Friends and Family*" dataset which contains rich data signals gathered from the smartphones of 140 adult members of a young-family residential community for over a year and is one of the most comprehensive mobile phone datasets gathered in academia to date.

Keywords: Machine Learning, Social Networks, Mobile Networks.

1 Introduction

Mobile phones, and increasingly smartphones, have become an integral part of many people's everyday lives. Users carry their smartphone almost everywhere and use it in order to perform many of their day-to-day communication and activities.

The pervasiveness of mobile phones has made them popular scientific data collection tools, as social and behavioral sensors of location, proximity, communications and context. Eagle and Pentland [1] coined the term ``Reality Mining'' to describe collections of sensor data pertaining to human social behavior. While existing work has demonstrated results for modeling and inference of social

A.M. Greenberg, W.G. Kennedy, and N.D. Bos (Eds.): SBP 2013, LNCS 7812, pp. 412–423, 2013.
© Springer-Verlag Berlin Heidelberg 2013

network structure and personal information out of mobile phone data, most are still mainly proofs of concept in a nascent field. The work of the "data scientist" is still that of an artisan, using personal experience, insight, and sometimes a "gut feeling" in order to extract meaning out of the plethora of data and noise.

As the field of computational social science matures, there is need for more structured methodology that would assist the researcher or practitioner in designing data collection campaigns, understanding the potential of collected datasets and estimating the accuracy limits of current analysis strategy vs. alternative ones. Such a methodology would facilitate the process of maturing from a field of "craft" into a field of science.

In this work, we present a first step in this direction. Specifically, we investigate the learning and prediction of social and individual models from raw phone-sensed data. We focus on social ties and individual descriptors that can be tied to social affiliation and affinity. For these prediction tasks, we look at the trade-off between the time period the data is collected in and the number of people that form the sample group. To do this, we use the *Friends and Family* dataset which contains rich data signals gathered from the smartphones of 140 adult members of a young-family residential community for over a year[2], in addition to self-reported personal and social-tie information. Preliminary results were described in [37] and [38].

We first build classifiers for predicting personal properties such as nationality or gender. We then proceeded to predict more complicated social links such as the subject's life-partner or "significant other". We demonstrated characteristics of the incremental learning of multiple social and individual properties from raw sensing data collected from mobile phones, as the information is accumulated over time, or alternatively, as the sample size is increased. We study the interplay between the change in time and the growth in sample size and present preliminary results that indicate that such a trade-off exists and that it reflects the network effect of the domain.

2   Related Work

In recent years, the social sciences have been undergoing a digital revolution, heralded by the emerging field of "computational social science". Lazer, Pentland, et al., [3] describe the potential of computational social sciences to increase our knowledge of individuals and groups with an unprecedented breadth, depth, and scale. Computational social sciences combine the leading techniques from network sciences [4-6] with new machine learning and pattern recognition tools specialized for the understanding of people's behavior and social interactions [7].

2.1   Mobile Phones as Social Sensors

The pervasiveness of mobile phones the world over has made them a premier data collection tool of choice and they are increasingly used as social and behavioral sensors of location, proximity, communications and context. Eagle and Pentland[1] coined the term "*Reality Mining*" to describe the collection of sensor data pertaining to human social behavior. They show that by using call records, cellular-tower IDs and Bluetooth proximity logs collected via mobile phones at the individual level, the subjects' regular patterns in daily activity can be accurately detected[1, 7]. Furthermore, mobile phone records from telecommunications companies have proven to be quite valuable in uncovering human level insights. For example, Gonzales et al. illustrate how cell-tower location information can be used to characterize human mobility, based on the observation that humans follow simple reproducible mobility patterns[8]. This approach has already expanded beyond academia with companies like Sense Networks [9] employing such tools in the commercial world to understand customer churn, enhance targeted advertisements, offer improved personalization and many other services.

2.2   Individual Based Data Collection

On one hand, data gathered through service providers includes information on a very large numbers of subjects, but on the other hand, this information is constrained to a specific domain (email messages, financial transactions, etc.) and there is very little, if any, contextual information on the subjects themselves. The alternative approach of gathering data at the individual level allows collection of many more dimensions related to the end user which are many times not available at the operator level. Madan et al.[10] follow up on Eagle and Pentland's work [1] and show that mobile social sensing can be used measure and predict the health status of individuals based on mobility and communication patterns. They also investigate the spread of political opinion within a community [11]. Other examples for using mobile phones for individual-based social sensing are those by Montoliu et al. [12], Lu et al. [13], and projects coming out of the CENS center, e.g., Campaignr by Joki et al. [14] as well as additional works as described in [15]. Finally, the *Friends and Family study*, which our paper uses as its data source, is probably the richest mobile phone data collection initiative to date with regards to the number of signals collected, study duration, and the number of subjects. The technical advancements in mobile phone platforms and the availability of mobile software development kits (SDKs) to any developer makes the collection of Reality Mining types of data easier to collect than ever before. Analysis of the security aspects of this trend can be found in [36].

In addition to mobile phones, a notable example for wearable sensor-based social data collection initiative is the *Sociometric Badge* by Olguin et al., capturing human activity and socialization patterns via a wearable sensor badge, that has been used mostly for in organizational settings [16]. The results of our work are applicable to these types of studies as well.

2.3   Learning and Prediction of Social and Individual Information

Many studies which involve predicting individual traits and social ties have been conducted in the recent years within the general context of social networking. Relevant works have been published by Liben-Nowell and Kleinberg [17],

Mislove [18] and Rokach et. al. [19], combining machine learning algorithms with social network data in order to build classifiers.

3   Data Collection and Analysis Methods

3.1   Friends and Family Dataset

To the best of our knowledge, the dataset generated from this study is probably the largest and richest dataset ever collected on a residential community to date. The accumulated size of the database files uploaded from the study phone devices adds up to *over 60 Gigabytes*. The data is composed of over *30 million* individual scan events (for all signals combined), where some events capture multiple data signals. Just as example, the dataset includes:

- 20 million wifi scans, which in turn accumulated 243 million total scanned device records.
- 5 million Bluetooth proximity scans, which in turn accumulated 16 million total scanned device records.
- 200,000 phone calls.
- 100,000 text messages (SMS).

In the analysis presented in this paper we give special focus to the data that was collected in November 2010 and April 2011, after the mobile platform was improved, new features, such as different call types where added, and several hardware problems where fixed. These two months were without a major holiday break in the academic schedule of the university and the bulk of participants were physically on campus.

In addition to the phone-based data, the study contained personal information on each participant, such as age, gender, religion, origin, current and previous income status, ethnicity, marriage information, and more.

3.2   Machine Learning Predictions

In order to evaluate learning over time, which is the main goal of our current work, we needed a set of learning and prediction models to work with. These are mostly illustrative models which enable us to conduct our main analysis. In order to achieve our final goal of predicting participants' personal and social information, we utilized two approaches; a machine learning approach, described in this section and a social network based prediction approach, described in the following section.

The first step in applying the machine learning methodology is to create feature vectors for each participant in the study. Each feature vector contains information on the participant's communication and phone usage patterns as they were collected during the study.

In order to cope with the huge amount of data collected during the study, we developed a code using C# and *Python's NetworkX* library [21]. Our code parsed the collected data and extracted feature vectors for each participant. We extracted 32 different features within a specified time interval. Namely, we collected the following features for each participant:

- **Internet usage features:** we calculated the number of distinct searches performed using the phone's browser and the number distinct bookmarks saved by the user.
- **Calls pattern features:** we computed the total number of calls, the number of unique phone numbers each user was in contact with and the total duration of all calls. We had also calculated the number of incoming/outgoing/missed calls and the total call duration, per call type.
- **SMS messages pattern features:** we computed the total number of SMS messages, the number of unique phone numbers each participant connected with via SMS and the of total incoming/outgoing SMS messages.
- **Phone applications related features:** we counted the number of applications installed and uninstalled on each device. We also computed the total number of currently running applications (originally sampled every 30 seconds).
- **Alarm features:** we counted the number of alarm-clock alarms and the number of "snooze" presses for each participant that used our alarm clock app.
- **Location features:** we calculated the number of different cellular cell tower ids and the number of different wifi network names seen by the smartphone. These features act as a rough indication of the number of different locations a participant visited during the time period.

Our next step was to extract all participant features for different time intervals. Using the extracted features, we can build different classifiers that are able to predict the participants' personal information. We used the *WEKA* software [22] in order to test different machine learning algorithms. In our experiments, we evaluated a number of popular learning methods: *WEKA*'s C4.5 decision trees, Naive-Bayes, Rotation-Forest, Random-Forest, and AdaBoostM1. Each classifier was evaluated using the 10-fold cross validation approach and in order to compare results between different classification algorithms, we used each classifier's Area Under Curve or AUC measure (also referred to as ROC Area) and F-measure results. In order to obtain an indication of the usefulness of various features, we analyzed their importance using *WEKA*'s information gain attribute selection algorithm.

Using the machine learning approach we built five different classifiers which predict the following: (1) the gender of the participant, (2) whether the participant is a student or not, (3) whether the participant has children or not, (4) whether the participant is above the age of 30, and (5) whether the participant is a native US citizen or not.

3.3 Social Network Predictions

Another method for predicting a participant's personal information details is by using the participants' different social networks. Using the data collected in the study, we can span different types of social networks between the participants according to different interaction modalities. Namely, we can define the following social networks:

- **SMS Social Network:** we constructed the community's SMS messages social network as a weighted graph $G_s = <V_s, E_s>$ according to the SMS messages the participants sent. Each weighted link $e = (u, v, w) \in E_s$ in this social network represents connections between two different phone numbers u, v ∈ V, while w is the strength of the link defined as the number of SMS message sent between the two phone numbers.

- **Bluetooth Social Network:** we constructed a weighted network graph $G_s = <V_B, E_B>$ of face-to-face interactions according to information collected about nearby Bluetooth devices. Each link $(u, v, w) \in E_s$ in this social network represent the fact that the two devices $u, v \in V_B$ encountered each other at least once, while w is the strength of the link, defined as the number of times the two devices meet.

- **Calls Social Network:** similar to the SMS social network, we can construct a network based on the participant's call graph $G_C = <V_C, E_C>$ according to the participants` phone calls. In this social network, each link $(u, v, w) \in E_C$ represents the fact that at least one call was made between two different phone numbers $,u, v \in V_C$, while w is the strength of the link, defined as the number of calls between u and v.

By using the social networks defined above, together with different graph theory algorithms, we can predict different types of personal and social information. In order to predict the participants' significant other we analyzed the Bluetooth social network. We predicted that each participant's significant other is the person that the participant spent the most time with during the measured interval. Namely, let $u \in V_B$ then:

$$significant\text{-}other(u)) = \{v | (u, v, w) \in E_B \text{ and } \forall (u`, v`, w`) \in E_b \ w > w`\}$$

In order to predict the subjects' ethnicity we used the SMS social network, using the *Louvain* algorithm for community detection [23], which separates the graph into disjoint groups.

For each iteration we assumed that we had information on the ethnicity of at least some of the nodes. We generated an ethnicity prediction for the members of each detected community based on the ethnicity of the majority of known nodes in that community (see more details in [24] or [18]).

3.4   Prediction Accuracy Evolution over Time

The goal of this work is to study and analyze the trade-off between the increased time given for the learning process of personal features and behavioral properties, and an increase in the sample size. For this analysis, we worry less about the specific learned models and their generalizability, and more about using them to study and benchmark the evolution of the learning process as the data accumulates. Understanding this process is of significant importance to researchers in a variety of fields, as it would provide an approximation of the amount of time that is needed in order to "learn" these features for some given accuracy, or alternatively, the level of accuracy that can be obtained for a given duration of time.

To generate the models presented here, we collected the performance results of the system using many combinations of learning time X sample sizes. Each predictor/classifier was executed on data gathered between November 1th and November 30th, 2010. Starting from an input of a single day (November 1st), in each consecutive execution, another day of data was added to the input so that iteration #1 was on data from November 1st, execution #2 had input of data for two days, November 1 and 2 together, and so on until the accumulation of 30 days in which the classifier ran on data from the entire month of November. This process was repeated for varying sizes of the community, starting from 2% of the users to 90%.

4   Results

Following are several examples of the evolution of the learning process of personal and social traits from mobile phone over time (namely, for growing segments of time used for collecting the learned data):

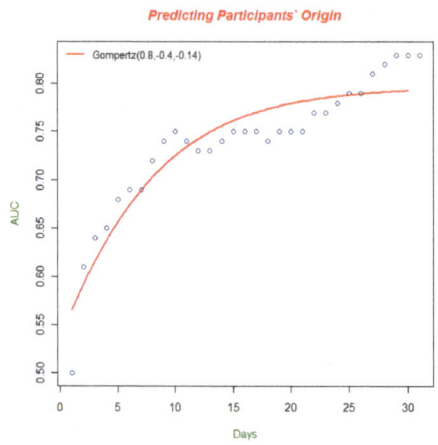

Fig. 1. Participants' origin Naïve-Bayes classifiers AUC results

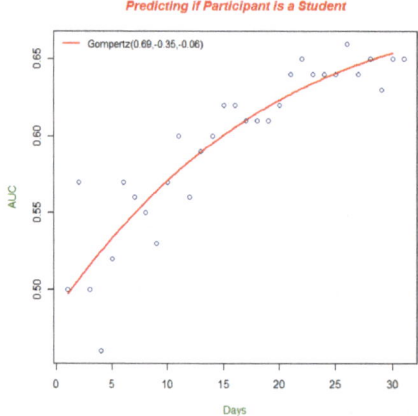

Fig. 2. Predicting if the Participant is a student over time: Rotation-Forest Classifier AUC results

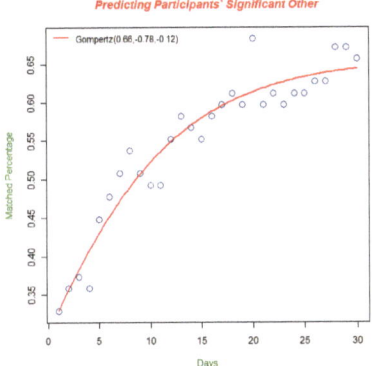

Fig. 3. Predicting significant other over time (the node with the maximum strength)

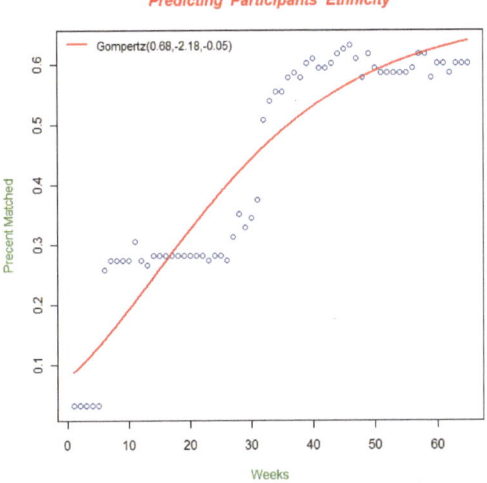

Fig. 4. Predicting ethnicity using SMS social network over time (65 weeks) using the Louvain Algorithm

The following charts illustrate the improvement in the prediction performance, as a function of the sample size (namely, the number of people whose data is used for the learning process):

After examining the way each of the properties affects the learning process separately, we can now examine the trade-off between an increase in the learning time and an increased sample size. The following charts represent the performance of the learning process (heights) as a function of the number of people of the sampling size (the longer axes) and the time in days. Notice that using the following charts the allocation of data collection and analysis resources can be empirically optimized. For example, given a data collection experiment that has X participants and that is going to last T days, the gradient of the appropriate $f(X,T)$ (corresponding to the trait that is to be analyzed) can be numerically calculated, hinting on the best use of any additional resources (i.e. shall we recruit additional participants, or increase the duration of the experiment).

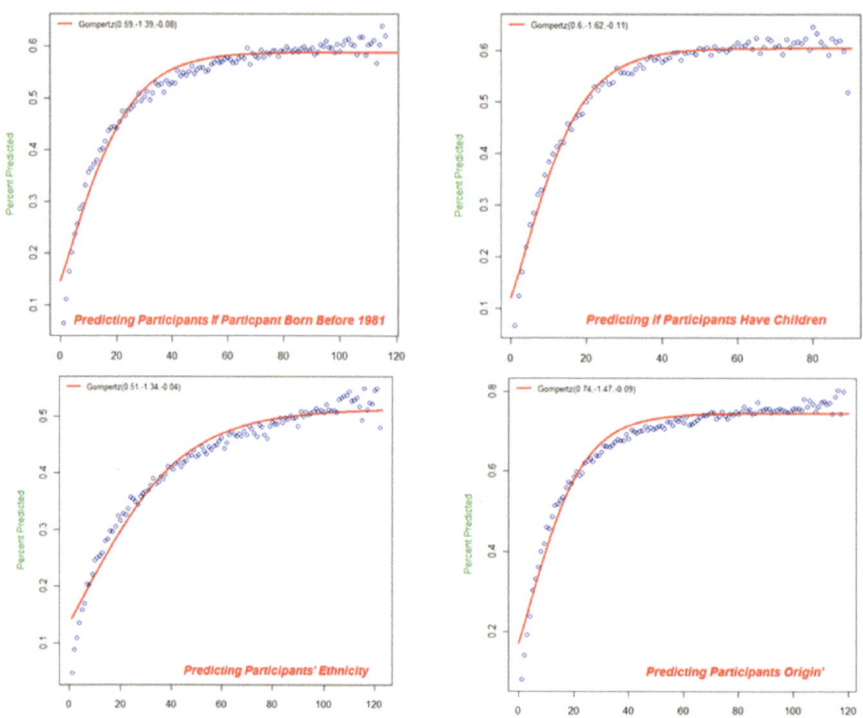

Fig. 5. The evolution of the learning process with the growth of the sample size (number of users as the X axes)

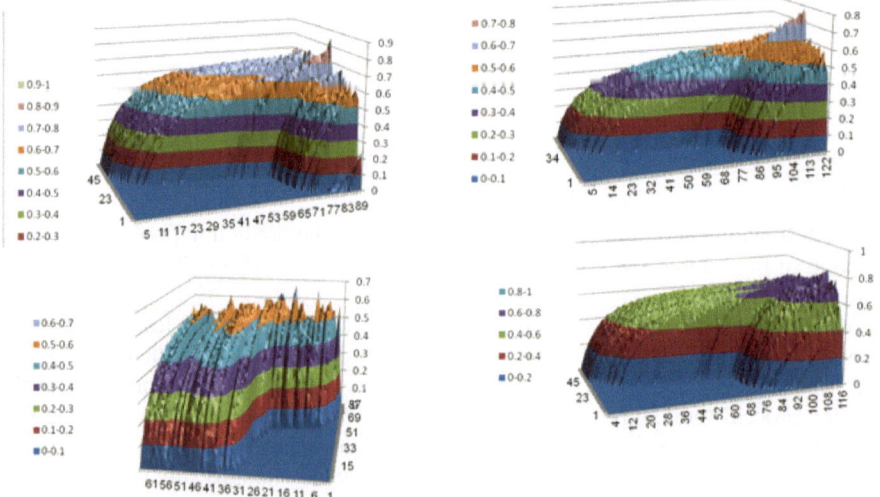

Fig. 6. An illustration of the trade-off in the learning performance of a mobile social network (height of the function) between the time axis (from 1 to 61 days) and the number of people in a sample group (from 1 to 120). These four charts correspond to the learning of age (top-left), parenting (top-right), ethnicity (bottom-left) and religion (bottom-right).

5 Conclusions

While there will always be the need for the expert and experienced "data artisan", the exponential increase in accumulated data and the rise of a big-data ecosystem creates an imperative need to design more accurate science and engineering of data collection, processing, and analysis. Our work is a building block towards this larger goal.

In this work we have discussed the trade-off between increased time (and data over time) and increased sample size, regarding the learning of personal features of mobile network users. For this, we have studied a comprehensive dataset of a mobile network containing phone calls, SMSs, and web activity, as well as self-reporting friendship queries.

We demonstrated the characteristics of incremental learning of multiple social and individual properties from raw sensing data collected from mobile phones as the information is accumulated over time. We then did the same for a systemic increase in the sample size used by the learning process. This was done through the use of state of the art techniques for machine learning using all the possible scenarios in terms of learning time and sample size.

We have presented the results of this analysis that hints to an inherent trade-off which is likely to be dominated by the "network-effect" of the domain the data is taken from. In the future stages of this research we intend to develop a mathematical model for the evolution of the learning process (with time and sample size) as well as suggest a mathematical model for the correlation and trade-off between the two.

In addition, it would be interesting to examine the correlation between the efficiency of the learning process and the way we select the identity of the network members whose activities we monitor. Crafting an optimal deployment scheme for monitoring agents throughout the network might provide an additional increase in the convergence rate of the learning mechanism, significantly enhancing our ability to achieve reliable predictions using a given amount of network access. For example, should we aim for symmetric deployment of our monitoring resources? Preliminary results [39] hint though that this might not necessarily be the case.

References

1. Eagle, N., Pentland, A.: Reality Mining: Sensing Complex Social Systems. Personal and Ubiquitous Computing 10, 255–268 (2006)
2. Aharony, N., et al.: Social fMRI: Investigating and shaping social mechanisms in the real world. In: Pervasive and Mobile Computing (2011)
3. Lazer, D., et al.: Life in the network: the coming age of computational social science. Science 323, 721 (2009)
4. Barabasiand, A.-L., Albert, R.: Emergence of scaling in random networks. Science (1999)
5. Newman, M.E.J.: The structure and function of complex networks.
6. Watts, D.J., Strogatz, S.H.: Collective dynamics of 'small-world' networks. Nature (1998)

7. Eagle, N., Pentland, A., Lazer, D.: From the Cover: Inferring friendship network structure by using mobile phone data. Proceedings of The National Academy of Sciences 106(36), 15274–15278 (2009)
8. Gonzalez, M.C., Hidalgo, A., Barabasi, A.-L.: Understanding individual human mobility patterns. Nature (2008)
9. Networks, S.: http://www.sensenetworks.com/
10. Madan, A., et al.: Social sensing for epidemiological behavior change. In: Ubiquitous Computing/Handheld and Ubiquitous Computing, pp. 291–300 (2010)
11. Madan, A., Farrahi, K., Gatica-Perez, D.: Pervasive Sensing to Model Political Opinions in Face-to-Face Networks (2011)
12. Montoliu, R., Gatica-Perez, D.: Discovering human places of interest from multimodal mobile phone data, pp. 1–10 (2010)
13. Lu, H., et al.: The Jigsaw continuous sensing engine for mobile phone applications, in Conference on Embedded Networked Sensor Systems, pp. 71–84 (2010)
14. Joki, A., Burke, J.A., Estrin, D.: Campaignr: A Framework for Participatory Data Collection on Mobile Phones (2007)
15. Abdelzaher, T.F., et al.: Mobiscopes for Human Spaces. IEEE Pervasive Computing 6(2), 20–29 (2007)
16. Olguín, D.O., et al.: Sensible Organizations: Technology and Methodology for Automatically Measuring Organizational Behavior. IEEE Transactions on Systems, Man, and Cybernetics 39(1), 43–55 (2009)
17. Liben-Nowell, D., Kleinberg, J.: The link-prediction problem for social networks. Journal of the American Society for Information Science and Technology 58(7), 1019–1031 (2007)
18. Mislove, A., et al.: You are who you know: inferring user profiles in online social networks. In: Web Search and Data Mining, pp. 251–260 (2010)
19. Rokach, L., et al.: Who is going to win the next Association for the Advancement of Artificial Intelligence Fellowship Award? Evaluating researchers by mining bibliographic data. Journal of the American Society for Information Science and Technology (2011)
20. Funf. Funf Project, http://funf.media.mit.edu
21. Hagberg, A.A., Schult, D.A., Swart, P.J.: Exploring Network Structure, Dynamics, and Function using Network X (2008)
22. Hall, M., et al.: The WEKA data mining software: an update. Sigkdd Explorations 11(1), 10–18 (2009)
23. Blondel, V.D., et al.: Fast unfolding of communities in large networks. Journal of Statistical Mechanics: Theory and Experiment 10 (2008)
24. Xie, J., Szymanski, B.K.: Community Detection Using A Neighborhood Strength Driven Label Propagation Algorithm. Computing Research Repository, abs/1105.3 (2011)
25. Rouvinen, P.: Diffusion of digital mobile telephony: Are developing countries different? Telecommunications Policy 30(1), 46–63 (2006)
26. Erickson, G.M.: Tyrannosaur Life Tables: An Example of Nonavian Dinosaur Population Biology. Science 313(5784), 213–217 (2006)
27. Donofrio, A.: A general framework for modeling tumor-immune system competition and immunotherapy: Mathematical analysis and biomedical inferences. Physica D-nonlinear Phenomena 208(3-4), 220–235 (2005)
28. Pan, W., Aharony, N., Pentland, A.: Composite Social Network for Predicting Mobile Apps Installation. In: Intelligence, AAAI 2011, San Francisco, CA (2011)
29. Krishnamurthy, B., Wills, C.E.: On the leakage of personally identifiable information via online social networks. Computer Communication Review 40(1), 7–12 (2009)

30. Binde, B.E., McRee, R., O'Connor, T.J.: Assessing Outbound Traffic to Uncover Advanced Persistent Threat, Sans Institute (2011)
31. Solutionary, White Paper: The Advanced Persistent Threat, APT (2011)
32. Brunner, M., et al.: Infiltrating Critical Infrastructures with Next-Generation Attacks. Fraunhofer-Institute for Secure Information Technology SIT Munich (2010)
33. Kalmijn, M.: Intermarriage and Homogamy: Causes, Patterns, Trends. Annual Review of Sociology 24(1), 395–421 (1998)
34. McPherson, M., Smith-Lovin, L., Cook, J.M.: Birds of a Feather: Homophily in Social Networks. Annual Review of Sociology 27(1), 415–444 (2001)
35. Dey, A.K., et al.: Getting Closer: an Empirical Investigation of the Proximity of Users to Their Smart Phones. In: Proc. of the 13th International Conference on Ubiquitous Computing, pp. 163–172 (2011)
36. Altshuler, Y., Aharony, N., Elovici, Y., Pentland, A., Cebrian, M.: Stealing Reality: When Criminals Become Data Scientists (or Vice Versa). IEEE Intelligent Systems 26(6), 22–30 (2011)
37. Altshuler, Y., Fire, M., Aharony, N., Elovici, Y., Pentland, A(S.): How Many Makes a Crowd? On the Evolution of Learning as a Factor of Community Coverage. In: Yang, S.J., Greenberg, A.M., Endsley, M. (eds.) SBP 2012. LNCS, vol. 7227, pp. 43–52. Springer, Heidelberg (2012)
38. Altshuler, Y., Fire, M., Aharony, N., Elovici, Y., Pentland, A.: Incremental Learning with Accuracy Prediction of Social and Individual Properties from Mobile-Phone Data, Arxiv preprint arXiv:1111.4645 (2011)
39. Altshuler, Y., Wagner, I.A., Bruckstein, A.M.: On Swarm Optimality in Dynamic and Symmetric Environments. Economics 7, 11–18 (2008)

Privacy Protection in Personalized Web Search: A Peer Group-Based Approach

Bin Zhou[1] and Jian Xu[2]

[1] University of Maryland, Baltimore County, USA
Department of Information Systems
bzhou@umbc.edu
[2] Yahoo! Labs, Beijing
xujian@yahoo-inc.com

Abstract. Privacy protection in web search engines is becoming more and more serious in recent days. In this paper, we study the problem of privacy protection in web search, with a special focus on IP-address based personalized web search. Our goal is to break the linkage between users' identities (e.g., IP address) and their issued queries so as to prevent privacy breaches. Our privacy model, which shares similar characteristics of l-diversity in privacy preserving data publishing of relational data, provides a strong privacy guarantee in web search. The central idea of our privacy model is to protect user's search activities within a social peer group. A social peer group contains a set of individual users. From search engines's perspective, search queries issued by users from the same peer group cannot be uniquely linked to individuals within the group. A framework based on grouping social peer users is proposed to achieve the privacy requirement. We also provide some experimental results to show that our methods achieve high efficiency in practice.

Keywords: Privacy, Personalized Web Search, Social, Peer Group.

1 Introduction

Web search engines nowadays have become an indispensable component for millions of users to search desired information on the web. Associated with this, however, are the increasing concerns that these search engine companies gather tremendous amounts of users' personal information. Although such information can be used to provide personalized web search which improves the accuracy of search results greatly, the intensive usage of users' personal information in web search engines also raises terrifying privacy threats to users.

Generally, information related to users' search activities such as IP address, search queries, and click-through data are all captured and maintained by web search engines using search logs. A few real-life examples (e.g., IP tracer [18] and AOL log data release [16]) indicate that detailed user profiles can be constructed from search logs. In such a case, search engine companies have to be trusted to not abuse their privileges, which may not always be desirable.

A.M. Greenberg, W.G. Kennedy, and N.D. Bos (Eds.): SBP 2013, LNCS 7812, pp. 424–432, 2013.

Privacy breach in web search engines has introduced more and more threats to individuals. There is an extremely high demand of effective privacy protection mechanisms in web search. Generally, there are two major questions related to privacy breach in web search. Firstly, who issued the search query? Secondly, what is the search query about? From the individual's point of view, if the answers to each of the two questions are identified by some malicious attackers, there is no big privacy concern. However, if a strong linkage between a user who issued the search query and the content of the search query is uniquely identified, the user's search activity is undoubtedly under risk.

Clearly, the linkage between a user and his/her search query should be well protected. If even the search engine companies could not correctly recover the linkage, user's privacy is strongly protected. In this paper, we focus on breaking the linkage between users' identities (e.g., IP address) and their issued queries so as to prevent private information to be disclosed by any parties. Several existing studies (e.g., onion routing [4] or anonymous proxy [17]) focus on hiding the true IP address of the users who issued a query. However, these methods cannot provide personalized searches which require the original IP address. For example, using IP address-based personalized web search, a query "weather" would return the detailed weather information for user's location as the top result. Once the true IP address is completely anonymized, the personalized search result cannot be returned. Thus, *can we develop techniques to provide strong privacy protection guarantees for search engine users without compromising the personalized search performance?*

The major contribution of our work is a novel privacy framework with guaranteed privacy protection in IP address-based personalized web search. The central idea of our privacy framework is to protect user's search activities within a social peer group. A peer group represents a social group of individuals who share similarities. The queries from the same peer group will be submitted to web search engines together. From search engines's perspective, search queries issued by users from the same peer group cannot be uniquely linked to individuals within the group. This framework consists of an online peer grouping step that dynamically constructs a peer group for each user, and an information obfuscation step which protects each individual user in the crowd (i.e., a peer group). We also provide a practical privacy model which shares similar characteristics of l-diversity in privacy preserving data publishing of relational data [9] to provide a strong privacy guarantee in personalized web search.

The rest of the paper is organized as follows. We review some related studies in Section 2. We present our privacy protection framework for personalized web search in Section 3. A practical privacy model with strong privacy protection guarantee is also discussed in this section. In Section 4, we discuss some strategies to efficiently formalize peer groups which serve as core foundations of our proposed privacy protection framework. A systematic empirical study conducted on the AOL search log data set is reported in Section 5. Section 6 concludes the paper.

2 Related Work

Privacy has become a more and more serious concern in many applications. One of the privacy related problems is publishing relational data for public use [12], which has been extensively studied in the recent years. The major objective of privacy preserving data publishing research is to hide sensitive knowledge from the data while maintaining the utility of data for various data analysis tasks [5]. Several privacy models, such as k-anonymity [14], l-diversity [9], and their variations have been proposed for the purpose of privacy protection. Other than relational data, some other types of data such as social networks and search log data also suffer from privacy breaching concerns. Recently, k-anonymity [14] and l-diversity [9] have been successfully extended to address privacy issues in social networks [19,8] and search logs [6].

Privacy in web search has also attracted much attention in the past several years. The *Private Information Retrieval* model [2] is considered to be the perfect private solution to address the privacy breaching issues in web search. However, due to its high complexity and the inability of personalized search, Private Information Retrieval does not have practical usage. Some recent studies [10,11] try to obfuscate the search query itself. Randomly generated keywords are injected into the actual query to hide the real search intent. However, these methods rely on a thesaurus for generating queries which is not practical in the web search scenario. In addition, linkages between users' identities and their queries are maintained in the search logs, which in fact still poses great privacy threats to associated individuals.

Our proposed privacy model is based on the concept of peer groups. Peer group, which represents a social group of individuals who share similarities, has been an important concept in social science research. The analysis of peer groups has been applied in many areas, such as stock analysis [7], collaborative information sharing [13], distributed computing [15] and cyber network structure [3]. However, the analysis of peer group has not been used for the purpose of privacy protection in web search.

3 Privacy Protection Framework and Privacy Model

In this section, we first discuss our framework for protecting user's privacy in personalized web search. Then, we discuss a practical privacy model with strong privacy protection guarantee in the web search scenario.

3.1 The Framework of Privacy Protection in Web Search

A web search activity usually involves interactions between a user (client) and a web search engine (server). Our methods address the problem of privacy preserving web search at the client side by formalizing a peer group for each web user.

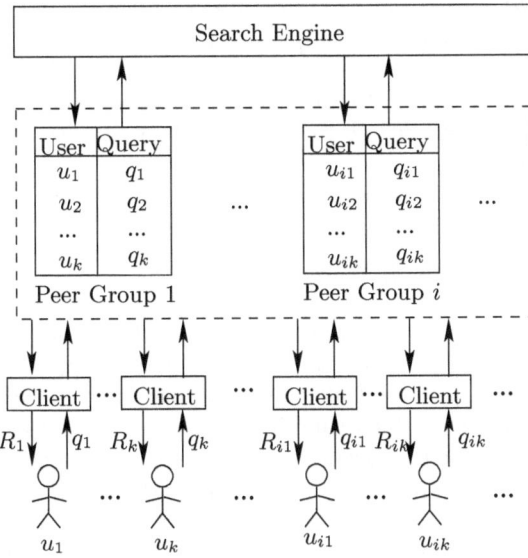

Fig. 1. The framework of privacy protection in personalized web search (u_i: user; q_i: query issued by u_i; R_i: ranked result for q_i)

Figure 1 presents the framework of our methods. While each user issues their own query as usual from their clients (e.g., web browsers), an automatic online grouping protocol will be applied to cluster users into peer groups. The queries from the same peer group will be submitted to web search engines together. The grouping protocol can be achieved by a software plug-in designed specifically for those web browsers.

While this framework does not completely hide users' actual search queries, however, from the search engine's point of view, it is equally plausible that an individual issues one of the queries in the same peer group. As a result, even the search engine cannot correctly infer which user issued what search query with 100% accuracy. Thus, the user's privacy in web search is well protected.

3.2   The Privacy Model in Web Search

According to the l-diversity model for relational data [9], if queries from the same peer group have the same search intent, a linkage is still able to be constructed between a user and a search intent. Thus, we define our privacy model by considering the diversity of queries.

Definition 1 (l-Diversity Search Privacy). *A user u_1 issuing a query q_1 has l-Diversity Search Privacy, if*

1. *The peer group G of u_1 has at least $l - 1$ other distinct users, denoted as $G = \{u_1, u_2, \ldots, u_l, \ldots\}$;*

2. Let $I(G)$ be the most frequent search intent of queries in G, thus, $\frac{I(G)}{|G|} \leq \frac{1}{l}$;

3. u_1 only appears in one peer group G at any time.

In general, l-Diversity Search Privacy has a similar property of l-diversity. If a user u satisfies l-Diversity Search Privacy, search engines could not determine a linkage between u's identity and u's search intent with a confidence higher than $\frac{1}{l}$. The larger the value of l, the stronger the privacy guarantee.

3.3 Discussions on the Privacy Framework

As the proposed privacy framework has an information obfuscation step to break the linkage between a user's identity and his/her search queries, it is necessary to consider whether this would affect the quality of web search performance. On the search engine side, each time it receives a group of user identities and search queries. If the size of a group is l, there exist l^2 different combinations of user identities and search queries. To ensure that the actual personalized search results are always generated, search engines need to conduct searches for all of these l^2 combinations. On the client side, the plug-in will only present to the users the personalized search results of the original IP address. Therefore, the quality of personalized search is not affected at all. It is interesting to see if a balance between the overheads on the search engine side and the search quality can be achieved. We leave this as a future research direction.

In the l-Diversity Search Privacy model, the grouping protocol knows all the search queries and their corresponding IP address mapping. The grouping protocol should be robust and reliable. Any mapping information should not be leaked. As an interesting future research direction, we plan to investigate practical security and encryption-based technique to enhance the security of the grouping protocol.

4   Online Construction of Peer Groups

As the formalization of peer groups serves as core foundations to protect individual's privacy in web search, an online formalization of peer groups is a necessity. Not surprisingly, in the current information era, millions of users are issuing queries to search engines at any time. An online grouping procedure needs to be conducted to form peer groups instantly. The details of the algorithms will be discussed in this section. In practice, users may stop issuing queries, and new users will start to issue queries. Thus, when a user does not satisfy the l-Diversity Search Privacy anymore, a reconstruction of peer groups is triggered automatically.

To construct peer groups online, we model users' search activities as a sequence of (u_i, q_i) pairs, where u_i is user's identification (e.g., IP address) and q_i is a query. The peer group construction problem then becomes a sequence partitioning problem such that each partition should satisfy the privacy requirements in Definition 1.

Algorithm 1. The GreedyAdd algorithm

Input: a stream of users' search queries $\mathcal{S} = \{(u_1, q_1), (u_2, q_2), \ldots, (u_i, q_i), \ldots\}$
Output: a user group $\mathcal{G}$;
1: let $G = \{u_1\}$;
2: let $pointer = 2$;
3: **while** $|G| < l$ **do**
4:     let $count = 0$;
5:     **for each** $u \in G$ **do**
6:       **if** $Sim(q, q_{pointer}) > \delta$ **then**
7:         $count = count + 1$;
8:       **end if**
9:     **end for**
10:    **if** $count = 0$ **then**
11:      let $G = G \cup \{u_{pointer}\}$;
12:      let $pointer = pointer + 1$;
13:    **end if**
14: **end while**
15: update $\mathcal{S}$;
16: **return** G;

To determine whether two queries have different search intents, a straightforward solution is to calculate a similarity score between them. We adopt a similarity measure based on the Vector Space model due to its popularity. That is, each query is regarded as a term vector. A cosine similarity is calculated to measure the similarity between two queries.

We develop a greedy solution to construct partitions from a sequence of (u_i, q_i) pairs. The major idea is to consider a variant of a traditional clustering problem: suppose n distinct users are issuing their own queries, we need to generate clusters of users (and their queries) $G = \{G_1, G_2, \ldots\}$ such that:

1. $\forall G_i \in G$, the size of G_i, denoted as $|G_i|$, satisfies $|G_i| \geq l$;
2. $\forall u_j, u_k \in G_i$, the similarity score of queries q_j, q_k issued by u_j, u_k (denoted as $Sim(q_j, q_k)$) satisfies $Sim(q_j, q_k) \leq \delta$, where δ is a parameter that determines whether two queries have different search intents.

The above problem is a variant of the *k-Gather Clustering* problem [1], which is NP-hard. However, in the web search scenario, the optimal solution is not necessary. In addition, millions of users may issue queries at the same time and new users and new queries will be issued continually. Taking the efficiency requirement into consideration, we develop the **GreedyAdd** algorithm.

The details of the **GreedyAdd** algorithm are summarized in Algorithm 1. The algorithm starts by picking the top-1 user in the sequence of (u, q) pairs. Then, it scans the remaining sequence and keeps adding users who issued diverse queries into the current peer group until the group size reaches l.

Once a peer group of l users is formalized, the queries are submitted to search engines together.

5   Experimental Results

We conduct some experiments using the well-known publicly released AOL search log data. The data set contains about 650,000 users over a 3-month period. We adopted this search log data for the simulation of users issuing queries to web search engines. Only user ids and their search queries are considered in the simulation experiment.

As discussed in Section 4, the quality of personalized web search is not affected at all using our proposed privacy protection framework. For the purpose of evaluation, one important efficiency measure we considered is the time delay for constructing the peer groups. Since a group of users and their queries are submitted together, some users who issued queries earlier may have to wait until the group is formed. To quantitatively evaluate this time delay, we use $p(u_i, q_i)$, the position of pair (u_i, q_i) in the sequence, as the time when u_i issued a query q_i. The largest position of a pair in the peer group G_j is denoted as p_{G_j}. Thus, the measure of time delay for G_j can be calculated as

$$Delay = \frac{\sum_{(u_i, q_i) \in G_j} |p(u_i, q_i) - p_{G_j}|}{|G_j|}. \tag{1}$$

Figure 2 shows the average time delay for all the users in the AOL search log data. The X-axis represents the value of l – the size of peer groups, and the Y-axis represents the delay as calculated using Equation 1. As a comparison, we also calculate the minimal time delay which refers to the case that each peer group contains a continuous set of (u_i, q_i) pairs in the sequence.

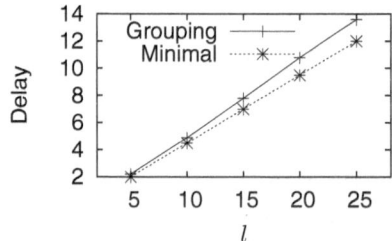

Fig. 2. The time delay in the simulation experiment

In general, the time delay due to grouping is quite small. When the value of l increases, the time delay increases as well. This is because the size of peer groups increases. Considering the fact that millions of users are issuing queries within a very short time period, the actual time delay for constructing peer groups in practice can be neglected.

6   Conclusion

In this paper, we proposed a practical privacy model for protecting user's privacy in personalized web search. The general idea is to hide individual's search activities in a social crowd. Thus, the linkages between user's identity and user's queries are disconnected.

There are several interesting future directions for our work, such as (1) how to extend the proposed privacy model to prevent privacy breaches which utilize individual's sequential search activities; (2) how to integrate users' click-through data to enhance the privacy model in web search. In addition, the proposed peer group formalization algorithm only considers whether user's queries share different search intents, it does not take into account whether users sharing similar social background should be grouped with high priority. We are also interested in exploring social profiles of users for more effective formalization of peer groups.

Acknowledgment. This research is supported in part by a UMBC Summer Faculty Fellowship grant and a UMBC Special Research Assistantship/Initiative Support grant. All opinions, findings, conclusions and recommendations in this paper are those of the authors and do not necessarily reflect the views of the funding agency.

References

1. Aggarwal, G., Feder, T., Kenthapadi, K., Khuller, S., Panigrahy, R., Thomas, D., Zhu, A.: Achieving anonymity via clustering. In: Proceedings of the 25th ACM SIGMOD-SIGACT-SIGART Symposium on Principles of Database Systems (PODS 2006), pp. 153–162. ACM, New York (2006)
2. Blundo, C.: Private information retrieval. In: Encyclopedia of Cryptography and Security, 2nd edn., pp. 974–976 (2011)
3. Bornhorst, N., Pesavento, M., Gershman, A.B.: Distributed beamforming for multiuser peer-to-peer and multi-group multicasting relay networks. In: ICASSP, pp. 2800–2803 (2011)
4. Clark, J., van Oorschot, P.C., Adams, C.: Usability of anonymous web browsing: an examination of tor interfaces and deployability. In: Proceedings of the 3rd Symposium on Usable Privacy and Security (SOUPS 2007), pp. 41–51. ACM, New York (2007)
5. Gkoulalas-Divanis, A., Verykios, V.S.: Hiding sensitive knowledge without side effects. Knowl. Inf. Syst. 20(3), 263–299 (2009)
6. Jones, R.: Privacy in web search query log mining. In: Buntine, W., Grobelnik, M., Mladenić, D., Shawe-Taylor, J. (eds.) ECML PKDD 2009, Part I. LNCS (LNAI), vol. 5781, p. 4. Springer, Heidelberg (2009)
7. Kim, Y., Sohn, S.Y.: Stock fraud detection using peer group analysis. Expert Syst. Appl. 39(10), 8986–8992 (2012)
8. Liu, K., Terzi, E.: Towards identity anonymization on graphs. In: Proceedings of the 2008 ACM SIGMOD International Conference on Management of Data (SIGMOD 2008), pp. 93–106. ACM Press, New York (2008)

9. Machanavajjhala, A., Gehrke, J., Kifer, D., Venkitasubramaniam, M.: L-diversity: Privacy beyond k-anonymity. In: Proceedings of the 22nd IEEE International Conference on Data Engineering (ICDE 2006). IEEE Computer Society, Washington, DC (2006)
10. Murugesan, M., Clifton, C.: Providing privacy through plausibly deniable search. In: Proceedings of the SIAM International Conference on Data Mining (SDM 2009), pp. 768–779. SIAM (2009)
11. Pang, H., Ding, X., Xiao, X.: Embellishing text search queries to protect user privacy. PVLDB 3(1), 598–607 (2010)
12. Samarati, P.: Protecting respondents' identities in microdata release. IEEE Transactions on Knowledge and Data Engineering (TKDE) 13(6), 1010–1027 (2001)
13. Shtykh, R.Y., Zhang, G., Jin, Q.: Peer-to-peer solution to support group collaboration and information sharing. Int. J. Pervasive Computing and Communications 1(3), 187–198 (2005)
14. Sweeney, L.: K-anonymity: a model for protecting privacy. International Journal on uncertainty, Fuzziness and Knowledge-based System 10(5), 557–570 (2002)
15. Tsuneizumi, I., Aikebaier, A., Enokido, T., Takizawa, M.: A scalable peer-to-peer group communication protocol. In: AINA, pp. 268–275 (2010)
16. http://en.wikipedia.org/wiki/AOL_search_data_leak
17. http://en.wikipedia.org/wiki/Proxy_server
18. http://www.ntia.doc.gov/legacy/ntiahome/privacy/files/smith.htm
19. Zhou, B., Pei, J.: Preserving privacy in social networks against neighborhood attacks. In: Proceedings of the 24th IEEE International Conference on Data Engineering (ICDE 2008), pp. 506–515. IEEE Computer Society, Cancun (2008)

Detecting Anomalous Behaviors Using Structural Properties of Social Networks

Yaniv Altshuler[1], Michael Fire[2], Erez Shmueli[1], Yuval Elovici[2], Alfred Bruckstein[3], Alex (Sandy) Pentland[1], and David Lazer[4]

[1] MIT Media Lab
{yanival,shmueli,sandy}@media.mit.edu
[2] Deutsche Telekom Lab, Department of Information Systems Engineering
Ben-Gurion University
{mickyfi,elovici}@bgu.ac.il
[3] Computer Science Department
Technion – Israeli Institute of Technology
freddy@cs.technion.ac.il
[4] College of Computer and Information Science & Department of Political Science
Northeastern University
d.lazer@neu.edu

Abstract. In this paper we discuss the analysis of mobile networks communication patterns in the presence of some anomalous "real world event". We argue that given limited analysis resources (namely, limited number of network edges we can analyze), it is best to select edges that are located around 'hubs' in the network, resulting in an improved ability to detect such events. We demonstrate this method using a dataset containing the call log data of 3 years from a major mobile carrier in a developed European nation.

Keywords: Mobile Networks, Anomalies Detection, Emergencies, Behavior Modeling.

1 Introduction

Analyzing the spreading of information in human networks has long been the focus in many studies of social networks [17, 20]. A main challenge in practical analysis of the way information flows between network's participants is the trade-off between the available analytic resources and the accuracy of the prediction they yield [3, 6].

Imagine a scenario where several people observe some extraordinary event, which triggers a cascading sequence of reports between "social neighbors". In this scenario, it is possible for an external observer to track the volume of network traffic, but not its content. How might that observer effectively make the inference that an extraordinary event has occurred?

This is in fact a plausible scenario, with the existence of communication systems where timing and volume of traffic is observed, but (typically) not content. Mobile phones are particularly notable in this regard, because of how pervasive they are.

A.M. Greenberg, W.G. Kennedy, and N.D. Bos (Eds.): SBP 2013, LNCS 7812, pp. 433–440, 2013.

Here we build on work examining detection of anomalous events in networks [10], but with the focus on how to aggregate those signals in a computationally efficient fashion. That is, if one cannot observe all nodes and edges, how best to sample the network? We argue that an efficient monitoring strategy is focusing on network edges that are located in vicinity to network hubs.

We demonstrate our approach using a comprehensive dataset, containing the entire internal calls as well as many of the incoming and outgoing calls within a major mobile carrier in a west European country, for a period of roughly 3 years. During this period that mobile users have made approximately 12 billion phone calls. We used the company's log files, providing all phone calls (initiator, recipient, duration, and timing) and SMS/MMS messages that the users exchange within and outside the company's network. The dataset also identifies the active cell towers of each call, thereby offering real time location (with tower resolution) data for each user. All personal details have been anonymized, and we have obtained IRB approval to perform research on it.

The rest of this paper contains related work that is discussed in Section 2, problem's definitions that is presented in Section 3, a demonstration of the proposed method using real world cellular data in Section 4 and concluding remarks appear in Section 5.

2   Related Work

Recent research around the use of mobile network data for detection of extraordinary events, had examined the question pertaining to the area where the event has occurred, and its nature: a bomb attack is narrowly localized in space, thus likely the anomalous calling activity will be limited to the immediate neighborhood of the event. This was observed in an analysis of mobile data in the vicinity of a bomb attack [10].

It has been recently shown that in trying to assess the societal changes and anomalous patterns that emerge in response to emergencies and security related events, it is crucial to understand the underlying social network [26, 27], the role of the link weights [16], as well as the response of the network to node and link removal [2].

Other works had examined the evolution of social groups, developing algorithms capable of identifying "new groups" – a certain kind of anomalous network pattern [23], or ways to cluster networks based on social and behavioral features [11]. In [19] the behavior and social patterns 2.5 million mobile phone users, making 810 million phone calls, were analyzed and resulted in clustering of the network to components showing striking resemblance to the geographical districts the users live in.

In the broader scope, this line of work aims for creating techniques for analyzing mobile phone data as ubiquitous and pervasive sensors networks. These techniques can be used to detect social relations [1, 13], evolving behavioral trends [7, 24], mobility patterns [15], environmental hazards [18, 25], socio-economical properties [14], and various security related features [4, 5].

Another question of interest with this regards is the specific way we utilize our sensors network. Namely, given a large number of sensors, what would be the best way to deploy them, in order to obtain maximal pervasiveness? Is the optimal deployment must always be symmetric (as some works actually argue against this approach [8, 9, 21])?

3   Problem Definitions

We denote the "global social network" as a graph $G = <V, E>$ where V is the set of all nodes and E is the set of directed edges over those nodes (an edge (u, v) exists if and only if there has been a reciprocal call between users u and v).

We assume that occasionally various anomalous events take place in the "real world", that are being directly observed by some of the network's users, that subsequently may (or may not) react by calling one of more of their friends (i.e. their neighbors in the social network).

Given a mobile carrier M, we denote its set of covered nodes (derived from its market share) as $V_M \subseteq V$, and its set of covered edges as $E_M \subseteq E$. An edge is covered by M if at least one of its nodes is covered by M, i.e. :

$$E_M = \{(u, v) | (u, v) \in E \wedge (u \in V_M \vee v \in V_M)\}$$

We assume that the operator M is interested to detect anomalous events such as emergencies, and to do so with as high accuracy rate as possible, and using as little resources as possible. We measure the amount of resources required by M as the overall number of edges being analyzed, or monitored. We denote the subset of edges processed by M as the "monitored edges", $S_M \subseteq E_M$.

Given an upper bound, ϵ on the size of $\frac{|S_M|}{|E|}$, we are interested in the highest detection performance obtainable by monitoring a portion of the edges smaller than ϵ.

Throughout this work we refer to the "1 ego-network" of a node v as the graph $G_1(V_1, E_1)$ such that V_1 contains all of the nodes u such that there exists an edge (v, u) in E, and that E_1 contains all the edges from v to the nodes of V_1. Furthermore, we denote by the "1.5 ego-network" the graph $G_{1.5}(V_{1.5}, E_{1.5})$ such that $V_{1.5} = V_1$, and that $E_{1.5} = E_1 \cup \Delta_{1.5}$ where $\Delta_{1.5}$ contains all the edges in E between nodes u_i and u_j such that $u_i, u_j \in V - v$. The definitions of 1 ego-network and the 1.5 ego-network are illustrated in Figure 1.

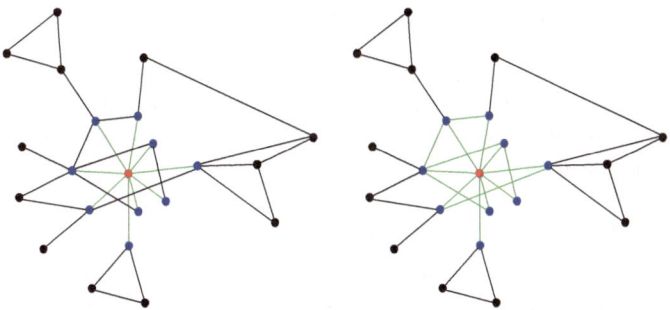

Fig. 1. An illustration of the 1 ego-network around (left chart) and the 1.5 ego-network (right chart) around the same node v (marked in red). The nodes $V_1 = V_{1.5}$ are marked in blue. The E_1 edges (left chart) and the $E_{1.5}$ edges (right chart) are marked in green.

3.1 Network Sampling

In this work we compare two alternatives for sampling a pre-defined number of network's edges. The basic method is simply to randomly select edges, with uniform distribution (this will serve as the baseline for comparison). The second method we propose in this work is to use the edges of the 1.5 ego-networks around network hubs – nodes with high traffic (either incoming or outgoing). The rationale behind the use of hubs is that hubs are highly likely to be exposed to new information, due to their high degree.

Specifically, given available resources ϵ, we select network nodes, $v_1, \ldots, v_n$, from V_M, such that those nodes have the highest degrees in V_M and the set $S_M = \bigcup_{1 \le i \le n} E_M^{1.5}(v_i)$ does not contain more than ϵ portion of the edges, where $E_M^{1.5}(v)$ denote the 1.5 ego-network around node v, that is – the edges between v and all of v's neighbors, as well as the edges between v's neighbors and themselves:

$$E_M^{1.5}(v_k) = E_M(v_k) \bigcup \{(u_1, u_2) | (u_1, u_2) \in E_M \wedge u_1 \in E(v_k) \wedge u_2 \in E(v_k)\}$$

3.2 Anomalies Detection

In order to detect anomalies in the dynamics of the social network around the network's hubs we use the Local-Outlier-Factor (LOF) anomaly detection algorithm. In other words, using the LOF algorithm for each network node we detected times where anomaly features occur. Using majority voting between all the hubs under monitoring we detected the times with the highest probability for anomalies.

We do so by ranking each day according to the number of hubs that reported it as anomalous. Then, for each day we look at the 29 days that preceded it, and calculate the final score of the day by its relative position in terms of anomaly-score within those 30 days. Namely, a day would be reported as anomalous (e.g., likely to contain some emergency) if it is "more anomalous" compared to the past month, in terms of the number of hubs-centered social networks influenced during it. Each day is given a score between 0 and 1, stating its relative "anomaly location" within its preceding 30 days.

4   Emergency Detection Using Real World Data

For evaluating our proposed monitoring method as an enhanced method for anomalies detection we have used a series of anomalous events that took place in the mobile network country, during the time where the call logs data was recorded. Figure 2 presents the events, including their "magnitude", in terms of the time-span and size of population they influenced.

We have divided the anomalies into the following three groups:

Concerts and Festivals. Events that are anomalous, but whose existence is known in advance to a large enough group of people. Those include events number 9-16, as appears in Figure 2.

"Small exposure events". Anomalous events whose existence is unforseen, and that were limited in their effect. Those include events number 1,2,5,6.

"Large exposure events". Anomalous events whose existence is unforseen, that affected a large population. Those include events number 3,4,7,8.

		Event	duration (hours)	$|G_0|$
Emergencies	1	*Bombing*	1.92	750
	2	*Plane crash*	2.17	2,104
	3	*Earthquake*	1.42	32,403
	4	*Blackout*	3.0	84,751
	5	*Jet scare*	1.67	3,556
	6	Storm 1	2.33	7,350
	7	Storm 2	2.0	14,634
	8	Storm 3	1.75	19,239
Non-emergencies	9	*Concert 1*	13.25	11,376
	10	Concert 2	6.67	3,939
	11	Concert 3	9.08	5,134
	12	Concert 4	12.08	2,630
	13	*Festival 1*	19.92	66,869
	14	Festival 2	2.17	1,453
	15	Festival 3	20.92	10,854
	16	Festival 4	11.25	3,117

Fig. 2. A detailed list of the anomalous events that were identified, including their duration (in hours) and the number of population that resided in the relevant region (denoted as G_P). Further details can be found in [10].

We rank each day between 0 and 1, according to its "anomalousness", based on the method explained above. This was done for increasingly growing number of monitored edges, in order to track the evolution of the detection accuracy. The result of this process was a series a numeric vectors pairs: $(\mathcal{V}_{BASE}, \mathcal{V}_{HUBS})_{|E|}$, corresponding to the two sampling methods used (e.g. the random network sampling for $\mathcal{V}_{BASE}$ and the hubs-sampling for $\mathcal{V}_{HUBS}$), for $|E|$ edges which were monitored. In addition, we created a binary vector $\hat{\mathcal{V}}$ having '1' for anomalous days and '0' otherwise.

For $|E|$ edges which were monitored we denote by $\delta_{|E|}$ the difference between the correlation coefficient of $\mathcal{V}_{HUBS}$ and $\hat{\mathcal{V}}$, and the correlation coefficient of $\mathcal{V}_{BASE}$ and $\hat{\mathcal{V}}$, namely :

$$\delta_{|E|} = CORR(\mathcal{V}_{HUBS}, \hat{\mathcal{V}}) - CORR(\mathcal{V}_{BASE}, \hat{\mathcal{V}})$$

for $(\mathcal{V}_{BASE}, \mathcal{V}_{HUBS})_{|E|}$, and for $CORR(x, y)$ the correlation coefficient function.

Figure 3 presents the values of $\delta_{|E|}$ for number of monitored edges ranging between 300 and 800. It can be seen how the hubs-sampling outperforms the basic random-sampling method. Furthermore, it can be seen that the positive delta increases with the increase in the amount of available resources (namely, number of monitored edges).

Figure 4 presents the values of $\delta_{|E|}$ for number of monitored edges between 300 and 800, for the three types of events.

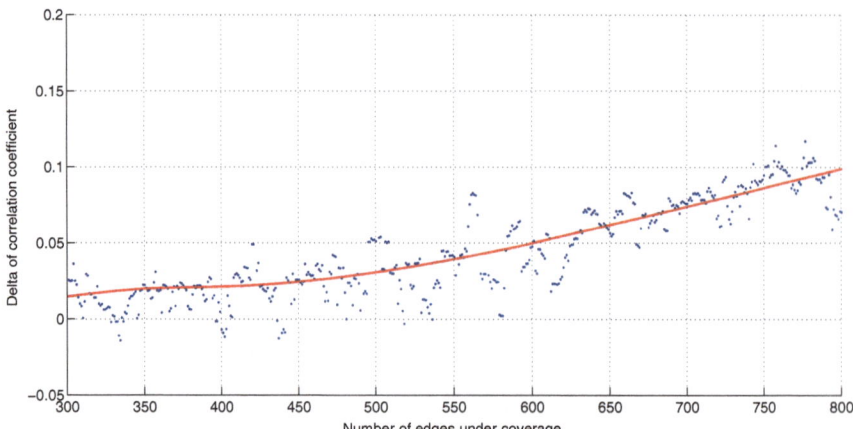

Fig. 3. The changes in the value of $\delta_{|E|}$ for growing numbers of edges being analyzed, evaluated using real anomalies and mobile calls data collected from a developed European country for a period of 3 years. Positive values indicate a higher detection efficiency of hubs-sampling compared to the basic random edge sampling, for the same number of monitored edges.

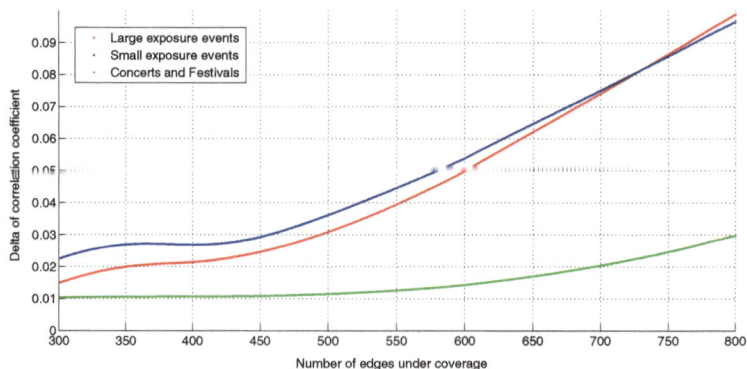

Fig. 4. The changes in the value of $\delta_{|E|}$ for growing numbers of edges being analysed, segregated by the type of event detected. Notice how concerts and festivals that have high exposure value a generate relatively lower values of $\delta_{|E|}$ (but still monotonously increase with $|E|$), while the small exposure events are characterized by the highest values of $\delta_{|E|}$, specifically for low values of $|E|$. It is important to note that a low value of $\delta_{|E|}$ does not imply that the accuracy of the detection itself is low, but rather that the difference in accuracy between the two methods is small.

5   Conclusions

In this paper we examined the problem of monitoring resources allocation for analyzing mobile networks, and have shown that by using social features of the network (namely, focusing on network hubs) prediction accuracy of anomalous events can be significantly increased. Specifically, we have shown that focusing on the neighborhood around a hub (the connections among the alters) will detect events external to the network that provoke spreading communication within the network. Hubs act as collectors (and as a result, amplifiers) of social information, through facilitating the spread of communication in their immediate neighborhood. Traces of small scale information diffusion processes are more likely to be revealed when tracking hubs' activities compared to randomly selected nodes. In this work we show that this effect is so intense that in many cases it outperforms the analysis of significantly larger amount of random nodes (in order to compensate of the fact that the analysis of a single hub requires coverage of much more edges than required for an arbitrary node).

We anticipate, however, that this methodology could be further extended and refined. For example, as hubs can sometimes be major bottlenecks, it is plausible that other neighborhoods within a large scale network would more efficiently act as social amplifiers. For example, it is possible that generally densely connected communities within a network would more efficiently disseminate observable changes in communication behavior, virtually acting as a kind of "distributed hub" (the dramatic effect of the network topology on the dynamics of information diffusion in communities was demonstrated in works such as [12, 22]). It is also possible that the incorporation of other kinds of information about the properties of nodes would greatly improve the model.

References

1. Aharony, N., Pan, W., Ip, C., Khayal, I., Pentland, A.: Social fmri: Investigating and shaping social mechanisms in the real world. Pervasive and Mobile Computing (2011)
2. Albert, R., Jeong, H., Barabási, A.L.: Error and attack tolerance of complex networks. Nature 406(6794), 378–382 (2000)
3. Altshuler, Y., Aharony, N., Fire, M., Elovici, Y., Pentland, A.: Incremental learning with accuracy prediction of social and individual properties from mobile-phone data. CoRR (2011)
4. Altshuler, Y., Aharony, N., Pentland, A., Elovici, Y., Cebrian, M.: Stealing reality: When criminals become data scientists (or vice versa). IEEE Intelligent Systems 26(6), 22–30 (2011)
5. Altshuler, Y., Elovici, Y., Cremers, A.B., Aharony, N., Pentland, A(S.): Security and privacy in social networks. Recherche 67, 2 (2012)
6. Altshuler, Y., Fire, M., Aharony, N., Elovici, Y., Pentland, A(S.): How many makes a crowd? On the evolution of learning as a factor of community coverage. In: Yang, S.J., Greenberg, A.M., Endsley, M. (eds.) SBP 2012. LNCS, vol. 7227, pp. 43–52. Springer, Heidelberg (2012)
7. Altshuler, Y., Pan, W., Pentland, A(S.): Trends prediction using social diffusion models. In: Yang, S.J., Greenberg, A.M., Endsley, M. (eds.) SBP 2012. LNCS, vol. 7227, pp. 97–104. Springer, Heidelberg (2012)
8. Altshuler, Y., Wagner, I.A., Bruckstein, A.M.: On swarm optimality in dynamic and symmetric environments, vol. 7, p. 11 (2008)

9. Altshuler, Y., Yanovsky, V., Wagner, I., Bruckstein, A.: Swarm intelligence-searchers, cleaners and hunters. Swarm Intelligent Systems, 93–132 (2006)
10. Bagrow, J.P., Wang, D., Barabási, A.L.: Collective response of human populations to large-scale emergencies. PloS One 6(3), e17680 (2011)
11. Carter, T.: Butts. The complexity of social networks: theoretical and empirical findings. Social Networks 23(1), 31–72 (2001)
12. Choi, H., Kim, S.H., Lee, J.: Role of network structure and network effects in diffusion of innovations. Industrial Marketing Management 39(1), 170–177 (2010)
13. Eagle, N., Pentland, A(S.), Lazer, D.: Inferring social network structure using mobile phone data. Proceedings of the National Academy of Sciences (PNAS) 106, 15274–15278 (2009)
14. Eagle, N., Macy, M., Claxton, R.: Network diversity and economic development. Science 328(5981), 1029–1031 (2010)
15. Gonzalez, M.C., Hidalgo, C.A., Barabasi, A.-L.: Understanding individual human mobility patterns. Nature 453(7196), 779–782 (2008)
16. Granovetter, M.S.: The strength of weak ties. American Journal of Sociology, 1360–1380 (1973)
17. Huberman, B.A., Romero, D.M., Wu, F.: Social networks that matter: Twitter under the microscope. First Monday 14(1), 8 (2009)
18. Kanaroglou, P.S., Jerrett, M., Morrison, J., Beckerman, B., Altaf Arain, M., Gilbert, N.L., Brook, J.R.: Establishing an air pollution monitoring network for intra-urban population exposure assessment: A location-allocation approach. Atmospheric Environment 39(13), 2399–2409 (2005)
19. Lambiotte, R., Blondel, V.D., de Kerchove, C., Huens, E., Prieur, C., Smoreda, Z., Van Dooren, P.: Geographical dispersal of mobile communication networks. Physica A: Statistical Mechanics and its Applications 387(21), 5317–5325 (2008)
20. Leskovec, J., Backstrom, L., Kleinberg, J.: Meme-tracking and the dynamics of the news cycle. In: Proceedings of the 15th ACM SIGKDD International Conference on Knowledge Discovery and Data Mining, pp. 497–506. Citeseer (2009)
21. Murray, A.T., Kim, K., Davis, J.W., Machiraju, R., Parent, R.: Coverage optimization to support security monitoring. Computers, Environment and Urban Systems 31(2), 133–147 (2007)
22. Nicosia, V., Bagnoli, F., Latora, V.: Impact of network structure on a model of diffusion and competitive interaction. EPL (Europhysics Letters) 94, 68009 (2011)
23. Palla, G., Barabasi, A.L., Vicsek, T.: Quantifying social group evolution. Nature 446(7136), 664–667 (2007)
24. Pan, W., Aharony, N., Pentland, A.: Composite social network for predicting mobile apps installation. In: Proceedings of the 25th Conference on Artificial Intelligence (AAAI), pp. 821–827 (2011)
25. Puzis, R., Altshuler, Y., Elovici, Y., Bekhor, S., Shiftan, Y., Pentland, A.S.: Augmented betweenness centrality for environmentally-aware traffic monitoring in transportation networks.
26. Waclaw, B.: Statistical mechanics of complex networks. Arxiv preprint arXiv:0704.3702 (2007)
27. Wassermann, S., Faust, K.: Social network analysis: Methods and applications, New York (1994)

Measuring User Credibility in Social Media

Mohammad-Ali Abbasi and Huan Liu

Computer Science and Engineering, Arizona State University
{Ali.abbasi,Huan.liu}@asu.edu

Abstract. People increasingly use social media to get first-hand news and information. During disasters such as Hurricane Sandy and the tsunami in Japan people used social media to report injuries as well as send out their requests. During social movements such as Occupy Wall Street (OWS) and the Arab Spring, people extensively used social media to organize their events and spread the news. As more people rely on social media for political, social, and business events, it is more susceptible to become a place for evildoers to use it to spread misinformation and rumors. Therefore, users have the challenge to discern which piece of information is credible or not. They also need to find ways to assess the credibility of information. This problem becomes more important when the source of the information is not known to the consumer.

In this paper we propose a method to measure user credibility in social media. We study the situations in which we cannot assess the credibility of the content or the credibility of the user (source of the information) based on the user's profile. We propose the *CredRank* algorithm to measure user credibility in social media. The algorithm analyzes social media users' online behavior to measure their credibility.

Keywords: Information Credibility, Behavior Analysis, Misinformation.

1 Introduction

Using social media, people easily can communicate and publish whatever they like. As a result, people are able to create huge amounts of data. For example, users on Twitter create 340 million tweets every day. Users on YouTube upload 72 hours of video every minute. In wordpress.com alone, bloggers submit 500,000 new posts and these posts receive more than 400,000 comments everyday[1].

People use social media either for communications or share newsworthy information. They use social media for almost every aspect of their lives. They use social media during disasters to report injuries, damage, or their needs. Examples of events in which social media was utilized include the recent tsunami in Japan, Hurricanes Irene, and Sandy, and the earthquake in Haiti [2]. In business and marketing, social media is also used for product and service review and recommendation. As a result, it is very common to read reviews and comments

---

[1] http://www.jeffbullas.com/2012/08/02/
blogging-statistics-facts-and-figures-in-2012-infographic/

A.M. Greenberg, W.G. Kennedy, and N.D. Bos (Eds.): SBP 2013, LNCS 7812, pp. 441–448, 2013.
© Springer-Verlag Berlin Heidelberg 2013

before purchasing a product or using a service. During the social events such as the *Arab Spring* and the *Occupy Wall Street (OWS)* movement, social media was effective to spread information about the movements[13]. In many cases, people report incidents and events almost instantly and their reports cover different aspects of the event (people act as social sensors). Social media provides first-hand data, but one pressing problem is to distinguish true information from misinformation and rumors. In many cases, social media data is user generated and can be biased, inaccurate, and subjective. Furthermore, some people use social media to spread rumor and misinformation[2]. Consequently, information in social media is not necessarily of equal value, and we need to assess the credibility of the data before using it for decision making.

Using credible information is a prerequisite for accurate analysis utilizing social media data. Non-credible data will lead to inaccurate analysis, decision making and predictions. Credibility is defined as "the quality of being trustworthy". In communication research, information credibility has three parts, message credibility, source credibility, and media credibility [14]. Comparing conventional media, assessing information credibility in social media is the more challenging problem. In the case of conventional media such as newspapers, the source and media are known; in addition the medium's owners take responsibility for the content. However in the case of social media, the source can be unknown thus no one takes responsibility about the content. In many cases a *username* is the only information we have about its source (e.g., an incomplete or even fake profile in Twitter or YouTube that publishes information about an incident).

Ranking social media users on their credibility is one approach to measure the credibility of the given piece of information. Twitter, for example, has a set of verified accounts. These accounts have a blue badge on their profiles. According to Twitter, "The verified badge helps users discover high-quality sources of information and trust that a legitimate source is authoring the account's tweets."[3] Neither Twitter nor other social media websites are able and want to verify all their users. In the best scenario, only a small portion of the users can be verified by the websites. Considering this and the fact that many users would prefer to remain unknown, it is expected that the majority of users in social media are unverified.

This anonymity is both an advantage and a disadvantage of social media. On one hand, people create content, and leave feedback or vote without being afraid of any negative side effects resulting from their activities. This is a great advantage especially for people in countries that lack the freedom of speech. On the other hand, people could also take advantage of openness and anonymity. Some would create many accounts in which to leave positive reviews in order to boost one product or negative reviews to downgrade another. During social and political movements (e.g., Arab Spring revolutions), one could observe many highly organized Twitter accounts that actively tweet against the revolution[1].

---

[2] http://personal.stevens.edu/~ysakamot/
726/paper/Grant/RAPIDdescription.pdf
[3] http://www.twitter.com/verified

We would see the same misbehavior in social bookmarking systems in which highly coordinated accounts would try to change the voting results in a specific direction. Such behavior could significantly decrease the quality of social media content.

Contributions. We propose to address user credibility to tackle the information credibility problem in social media. We propose the *CredRank algorithm*, which measures the credibility of social media users based on their online behavior.

Paper Organization. The rest of the paper is organized as follows. Section 2, *Literature Review*, reviews the related work on information credibility and credibility in social media. Section 3, *Problem Statement*, discusses the credibility problem in social media. Section 4, *A Proposed Solution*, introduces the CredRank algorithm as one solution to measure user credibility. Section 5, *Experiments*, shows the use of the proposed method on the U.S. Senate voting record data. Section 6, *Discussion*, summarizes findings and describes future work.

2   Literature Review

Assessing credibility is an important part of research on mass communication. Seminal work on credibility concentrated on source credibility as well as credibility attributed to different media channels [9]. In traditional media as well as social media, the credibility of the source has a great effect on the process of acquiring the content and changing audience attitudes and beliefs [5]. Studies confirm that people consider Internet information as credible as traditional media such as television, radio, and magazines but not as credible as newspapers [11].

Castillo et al. in [6] discussed the information credibility of news propagated through Twitter. They used users' profile information, network information, and users' behavior (tweets and retweets) to assess the credibility of tweets. Barbier and Liu in [4] proposed a method to find provenance paths leading to sources of the information to evaluate its credibility. Researches have used trust information to evaluate online content. Trust is widely exploited to help online users collect reliable information in applications such as high-quality reviews detection and product recommendations [10]. Guha et al. [8] studied the problem of propagating trust and distrust among Epinions' [4] users, who may assign positive (trust) and negative (distrust) ratings to one another. They used trust information to rank users and using that, rank the content generated by the users. Castilo et al. [6], used features from users' posting behavior (tweet and retweet), text, and the network (# of friends and # of followers) to distinguish credible from not credible tweets. Agichtein et al. [3], used community information to identify high quality content in question and answering portal (Yahoo! Answers[5]). They used features from *answers, questions, votes,* and *users' information and relationship* to build a model to measure quality of the content.

---

[4] http://www.epinions.com
[5] http://answers.yahoo.com

Popularity is the most accepted measure of assessing credibility of users and content in social media. Usually popularity and credibility are used interchangeably. For example many users would trust a Twitter user who has many followers. Similarly one might trust a piece of information (a video clip on YouTube) if many people had already watched it. Using popularity idea, there are some work that used link based information (e.g., PageRank and HITS) to rank the users and evaluate the content based on the source's rank. [12] used HITS to rank users and find experts and high quality answers in the question and answering communities. Using number of inlinks (# of friends on Facebook or # of followers on Twitter) is well-accepted feature to measure the importance (credibility or influence in different concepts) or users. Cha et al. [7] use three approaches (*indegree*, *retweet*, and *mention*) to measure users' importance in Twitter. The study shows that although *indegree* measures the popularity of a user, it does not necessarily reflect the importance (or influence in some domains) of the user.

3   Non-credible Information and the Need for Detection

Non-credible users are responsible for part of the non-credible information in social media. In this section, using real examples from the Arab Spring movements in social media sites, we show how users can generate and spread misinformation or prevent the spread of trustworthy information.

Twitter. Usually for disasters or social events, the most useful and novel information is distributed by unknown or unpopular social media users. In this situation, the receiver of the information does not have adequate time and/or resources to assess source's credibility. Since they cannot assess the credibility of the information or sources, ordinarily the user relies on the popularity of the source or the content (# of followers or # of retweets). However ordinary Twitter users do not have many followers; therefore, their content would not get attention and would be lost among many other tweets. On the other hand some organized users take advantage of this situation. For example, during Arab Spring there were many coordinated users tweeting against the revolutionists. Many of them could be government-supported accounts. These users were very organized and most of them followed one another and retweeted one another's tweets. Therefore, their content could easily be noticed in Twitter.

YouTube. In some cases, coordinated users targeted a video clip on YouTube to take it down. They frequently submitted false reports against the video and in many cases, these false reports led to the video removal from YouTube.

Voting Systems. In voting systems (social bookmarking systems), users who have more supporters can get more votes and publicize their content more easily than others. It is common in this kind of system that many users create a clique, usually vote for one another's content, and against other users' contents. This enables them to publicize their content and deter other users who cannot collect enough votes for their content. These highly connected users also easily can degrade others' content by awarding them negative votes.

In all of these cases *coordinated users* can easily suppress independent users and prevent their content from spreading in social media. They also would be able to spread misinformation. These *coordinated users* have highly correlated behavior (e.g. their tweets are very similar or their votes have similar patterns). The next section shows our attempt on detecting these users by monitoring their online behavior. These are non-credible users who generate and spread misinformation or prevent other users from spreading their content. The main properties of non-credible users are:

- Creating a large number of accounts and using the accounts to spread the word.
- Voting, regardless of content, for other users in their group. As the votes go for the user, not the content, the number of votes coming from the group members does not represent the quality of the content[6].

4   Proposed Solution

Our solution gives each independent person an equal vote and a chance to publicize his/her content. We perform the following steps: (1) *detect and cluster* coordinated users (dependent users) together and (2) *weight* each cluster based on the size of the cluster. We design the *CredRank* algorithm to perform these two steps.

4.1   CredRank Algorithm

This algorithm finds users with similar behavior and clusters them. CredRank uses a hierarchical clustering method to cluster similar users into clusters. We measure similarity of behaviors to calculate the similarity between users.

$$Sim(u_i, u_j) = \frac{1}{t_n - t_0} \sum_{t=t_0}^{t_n} \sigma(B(u_i, t), B(u_j, t)) \tag{1}$$

where $B(u_i, t)$ is user u_i's behavior in timestamp t and $\sigma(B(u_i, t), B(u_j, t))$ is a function that measures the similarity of two users' behavior in the given timestamp t.

For each domain we use a specific function to measure similarity. For example, to measure Twitter users' similarity we calculate the similarity of their tweets. In social bookmarking systems, we measure the similarity of their votes. We can use various similarity measure approaches such as *edit-distance, tf-idf,* or *Jaccard's coefficient*. In our experiments we use Jaccard's coefficient to calculate behaviors' similarity.

$$\sigma(B_i, B_j) = \frac{|B_i \cap B_j|}{|B_i \cup B_j|} \tag{2}$$

---

[6] This problem also exists in political parties. In many cases, regardless of the topic, legislators vote for their party

---

Algorithm 1. CredRank Algorithm

1: Measure the pairwise similarity between users based on their behavior $(Sim(u_i, u_j))$.

2: Cluster users together if their similarity exceeds the threshold τ.

3: Assign $\omega_{C_i} = \frac{\sqrt{|C_i|}}{\sum_j \sqrt{|C_j|}}$ to each cluster, which is the cluster's weight. Each member

in the cluster C_i, has a weight of $\frac{\sqrt{|C_i|}}{|C_i|}$ which is the credibility assigned to the member.

$\sigma(B_i, B_j) = 1$ shows that two users' behaviors are completely similar and $\sigma(B_i, B_j) = 0$ shows that their behaviors are different.

After calculating the similarity between users, we cluster users together if their similarity exceeds the threshold τ. The value of τ varies for different domains.

In the next step, using the following formula, we assign clusters' weights.

$$\omega_{C_i} = \frac{\sqrt{|C_i|}}{\sum_j \sqrt{|C_j|}} \tag{3}$$

where ω_{C_i} is the weight assigned to the cluster C_i with $|C_i|$ members. Each member in cluster C_i, has a weight of $\frac{\sqrt{|C_i|}}{|C_i|}$. This value show the amount of the credibility associates with the member.

5   Experiments

We use US Senate voting history data to show how the proposed algorithm helps us to detect coordinated collective behavior. In this case we consider the highly coordinated voting as non-credible behaviors. In this section we show that how we can detect these coordinated behavior. Then we use these coordinated behaviors to detect senators with similar voting history. Then we cluster senators with similar voting history in the same groups. Our analysis show that usually votes in each cluster are highly correlated.

Dataset. We crawled United States Senate official websites [7] to collect senators' votes records. The website provides Senate "Roll Call Vote" results for the current and several prior Congresses. We crawled voting history from 1989 to 2012. For each issue, we collected each Senator's vote (yea or nay) and the vote result (rejected, agreed to, passed, and confirmed).

To analyze the correlation between votes, we use CredRank algorithm idea and calculate top eigenvalues as we report them in table 1.

---

[7] http://www.senate.gov

Table 1. Top eigenvalues of senators' voting history

Eig 1	Eig 2	Eig 3	Eig 4	Eig 5	Eig 6
70.3542	12.2152	1.6815	0.9427	0.7385	0.7013

By using k-means, with different values of k, in most cases Democratic and Republican senators cluster into different clusters. In almost all cases senators' votes depend on their party. In only a few cases the votes really represent senator's own opinion. Oftentimes votes of a few independent senators are highly influential on the result of votes and usually the result is highly dependent on their votes.

Referring to Table 1, the results show that Senators' votes are highly correlated. By using CredRank we can cluster senators into 6 clusters and rank them based on the number of senators in each group. If we pick one representative for each group, with the weight calculated by step 2 of the algorithm, we would be able to generate the same vote results as of votes from all of the senators.

Despite the real-world voting systems, we do not expect to observe coordinated voting behavior in social media. If two users in a rating system have very similar votes for many products, we consider this voting behavior as non-credible behavior. Therefore the users considered as non-credible users.

6 Discussion

In this paper, we propose a method to detect coordinated behavior in social media and assign a lower credibility weight to users who are involved in the coordinated behavior. In this process, we are able to prevent the spread of misinformation generated by these users, which is an attempt to increase the quality of information in social media. The proposed algorithm helps us to detect individuals who use many social media accounts and do so in a way to diffuse their content. The CredRank algorithm can be used in many cases such as: preventing the distribution of rumors, averting coordinated activities, and thwarting fake product reviews. In the future work, we will improve the algorithm to solve two drawbacks we mention them next. Using the method might prevent true diffusion of information. In addition, calculating similarity among all users' behaviors in real time might be computationally expensive.

In order to achieve better results, we must consider all three parts involved in information credibility, including message credibility, source credibility, and media credibility. Focusing on source credibility and considering more features of sources, such as network and profile, to assess user credibility is an extension to this work. By considering behavior, network, and profile we expect to construct a reliable model to assess source credibility in social media.

Acknowledgments. This research is sponsored, in part, by Office of Naval Research (Grant number: N000141110527).

References

1. Abbasi, M.-A., Chai, S.-K., Liu, H., Sagoo, K.: Real-world behavior analysis through a social media lens. In: Yang, S.J., Greenberg, A.M., Endsley, M. (eds.) SBP 2012. LNCS, vol. 7227, pp. 18–26. Springer, Heidelberg (2012)
2. Abbasi, M.-A., Kumar, S., Filho, J.A.A., Liu, H.: Lessons learned in using social media for disaster relief - ASU crisis response game. In: Yang, S.J., Greenberg, A.M., Endsley, M. (eds.) SBP 2012. LNCS, vol. 7227, pp. 282–289. Springer, Heidelberg (2012)
3. Agichtein, E., Castillo, C., Donato, D., Gionis, A., Mishne, G.: Finding high-quality content in social media. In: Proceedings of the International Conference on Web Search and Web Data Mining, pp. 183–194. ACM (2008)
4. Barbier, G., Liu, H.: Information provenance in social media. In: Salerno, J., Yang, S.J., Nau, D., Chai, S.-K. (eds.) SBP 2011. LNCS, vol. 6589, pp. 276–283. Springer, Heidelberg (2011)
5. Burgoon, J., Hale, J.: The fundamental topoi of relational communication. Communication Monographs 51(3), 193–214 (1984)
6. Castillo, C., Mendoza, M., Poblete, B.: Information credibility on twitter. In: Proceedings of the 20th International Conference on World Wide Web, pp. 675–684. ACM (2011)
7. Cha, M., Haddadi, H., Benevenuto, F., Gummadi, K.: Measuring user influence in twitter: The million follower fallacy. In: 4th International AAAI Conference on Weblogs and Social Media, ICWSM (2010)
8. Guha, R., Kumar, R., Raghavan, P., Tomkins, A.: Propagation of trust and distrust. In: Proceedings of the 13th International Conference on World Wide Web, pp. 403–412. ACM (2004)
9. Hovland, C., Weiss, W.: The influence of source credibility on communication effectiveness. Public Opinion Quarterly 15(4), 635–650 (1951)
10. Jamali, M., Ester, M.: Trustwalker: a random walk model for combining trust-based and item-based recommendation. In: Proceedings of the 15th ACM SIGKDD International Conference on Knowledge Discovery and Data Mining, pp 397–406 ACM (2009)
11. Johnson, T., Kaye, B.: Cruising is believing?: Comparing internet and traditional sources on media credibility measures. Journalism & Mass Communication Quarterly 75(2), 325–340 (1998)
12. Jurczyk, P., Agichtein, E.: Discovering authorities in question answer communities by using link analysis. In: Proceedings of the Sixteenth ACM Conference on Conference on Information and Knowledge Management, pp. 919–922. ACM (2007)
13. Kwak, H., Lee, C., Park, H., Moon, S.: What is twitter, a social network or a news media? In: Proceedings of the 19th International Conference on World Wide Web, pp. 591–600. ACM (2010)
14. Metzger, M., Flanagin, A., Eyal, K., Lemus, D., McCann, R.: Credibility for the 21st century: Integrating perspectives on source, message, and media credibility in the contemporary media environment. Communication Yearbook 27, 293–336 (2003)

Financial Crisis, Omori's Law, and Negative Entropy Flow

Jianbo Gao[1,2] and Jing Hu[1]

[1] PMB Intelligence LLC,
P.O. Box 2077, West Lafayette, IN 47996, USA
jbgao.pmb@gmail.com
http://www.gao.ece.ufl.edu
[2] Mechanical and Materials Engineering
Wright State University, Dayton, OH 45435, USA

Abstract. Understanding the mechanism of financial crises is an important issue, especially in a time of profound economic difficulty worldwide. To gain insights into how economic crises develop, we examine the exposure network associated with Fannie Mae/Freddie Mac, Lehman Brothers, and American International Group, and show that the losses associated with them can be modeled by an Omori-law-like distribution for earthquake aftershocks. Under certain conditions, Omori's law leads to Pareto distribution. Positive Pareto incomes, together with Omori's law, motivate us to examine whether distributions of negative incomes during crises may also be modeled by Pareto distributions. We find that during crises, negative incomes not only may indeed be modeled as Pareto-like distributions, but actually have heavier tails than those for positive incomes. As a result, entropy flow associated with losses or negative incomes provides an excellent technique for predicting economic downturns.

1 Introduction

There have been a growing number of financial crises in the world, according to the International Monetary Fund [1]. Important lessons may be learned from previous major financial crises [2], such as (i) globalization has increased the frequency and spread of financial crises, but not necessarily their severity, (ii) early intervention by central banks is more effective in limiting their spread than later moves, (iii) it is difficult to tell at the time whether a financial crisis will have broader economic consequences, and (iv) regulators often cannot keep up with the pace of financial innovation that may trigger a crisis. While financial crises can be studied through theoretical modeling [3,4] and analysis of individual companies [1,5,6,8], it is still an important open question whether financial crises in general and the recent gigantic economic crisis in US in particular, can be analyzed using analogies from the physical world. In this work, we focus on the collective economic dynamics associated with losses or negative incomes. Specifically, we shall start from the exposure networks of Fannie Mae/Freddie

A.M. Greenberg, W.G. Kennedy, and N.D. Bos (Eds.): SBP 2013, LNCS 7812, pp. 449–457, 2013.
© Springer-Verlag Berlin Heidelberg 2013

Mac (FNM/FRE), Lehman Brothers (LEH), and American International Group (AIG), then examine the generic distributions associated with negative incomes, and finally examine the entropy flow associated with losses or negative incomes and prediction of economic downturns.

2   Distribution of Losses in Crisis Exposure Networks

The 2007 - 2009 financial crisis was a truly gigantic one, as it claimed Bear Stearns, Fannie Mae (FNM), Freddie Mac (FRE), Lehman Brothers (LEH), American International Group (AIG), and Washington Mutual (WaMu) in the United States, and spiraled to the entire world and consumed other businesses, including Japan's Yamato Life Insurance. It even has much to do with the current dire economic conditions in many countries in the European Unions. Here, we report an effort of constructing networks exposed to AIG, LEH, and FNM/FRE. We find that the distribution for the investments in those exposure networks is strikingly similar to the Omori's law for the rate of occurrence of aftershocks triggered by a main earthquake. The similarity stimulates us to take an evolutionary point of view and examine the entropy production associated with the formation of exposure networks.

By searching on the Internet daily list of companies reporting AIG, LEH, and FNM/FRE exposures in 2008, we found respectively, 34, 151, and 146 companies worldwide, exposed to AIG, LEH, and FNM/FRE, and constructed three separate exposure networks. Based on the foreign exchange rate data on October 8, 2008, the amount of investments ranges from less than 1 million to hundreds of millions or even tens of billions of US dollars in each network. AIG, LEH, and FNM/FRE, being connected to all the companies exposed to them, may be called hub-nodes, just as hubs of major airlines. In contrast, other companies may be called end-nodes, and the amount of their exposures to AIG, LEH, and FNM/FRE determines the strength of the links connecting them to the hub-nodes. To assess the severity of the collapse of hub-nodes, it is most important to find the number of companies with big investments on those hub-nodes. Mathematically, this amounts to estimating the complementary cumulative distribution function (CCDF), $P(X \geq x) = $ Probability that $X \geq x$ million. The huge range of capitals involved in each network makes such a task feasible and meaningful.

To reliably estimate distributions from sparse data with wide range, we have used the technique of equal-log-bin [8] by first taking logarithm of the data, then estimating the CCDF. The results are shown in Fig. 1. We observe that the CCDFs for AIG and LEH are concave to the origin, while that for FNM/FRE is a power-law. The similarity between AIG and LEH suggests that they may have very similar business models/strategies. The power-law behavior for FNM/FRE suggests that the FNM/FRE exposure network is a type of power-law or scale-free network [9]. The difference between FNM/FRE and AIG/LEH might partly be due to the fact that many companies had greatly reduced their investments on FNM/FRE before FNM/FRE were taken over by the U.S. government.

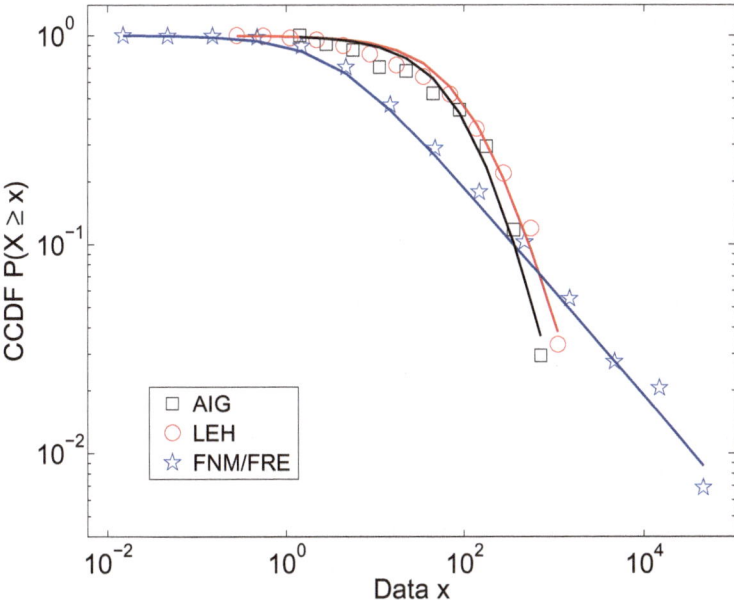

Fig. 1. CCDF for the investment distribution of the 3 exposure networks. (α, β) are $(2, 118), (1.6, 116)$, and $(0.5, 2)$, for AIG, LEH, and FNM/FRE, respectively.

Can we find a simple closed-form formula to fit all the CCDFs found here? The task sounds challenging, since the CCDFs for AIG and LEH are very different from that for FNM/FRE. Surprisingly, it turns out that an Omori-like law for earthquake aftershock activity solves the problem neatly. The Omori law for aftershock activity states that [10],

$$r(t) = \frac{1}{\tau[1 + t/c]^p} \tag{1}$$

where $r(t)$ is the rate of occurrence of aftershocks with magnitudes greater than m at time t after a major earthquake, τ and c are certain characteristic times, and $p > 0$ is the scaling parameter. The Omori law has been used to describe volatility return intervals in stock markets [11,12]. It has an interesting property that when $p \neq 1$, it essentially keeps its functional form with integration or differentiation. This motivates us to fit our CCDFs by the following formula,

$$P(X \geq x) = \left(1 + \frac{x}{\beta}\right)^{-\alpha}, \tag{2}$$

where $\alpha > 0$ and $\beta > 0$ are parameters. The fitting is shown in Fig. 1 as red smooth curves, where (α, β) are $(2, 118), (1.6, 116)$, and $(0.5, 2)$, for AIG, LEH, and FNM/FRE, respectively. In all three cases, the fitting is remarkably good. This suggests that cascading of failures of financial institutions after the collapse of a major financial player is remarkably similar to the triggering of a sequence of aftershocks by a major earthquake.

Note that when $x \gg \beta$, Eq. (2) becomes a power-law,

$$P(X \geq x) \sim x^{-\alpha} \tag{3}$$

When $\alpha < 2$, such a distribution is heavy-tailed having infinite variance. When $\alpha \leq 1$, the mean also becomes infinite.

3   Distribution of Losses around Recession Times

It is well-known that positive incomes often follow a Pareto-like distribution. In good economic times, losses are rare, and in general, one would not anticipate Pareto-law to be relevant to the distribution of losses. However, the situation drastically changes right before or during economic recessions. Since the Omori-like law described by Eq.(2) has provided an excellent model for the loss distribution in exposure networks, we thus ask whether Omori's Law in general, and Pareto-law in particular, may be relevant for the characterization of losses around recession times.

To show this, we examined pretax quarterly incomes, from the beginning of 1990 to the end of 2011, of thousands of U.S. companies in 9 sectors: Financial, Consumer Goods, Consumer Services, Basic Materials, Health Care, Industrials, Oil/Gas, Tech-Telecommunications, and Utilities (See Data Source). Prior to 1990, income data for many companies in these sectors are incomplete, and thus meaningful analysis is not feasible. In the analyses below, we consider negative and positive incomes as separate clusters; when the number of companies in each category or cluster is sufficiently large, distributional analysis is feasible. The black circles in Fig. 2 illustrate the positive income distributions in general, and the straight lines in the tail region of the log-log plots clearly indicate a heavy-tailed distribution. While the shape of the positive income distribution is largely independent of the economic health of the sector, the parameter α is not universal; instead, it correlates well with the health of the economy — it attains a larger value when a recession is more serious.

The shape of distribution of negative incomes, in contrast, strongly depends on the economic health of the sector. During healthy periods, negative incomes occur because of nonsystemic reasons (e.g., poor management) and are rare and small. When losses are very few, the loss distribution may not be well-defined; when losses become slightly more numerous, $X = -Y$ (where $Y < 0$ denotes negative incomes) may roughly follow an exponential distribution, or, with even more numerous losses, resemble a heavy-tailed distribution.

For the 2008 Financial Crisis, prior to the third quarter of 2007, the distribution of losses in any quarter is thinner than that of positive incomes in the same quarter (i.e., the tail of the loss distribution lies beneath the tail of the profit distribution). However, near the onset of and during a crisis, the distributions of negative incomes not only have also become Pareto, but have even heavier tails

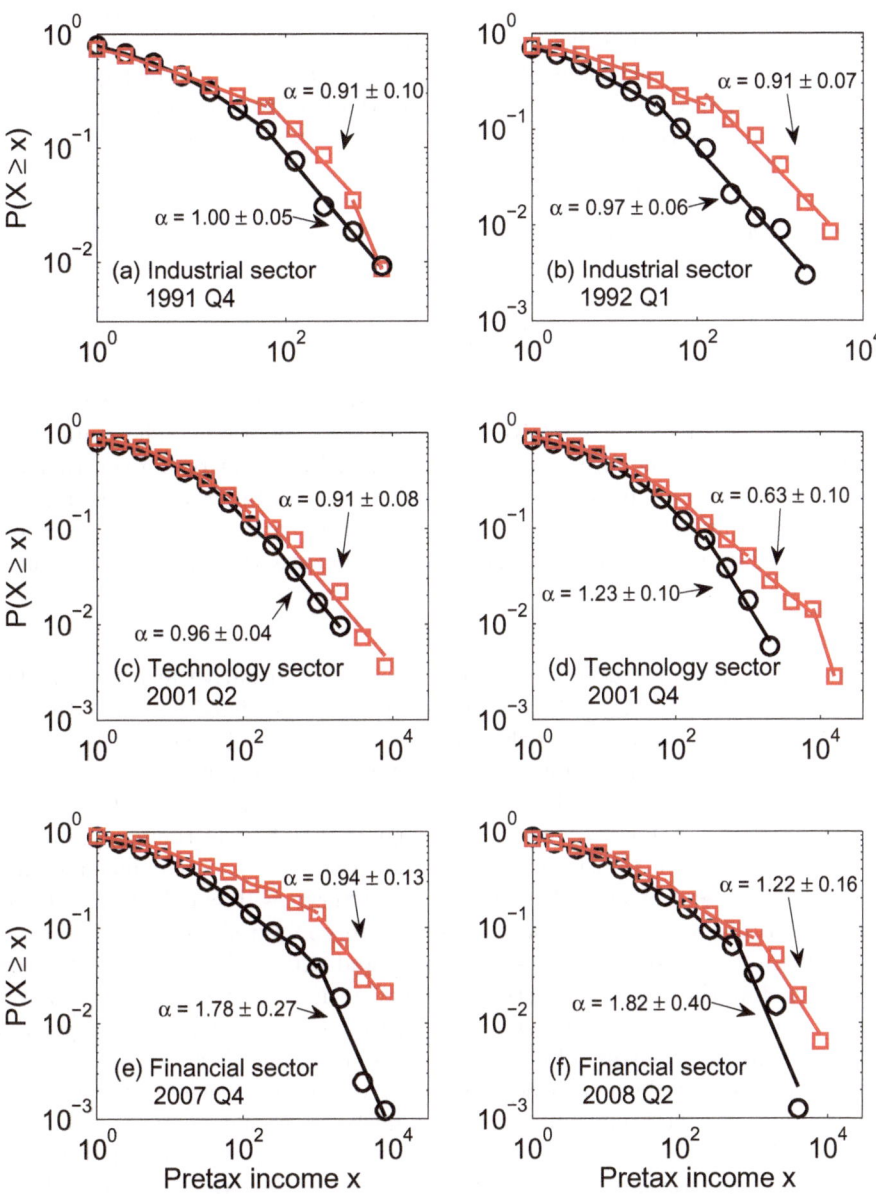

Fig. 2. CCDF (log-log scale) for negative (red square) and positive (black circle) pretax incomes amongst U.S. companies: (a) industrial sector, fourth quarter of 1991, 116 and 327 negative and positive pretax incomes; (b) industrial sector, first quarter of 1992, 118 and 335 negative and positive pretax incomes; (c) technology sector, second quarter of 2001, 275 and 418 negative and positive pretax incomes; (d) technology sector, fourth quarter of 2001, 356 and 345 negative and positive pretax incomes; (e) financial sector, fourth quarter of 2007, 141 and 823 negative and positive pretax incomes; (f) financial sector, second quarter of 2008, 157 and 800 negative and positive pretax incomes.

than those of positive incomes during the same period (Fig. 2(e,f)). Again, such behaviors is also seen in other recessions (Fig. 2(a–d)).

Two important conclusions can be deduced from Fig. 2: (i) the shape of the distribution for the losses indicates the severity of the losses; and (i) a recession may be defined as the instance that the distribution for the negative incomes is on top of that for the positive incomes. Guided by these understanding, we have found that the recent US recession ended around Q2 of 2009. Unfortunately, the losses, especially for financial institutions, have still been significant, at least up to Q2 of 2011.

4   Entropy Flow Associated Losses

As with income distribution analysis, it is more convenient to consider the entropy of the positive and negative income clusters separately. The stability or strength of an income cluster may be quantified by its entropy, given that the second law of thermodynamics can be equivalently restated as saying that the most stable configuration is the one with the highest entropy. For discrete probabilities estimated from data, the following formula for entropy is most convenient

$$H = - \sum P_i \log P_i, \tag{4}$$

where P_i are the probabilities that the positive or negative incomes will fall within a prescribed bin i (where a bin is an interval of fixed length). We shall take 2 as the base of the logarithm so that the unit of the entropy is the bit. Note that when all the probabilities are equal, Shannon entropy attains its largest value; such a situation may be associated with the discretization of a uniform distribution. Fig. 3 shows H from the first quarter of 2006 to the fourth quarter of 2008, for positive (black circles) and negative (red squares) incomes in 5 sectors (Financial, Consumer Goods, Consumer Services, Technology, and Health Care). The discrete probabilities in Fig. 3 are computed using a bin size of $15 million. Tests using bin sizes of $5, $10, and $20 million shifted the curves vertically, but the difference between H for positive and negative incomes is largely independent of the bin size. The entropy of the distribution of positive incomes is almost constant, for all five sectors examined here. In contrast, the entropy of the distribution of negative incomes varies considerably with time. For example, for the Financial sector, it is markedly smaller than the entropy of positive incomes until the third quarter of 2007, when the entropy rises sharply. By the third quarter of 2007, the difference between the two entropies is almost zero, suggesting that the cluster of financial companies with losses is almost as strong (i.e., this configuration is nearly as stable) as the cluster of profitable financial companies, and signals weakness in the sector. From the third quarter of 2007 on, the entropy of the distribution of negative incomes is noticeably larger than for positive incomes, indicating that the cluster of financial companies with negative incomes is well-established and stronger than the cluster of financial companies with positive incomes, consistent with the progression of the 2008 Financial Crisis.

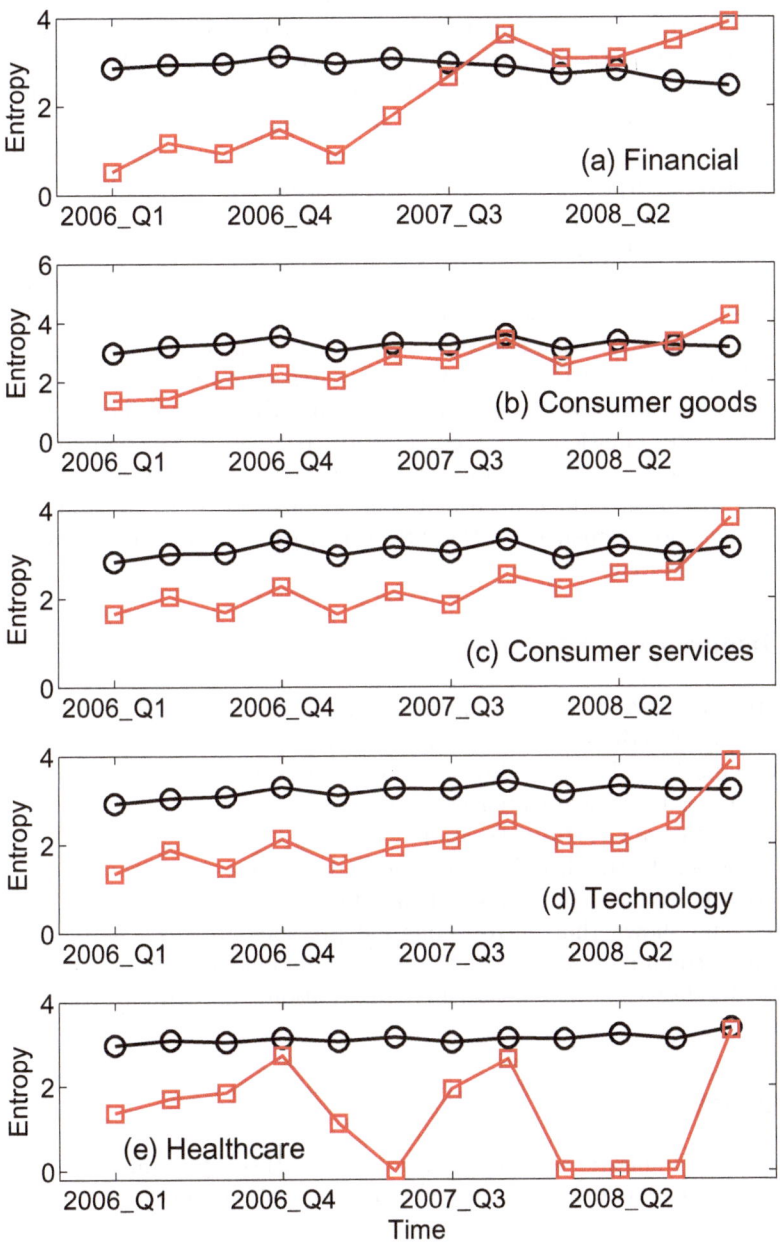

Fig. 3. Entropies (in units of bits) for the distribution of positive (black circles) and negative (red squares) incomes from the first quarter of 2006 to the fourth quarter of 2008 for 5 sectors

The time series of entropy H for other sectors in the period preceeding and during the 2008 Financial Crisis show behavior consistent with the leading role of the Financial sector in the crisis, and propagation of the crisis from that sector into other sectors of the economy. Consumer Services (Fig. 3(c)) and Technology (Fig. 3(d)), for instance, do not show noticeable weakness until after third quarter of 2008. The temporal variations of entropy for Consumer Goods (Fig. 3(b)) shows similar behavior, though in second quarter of 2007 the negative income entropy nearly surpasses positive income entropy, suggesting the beginning of sector weakness at that time.

Note that the propagation of weakness from one sector to another during the 2008 Financial Crisis, as revealed in positive and negative income entropy, is also seen during other periods of market decline, such as the 1999–2003 technology stock fueled decline and the early 1990 recession. These results compare well with the National Bureau of Economic Research (NBER)'s business cycle contraction onset dating determinations during this period. The NBER business cycles determinations are, however, retrospectives. If we compare the dates of our entropy-based downturn onset identifications with the dates the NBER *announced* their onset identifications, our entropy-based identifications preceed the NBER announcements; for the 2008 crisis, the entropy-based identification occurs 1 year earlier.

5   Discussions

To better understand the mechanism of financial crises, in this paper, we have examined the distribution of losses associated with exposure networks of Fannie Mae/Freddie Mac, Lehman Brothers, and American International Group, and show that the losses associated with them can be modeled by an Omori-law-like distribution for earthquake aftershocks. We also have shown that this law is very relevant to the distribution of negative incomes around recession times. Furthermore, we have considered entropy flow associated with losses or negative incomes, and shown that entropy method provides an excellent technique for predicting economic downturns. A remaining significant question would be to analytically derive the Omori-law-like distribution, through an evolution point of view.

References

1. Aziz, J., Caramazza, F., Salgado, R.: Currency crises. In: Search of common elements. IMF Working Paper, vol. 67. International Monetary Fund, Washington, DC (2000)
2. BBC, Financial crises: Lessons from history (2007),
   http://news.bbc.co.uk/2/hi/business/6958091.stm
3. Jorge, A.C.-L., Chen, Z.H.: A theoretical model of financial crisis. Review of International Economics 10, 53 (2002)
4. Kindleberger, C.: Manias, panics, and crashes: A history of financial crises, 3rd edn. Wiley, New York (1996)

5. Ozkan-Gunay, E.N., Ozkan, M.: Prediction of bank failures in emerging financial markets: an ANN approach. The Journal of Risk Finance 8, 465 (2007)
6. Yeh, Y.H., Woidtke, T.: Commitment or entrenchment? Controlling shareholders and board composition. Journal of Banking & Finance 29, 1857 (2005)
7. Niemiraa, M.P., Saaty, T.L.: An Analytic Network Process model for financial-crisis forecasting. International Journal of Forecasting 20, 573 (2004)
8. Gao, J.B., Cao, Y.H., Tung, W.W., Hu, J.: Multiscale Analysis of Complex Time Series — Integration of Chaos and Random Fractal Theory, and Beyond, pp. 38–40. Wiley, Hoboken (2007)
9. Albert, R., Jeong, H., Barabasi, A.-L.: Error and attack tolerance of complex networks. Nature 406, 378–482 (2000)
10. Utsu, T., Ogata, Y., Matsu'ura, R.S.: The centenary of the Omori formula for a decay law of aftershock activity. J. Phys. Earth 43, 1–33 (1995)
11. Lillo, F., Mantegna, R.N.: Power-law relaxation in a complex system: Omori law after a financial market crash. Phys. Rev. E 68, 016119 (2003)
12. Weber, P., Wang, F., Vodenska-Chitkushev, I., Havlin, S., Stanley, H.E.: Relation between volatility correlations in financial markets and Omori processes occurring on all scales. Phys. Rev. E 76, 016109 (2007)

Trust Metrics and Results for Social Media Analysis

Eli Stickgold, Corey Lofdahl, and Michael Farry

Charles River Analytics
625 Mt. Auburn Street
Cambridge, MA 02138
{estickgold,clofdahl,mfarry}@cra.com

Abstract. Social media has changed the information landscape for a variety of events including natural disasters, demonstrations, and violent crises. During these events, people use a variety of social media, such as Twitter, to share information with the world. Given the massive amount of data generated, it is difficult to identify the valuable information in a sea of noise. In this study, we focus on a universal contributing factor to information value—trust—which is analyzed in two steps. Leveraging the theory of *trust in information*, a set of metrics is developed that focus on trusted relationships and behavioral indicators of trustworthiness within social media. Second, these trust metrics are tested on an anonymized data set and their results presented.

Keywords: trust theory, trust metrics, social networks, graph theory, automated social media analysis.

1 Introduction

The proliferation of social media has changed the information landscape for a variety of events including natural disasters, demonstrations, and violent crises. During these events, people use a variety of social media (e.g., Twitter) to share information. Given the massive amount of data that can be generated in a short period of time by social media, it is difficult to identify valuable information in a sea of noise. That analysis is dependent on careful definitions of information value, which may shift according to the type of data considered, who is inspecting the data, and for what purposes the data is being used. To accommodate that broad range of definitions, many current attempts to exploit social media are highly manual, using "mechanical Turk" approaches to synthesize information. For example, the disaster response information sharing service Ushahidi (http://www.ushahidi.com) used a team to manually categorize and store responses to the Haiti earthquake in January 2010. However, reliance on manual synthesis impedes scaling. As a result, approaches like Ushahidi are not well-suited in all situations to determine information value. A more robust solution, one that explicitly models information value in more automated and decision-relevant terms, is clearly required.

This paper examines the notion of trust as a way to determine key information within social media using automated tools, which is accomplished in two steps. Leveraging the theory of "trust in information" as articulated by Henry and Dietz

A.M. Greenberg, W.G. Kennedy, and N.D. Bos (Eds.): SBP 2013, LNCS 7812, pp. 458–465, 2013.

(2011), a set of metrics is developed that focus on trusted relationships within social media data sets. These metrics focus on the structure of relationships rather than the content of messages passed within social media, which we posit as a more reliable indicator of trust. Second, these trust metrics are then tested on an anonymized Twitter data set from the Occupy Wall Street (OWS) movement. The results of this test indicate that trust relationships can be found within social media using automated tools.

2   Metric Development

We now present structure-based metrics that focus on the identification of trustworthy individuals. Our concept of trust in information considers network structural elements rather than message content, which is outside the scope of this study.

We began by formalizing a conceptual hypothesis: that a node (or social media user) that served as a source of information to high-profile and widely-distributed users (e.g., news sources, celebrities) could provide high-value and potentially trustworthy information. This hypothesis leads to two graph-theoretic definitions. We define *first-order influence* as simply the degree to which a user receives attention from others in a social network. First-order influencers are not necessarily inherently trustworthy— news sources may be biased or untrustworthy in many areas of the world, and the same is true of other central figures such as celebrities. Rather, it is the second concept that allows for the identification of trust relationships. We define *second-order influence* as the degree to which a user *indirectly* receives attention in a social network. Second-order influence is measured by the propagation of a user's ideas into a social network by means other than direct messages. We hypothesized that second-order influencers are more likely to provide trustworthy information, since first-order influencers look to them for local, timely, and detailed knowledge. In our OWS Twitter data set, we had access to the messages, but not to follower data for the users in question. In situations like this, a simple metric like degree centrality performed on a directed unweighted "mention network" in tweets can identify users with high first-order influence. Without resorting to computationally-intensive and unreliable content analysis techniques, we exploited a simple Twitter social convention, retweeting, to track second-order influence. While this specific mechanism is unique to Twitter, the general concept applies to most social media (e.g., resharing on Facebook), email systems (forwarding or quoting), and collaborative data sharing and office productivity software (through mechanisms as simple as cutting and pasting).

In our test data, we considered users to have high second-order influence if they were frequently retweeted by users with high first-order influence. Several metrics, such as Katz centrality or the PageRank metric, can identify users with high second-order influence, but our desired property is slightly more specific. Due to frequently dense connections between high first-order influence users, these users also have very high second-order influence, usually the highest in the network. For this study, we wanted to locate users with high second-order and low first-order influence. This distinction is not as simple as eliminating users with high first-order influence from the result set, as we will see in the discussion below. These users, we theorized, would include on-the-scene local reports. These might be as diverse as trusted family members or news reporters reporting on natural disasters, to the man from Abbottabad who tweeted about helicopter noise, indicating the raid that successfully targeted Osama Bin Laden. Identifying and tracking these users would remove the delay associated with them being retweeted by high-profile news sources and would remove any filtering being done by those sources. To locate these users, we developed and tested a suite of potential metrics.

The first two metrics we tested tried different approaches to isolating these users. Our first metric, Decremented Second-Order Centrality (see Equation 1), is a straightforward naïve approach. We computed a simple approximation of second-order influence for each user by summing the in-degree of all users that mentioned the user in question (in the future we will refer to this value as the unweighted second-order degree centrality), then subtracted the degree of the user. This proved insufficient, for several reasons. First, the natural degree distribution of the network was broad enough that the average degree of the network was significantly greater than 1 (we considered subtracting the degree of the central user multiplied by a coefficient equal to the average degree of the network, but rejected this as too closely associated with a single data set, making it harder to generalize). Second, we found that users with high first-order influence often had considerably higher second-order influence than the users we were looking for, leading to them continuing to outperform our target users even in this metric.

$$\left(\sum_{user \in adj(x)} \deg(user) \right) - \deg(x)$$

Equation 1: Decremented Second-Order Centrality

Our second metric, Scaled Second-Order Centrality (see Equation 2), took a different tack. Rather than the linear difference of second-order and first-order degree centralities, we took the ratio of the two (or equivalently, we calculated the average degree of users who referred to the central user). This metric proved much more successful. The comparatively larger second-order influence of high-profile users was reduced, effectively, to simply the number of high-profile users referred to them, which was considerably lower than the values for low-profile users referred to by high-profile ones. This metric provided good progress towards our goals, but had several drawbacks, most notably that it did not provide higher values for users that were mentioned repeatedly by high-profile users or for users that were mentioned by multiple high-profile users.

$$\sum_{user \in adj(x)} \deg(user) \deg(x)$$

Equation 2: Scaled Second-Order Centrality

Based on the results of these two metrics, we iterated by developing a new pair of metrics, each a refined version of one of the previous two. We began with a refined version of the Decremented Second-Order Centrality metric, which we called Logarithm-Based Second-Order Centrality, illustrated in Equation 3. Rather than effectively subtracting 1 from every adjacent user's degree before adding it to a running sum, we instead took the logarithm of each user's centrality before adding it to the sum. This still had the effect of removing all nodes with degree 1 from the sum, but also reduced the impact of the medium-profile users. Ultimately, however, even this proved ineffective, as it failed to resolve the problems that stemmed from the comparatively higher second-order influence of high-profile users over the users for which we were looking.

$$\sum_{user \in adj(x)} \log(\deg(user))$$

Equation 3: Logarithm-Based Second-Order Centrality

Our refined version of the Scaled Second-Order Centrality metric was called Edge-Weighted Second-Order Centrality, illustrated in Equation 4. To calculate it, we added new information to the graph by giving edges (based on mentions) a weight equal to the log of the number of mentions between the relevant pair of users. From this, we calculated a weighted degree average by multiplying each adjacent user's degree by the weight of the edge between them and the central user before adding them to the sum (we call this sum, before it is divided by the central node's degree, the weighted second-order degree centrality). This successfully promoted users repeatedly mentioned by the same high-profile source over those who were only mentioned a single time. This approach still resulted in the highest scores being achieved by users only ever referenced by a single high-profile user, but this was considered acceptable, as ultimately any metric that does not suffer from this problem requires a way to categorize nodes into "low-profile" and "high-profile," and as both iterations of the linear difference metric proved, this ends up being a highly network-specific question.

$$\frac{\sum_{user \in adj(x)} \deg(user) \cdot edgeweight(user, x)}{\deg(x)}$$

Equation 4: Edge-Weighted Second-Order Centrality

After comparison of all four examined metrics, we determined that Edge-Weighted Second-Order Centrality generally was most successful at identifying users with high second-order influence and low first-order influence. These metrics are tested in the following section.

3   Metric Results

We applied the trust metrics to the anonymized OWS data set, which contained 600MB worth of tweets, and compared the metrics against one another to test their ability to identify promising second-order influencers. The running time to analyze the data set was under five minutes. Each of the algorithms rely upon a summation of the degree of surrounding users, meaning that running an algorithm on a network of n nodes will require an analysis of each node's degree, and for each node, those degrees must be operated on and summed. This process scales linearly, meaning that these metrics are likely to prove effective for all conceivable applications.

We present the results here in four tables with each corresponding to the four algorithms developed: (1) Decremented Second-Order (DSO) Centrality, (2) Scaled Second-Order (SSO) Centrality, (3) Log-Based Second-Order (LBSO) Centrality, and (4) Edge-Weighted Second-Order (EWSO) Centrality. Each table focuses on a separate metric, with the highest value nodes from that algorithm presented in order. The node IDs and values for the other algorithms are provided as well to allow for comparison of results across algorithms and to identify correlations among their results. Table 1 demonstrates the results from the naïve metric, Decremented Second-Order Centrality. The table lists the nodes with the highest degree, decremented by one (the highest-degree node in the OWS data set was node 19 with 561 links). Little insight is provided by the first table alone, but having this metric provides insight for the subsequent analyses.

Table 1. Trust metrics ordered by Decremented Second-Order (DSO) Centrality

Node ID	DSO	SSO	LBSO	EWSO
19	560	1.43	12.2	2.77
1818	391	1.18	5.26	1.74
62	386	2.018	10.9	3.32
367	236	3.19	12.0	5.86
2035	207	0.207	0.0414	0.207
259	195	1.18	2.50	1.56
20	187	3.39	6.80	4.23
826	181	2.58	3.96	3.39
212	171	2.06	3.90	3.29
1194	163	3.57	9.74	6.41

The top results for Scaled Second-Order Centrality are illustrated in Table 2. The top node, 14043, also correlates with high values for Log-Based and Edge-Weighted Second-Order Centrality. Of note, all of the top Scaled nodes have a value of zero for their Decremented cases. Reviewing this conclusion, we see that the Scaled Second-Order Centrality algorithm is working as designed: it focuses its attention on individuals that do not have high degree, but nonetheless have high influence. However, as noted in Section 4.4, this algorithm does not provide higher values for users that are mentioned repeatedly by high-profile users or for users that are mentioned by multiple high-profile users. This point will not become fully clear until we study Edge-Weighted Second-Order Centrality.

Table 2. Trust metrics ordered by Scaled Second-Order (SSO) Centrality

Node ID	DSO	**SSO**	LBSO	EWSO
14043	0	188	2.68	394
14006	0	188	1.28	188
5316	0	145	1.16	145
16907	0	123	1.09	123
14627	0	123	1.09	123
14314	0	123	1.09	123
14088	0	123	1.09	123
14087	0	123	1.09	123
13525	0	123	1.09	123
12827	0	123	1.09	123

Table 3 provides an overview of results ordered by Log-Based Second-Order Centrality. The intent of this algorithm was to "filter out" or lessen the impact of linkages to other nodes with low first-order influence by summing the logarithms of neighbors' first-order influences. This method was generally successful, but looking at the top results, many top Log-Based influencers also have high Decremented results. This result means that the Log-Based algorithm is working as advertised, but fails to perform a second step of filtering out users who are high first-order influencers in addition to second-order influencers. In practice, additional predicate logic can filter out first-order influencers to truly reveal the non-obvious trusted sources, but that approach may be subject to scalability issues.

Table 3. Trust metrics ordered by Log-Based Second-Order (LBSO) Centrality

Node ID	DSO	SSO	**LBSO**	EWSO
344	122	5.21	14.0	11.1
19	560	1.43	12.2	2.77
367	236	3.19	12.0	5.86
62	386	2.01	10.9	3.32
4452	63	11.8	10.6	17.8
385	81	6.13	10.6	15.9
1194	163	3.57	9.74	6.41
2206	144	5.10	9.71	7.41
436	133	4.09	9.20	6.26
1309	66	5.45	8.70	12.0

Table 4 provides an overview of results ranked by Edge-Weighted Second-Order Centrality. Results are similar to the Scaled results (see Table 2), as one might expect since the two algorithms are related. Similar to the Scaled results, the highest Edge-Weighted results contain exclusively nodes with minimal first-order influence, which is a positive result. The intention of adding edge weighting in this algorithm was to promote users who were mentioned frequently by individuals with high influence. The results indicate that we succeed in promoting those users, since the table reveals that users must satisfy three general criteria simultaneously: (1) they must have low first-order influence; (2) they must have high Scaled influence (a direct but naïve indicator of second-order influence); and (3) they must have high Log-Based influence (a more sensitive metric of influence). Based on these results, we posit that Edge-Weighted Second-Order Centrality is a very viable metric to identify non-obvious sources of trustworthy and high-value information.

Table 4. Trust metrics ordered by Edge-Weighted Second-Order (EWSO) Centrality

Node ID	DSO	SSO	LBSO	EWSO
14043	0	188	2.68	395
11455	0	123	2.61	294
11454	0	123	1.86	208
2778	0	123	1.86	208
1062	0	123	1.86	208
14006	0	188	1.28	188
5316	0	145	1.16	145
16907	0	123	1.09	123
14627	0	123	1.09	123
14314	0	123	1.09	123

4   Conclusion

The proliferation of social media has changed the information landscape for a variety of events. In this study, we focus on a universal contributing factor to information value: trust. To support analysts in rapidly assessing the trustworthiness of social media data, we designed a set of metrics based on the theory of trust in information and tested those metrics on anonymized Twitter data, which was described in two sections. Leveraging the theory of "trust in information" that stresses structural network relationships, we developed a set of trust metrics. We then presented initial results when they were tested against an anonymized Twitter data set and demonstrated that our algorithms accurately assessed the trustworthiness of social media data. Our algorithm design indicates that while the Edge-Weighted Second-Order Influence metric is the most accurate in assessing information value and trustworthiness in general cases, the other metrics designed under this program also yield insight into network dynamics that are useful to analysts assessing trustworthiness. These metrics are likely to help analysts identify trustworthy information more quickly than manual inspection. Future work may consider the application of these algorithms to various domains, including responses to natural disasters and crises that result from violent crowds. It would also be instructive to investigate how advanced content analysis techniques could be combined with these metrics.

Acknowledgements. This work was performed under US Navy contract number N00014-12-M-0259. The authors thank Dr. Rebecca Goolsby for her significant technical support and eager engagement on this project. This work was funded in its entirety by the Office for Naval Research (ONR). We also acknowledge the contribution of Professor Adam Henry of the University of Arizona who introduced the authors to the concept of trust in information.

Reference

1. Henry, A.D., Dietz, T.: Information, networks, and the complexity of trust in commons governance. International Journal of the Commons 5(2) (2011)

Predicting Mobile Call Behavior via Subspace Methods

Peng Dai, Wanqing Yang, and Shen-Shyang Ho

School of Computer Engineering,
Nanyang Technological University, Singapore
{daipeng,wyang006,ssho}@ntu.edu.sg

Abstract. We investigate behavioral prediction approaches based on subspace methods such as principal component analysis (PCA) and independent component analysis (ICA). Moreover, we propose a personalized sequential prediction approach to predict next day behavior based on features extracted from past behavioral data using subspace methods. The proposed approach is applied to the individual call (voice calls and short messages) behavior prediction task. Experimental results on the Nokia mobility data challenge (MDC) dataset are used to show the feasibility of our proposed prediction approach. Furthermore, we investigate whether prediction accuracy can be improved (i) when specific call type (voice call or short message), instead of the general call behavior prediction, is considered in the prediction task, and (ii) when workday and weekend scenarios are considered separately.

Keywords: Sequential Prediction, Eigenbehavior, Principal Component Analysis, Independent Component Analysis, Behavior Prediction.

1   Introduction

To make accurate prediction on individual activities and behavioral patterns are new research directions in data mining, machine learning, and pervasive computing research communities. Based on data collected from mobile devices such as smart phones, one can predict and understand an individual's behavior and provide useful services or information to the individual. The industry takes a serious interest in these research topics with their game-changing potential in the highly competitive mobile device market [6]. Eagle and Pentland [2] introduced the eigenbehavior to represent repeating structures in an individual's behavior using principal components similar to those for eigenface [7]. They further claimed that "dimensionality reduction techniques [...] will play an increasingly important role in behavioral research".

In this paper, the two main contributions are (i) our investigation on whether independent components can be as useful as principal components in their representation of individual behavior and (ii) a sequential prediction approach to predict daily personal behavior modeled at hourly intervals. Our proposed approach assumes that the behavior of interest represented by primary

A.M. Greenberg, W.G. Kennedy, and N.D. Bos (Eds.): SBP 2013, LNCS 7812, pp. 466–475, 2013.
© Springer-Verlag Berlin Heidelberg 2013

(either principal or independent) components remain (almost) unchanged in the near future (e.g., the next few days). We demonstrate the feasibility of our proposed sequential prediction approach on the individual mobile call behavior prediction task. For this task, the objective is to predict whether an individual will call (voice call or/and short message) within some hour interval on the next day. Moreover, we investigate (i) whether predicting specific call type (voice call or short message) is a better problem setting than the general call behavior prediction setting; and (ii) whether splitting the training data to workday and weekend data can improve the prediction performance.

2   Dataset and Data Preprocessing

In Section 2.1, we briefly describe the Nokia Mobility Data Challenge (MDC) dataset that is used in this paper. In Section 2.2, we describe how we process the MDC data for the mobile call prediction task.

2.1   Nokia MDC Dataset

The MDC dataset consists of smartphone data collected in the Lake Geneva region from October 2009 to March 2011. Data types related to location (GPS, WLAN), motion (accelerometer), proximity (Bluetooth), communication (phone call and SMS logs), multimedia (camera, media player), and application usage (user-downloaded applications in addition to system ones) and audio environment (optional) were collected [6]. A total of 185 participants were involved. 38% of the participants are females and the rest are males. About two thirds of the participants are of age ranging from 22 to 33. Individual data was collected using the Nokia N95 smartphone and a client-server architecture. The open challenge data subset from the MDC dataset consisting of data from 38 participants for 8154 days are used in this paper. We focus on the voice calls, short messages, and the time they occurred.

2.2   Data Preprocessing

The call log data, consisting of call time, call duration, call type (short message or voice call), and etc. Call time and call type are used in our investigation. We, first, categorize about 2 years of daily call information for all participants into valid and invalid days. A valid day is a day where there are some phone activities (either voice calls or short messages). Otherwise, when there is no phone activity, it is a invalid day. Invalid days are ignored in the construction of the call behavior matrix for a participant so that there can be no row of zeros (i.e. no phone activity). Hence, we may not have consecutive days of call behavior vectors in the matrix. This call behavior matrix construction assumes that a person must have daily phone activity. Towards this end, the trivial prediction of no phone activity is not possible for our approach.

The call behavior of a participant is characterized by a $D_i \times 24$ matrix M_i, where i is the unique index for a participant and D_i is the total number of valid day used to construct our matrix for participant i. The call behavior matrix consist of binary values, one and zero, representing the existence or the non-existence of phone activity, respectively. Figure 1(a) shows the first 60 consecutive valid days of the phone activities for participant 2. Figure 1(b) shows the total number of valid days for the thirty-eight participants. Data from participant 7 include only 25 valid days and hence his data are not used in our experiments.

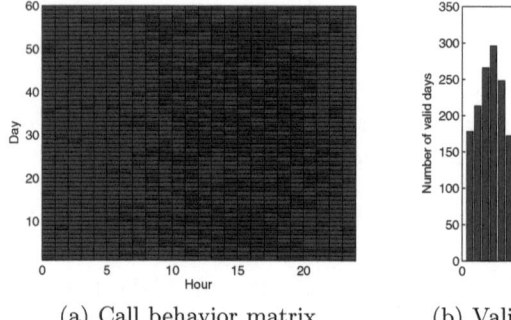
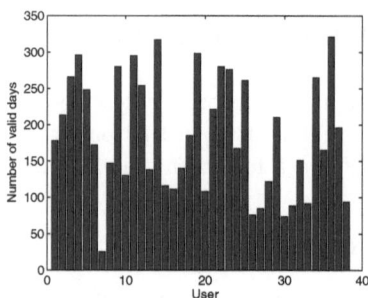

(a) Call behavior matrix   (b) Valid days for all participants

Fig. 1. The MDC data

3   Behavioral Representations

In Section 3.1, we introduce eigenbehavior and its implementation using principle component analysis (PCA). In Section 3.2, we introduce independent component analysis (ICA) as an alternative behavioral representation.

3.1   Eigenbehavior and Principal Component Analysis

Eigen representations have become one of the most popular techniques in pattern recognition (e.g. face recognition [7]) because of its strong discriminative ability. Eagle and Pentland [2] proposed using the so-called eigenbehavior to measure the distance between people, which is then used for the construction of a social network. They also apply the eigenbehavior for individual location prediction [1]. Eigenbehavior is based on the application of principal component analysis [5,2] on task-dependent daily individual behavioral representations.

Given an individual's daily m-dimensional behavior vectors, $\Gamma_1, \Gamma_2, \ldots, \Gamma_i$, $\ldots, \Gamma_D$, for a total of D days. Based on the convention used in [1], the average behavior of the individual is

$$\Psi = \frac{1}{D} \sum_{i=1}^{D} \Gamma_i. \tag{1}$$

The behavior deviation for a particular day from the mean behavior is

$$\Phi_i = \Gamma_i - \Psi. \tag{2}$$

Principal components analysis (PCA) is then performed on these vectors generating a set of m orthonormal vectors that can be linearly combined that best describe the distribution of the set of behavior vectors. The vectors and their corresponding scalars computed from PCA are the eigenvectors and eigenvalues of the covariance matrix

$$C = \frac{1}{D} \sum_{i=1}^{D} \Phi_i \Phi_i^T \tag{3}$$

3.2   Representing Behavior Using Independent Components

The goal of PCA is to find a set of orthogonal components that minimize the error in the reconstructed data. In fact, PCA seeks a transformation of the original data into a new frame of reference with as little error as possible, using fewer factors (i.e., principal components) than the original data. In particular, PCA is a popular approach to perform dimensionality reduction [7].

Here, we investigate whether independent components derived from independent component analysis (ICA) can be used to obtain behavior representation as useful as eigenbehavior for prediction tasks. In contrast to PCA, ICA seeks, not a set of orthogonal components, but a set of independent components. Two components are independent if any knowledge about one implies nothing about the other.

Again, given an individual's daily m-dimensional behavior vectors, $\Gamma_1, \Gamma_2, \ldots, \Gamma_i, \ldots, \Gamma_D$, for a total of D days. Each behavior vector

$$\Gamma_i = \sum_{i=1}^{n} w_i s_i \tag{4}$$

is assumed to be generated by the set of independent components $s_i, i = 1, \ldots, n$ and $w_i, i = 1, \ldots, n$, are the corresponding weights.

Our ICA representation is constructed using the InfoMax algorithm [3]. It is based on maximizing the output entropy (or information flow) of a neural network with non-linear outputs. Assume that $\mathbf{x}$ is the input to the neural network whose outputs are of the form $\phi_i \left(\mathbf{w}_i^T \mathbf{x} \right)$, where the ϕ_i are some non-linear scalar functions, and the $\mathbf{w_i}$ are the weight vectors of the neurons [3]. Then ICA model can be obtained by maximizing the entropy

$$H \left[\phi_1 \left(\mathbf{w}_1^T \mathbf{x} \right), \cdots, \phi_n \left(\mathbf{w}_n^T \mathbf{x} \right) \right] \tag{5}$$

of the outputs [4]. The MATLAB implementation of InfoMax algorithm is publicly available in DTU toolbox [8].

4   Behavior Prediction Approaches

In Section 4.1, we introduce the approach proposed by Eagle and Pentland [2] that predicts the later part of the day based on information on the earlier part of the day. In Section 4.2, we describe our proposed sequential prediction approach for the next day(s) based only on data from previous days.

4.1   Single-Day Method

Both PCA (or eigenbehavior) and ICA share the same idea that the daily behavior vector obtained in Section 2.2 can be treated as a combination of several primary daily behavior components generated by either approach. An individual's primary daily behavior components represent a space upon which all of his daily behavior can be projected with different levels of accuracy. Using the primary behavior components, it is possible to predict the future behavior for an individual.

One straightforward way to predict the future behavior for an individual at the later part of a particular day is to reconstruct an entire daily behavior vector using only behavior information from an earlier part for that day [2]. Let

$$\mathbf{A} = [\mathbf{\Phi}_1, \mathbf{\Phi}_2, \ldots, \mathbf{\Phi}_i, \ldots, \mathbf{\Phi}_M] \tag{6}$$

denotes the primary behavior matrix calculated from N days of behavior data such that each column contains one principal/independent component $\mathbf{\Phi}_i$ that is a $24-$dimensional vector corresponding to the ith primary behavior component. Assuming the first p hours behavior for that day, $\mathbf{\Gamma}_{1:p}$, are known. Hence,

$$\mathbf{A}_s \mathbf{v} = \mathbf{\Gamma}_{1:p} \tag{7}$$

where $\mathbf{A}_s$ is a $p \times M$ matrix corresponding to the first p row of $\mathbf{A}$. Then one obtains a M-dimensional reconstruction vector

$$\mathbf{v} = \mathbf{A}_s^{-1} \mathbf{\Gamma}_{1:p} \tag{8}$$

where $\mathbf{A}_s^{-1}$ is the pseudo inverse matrix of $\mathbf{A}_s$. To predict the rest of the day, i.e., $p+1$ to 24 hours in $\mathbf{\Gamma}$, one reconstructs the entire behavior vector using

$$\mathbf{\Gamma} = \mathbf{A}\mathbf{v}. \tag{9}$$

The above predictive model assumes dependency of behavior within the same day, and the relationship stays relatively stable. We refer to this prediction approach as the **single-day method**.

4.2   Multiple-Day Method

An alternative prediction approach is to model the daily behavior as a whole day event and then predict the next day(s). Assuming a sequence of D days

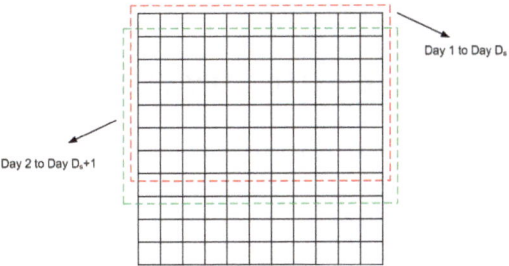

Day 1 to Day D_s

Day 2 to Day D_s+1

Fig. 2. Multiple-day method for generic future behavior prediction

of 24-dimensional behavior vectors in our prediction scenario, one predicts the behavior for day D_s+1 given behavior information from previous D_s days. Based on the assumption that there cannot be too much changes in a person's behavior within a short time interval, our proposed approach models daily behavior within a fixed temporal window of D_s days (see Figure 2) and predict the next day's (day $D_s + 1$) behavior based on this daily behavior model. We first obtain the primary behavior matrix representing the behavior from day 1 to D_s, denoted as $\Gamma_1^{D_s}$ corresponding to the red bounding box in Figure 2. According to our assumption, $\Gamma_2^{D_s+1}$ (denoted as the green bounding box in Figure 2) share the same primary behavior matrix as $\Gamma_1^{D_s}$. Note that row $D_s + 1$ represents the unknown next day behavior that we want to predict.

Using the D_s days of daily behavior vectors, one obtains the daily behavior model as a set of D_s-dimensional primary components, $\Phi'_i, i = 1, \ldots, 24$. The first M primary components are chosen to construct the primary behavior matrix

$$\mathbf{C} = [\Phi'_1, \Phi'_2, \ldots, \Phi'_M] \tag{10}$$

where each column of C correspond to a primary vector. Then one obtains the $M \times 24$ reconstruction matrix

$$\mathbf{V} = \mathbf{C_s}^{-1}\Gamma_2^{D_s} \tag{11}$$

where $\mathbf{C_s}^{-1}$ is the pseudo inverse matrix of the $(D_s - 1) \times M$ matrix $\mathbf{C_s}$ corresponding to the first $D_s - 1$ rows of $\mathbf{C}$, since

$$\Gamma_2^{D_s} = \mathbf{C_s}\mathbf{V} \tag{12}$$

Then, the prediction of day $D_s + 1$ can be obtained as the last row of

$$\Gamma = \mathbf{CV}. \tag{13}$$

This method makes use of the relationship embedded in the historical data from the previous D_s days. We refer to this prediction approach as the **multiple-day method**. Note that the multiple-day method can be used to predict not only the behavior for the next day (i.e., $D_s + 1$) but also the next $n(<< D_s)$ day's behavior. The modified prediction scheme for day $D_s + n$ is

$$\Gamma = \mathbf{C} \left(\mathbf{C'_s}^{-1} \Gamma^{D_s}_{n+1} \right) \tag{14}$$

where $\mathbf{C'_s}$ is a $(D_s - n) \times M$ matrix corresponding to the first $D_s - n$ rows of $\mathbf{C}$.

5   Experimental Results

First, we study the prediction performance of two subspace approaches, PCA and ICA, utilizing different number of daily behavior vectors via the single-day and multiple-day methods. Then, we investigate whether the prediction performance can be improved by (i) considering the call types: short messages and voice calls; and (ii) then further splitting the data into workday and weekend observations. For illustration purposes, we apply only the PCA-based single-day and multiple day methods for this investigation. For all our empirical results, we use 4 primary components for either PCA or ICA. Since the objective of the prediction task is to predict whether an individual will or will not call (i.e., 1 or 0) within some hour interval the next day, a threshold is required to decide on the final prediction. Here, the threshold is set to 0.5.

From Figure 3, we observe that both PCA-based and ICA-based multiple-day methods perform better than the single-day methods. Moreover, their prediction performance are comparable. PCA-based single-day method performs slightly better than ICA-based single-day method. One thing to note is that the single day approach has to use the first p (here, $p = 12$, i.e., using first half of the day to predict the second half) hourly observations to calculate the reconstruction vector, $\mathbf{v}$ in (8). Thus, it is impossible to predict the whole day. On the other hand, the multiple-day method uses the previous days' observations to calculate the reconstruction matrix $\mathbf{V}$ in (11). Therefore, it can predict an individual's behavior for the entire day.

From Figure 3, we see that the number of daily behavior vectors used for optimal prediction performance is seventy for the PCA-based single-day method and the multiple-day methods. However, PCA-based single-day method has comparable prediction performance when the number of days used is between 10 and 80. The performance of ICA-based single-day methods degrades as the number of days used increases.

Using 20 days and 70 days of data to build the behavior matrices, we investigate the distribution of participants at various level of average prediction accuracy shown in Figure 4. Considering using 70 days of data, we observe that the multiple-day method is very competitive due to its use of information from previous 70 days when prediction is made. In particular, 21 out of the 38 participants achieve prediction accuracy of more than 80% for each multiple-day method. Furthermore, one observes that prediction performance for single methods can go as low as 55% for a user while multiple-day methods achieves a minimum of 65% accuracy for the participants. Considering using only 20 days of data, PCA-based multiple-day method performs the best with 19 out of 38 participants achieve 80% or more prediction accuracy. While PCA-based single day performs relative well with 17 participants achieving accuracy of 80% or more, we observe from the Figure 4 that prediction performance for 6 participants are 65% or below. Compared to the other approaches, the number of participants with poor prediction performance is significantly higher. One notes that when a small number of days of data are used, ICA-based methods have average prediction performance with respect to the number of participants. Again, readers are reminded that multiple-day and single-day methods can be considered to be solutions for two different prediction tasks or problem settings.

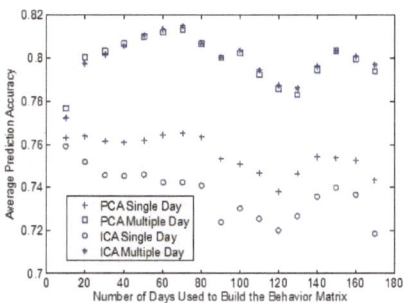

Fig. 3. Effect of different number of daily behavior vectors on prediction performance

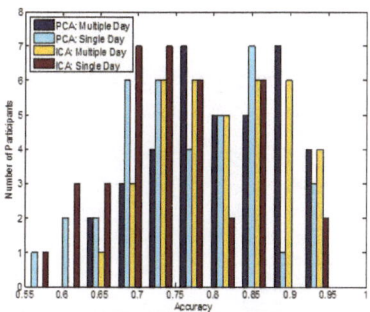 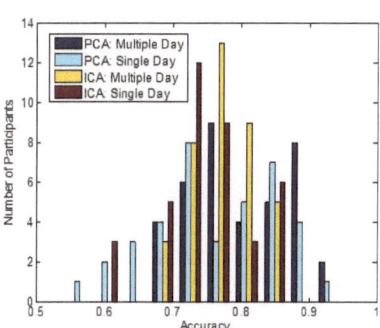

Fig. 4. The number of participants achieving various average prediction accuracy for the different approaches when the number of days to build the behavior matrix are 70 (left) and 20 (right), respectively

From Figure 5, we observe that prediction performances are improved for both methods when call types: short messages and voice calls, are considered. Hence, specific (call) behaviors are more predictable. One notes that the number of days of data used has a significant effect on the PCA-based single-day method for short messaging prediction.

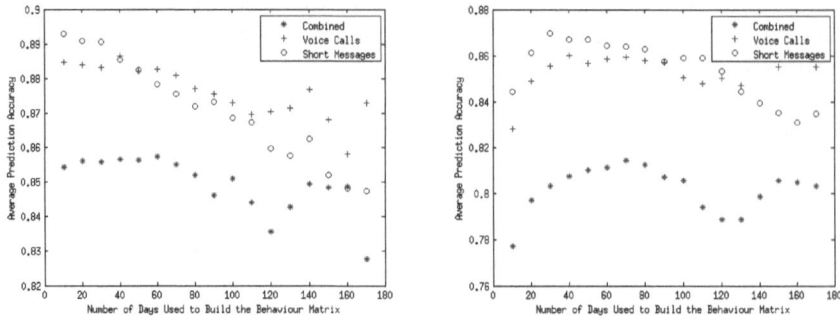

Fig. 5. Prediction accuracy for PCA-based Single-Day Method (left) and PCA-based Multiple-Day Method (right) when call behavior is further categorized into sending short messages and making voice calls

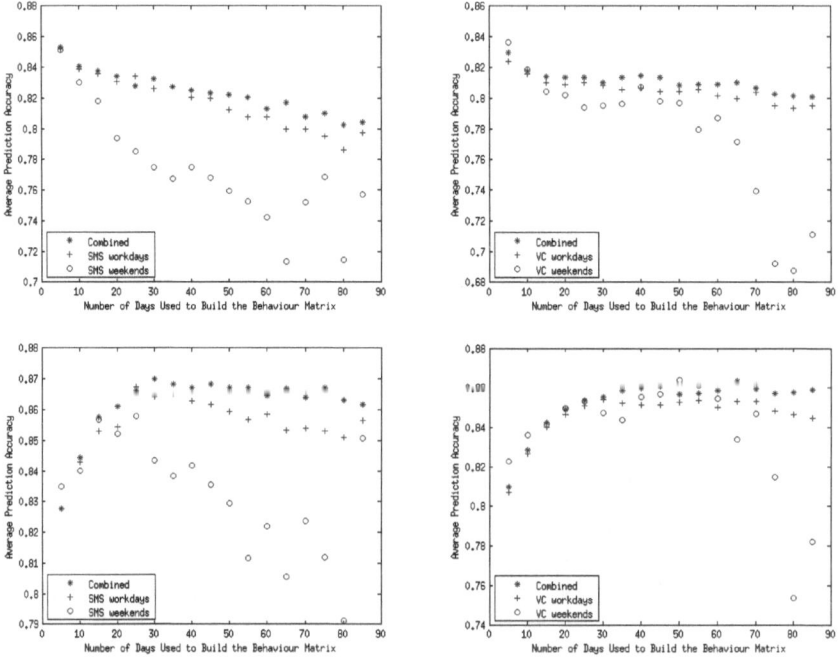

Fig. 6. Prediction accuracy for PCA-based Single-Day Method (top) and PCA-based Multiple-Day Method (bottom) when short message data (left) and voice call data (right) are split based on whether they occurred on weekends or workdays

From Figure 6, we observe that in general the two methods have better prediction performance for workday call behavior than for weekend call behavior. It is particularly significant that short messaging behavior is less predictable during weekends. The significant drop and haphazardness in the prediction performance during weekends as the number of days of observation used increases is most probably due to shortage of data for testing purposes. Hence, the empirical results using more than 60 days of data should be ignored. Towards this end, we conclude that PCA-based multiple-day method predicts well on voice call behavior for both weekend and workday, and occasionally even slightly better than using weekend and workday data together (see bottom right graph in Figure 6).

6   Conclusions

In this paper, we investigate whether independent components can be as useful as principal components in their representation of individual behavior. Moreover, we propose a sequential prediction approach to predict daily personal behavior modeled at hourly intervals. We demonstrate the feasibility of our proposed sequential prediction approach on the individual mobile call behavior prediction task using the Nokia MDC dataset. We observe that formulating a specific (voice call or short message) behavior prediction problem is better than a general (call) behavior prediction problem as one can obtain better prediction accuracy in the former task. Also in general, we observe that workday behavior is more predictable than weekend behavior.

References

1. Eagle, N., Pentland, A., Lazer, D.: Inferring Social Network Structure using Mobile Phone Data. Proceedings of the National Academy of Sciences 106(36), 15274–15278 (2009)
2. Eagle, N., Pentland, A.S.: Eigenbehaviors: Identifying structure in routine. Behavioral Ecology and Sociobiology 63, 1057–1066 (2009)
3. Hyvarinen, A., Oja, E.: Independent component analysis: algorithms and applications. Neural Netw. 13(4-5), 411–430 (2000)
4. Bell, A., Sejnowski, T.J.: An Information-Maximization Approach to Blind Separation and Blind Deconvolution. Neural Computation 7, 1129–1159 (1995)
5. Jolliffe, I.T.: Principal Component Analysis. Springer-Verlag New York, Inc. (1997)
6. Laurila, J.K., Gatica-Perez, D., Aad, I., Blom, J., Bornet, O., Do, T.-M.-T., Dousse, O., Eberle, J., Miettinen, M.: The mobile data challenge: Big data for mobile computing research. In: Proc. on Mobile Data Challenge by Nokia Workshop in Conjunction with Int. Conf. on Pervasive Computing, Newcastle (June 2012)
7. Turk, M., Pentland, A.S.: Eigenfaces for Recognition. Journal of Cognitive Neuroscience 3(1), 71–86 (1991)
8. ICA:DTU Toolbox, http://cogsys.imm.dtu.dk/toolbox/ica/

Modeling the Interaction between Emergency Communications and Behavior in the Aftermath of a Disaster

Shridhar Chandan, Sudip Saha, Chris Barrett, Stephen Eubank,
Achla Marathe, Madhav Marathe, Samarth Swarup,
and Anil Kumar S. Vullikanti

Network Dynamics and Simulation Science Laboratory,
Virginia Bioinformatics Institute,
Virginia Tech, 1880 Pratt Drive, Blacksburg, VA 24060, USA
{shridhar,ssaha,cbarrett,seubank,amarathe,
mmarathe,swarup,akumar}@vbi.vt.edu

Abstract. We describe results from a computer simulation-based study of a large-scale, human-initiated crisis in a densely populated urban setting. We focus on the interaction between human behavior and the communication infrastructure in the aftermath of the crisis. We study the effects of sending emergency broadcasts immediately after the event, advising people to shelter in place, and show that this relatively mild intervention can have a large beneficial impact.

Keywords: synthetic information, computer simulations, disaster modeling, nuclear terrorism.

1   Introduction

Increasing threats from natural disasters (e.g., hurricanes Katrina and Sandy) and human-initiated attacks/failures (especially, nuclear threats) on urban populations and infrastructures have led to significant interest in the analysis of potential impacts and contingency planning, e.g. [1, 2]. Government agencies at all levels–federal, state and local–routinely plan for such disasters. There has been a lot of work on modeling the effects of nuclear disasters [1–5], primarily focused on health effects and "evacuation" vs. "shelter in place" policies. They capture the physical, geographical and radiological aspects in great detail, but ignore the interdependencies between infrastructures and the effect of behavior. Failures (and methods to prevent them) have been studied in inter-dependent critical infrastructures in the case of non-nuclear disasters [6–9]. Some of these, e.g., [8, 10, 11] have studied the impact of the communication network, but primarily in the context of evacuation.

A common limitation in most of this research is that effects of individual behavior are not adequately modeled. It has been observed repeatedly that people do not always follow emergency plans. This is especially important in disasters

A.M. Greenberg, W.G. Kennedy, and N.D. Bos (Eds.): SBP 2013, LNCS 7812, pp. 476–485, 2013.

that might be caused by nuclear explosions, where exposure in the first few hours of the event can cause significant health hazards. A common policy recommendation in such events is to shelter in place; however, many studies from past disasters (both nuclear and non-nuclear) have shown that people do not adhere to such recommendations, e.g., [12–14]. Some of the key reasons include: lack of information, need to coordinate with their family (e.g., ensuring the safety of children), lack of resources. It suggests that the availability of the communication network would have a significant impact on the response to a disaster, which is borne out from past experience. There has been very limited study of the impact of failures of the communication network, because of the modeling and computational challenges involved. Therefore, modeling individual behavior, and interdependent infrastructures is crucial for analyzing and planning for large disasters in urban regions (see, e.g., [8, 10]).

In this paper, we use an agent-based simulation approach to evaluate the impact of failures in the communication infrastructure on individual behavior and their health. We study a hypothetical nuclear detonation scenario in Washington DC, in which parts of the cellular infrastructure (as well as other infrastructures, such as roads and the power network) are partially destroyed. We study a specific aspect of the communication network, namely, the support for Emergency Broadcasts in the affected region in the aftermath of the event, and its impact on panic behavior, and the overall health effects. We find that even a simple intervention of emergency broadcast (EBR) which prompts people to take shelter-in-place, sent every two hours, can save thousands of lives and reduce panic. More specifically, we find that without EBR, 5-7% of people seek shelter whereas with EBR 15-20% people seek shelter, in our simulated scenario. This, in turn, leads to people who received the EBR having less than a quarter of the exposure to radiation of those who do not receive an EBR. This suggests that restoring the communication network might be a very useful strategy. Further information on this topic can be found at [15].

2   Scenario and System Model

Scenario Description: The scenario assumes a nuclear detonation at a ground level on a working day at 10 a.m. in the heart of Washington DC. The prompt radiation (gamma and neutron) effects of this blast cover a circular area from ground zero and the degree of damage subsides with distance from ground zero.

Most of the human casualties occur from thermal burns, exposure to radiation, pressure wave, impact of projectiles and fragments including glass created by the blast, and radioactive fallout. Due to environmental contamination, the occurrence of injuries and fatalities continue even after the burning and blast effects have tapered off.

In case of a nuclear detonation, limited resources need to be used to ensure that people shelter-in-place instead of evacuating to minimize exposure to radiation. This is especially important for individuals who are likely to be in the fallout path as they evacuate. Another critical step is to communicate to the exposed population how to efficiently decontaminate. Timely decontamination of highly contaminated individuals has been proven to be effective in reducing casualties [4].

Given the importance of disseminating timely and appropriate instructions, we focus on understanding the role of communication in influencing human behavior and its further effect on their exposure to radiation. In such a scenario, there will be widespread power outages and heavy losses to the cellular infrastructure and electronic devices. We assume that the planning authorities will try to send "Emergency Broadcasts" (EBR) every two hours to communicate instructions to the people. However due to power outage, only a small subset of base stations and cell phone devices are operational and hence only 10% of the individuals in the affected study area receive the EBR. We further assume that individuals who receive the EBR have a 50% chance of sheltering-in-place because in spite of understanding the benefits of sheltering, some people among the ones who received the EBR, will still try to evacuate or reconstitute family or visit a hospital etc.

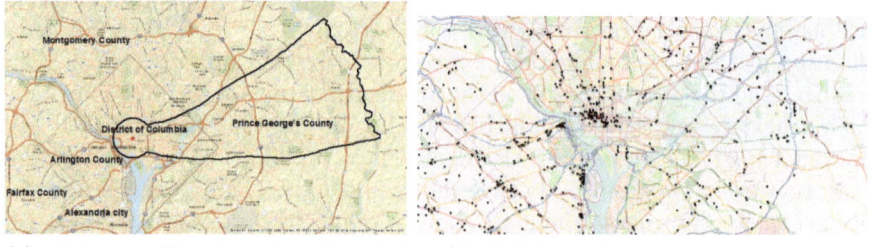

(a) Polygon Showing the Detailed Study Area and the fallout path

(b) Cell tower locations in the region.

Fig. 1. Simulation region and communication infrastructure

Modeling the Communication Infrastructure: Our communication network model builds on the framework of [16], which develops a first-principles based approach for modeling mobile networks and spectrum demand. Most cellular infrastructure data is proprietary, and is difficult to obtain. Hence we construct a partial representation of the cellular infrastructure in this region by integrating cell tower data from TowerMaps [17], and by queries to the APIs from Sprint, AT&T and the FCC Antenna Registration System; let $\mathcal{C}$ denote the set of all the inferred cell tower locations. For our region, we have $|\mathcal{C}| = 5187$. Since we do not have information about sharing agreements between providers, we assume that all the cell towers for different providers are co-located (i.e., each tower in $\mathcal{C}$ hosts antennas for all the providers).

Further, since the antenna characteristics are not known, we assume the coverage region for a tower $C \in \mathcal{C}$ is part of the region associated with C in the voronoi decomposition for $\mathcal{C}$, within a given service range (which varies from 0.75 miles to 1.5 miles, depending on the scenario). Each tower is assumed to have a maximum capacity of 50 Mbit/sec. and the cell tower battery is expected to last for one hour; therefore, the region which loses power also loses cell phone coverage after 1 hour, except for those parts which are within the service range of other cell towers with power. Also, all cell towers within 0.6 miles of the blast site get destroyed by the blast.

Individuals in the population are assigned cellular devices based on their demographics, using a CDC survey [18]. The survey data provides household's size, income and ages, number of workers in the household, number of cells phones and a device penetration rate of 53.44%. We construct regression trees based on the demographic information that best describes the average number of cell phones among households, and use them to assign device ownership [16]. This method leads to a cell phone assignment that closely matches the survey data.

We assign wireless providers to device owned households randomly, in proportion to their market shares as given in [19]. Initially, we assume phones have battery levels chosen randomly; these drain over time, depending on the number of calls made. The communication network supports normal voice and digital calls, as well as Emergency Broadcasts from the cell towers.

We assume in the event of such a disaster, normal traffic is disrupted, and people call only for reconstituting their families, in order to respond to the disaster. Therefore, instead of using measured arrival rate distributions for determining calls (as in [16]), the *Behavioral Module* selects the calls to be made. More details of the communication model e.g. call sessions on a normative day, mobility of the people, load calculation, channel allocation, etc. can be found in [20–22].

The Emergency Broadcasts (EBR) are done every two hours at locations where cell tower coverage is available. We assume that EBR messages can be received by individuals even if they do not have functioning phones, through sharing–as a user who receives an EBR can inform others who might not have phones at that location. Therefore, the following simple model works for EBR: each time the EBR is made, a randomly selected 10% of the population in the cell coverage area receives the EBR. Cell tower coverage is available where cell towers either have power or backup battery. Given the random selection and the movement of the population, every time the EBR is sent, a partially different set of population receives it.

A related study on human-initiated cascading failures in societal infrastructures which uses this communication model is available in [10]. Also see [23,24].

Modeling the Population and Human Behavior: We model a *synthetic population* of the Washington DC Metro Area using techniques described in Barrett et al. [25] and Beckman et al. [26]. This generated synthetic population consists of ~4.1 million agents, endowed with demographics, home locations, and daily activity patterns. These daily activities are geo-located, which allows us to pinpoint the geographic location of each individual at each instant in a typical weekday.

From this population, we sub-select the individuals who are present in the study area at the time of the detonation. This is a population of 730,833 agents, which includes residents of the area, residents of surrounding areas whose activities have brought them into the study area at the moment of the detonation, as well as a set of transients who happen to be visiting the region (tourists and

Synth. pop.	Data sources
Base US pop.	Amer. Community Survey
	TIGER/Line shapefiles
	Nat. Center for Edu. Stat.
	Nat. Household Travel Sur.
	Navteq
	Dun & Bradstreet
Transient pop.	Destination DC
(additional)	Smithsonian visit counts
Dorm students	CityTownInfo
(additional)	DC public access -
	online Data Catalog

(a)

Behavior	High-level description
Household	Call, move towards
reconstitution	household members
Evacuation	Move outside region
Shelter-seeking	Shelter in place,
	or move towards shelter
Healthcare-	Call 911,
seeking	move towards hospital
Panic	Call 911, run outdoors,
	move towards hospital
Aid & Assist	Transport hurt
	individuals to hospital

(b)

Fig. 2. Datasets and Behaviors for the synthetic population used in the simulation

business travelers) [27]. The data sets used and the populations generated are summarized in table 2a.

Agent behavior is modeled using a decentralized semi-Markov decision process formalism using the framework of options [28]. An option is a behavior that is described as a policy over low-level actions, together with conditions for starting and stopping the execution of the option. The low-level actions in our simulation are just moving (towards a chosen destination) and/or calling (a family member or 911). The high-level behaviors, or options, are described in table 2b. More details are available elsewhere [27].

This study is computationally very challenging, and we summarize the resources used here. The total compute time for one complete simulation run is about 35 hours, on a large 60 node multi-core cluster. The computation is very data intensive, and in order to facilitate the data exchange, our system is tightly coupled with an oracle database. Each module uses the database for storing the inputs and outputs, as well as intermediate data, in some cases. Each simulation run requires about 27GB of space for storing all the data, which leads to a few TB of space for a complete experiment.

3   Results and Discussion

In order to understand the impact of timely information dissemination, and its consequent effect on radiation exposure, we set up the following experiment. An emergency broadcast (EBR) is made every two hours after the detonation, through which authorities communicate relevant information to the people. We assume that only 10% of the population receives EBR and those who receive EBR have a 50% probability of taking shelter. However people who do not receive EBR have only 10% probability of sheltering. The higher conditional probability of sheltering by those who receive EBR results in lower exposure to radiation because more of these individuals choose to shelter as opposed to engaging in other behavioral options such as family reconstitution, evacuation, panic etc. which are likely to keep them outdoors and exposed.

The results highlight the changes in people's behavior due to EBRs and the resulting changes in health outcomes due to differences in the speed of communication of safety information. Figure 3a shows the behaviors adopted by the subpopulation which never receives EBR. It also shows the proportion of those (among the ones who never receive EBR) who have died and have moved out of the study area. Iteration 1-35 correspond to hour 0-16; hour 0 being the time of detonation. It shows that by hour 16, about 75% of the people with no communication have died.

A very small fraction, i.e. approximately 5%, have moved out of the study area whereas a large fraction of people are engaged in household reconstitution right after the event. Unaware of the existence of radiation and the importance of sheltering, it is natural to try and find family members in a disaster. This drives people to go outdoors, resulting in exposure and ultimately casualties for some. The exposure to radiation is maximum in the early part of post-detonation and this is also the time when people are confused, panicked, worried about family and unsure of what just occurred. The figure shows that there is high level of

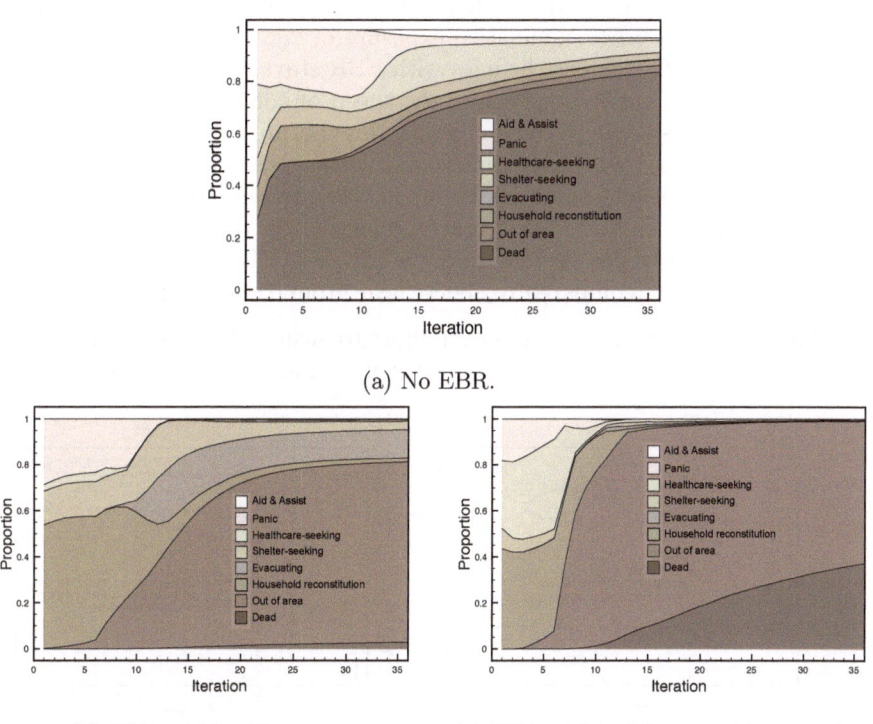

(a) No EBR.

(b) EBR within 4 hours.   (c) EBR within 4 hours, within 1 mile.

Fig. 3. Proportions of behaviors/state for two different sub-populations

panic and family constitution in the early part of the graph, resulting in higher deaths in the later part as exposure accumulates.

Figure 3 focuses on individuals who received at least one emergency broadcast during the study time. Figure 3b shows the behavior or state of the people who received an EBR within the first 4 hours of the detonation whereas figure 3c spatially restricts it to people who were within 1 mile of ground zero.

Lets first compare figure 3b with figure 3a to observe the difference in behavior between people who received an EBR versus those who did not receive an EBR. Important observations include: (1) Less than 5% of the people who receive an EBR die as opposed to 75% in the "no EBR" case. (2) Without EBR 5-7% of the people are seeking shelter whereas with EBR 15-20% people are seeking shelter. (3) After iteration 10 (hour 3), when the radiation has subsided, lot more people with EBR are evacuating to safer grounds and out of area.

Now lets compare figure 3b and 3c. Both figures consider the sub-population that received EBR within 4 hours however 3c constraints it further by spatially restricting it to within 1 mile of ground zero. The plots show that the proportion of dead and panic near ground zero is much higher within 1 mile compared to the rest of the study area.

Figures 4 shows the average cumulative exposure levels of the people over time. Figure 4a shows radiation exposure of individuals who never received an EBR and those who received within 4 hours and were located within a mile of ground zero. The average cumulative exposure of "no EBR" population grows to more than 800cG whereas the ones with EBR stays constant around 200cG.

Figure 4b shows the importance of the speed of communication in the time of crisis. It contrasts the average cumulative exposure of people who received EBR within 4 hours; within 4-6 hours and within 6-8 hours since the time of detonation. Results find that the longer it takes to communicate, the higher is the average exposure level. Since the population in the plot is not spatially restricted and is averaged over the entire study area, the exposure level is much lower at up to 50cG compared to 200cG near ground zero in figure 4a.

Validation. We have made significant effort to ensure that the models used in the study are validated. We discuss this briefly below. Additional work needs to

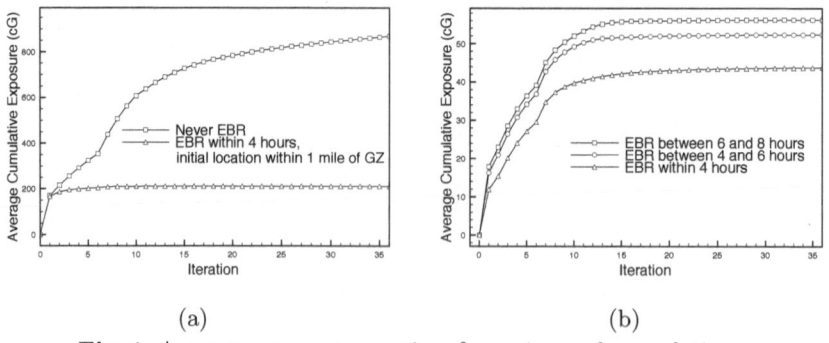

(a)   (b)

Fig. 4. Average exposure over time for various sub-populations

be done in this area. The advent of social media and data collected during recent large-scale crisis (e.g. Fukashima disaster) can be very useful in this regard. Additional discussion on this subject can be found in [15, 29, 30].

1. A range of public and commercial data sets (see Table 2) were used to ensure that the synthetic population and the built infrastructure is accurately represented at the time of the detonation. Specifically, our methods ensure that the statistical distribution of our synthetic population agrees with measured census data. Data regarding built infrastructure has been obtained via a number of open source and commercial sites. Information regarding cellular infrastructure is also obtained from diverse sources to ensure that cell towers are located accurately. See [16, 31–33] for more information on this topic.
2. Structural validation was performed to ensure that emergent system level properties exhibit observed structural features. This includes, the spatial and temporal density of the populations, distribution of digital devices and movement patterns of individuals.
3. For certain parameters, e.g. probability of sheltering and EBR fraction, there is no published data. Given this, we do a sensitivity analysis study to understand their impact on the system-level behavior. We refer the reader to [20, 30, 32, 33] for more details on the population model and validation.

4   Conclusions

This research uses an agent-based simulation approach to understand the impact of timely communication, after an improvised nuclear device detonation, on individuals' behavior and health. Results show that emergency broadcast which prompts people to take shelter-in-place, sent every two hours, can have significant impact on reducing casualties and adverse health effects. They can also reduce fear and panic in the population. Our work is a first step in evaluating important policy recommendations in [1,34] as they relate to use of tele-communication networks for timely information dissemination, using realistic large scale computer simulations.

Acknowledgements. The authors would like to thank members of the Network Dynamics and Simulation Science Laboratory (NDSSL) for their suggestions and comments. This work has been partially supported by NSF NETS Grant CNS-0831633, NSF NetSE Grant CNS-1011769, NSF SDCI Grant OCI-1032677, DTRA R&D V&V Grant HDTRA1-11-1-0016, DTRA CNIMS Contract HDTRA1-11-D-0016-0001, NIH MIDAS Grant 2U01GM070694-09, NSF PetaApps Grant OCI-0904844 and NSF ICES Grant CCF-1216000.

References

1. Buddemeier, B.R., Valentine, J.E., Millage, K.K., Brandt, L.D.: National Capital Region: Key response planning factors for the aftermath of nuclear terrorism. Technical Report LLNL-TR-512111, Lawrence Livermore National Lab. (2011)

2. Wein, L.M., Choi, Y., Denuit, S.: Analyzing evacuation versus shelter-in-place strategies after a terrorist nuclear detonation. Risk Analysis 30(6) (2010)
3. Dombroski, M.J., Fischbeck, P.S.: An integrated physical dispersion and behavioral response model for risk assessment of radiological dispersion device (RDD) events. Risk Analysis 26(2), 501–514 (2006)
4. Homeland Security Council Interagency Policy Coordination Subcommittee: Planning guidance for response to a nuclear detonation (2009)
5. Millage, K.: Modeling the effects of nuclear weapons in an urban setting. In: Radiation Countermeasures Symposium an AFRRI 50th Anniversary Event (2011)
6. Panzieri, S., Setola, R.: Failure propagation in critical interdependent infrastructures. Int. J. Modeling, Identification and Control, 69–78 (2008)
7. Peeta, S., Hsu, Y.T.: Integrating Supply and Demand Aspects of Transportation for Mass Evacuation under Disasters. Technical Report NEXTRANS Project No. 019PY01, Purdue University (October 2010)
8. Barrett, C., Beckman, R., Channakeshava, K., Huang, F., Kumar, V.A., Marathe, A., Marathe, M.V., Pei, G.: Cascading failures in multiple infrastructures: From transportation to communication network. In: The Fifth International IEEE CRIS Conference on Critical Infrastructures, Beijing, China (2010)
9. Little, R.: Controlling cascading failure: Understanding the vulnerabilities of interconnected infrastructures. Journal of Urban Technology, 109–123 (2002)
10. Barrett, C., Beckman, R., Channakeshava, K., Huang, F., Kumar, V.A., Marathe, A., Marathe, M.V., Pei, G., Saha, S.: Human initiated cascading failures in societal infrastructures. PLoS ONE (2012)
11. Liu, H.X., Ban, J.X., Ma, W., Mirchandani, P.B.: Model reference adaptive control framework for real-time traffic management under emergency evacuation. Journal of Urban Planning and Development 133(1), 43–50 (2007)
12. Drabek, T.E., Boggs, K.S.: Families in disaster: reactions and relatives. Journal of Marriage and the Family 30, 443–451 (1968)
13. Gerber, B.J., Ducatman, A., Fischer, M., Althouse, R., Scotti, J.R.: The potential for an uncontrolled mass evacuation of the DC metro area following a terrorist attack: A report of survey findings. Technical report, West Virginia University, Homeland Security Programs (2006)
14. Guterbock, T.M., Lambert, J.H., Bebel, R.A., Parker, M.W.: NCR behavioral survey 2011: Work, school or home? Issues in sheltering in place during an emergency. Technical report, Center for Survey Research, University of Virginia (August 2011)
15. NDSSL, Virginia Tech: Computational analysis of behavior and urban disaster resilience, http://ndssl.vbi.vt.edu/projects/disaster-resilience/
16. Beckman, R., Channakeshava, K., Huang, F., Kumar, V., Marathe, A., Marathe, M., Pei, G.: Synthesis and analysis of spatio-temporal spectrum demand patterns: A first principles approach. In: IEEE DySPAN (2010)
17. TowerMaps: Wireless antenna facility location data, http://www.towermaps.com/
18. Center for Disease Control: National Health Interview Survey (NHIS), http://www.cdc.gov/nchs/about/major/nhis/nhis_2007_data_release.htm
19. Beckman, R., Channakeshava, K., Huang, F., Kumar, V.A., Marathe, A., Marathe, M., Pei, G.: Implications of dynamic spectrum access on the efficiency of primary wireless market. IEEE Dynamic Spectrum Access Networks, DySPAN, 2–12 (April 2010)
20. Kim, J., Kumar, V., Marathe, A., Pei, G., Saha, S., Subbiah, B.: Modeling cellular network traffic with mobile call graph constraints. In: Proceedings of the 2011 Winter Simulation Conference, WSC, pp. 3165–3177 (December 2011)

21. Saha, S., Kumar, V., Marathe, A., Pei, G., Subbiah, B., Kim, J.: Clearing secondary spectrum market with spatio-temporal partitioning. IEEE Dynamic Spectrum Access Networks, DySPAN (October 2012)

22. Kim, J., Kumar, V., Marathe, A., Pei, G., Saha, S., Subbiah, B.: Analysis of policy instruments for enhanced competition in spectrum auction. IEEE Dynamic Spectrum Access Networks, DySPAN (October 2012)

23. Parikh, N., Swarup, S., Stretz, P., Rivers, C., Lewis, B., Marathe, M., Eubank, S., Barrett, C., Lum, K., Chungbaek, Y.: Modeling human behavior in the aftermath of a hypothetical improvised nuclear detonation. In: The 12th International Conference on Autonomous Agents and Multiagent Systems, AAMAS, Minnesota, USA (May 2013)

24. Lewis, B., Swarup, S., Bisset, K., Eubank, S., Marathe, M., Barrett, B.: A Simulation Environment for the Dynamic Evaluation of Disaster Preparedness Policies (2013)

25. Barrett, C., Bisset, K., Leidig, J., Marathe, A., Marathe, M.: An integrated modeling environment to study the co-evolution of networks, individual behavior, and epidemics. AI Magazine 31(1), 75–87 (2010)

26. Beckman, R.J., Baggerly, K.A., McKay, M.D.: Creating synthetic baseline populations. Transportation Research Part A: Policy and Practice 30(6), 415–429 (1996)

27. Network Dynamics and Simulation Science Laboratory: Social, Health and Sociotechnical effects of an IND in the National Capitol (2012)

28. Sutton, R., Precup, D., Singh, S.: Between MDPs and semi-MDPs: A framework for temporal abstraction in reinforcement learning. Artificial Intelligence 112(1-2), 181–211 (1999)

29. Barrett, C., Eubank, S., Marathe, A., Marathe, M., Pan, Z., Swarup, S.: Information integration to support policy informatics. The Innovation Journal 16(1), article 2 (2011)

30. Barrett, C., Beckman, R., Berkbigler, K., Bisset, K., Bush, B., Campbell, K., Eubank, S., Henson, K., Hurford, J., Kubicek, D., Marathe, M., Romero, P., Smith, J., Smith, L., Speckman, P., Stretz, P., Thayer, G., Eeckhout, E., Williams, M.D.: TRANSIMS: Transportation analysis and simulation system. Technical Report LA-UR-00-1725, Los Alamos National Laboratory (2001)

31. Barrett, C., Beckman, D., Khan, M., Kumar, V.A., Marathe, M., Stretz, P., Dutta, T., Lewis, B.: Generation and analysis of large synthetic social contact networks. In: Winter Simulation Conference (2009)

32. Beckman, R., Channakeshava, K., Huang, F., Marathe, A., Marathe, M., Pei, G., Saha, S., Vullikanti, A.: Integrated multi-network modeling environment for spectrum management. NDSSL Technical Report (2013)

33. Eubank, S.G., Guclu, H., Kumar, V.S.A., Marathe, M.V., Srinivasan, A., Toroczkai, Z., Wang, N.: Modelling disease outbreaks in realistic urban social networks. Nature 4, 180–184 (2004)

34. Human Behavior and WMD Crisis /Risk Communication Workshop: Defense Threat Reduction Agency, Federal Bureau of Investigation, U.S. Joint Forces Command (March 2001)

Modeling the Dynamics of Dengue Fever

Kun Hu[1,*], Christian Thoens[2], Simone Bianco[3], Stefan Edlund[1], Matthew Davis[1],
Judith Douglas[1], and James Kaufman[1]

[1] IBM Almaden Research Center, San Jose, California, USA
{khu,sedlund,mattadav,jvdougla,jhkauf}@us.ibm.com
[2] Federal Institute for Risk Assessment, Biological Safety, Berlin, Germany
christian.thoens@bfr.bund.de
[3] University of California, San Francisco, Bioengineering and Therapeutic Sciences,
San Francisco, California, USA
Simone.Bianco@ucsf.edu

Abstract. Dengue is a major international public health concern that impacts one-third of the world's population. There are four serotypes of the dengue virus (DENV). Infection with one serotype affords life-long immunity to that serotype but only temporary cross immunity (CI) to other serotypes. The risk of lethal complications is elevated upon re-infection, possibly because of the effect of antibody-dependent enhancement (ADE). In this paper we propose a system dynamics model that captures both host and vector populations, latency, and four dengue serotypes. This model allows one to study both CI and ADE. Modeling the *Aedes* vector adds complexity, but we consider this to be important because combating the mosquito vector may be the most practical intervention in the absence of an effective vaccine. Our results support the need to model the vector population and ADE to explain the observed epidemiological data.

Keywords: Dengue, cross immunity, antibody-dependent enhancement, system dynamics model, dynamic behaviors.

1   Introduction

Dengue is now endemic in half the world's nations and impacts more than one-third of the world's population [1]. Countries in Southeast Asia, the Americas, Africa, and the Western Pacific have become focal points for this mosquito-borne plague [1]. Dengue is a leading cause of childhood hospitalization in Thailand [2] and death in Southeast Asia [3]. Officials report that over 30,000 people in India had been sickened with dengue fever (hospitalizations and confirmed by laboratories) through October 2012, a 59% jump from the 2011 record of 18,860. In countries where dengue is not endemic it can be introduced through international travel, migration, and trade [4]. Growing numbers of Western tourists return from warm-weather vacations with the disease, which has now reached the United States and Europe.

---

* Corresponding author.

A.M. Greenberg, W.G. Kennedy, and N.D. Bos (Eds.): SBP 2013, LNCS 7812, pp. 486–494, 2013.

As a vector-borne viral disease, the spread of dengue is attributed to the expansion of the geographic distribution of the four serotypes of dengue viruses (DENV1-4) and their vector, *Aedes* mosquitoes. Changing climate and a rapid rise in urban mosquito population are bringing ever greater numbers of people into contact with DENV [5]. The four dominant strains have progressively spread to virtually all tropical countries around the globe [6]. Infection with one serotype affords life-long immunity to that serotype [1], but only temporary partial immunity to the other three. Patients re-infected with a second serotype experience an increased risk of developing dengue hemorrhage fever (DHF) and dengue shock syndrome (DSS) [7]. This increased risk has been attributed to the effects of antibody-dependent enhancement (ADE) [7-8]. Development of a multi-serotype vaccine is complicated by the complexity of between-serotype interactions.

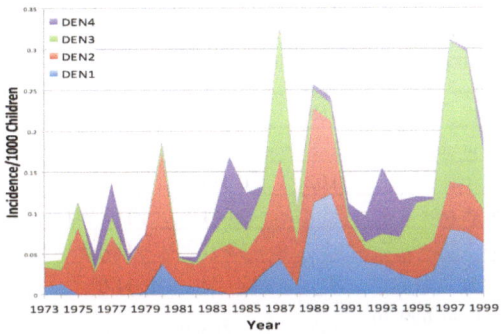

Fig. 1. Eight dengue outbreaks in Bangkok, Thailand, over 30 years (a periodicity of 3.75 years) derived from data by Nisalak [2]

A stacked area chart in Fig. 1 shows the incidence of each serotype in relation to the total incidence by year in Bangkok, Thailand, over 30 years. Dengue epidemics have occurred with an approximate period of 3-4 years [2-3]. The effect of ADE on the transmission dynamics of dengue was first studied by Ferguson [9] who showed that ADE could promote the co-existence of strains and lead to large oscillations in dengue dynamics. Recker et al. [10] have argued that ADE alone can produce the observed periodicities and the desynchronized oscillations. Wearing and Rohani [11] observe that "contrary to perceived wisdom," the dynamics of dengue epidemics can be generated entirely by a short-lived period of (partial) cross-immunity without ADE.

Several mathematical models have been proposed to describe the dynamic behavior of dengue transmission (see: review paper [12]). The models differ in complexity and details [11, 13-14]. In some cases, latency is neglected; in others the mosquito vector is omitted and replaced with a seasonality-modulated (sinusoidal) human-to-human transmission pathway [11, 14-15].

In this study, we proposed a model of dengue that enables evaluation of several important epidemiological factors. The model captures *both* incubation period (latency) and explicitly captures disease transmission through a vector population [16]. We attempt to provide new insights into the factors that most influence the dynamics of this complicated multi-serotype virus. In particular, the observed quasi-periodic outbreaks of dengue have been attributed to two critical factors: antibody-dependent enhancement (ADE) and/or the strength of cross strain immunity (CI).[1]

2   Description of the Models and Experiments

In order to evaluate the effects of CI and ADE, it is necessary to explicitly model all four strains of the virus. Nisalak et al. [2] showed that the majority of dengue cases are either primary or secondary infections (with no evidence for tertiary infection) [17]. We defined a set of differential equations to describe the disease state in both host and vector populations for four DENV serotypes and up to two sequential infections. Epidemiological parameters in the model were chosen based on a review of the literature and listed in Table 1.

Table 1. Summary of model inputs for mathematical models designed to simulate the transmission of dengue in predefined populations

Model input / Parameter	Value	Range	Sources (ref.)	Unit
R_0, basic reproductive number	2	2-4	[18-20]	dimensionless
μ^h, host birth and death rate	0.02	0.03-0.01	[9]	year^{-1}
μ^v, vector birth and death rate	35	24-61	[19, 21]	year^{-1}
γ^h, host incubation rate	65	52-91	[19]	year^{-1}
γ^v, vector incubation rate	30	23-33	[19]	year^{-1}
β^h, transmission rate from host to host (no vector in model)	200	100-300	[22]	year^{-1}
β^{vh}, transmission rate from vector to host[*]	15	12-21	[19]	year^{-1}
β^{hv}, transmission rate from host to vector[*]	530	136-1000	[19]	year^{-1}
σ^h, host recovery rate	100	50-200	[9, 23]	year^{-1}
θ^h, cross immunity loss rate of host population	2	2-9	[11]	Month
m, vector per host	1.9[‡]	1-6	[19]	dimensionless
ϕ, ADE factor	n.a.[§]	1-3	Range explored	dimensionless
ε, strength of cross immunity	n.a.[§]	0-1	Range explored	dimensionless

* We obtained the values of transmission rates in the model when integrating vector population from the biting rate (unit: bites per mosquito per day) used in [19]. In Table 1 we adjusted the value from the literature so that the units of transmission rate are year^{-1}.

‡ The value of m is calculated to be 1.9 to keep host-to-host transmission constant even after adding vector component (since the model in this study is based on the base model by Bianco [16]).

§ Not applicable.

---

[1] The model reported here was developed using the Eclipse Spatio-Temporal Epidemiological Modeler (STEM) and is available free (source code and executable) as open source through the Eclipse Foundation at http://www.eclipse.org/stem/.

The basic reproductive number, R_0, is defined as the number of secondary host infections caused by one primary host infection introduced to a fully susceptible host population at a demographic steady state [24-26]. The literature reports a wide range of values of R_0 for dengue fever [27]. In most cases, however, R_0 is generally thought to be near 2.0 for dengue, which we adopted in this study. It is noted that R_0 itself is not an independent parameter. The formulation for R_0 shown below is slightly different from the standardized formulation for host-host transmission [16] in that it includes host-vector transmission as well as the exposed (E) host state. Specifically, the number of exposed vectors generated by one primary infectious host is $m\beta^{hv}/(\sigma^h+\mu^h)$, the fraction of $\gamma^v/(\mu^v + \gamma^v)$ of which become infectious, where m is the number of vectors per host (female mosquitoes per human). In addition, $\gamma^h/(\gamma + \mu^h)$ stands for the probability that the exposed host survives the incubation period. We note that this expression does not depend on the ADE factor, ϕ or the strength of cross immunity, ε.

$$R_0 = \frac{m\beta^{hv}\gamma^v\beta^{vh}\gamma^h}{(\sigma^h+\mu^h)(\mu^v+\gamma^v)(\gamma^h+\mu^h)\mu^v} \tag{1}$$

2.1   Model Structure and Formulations

The full compartment diagram for the dengue model reported here is shown in Fig. 2. The model includes ordinary differential equations for the disease state (four serotypes) in both human and mosquito populations, and includes exposed state to capture the known period of incubation. The effects of partial cross strain immunity CI and ADE can be explored by varying the corresponding epidemiological parameters, effecting the state transition rates shown in Fig. 2.

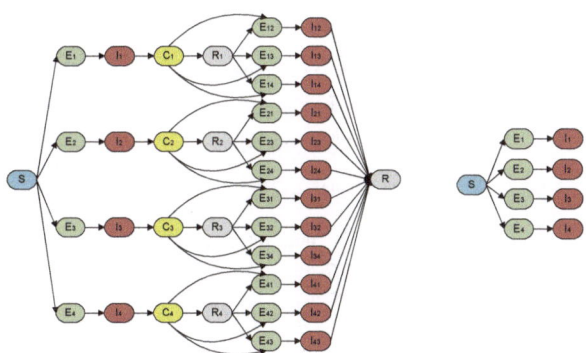

Fig. 2. The structure of the full model composed by 51 compartments including both host and vector populations

In the figure, population sizes are constant and normalized to unity, so each state represents a fraction of the total population. Populations are divided into the following states: S denotes the fraction of the population that has not yet been infected by any of the serotypes and is thus totally susceptible; E_i is the proportion of the exposed population infected by serotype i without shedding any viruses (latency); I_i is the proportion of the population infectious due to a primary infection with serotype i; C_i

is the proportion of the population recovered from a primary infection and temporarily immune to all serotypes; R_i is the proportion of the population recovered from a primary infection with serotype i and susceptible to serotypes other than i; E_{ij} is the proportion of the exposed population recovered from serotype i and currently infected by serotype j who have not shed any virus of j yet ($i \neq j$); I_{ij} is for secondary infection, in which individuals are infectious with serotype j, but recovered from infection with serotype i. Finally, R is the proportion of population who are completely immune because of historical exposure to two serotypes [2].

Female mosquito population is categorized as susceptible (S), exposed (E_i) and infectious (I_i) states. Once infected by a blood meal, a mosquito is assumed to experience a fixed extrinsic incubation period gv for about 8–12 days in the exposed (E_i) state [19]. An infected mosquito never recovers from dengue fever so there is no "recovered" compartment for the vector component. The life span of the adult *Aedes* mosquito is relatively short, roughly around 8–10 days [1].

The structure of the full model requires 51 differential equations. To simplify the notation, we define two infection rates in terms of two populations: λ_i^h is the infection rate of susceptible hosts for serotype i; λ_i^v is the infection rate of susceptible vectors for serotype i. The differential equations are then given by Eqs (2):

Vector:

$$\frac{dS^v}{dt} = \mu^v - S^v \sum_{i=1}^{4} \lambda_i^v - \mu^v S^v$$

$$\frac{dE_i^v}{dt} = S^v \lambda_i^v - \gamma^v E_i^v - \mu^v E_i^v$$

$$\frac{dI_i^v}{dt} = \gamma^v E_i^v - \mu^v I_i^v$$

Host:

$$\frac{dS^h}{dt} = \mu^h - S^h \sum_{i=1}^{4} \lambda_i^h - \mu^h S^h$$

$$\frac{dE_i^h}{dt} = S^h \lambda_i^h - \gamma^h E_i^h - \mu^h E_i^h$$

$$\frac{dI_i^h}{dt} = \gamma^h E_i^h - \sigma^h I_i^h - \mu^h I_i^h$$

$$\frac{dC_i^h}{dt} = \sigma^h I_i^h - (1-\epsilon)C_i^h \sum_{j \neq i} \lambda_j^h - \theta^h C_i^h - \mu^h C_i^h \qquad (2)$$

$$\frac{dR_i^h}{dt} = \theta^h C_i^h - R_i^h \sum_{j \neq i} \lambda_j^h - \mu^h R_i^h$$

$$\frac{dE_{ij}^h}{dt} = R_i^h \lambda_j^h + (1-\epsilon)C_i^h \lambda_j^h - \gamma^h E_{ij}^h - \mu^h E_{ij}^h$$

$$\frac{dI_{ij}^h}{dt} = \gamma^h E_{ij}^h - \sigma^h I_{ij}^h - \mu^h I_{ij}^h$$

$$\frac{dR_{ij}^h}{dt} = \sigma^h I_{ij}^h - \mu^h R_{ij}^h$$

where,

$$\lambda_i^h = m\beta^{vh} I_i^v$$

$$\lambda_i^v = \beta^{hv} \left(I_i^h + \phi \sum_{i \neq j} I_{ji}^h \right)$$

3   Result and Analysis

3.1   Dominant Period

Measurement of the bifurcation diagram from a simple time series (e.g., of the fraction of susceptible) is a useful way to determine if a system leads to sustained oscillations, but does not provide insight into the dominant period or relevant timescale of these oscillations. To determine this in one measurement, we measured the power spectrum of the susceptible fraction of total dengue cases for over 20,000 simulations with $1 \leq ADE \leq 3$, and $0 \leq CI \leq 1$. The power spectrum of the susceptible (S state) provides a measure of the dominant period. Phase diagrams were then created from this data with color indicating the dominant period, and intensity the amplitude of the primary peak in each power spectrum. The results are shown for two values of the host cross immunity loss rate (CILR) spanning the accepted range. The black regions indicate the system has reached a disease free or endemic equilibrium state (infinite period). In regions where the system exhibits sustained oscillations, the colors show the dominant time scale measured in the power spectrum (refer to the legend on the right hand side of Fig. 3). Solid color indicates a transition to periodic behavior (with one or more frequency components) whereas speckle indicates frequency instability or chaos (e.g., in Fig. 3 (a), when $CI > 0.7$ and $1.5 < ADE < 2.5$).

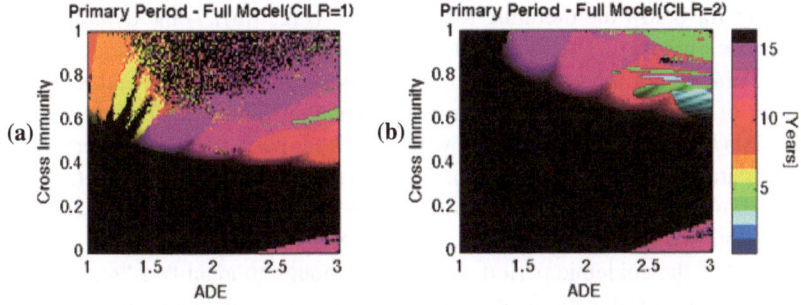

Fig. 3. Periodicity plot for full model with CILR=1 and 2[2]

Analysis of the bifurcation phase diagrams as a function of ADE and CI can provide insight into the range of parameters likely to produce the periodicities reported for real dengue outbreaks (approximately 3-4 years in Bangkok, Thailand. See: Fig. 1). As shown in Fig. 3, a transition from steady state to periodic or even chaotic behavior is observed as a function of both CI and ADE. For a CILR=2, the model exhibits no transition at ADE=1 (i.e., increasing only CI). If the vector is not included, the model [16] exhibits a very high degree of frequency instability at high CI. Including the exposed state, E, also quenches this frequency instability.

---

[2] With the largest amplitude scaled to maximum brightness for the pixel color.

3.2   Sensitivity Analysis

Fig. 4 shows the sensitivity of the periodicity plots in Fig. 3 to variation of the basic reproductive number, R_0. The consensus range of R_0 from the literature (Table 1) is $2.0 \leq R_0 \leq 4.0$. A base value of $R_0 = 2.0$ was chosen for consistency with Billings et al. [18] and Bianco et al. [16]. Fig. 4(a) show the periodicity plots for $R_0 = 1.2$ (40% below the literature minimum), and Fig. 4(b) shows the corresponding plots for $R_0 = 4$. The same protocol used for base case experiment was adopted. R_0 can be changed by varying β^{vh}, β^{hv}, or m. They are mathematically equivalent (refer to Eq. 2).

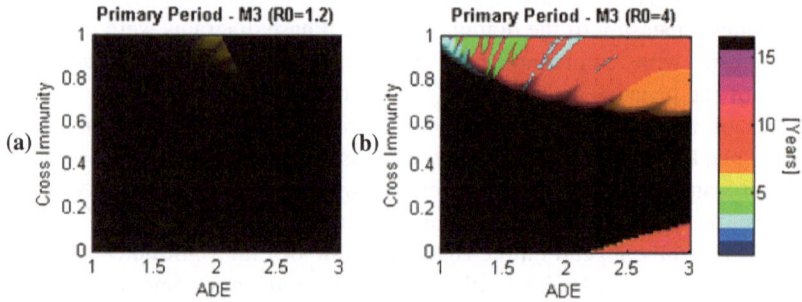

Fig. 4. Sensitivity analysis for the primary period obtained by increasing and decreasing the basic reproductive number, R_0, by +/- 40%

Lowering R_0 below the literature range by 40% (Fig. 4(a)), some dynamic regions disappear (see Fig. 4(a)). This is expected as R_0 approaches 1. Black regions indicate epidemic extinction or a fixed point of endemic infection. At higher values of R_0, the basic form of the frequency plots is robust throughout the literature range. At the upper bound ($R_0 = 4$), coherent regions with a 3-4 year period are still observed, albeit shifted in parameter space (Fig. 4(b)). Regions with "realistic" period (3-4 years) observed above ADE=2.5 in Fig. 3(b) shift with ADE as R_0 is increased. A new narrow region with this period appears at ADE=1 with CI=1.0 (see Fig. 4(b)). Measurement of the epidemic period alone is not enough to identify a "realistic" point in parameter space for dengue fever. It is also necessary to examine the detailed time series data. Given that the dominant epidemic period of dengue fever shifts with variation in R_0 and other parameters, future work will be required to establish the precise range of epidemiological parameters describing dengue. This work suggests nonzero values of both CI and ADE will be required.

4   Conclusion

In this paper, we proposed a deterministic model for dengue fever, including mosquito vector and the effects of host incubation, in a four serotype model extending Bianco [16]. The model makes it possible to explore the interacting effects of ADE and CI on the dynamics of dengue epidemics. As these important factors are varied, a systematic change in dynamics is observed. The vector model exhibits a transition from steady state to periodic dynamics as a function of both CI and ADE. No dynamic behavior is observed by including CI alone (ADE=1).

We also measured the dominant timescale of the dynamics from the peak in the power spectrum of the epidemic time series. It is possible to find regions of parameter space with a realistic period for dengue outbreaks (3-4 years). Regions with "realistic" dynamics and desynchronization between serotypes are observed at nontrivial levels of ADE (>2). The epidemics for different serotypes are phase locked, even though one can obtain a 3-4 year period at ADE=1. Also, it is inevitable that partial cross strain immunity plays a very important and necessary role.

The model reported here is not intended to represent a "perfect" model of dengue fever. Indeed, the model is analyzed in a deterministic setting and lacks a stochastic component. Stochasticity, realistic seasonal forces, demographic dynamics, and longer term environmental and climate changes may all influence dengue burden and its dynamical signatures [27]. More work is required to evaluate these factors.

Without inclusion of the mosquito vector [16], a large degree of frequency instability or "chaotic" behavior is observed. Such instability makes future creation of a predictive model problematic (even for short term predictions). Explicit inclusion of the vector may resolve this problem as it quenches the instability, stabilizing the epidemic dynamic timescales over extended regions of parameter space. This could be quite important to future attempts to calibrate a model of dengue fever based on surveillance data, and to derive more accurate estimates of the epidemiological parameters for the known serotypes. Models including the vector component are also more robust with respect to large seasonal changes in R_0.

References

1. World Health Organization, http://www.who.int/mediacentre/factsheets/fs117/en/#
2. Nisalak, A., Endy, T.P., Nimmannitya, S., Kalayanarooj, S., Thisayakorn, U., Scott, R.M., Burke, D.S., Hoke, C.H., Innis, B.L., Vaughn, D.W.: Serotype-Specific Dengue Virus Circulation and Dengue Disease in Bangkok, Thailand from 1973 to 1999. Am. J. Trop. Med. Hyg. 68(2), 191–202 (2003)
3. Cummings, D.A.T., Iamsirithaworn, S., Lessler, J.T., McDermott, A., Prasanthong, R., Nisalak, A., Jarman, R.G., Burke, D.S., Gibbons, R.V.: The Impact of the Demographic Transition on Dengue in Thailand: Insights from a Statistical Analysis and Mathematical Modeling. PLoS Med. 6(9) (2009)
4. Semenza, J.C., Menne, B.: Climate Change and Infectious Diseases in Europe. Lancet Infect. Dis. 9(6), 365–375 (2009)
5. Hopp, M., Foley, J.: Global-Scale Relationships between Climate and the Dengue Fever Vector, Aedes Aegypti. Climatic Change 48(2), 441–463 (2001)
6. Halstead, S.B., O'Rourke, E.J.: Antibody-Enhanced Dengue Virus Infection in Primate Leukocytes. Nature 265, 739 (1977)
7. CDC website (2012), http://www.cdc.gov/dengue/
8. Kawaguchi, I., Sasaki, A., Boots, M.: Why Are Dengue Virus Serotypes So Distantly Related? Enhancement and Limiting Serotype Similarity between Dengue Virus Strains. Proc. R. Soc. London [Biol.] 270, 2241–2247 (2003)

9. Ferguson, N.M., Anderson, R.M., Gupta, S.: The Effect of Antibody-Dependent Enhancement on the Transmission Dynamics and Persistence of Multiple-Strain Pathogens. Proc. Natl. Acad. Sci. U.S.A. 96, 790 (1999)

10. Recker, M., Blyuss, K.B., Simmons, C.P., Tinh Hien, T., Wills, B., Farrar, J., Gupta, S.: Immunological Serotype Interactions and Their Effect on the Epidemiological Pattern of Dengue. Proc. Biol. Sci. 276(1667), 2541–2548 (2009)

11. Wearing, H.J., Rohani, P.: Ecological and Immunological Determinants of Dengue Epidemics. Proc. Natl. Acad. Sci. U.S.A. 103, 11802–111807 (2006)

12. Johansson, M.A., Hombach, J., Cummings, D.A.T.: Models of the Impact of Dengue Vaccines: A Review of Current Research and Potential Approaches. Vaccine 29(35), 5860–5868 (2011)

13. Cummings, D.A.T., Schwartz, I.B., Billings, L., Shaw, L.B., Burke, D.S.: Dynamic Effects of Antibody-Dependent Enhancement on the Fitness of Viruses. Proc. Natl. Acad. Sci. U.S.A. 102, 15259–15264 (2005)

14. Nagao, Y., Koelle, K.: Decreases in Dengue Transmission Act to Increase the Incidence of Dengue Hemorrhagic Fever. Proc. Natl. Acad. Sci. U. S. A. 105, 2238–2243 (2008)

15. Adams, B., Boots, M.: Modelling the Relationship between Antibody-Dependent Enhancement and Immunological Distance with Application to Dengue. J. Theor. Biol. 242, 337–346 (2006)

16. Bianco, S., Shaw, L.B., Schwartz, I.B.: Epidemics with Multistrain Interactions: The Interplay between Cross Immunity and Antibody-Dependent Enhancement. Chaos 19(4), 9 (2009)

17. Halstead, S.B.: Dengue Virus-Mosquito Interactions. Annu. Rev. Entomol. 53, 273–291 (2008)

18. Billings, L., Schwartz, I.B., Shaw, L.B., McCrary, M., Burke, D.S., Cummings, D.A.T.: Instabilities in Multiserotype Disease Models with Antibody-Dependent Enhancement. J. Theor. Biol. 246, 18–27 (2007)

19. Chowell, G., Diaz-Duenas, P., Miller, J.C., Alcazar-Velazco, A., Hyman, J.M., Fenimore, P.W., Castillo-Chavez, C.: Estimation of the Reproduction Number of Dengue Fever from Spatial Epidemic Data. Math. Biosci. 208(2), 571–589 (2007)

20. Koopman, J.S., Prevots, D.R., Marin, M.A.V., Dantes, H.G., Aquino, M.L.Z., Longini, I.M., Amor, J.S.: Determinants and Predictors of Dengue Infection in Mexico. Am. J. Epidemiol. 133(11), 1168–1178 (1991)

21. Muir, L.E., Kay, B.H.: Aedes Aegypti Survival and Dispersal Estimated by Mark–Release–Recapture in Northern Australia. Am. J. Trop. Med. Hyg. 58(3), 277–282 (1998)

22. Ferguson, N.M., Donnelly, C.A., Anderson, R.M.: Transmission Dynamics and Epidemiology of Dengue: Insights from Age-Stratified Sero-Prevalence Surveys. Philos. Trans. R. Soc. London [Biol.] 354(1384), 757–768 (1999)

23. Gubler, D.J., Suharyono, W., Tan, R., Abidin, M., Sie, A.: Viraemia in Patients with Naturally Acquired Dengue Infection. Bull. W. H. O. 59, 623–630 (1981)

24. Anderson, R.M., May, R.M.: Infectious Diseases of Humans: Dynamics and Control, 2nd edn. Oxford University Press (1991)

25. MacDonald, G.: The Epidemiology and Control of Malaria. Oxford University Press (1957)

26. MacDonald, G.: The Dynamics of Helminth Infections, with Special Reference to Schistosomes. Trans. R. Soc. Trop. Med. Hyg. 59(5), 489–506 (1965)

27. Kamo, M., Sasaki, A.: The effect of cross-immunity and seasonal forcing in a multi-strain epidemic model. Physica D: Nonlinear Phenomena 165, 228–241 (2002)

Improving Markov Chain Monte Carlo Estimation with Agent-Based Models

Rahmatollah Beheshti and Gita Sukthankar

Department of EECS
University of Central Florida
Orlando, Florida 32816
{beheshti@knights,gitars@eecs}.ucf.edu

Abstract. The Markov Chain Monte Carlo (MCMC) family of methods form a valuable part of the toolbox of social modeling and prediction techniques, enabling modelers to generate samples and summary statistics of a population of interest with minimal information. It has been used successfully to model changes over time in many types of social systems, including patterns of disease spread, adolescent smoking, and geopolitical conflicts. In MCMC an initial proposal distribution is iteratively refined until it approximates the posterior distribution. However, the selection of the proposal distribution can have a significant impact on model convergence. In this paper, we propose a new hybrid modeling technique in which an agent-based model is used to initialize the proposal distribution of the MCMC simulation. We demonstrate the use of our modeling technique in an urban transportation prediction scenario and show that the hybrid combined model produces more accurate predictions than either of the parent models.

Keywords: Markov Chain Monte Carlo, agent-based models.

1   Introduction

Markov chain Monte Carlo (MCMC) simulation is a simple, easily parallelizable methodology for estimating the summary statistics of a population from minimal information. The aim of the process is to approximate the posterior distribution of the model parameters based on the observed data. By using Monte Carlo simulations to perform the high-dimensional integrations necessary to calculate marginal and posterior distributions, algorithms such as Metropolis-Hastings can make the Bayesian inference process tractable. MCMC has been used as a key component in the model fitting process in many types of social modeling and prediction problems. For instance, Cauchemez et al. use a Bayesian MCMC approach to examine the main characteristics that affect influenza disease transmission between households [1]. Similarly, the effect of spatial influences on geopolitical conflicts has been modeled using an MCMC formulation in which the likelihood of war involvement for each nation is conditioned on the decisions of proximate states [2].

A.M. Greenberg, W.G. Kennedy, and N.D. Bos (Eds.): SBP 2013, LNCS 7812, pp. 495–502, 2013.
© Springer-Verlag Berlin Heidelberg 2013

Although the MCMC methodology has many advantages, many of the commonly used MCMC algorithms are strongly dependent upon good initialization of the proposal distribution. In cases where the proposal distribution is far from the desired posterior distribution the algorithm may converge to a poor local minimum or require a long time to achieve convergence. In this paper, we focus on the question of how to select a good proposal distribution for MCMC algorithms. To address this problem, we turn to another modeling technique, agent-based modeling (ABM), to generate simulated data which is then used to initialize the proposal distribution of the MCMC. The combination of the two models, agent-based and MCMC, produces a more accurate result than either of the parent models and facilitates the MCMC convergence. To demonstrate the strengths of this approach, we present a case study on modeling and predicting transportation patterns and parking lot usage on a large university campus.

2   Related Work

Markov Chain Monte Carlo describes a family of methods for performing Bayesian inference through stochastic simulations of a Markov process. In the domain of social modeling and prediction, MCMC is well suited for studying the effect of long-term influences on dynamic systems of social agents. For instance, SIENA (Simulation Investigation for Empirical Network Analysis) uses MCMC for analyzing longitudinal data of networks and behavior [3]. SIENA is a powerful toolkit that can be used to test hypotheses about the effects of actor and tie covariates on network structure and actor behavior [4]. However, for large and complicated datasets, it can be challenging to get the MCMC component of SIENA to converge in a reasonable period of time. Since our proposed method initializes the proposal distribution at a point closer to the target distribution, it improves the convergence rate of MCMC.

MCMC is an alternative to two other commonly used approximation methods: 1) importance sampling—samples are drawn from a distribution other than the target one, then reweighted to account for differences between the two distributions, and 2) variational inference—the original integration problem is transformed into an optimization problem [5]. Effectively MCMC allows us to draw samples from a distribution $\pi(x)$ without having to know its normalization. With these samples, it is possible to compute any quantity of interest about the distribution of x, such as means, confidence regions, or covariance [6]. In this paper, MCMC is used as a simulation technique, and the sample set used to characterize the posterior distribution is simply compared against the output of other simulation techniques such as agent-based modeling, rather than used to perform Bayesian inference over model parameters.

This paper focuses on improving the performance of the Metropolis-Hastings algorithm (MH) [7] which is relatively sensitive to the initial proposal distribution. It is because of this sensitivity that researchers sometimes opt to use alternative MCMC algorithms, such as Gibbs sampling [8]. Our proposed method is a variation on the idea of using suboptimal inference and learning algorithms to generate data-driven proposal distributions for the MH algorithm [9]. Eaton et al. [10] used dynamic programming to create a proposal distribution for MCMC in the space of directed acyclic graphs. They showed that this hybrid technique converges to the posterior faster than other methods, resulting in more accurate structure learning of graphical models and higher predictive likelihoods on test data.

In [11], de Freitas et al. introduce two different methods to overcome the problem of finding a good proposal distribution. In the first approach, a mixture of two kernels is used to drive the search process: 1) a variational kernel to broadly explore the problem domain and locate regions of high-probability and 2) a Metropolis kernel to explore the local regions. One drawback with this method is that finding a good variational kernel can be difficult to do. To combat this issue, the authors propose a second technique called adaptive MCMC in which the proposal distribution is updated at run-time based on the behavior of Markov chain. Adaptive methods generally seek to construct a better proposal distribution through the combination of stochastic approximation and MCMC [12]. One issue with this class of adaptive techniques is that they often rely on certain mathematical assumptions being valid, and thus can only be used in a limited set of conditions unlike our proposed approach. Reversible jump MCMC is a different form of run-time modification in which the dimensionality of proposal distribution is changed; this technique can be used even in cases that the number of parameters is not known [13]. Brooks et al. introduced a new methodology for constructing efficient reversible jump MCMC proposal distributions [14].

Agent-based modeling can be an effective way of modeling complex systems that are not easy to characterize analytically. Typically, each agent in the simulation operates according to a set of simple rules representing the decision-making process of a human, or a group of humans. Simulating the social system reveals emergent interactions between the agents, which are often not immediately obvious from the rules of the system. For a more comprehensive overview of agent-based modeling approaches and applications, the reader is referred to [15]. Although agent-based systems are a powerful simulation and modeling tool in the hands of a domain expert, it is generally difficult to reproduce or verify conclusions drawn from more complicated ABMs since it rarely possible to exhaustively describe all the interactions which occur within the ABM or to quantify the impact of software modifications to the simulation. In this paper, since the ABM is used exclusively to shape the proposal distribution, it is easy to quantify the contribution of the ABM and reproduce the results.

3   Method

Figure 1 provides an overview of our proposed hybrid modeling technique in which an agent-based model is used to generate the proposal distribution used by the Markov Chain Monte Carlo algorithm. For this paper, we present a case study illustrating the usage of our technique as part of modeling effort to understand transportation patterns and parking lot occupancy on the campus of the University of Central Florida.

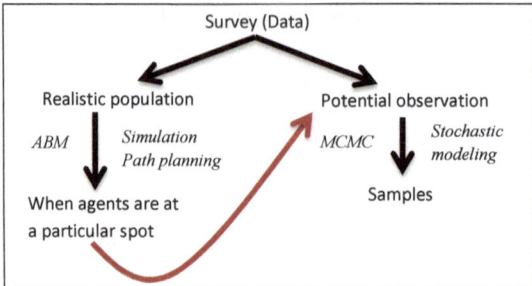

Fig. 1. Overview of proposed method

First, we distributed an online survey to the population of interest on campus-wide email lists; the results of the survey were then used to initialize an ABM model with reasonable parameters. The ABM creates 1) a realistic simulated campus population according to parameters fitted with the survey data 2) schedules for the simulated population members using an activity-based microsimulation 3) paths for the agents to move between their scheduled activities. General trends of student movement can be viewed using the ABM. To estimate a specific quantity of interest, such as usage for a specific parking lot at a particular time and day, the MCMC is used. The Metropolis-Hastings algorithm is initialized using a proposal distribution based directly on the output of the ABM and run to convergence.

In this paper, we compare the prediction performance of three different modeling techniques: 1) ABM only 2) an MCMC using a standard proposal distribution combined with observed data based directly on the surveys 3) our hybrid method in which the classic MCMC sampling is done in two separate phases. In the first phase, data from the agent-based population is sampled to create a proposal distribution, and in the second phase the Markov chain is repeatedly sampled to obtain the target distribution.

3.1 Agent-Based Transportation Model

To perform transportation forecasting on the UCF campus, we created an agent-based model for simulating the common activities (transportation, dining, recreation, and building occupancy) performed by the 47,000 students on the main campus. 1003 students responded to our online survey posted on `KwikSurveys` which was advertised on various campus email lists. The questions on the survey were grouped into different categories, related to possible places that could be visited on the main campus, and students were specifically queried about their visitation frequencies. Based on this data, we created the activity-based microsimulation described in [16].

Each agent in the model represents an individual student and has a unique set of parameters that govern his/her activity profile. An agent's defining parameters are: *entrance, dormitory, department, class building, arrive, depart, lunch, dinner, beverage, recreation and wellness, parking, shuttle,* and *miscellaneous*. The

first four parameters designate the single (most common) value of the agents'
entry point to the campus, housing situation, home department, and main class
building. *Arrive* and *depart* are lists showing the times the agent enters the campus and leaves it. The remaining parameters are lists of locations for the agent's
dining, recreation, and commuting. Additionally, each parameter that includes
a location has another matching parameter that shows the time or frequency of
visiting that location.

Rather than directly mapping the survey data to simulated entities that match
the exact preferences of one of the survey respondents, we attempt to learn a
general model of the population by fitting a set of distributions to the answers
of every question. When the simulation commences, all the agents are initialized
with parameters that remain constant over the lifetime of the agent and are used
to create daily activity profiles. Our simulation is implemented in the Netlogo [17]
environment and is freely available at: `http://code.google.com/p/ucf-abm/`.

3.2 MCMC

To benchmark the performance of our ABM MCMC model, we created a Markov
Chain Monte Carlo simulation with a standard proposal distribution for making
a limited set of forecasts based on the survey data. We use the Metropolis-
Hastings algorithm as follows:

- Select a proposal distribution Q
- Initialize the starting point, x_0
- Do
  - Generate a candidate point x_c, according to the probability $Q(x_c|x_i)$
  - Calculate the acceptance probability:

$$\alpha(x_i, x_c) = min(1, \frac{\pi(x_c)q(x_i|x_c)}{\pi(x_i)q(x_c|x_i)})$$

  - Choose $x_{i+1} = x_c$ with probability α, $x_{i+1} = x_i$ with probability $(1 - \alpha)$

This procedure is executed until the Markov chain has reached its stationary
distribution according to a convergence diagnostic. To validate the simulation,
MH is used to estimate the number of cars entering the parking lots at different
times of a day. One can envision this as a two dimensional diagram with the horizontal axis corresponding to the time of a day, and the vertical one showing the
number of cars entering a specific parking lot. The survey data from the questions about the attendance pattern and frequency of parking lot usage is used to
initialize observed data used by the MCMC model. Our MCMC model assumes
the unnormalized distribution, $\pi(x)$, is of the form of a Poisson distribution, and
a standard multivariate Gaussian is used for the proposal distribution.

3.3 ABM MCMC

In our proposed method, the samples produced by the ABM are used to construct
the proposal distribution. Then this distribution is employed by the MCMC

method to find the target distribution. In this case study, the goal of the campus modeling problem is to build a model describing the transportation patterns of students, hence the distribution that we are seeking (the target distribution) should represent the location of students at different times. The samples that are collected from the agent-based model include x and y coordinates of agents at each hour. This produces a population of samples containing x, y and *time*. The proposal probability of each vector is set equal to the number of times the vector exists in the dataset divided by the total number of dataset records. This makes the implicit assumption that the agent-based model has produced an evenly distributed set of samples from the population domain.

4   Results

One of the main applications of our microsimulation is analyzing pedestrian movement and car traffic on campus. Figure 2 shows the average visitation frequency for UCF campus locations (junctions, roads, and buildings) as predicted by the ABM MCMC simulation. The darkness of the circles in Figure 2 is proportional to the number of the students who passed or visited these places.

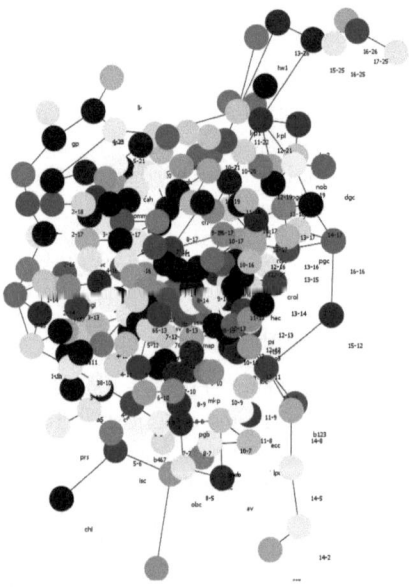

Fig. 2. Average traffic through different locations on the campus as predicted by ABM MCMC estimation with darker circles showing more probable locations. The simulation clearly shows several campus usage trends that are easily verified, including high student union usage (center) and high traffic at main campus entrances (bottom left and up left).

A question of daily interest for most students is parking lot usage: which lots have vacancies and where can the best parking spots be found? UCF Parking

Service performed a visual survey of lot usage in Fall 2011 and created a data set which we compared to our hourly forecasts of student lot usage. Figures 3a and 3b show the microsimulation forecasts for the different student parking lots as predicted by: 1) **ABM**: the agent-based model; 2) **MCMC**: the Markov Chain Monte Carlo with standard proposal distribution 3) **ABM MCMC**: the proposed hybrid method. The horizontal axis shows the names of the parking lots and the vertical the difference between the model predictions and the actual parking lot occupancy tallied by UCF Parking Office. The ABM is much better at predicting parking lot usage, compared to the MCMC (standard proposal distribution). However, the hybrid method produces estimations of parking lot usage that are virtually identical to the actual parking lot survey, with improved convergence rates.

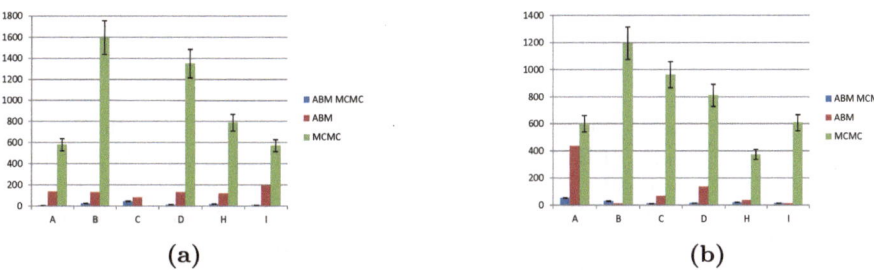

(a)   (b)

Fig. 3. The absolute value of the prediction error of the MCMC simulation with standard proposal distribution (**MCMC**), the agent-based modeling method (**ABM**), and our proposed method (**ABM MCMC**). Shorter bars represent predictions that diverge less from the actual observed Parking Services data. Our proposed method accurately forecasts the parking lot usage across all the parking lots (A-I) at noon (3a) and 4 pm (3b).

5   Conclusions

This paper introduces a new hybrid modeling method for combining agent-based models with MCMC. We demonstrate that the proposed method for initializing the MCMC proposal distribution with ABM data significantly reduces the prediction error over standard MCMC and also improves upon the ABM alone. We hypothesize that the combined ABM MCMC finds a more general model of the the posterior distribution than the ABM alone. Although agent-based models are often difficult to formally specify and reproduce exactly, the contribution of the ABM can be entirely quantified by the single proposal distribution, which makes it possible to reproduce the results without replicating the entire ABM.

Acknowledgments. This research was supported in part by NSF award IIS-0845159.

References

1. Cauchemez, S., Carrat, F., Viboud, C., Valleron, A.J., Boëlle, P.Y.: A Bayesian MCMC approach to study transmission of influenza: application to household longitudinal data. Statistics in Medicine 23(22), 3469–3487 (2004)
2. Ward, M.D., Gleditsch, K.S.: Location, location, location: An MCMC approach to modeling the spatial context of war and peace. Political Analysis 10(3), 244–260 (2002)
3. Snijders, T.: Markov Chain Monte Carlo estimation of exponential random graph models. Journal of Social Structure 3 (2002)
4. Snijders, T.: Models and methods in social network analysis. Cambridge University Press, New York (2005)
5. Carbonetto, P., King, M., Hamze, F.: A stochastic approximation method for inference in probabilistic graphical models. In: NIPS, vol. 22, pp. 216–224 (2009)
6. Press, W., Teukolsky, S., Vetterling, W., Flannery, B.: Numerical Recipes: The Art of Scientific Computing. Cambridge University Press (2007)
7. Metropolis, N., Rosenbluth, A., Rosenbluth, M., Teller, A., Teller, E.: Equation of state calculations by fast computing machines. Journal of Chemical Physics 21, 1087–1093 (1953)
8. Geman, S., Geman, D.: Stochastic relaxation, Gibbs distributions, and the Bayesian restoration of images. IEEE Transactions on Pattern Analysis and Machine Intelligence (6), 721–741 (1984)
9. Andrieu, C., De Freitas, N., Doucet, A., Jordan, M.: An introduction to MCMC for machine learning. Machine Learning 50(1), 5–43 (2003)
10. Eaton, D., Murphy, K.: Bayesian structure learning using dynamic programming and MCMC. In: Proceedings of the Conference on Uncertainty in Artificial Intelligence, pp. 101–108 (2007)
11. De Freitas, N., Højen-Sørensen, P., Jordan, M., Russell, S.: Variational MCMC. In: UAI, pp. 120–127 (2001)
12. Andrieu, C., Moulines, É.: On the ergodicity properties of some adaptive MCMC algorithms. The Annals of Applied Probability 16(3), 1462–1505 (2006)
13. Yeh, Y., Yang, L., Watson, M., Goodman, N., Hanrahan, P.: Synthesizing open worlds with constraints using locally annealed reversible jump MCMC. ACM Transactions on Graphics (TOG) 31(4), 56:1–56:11 (2012)
14. Brooks, S., Giudici, P., Roberts, G.: Efficient construction of reversible jump Markov chain Monte Carlo proposal distributions. Journal of the Royal Statistical Society: Series B (Statistical Methodology) 65(1), 3–39 (2003)
15. Macal, C., North, M.: Tutorial on agent-based modelling and simulation. Journal of Simulation 4(3), 151–162 (2010)
16. Beheshti, R., Sukthankar, G.: Extracting agent-based models of human transportation patterns. In: Proceedings of the ASE/IEEE International Conference on Social Informatics, Washington, D.C., pp. 157–164 (December 2012)
17. Wilensky, U.: NetLogo. Evanston, IL: Center for Connected Learning and Computer-Based Modeling, Northwestern University (1999), http://ccl.northwestern.edu/netlogo/ (retrieved)

Geographic Profiling of Criminal Groups for Military Cordon and Search*

Samuel H. Huddleston, Matthew S. Gerber, and Donald E. Brown

Department of Systems and Information Engineering, University of Virginia
{shh4m,msg8u,brown}@virginia.edu

Abstract. In the course of counter-insurgency campaigns, military forces expend considerable resources and time conducting cordon and search operations in an effort to interdict and suppress criminal groups. However, these operations have a low success rate, with most operations yielding little intelligence or marginal tactical gains while simultaneously angering the local populace. This paper demonstrates methods for improving the success rate of cordon and search operations by leveraging Criminal Site Selection (CSS) models for geographic profiling. This new modeling approach provides statistically significant performance improvements over the current best method for geographic profiling and provides geographic profiles that are often accurate enough to facilitate tactical success, with the modeled criminal group's anchor point falling within the search profile for military unit cordon and search operations.

1 Introduction

Military forces engaged in counter-insurgency campaigns perform many of the same security and crime suppression activities as domestic police forces. However, they are much more focused on capturing or suppressing groups of offenders than capturing and prosecuting individual serial criminals. These criminal groups, organized into "cells" or small paramilitary elements, cooperate to coordinate attacks on the local population, the government, and security forces in an effort to destabilize the government. Cordon and search operations are one of the most frequently employed techniques military forces use in targeting these criminal groups [20]. Military forces perform cordon and search operations by establishing an impermeable outer security perimeter (the cordon) and then systematically searching the target area in an effort to locate enemy combatants or equipment. Effective geographic profiling techniques could provide a method for determining high probability search zones for military cordon and search.

Geographic profiling is an investigative technique used by police that uses the known locations of a crime series to determine a serial offender's anchor point, usually a residence or workplace. However, despite several high-profile successes, geographic profiling models [9,13,15,17] have not yet been developed that are demonstrably more accurate than simple centrographic techniques such

---

* The rights of this work are transferred to the extent transferable according to title 17 U.S.C. 105.

A.M. Greenberg, W.G. Kennedy, and N.D. Bos (Eds.): SBP 2013, LNCS 7812, pp. 503–512, 2013.
© Springer-Verlag Berlin Heidelberg 2013

as calculation of the Fermat-Weber point, more commonly known as the Center of Minimum Distance (CMD) [9, 14, 15]. Effective solutions to this problem are in high demand because a critical function in the criminal investigative process is locating unknown serial offenders [17].

This paper demonstrates the use of Criminal Site Selection (CSS) models to develop geographic profiles in support of military cordon and search operations against the anchor points (addresses) for the criminal activities of known criminal groups. CSS models are a technique used to develop hot-spot maps to predict future criminal activity. CSS models have been shown to significantly improve predictive performance over traditional kernel density hot-spot methods for predicting crimes such as burglaries [10, 21], terrorist events [3], suicide bombings [18], and criminal activity by street gangs [5, 6]. However, CSS models have not yet been applied to geographic profiling.

2   Data Set

Many researchers have noted the similarities between criminal street gangs and insurgent groups operating in urban environments [1, 4, 11]. In our investigation, we used crime data for criminal street gangs in Santa Ana, California as a substitute for sensitive datasets from ongoing military operations. The data came from three sources: the Gang Incident Tracking System (GITS) crime dataset for the city of Santa Ana [12], the 2000 US Census, and a gang intelligence map provided by the Santa Ana Police Department. The gang intelligence map detailed known gang territories and point locations (addresses) for many of the criminal gangs active in the city during the study period. We were able to identify crime series containing more than 3 criminal events from the GITS database for 17 of the gangs for which there was a defined anchor point (address) in the gang intelligence map. The US Census provided socio-economic and demographic information at the census block group level for Santa Ana.

3   Methodology

The formal definition of this problem is to identify the anchor point $z_j \in \Re^2$ for criminal group j from a crime series of size N_j committed by that group at locations $S_j = \{s_{j1}, s_{j2}, ..., s_{jN_j}\}$. This anchor point will be probabilistically assigned to a grid cell of possible locations $i \in \Re^2$ where i indexes a series of 50 meter x 50 meter grid cells in the domain of interest. The variable I represents the total count of approximately 30,000 grid cells mapped within the city limits of Santa Ana, California. Figure 1 provides an example of a geographic profile (the mapped model outputs over the index i) for a criminal anchor point using one of the methods developed below.

3.1   Data Set Preparation

Developing a CSS geographic profiling model requires a training dataset containing (solved) crime series linked to serial offenders and their known anchor points.

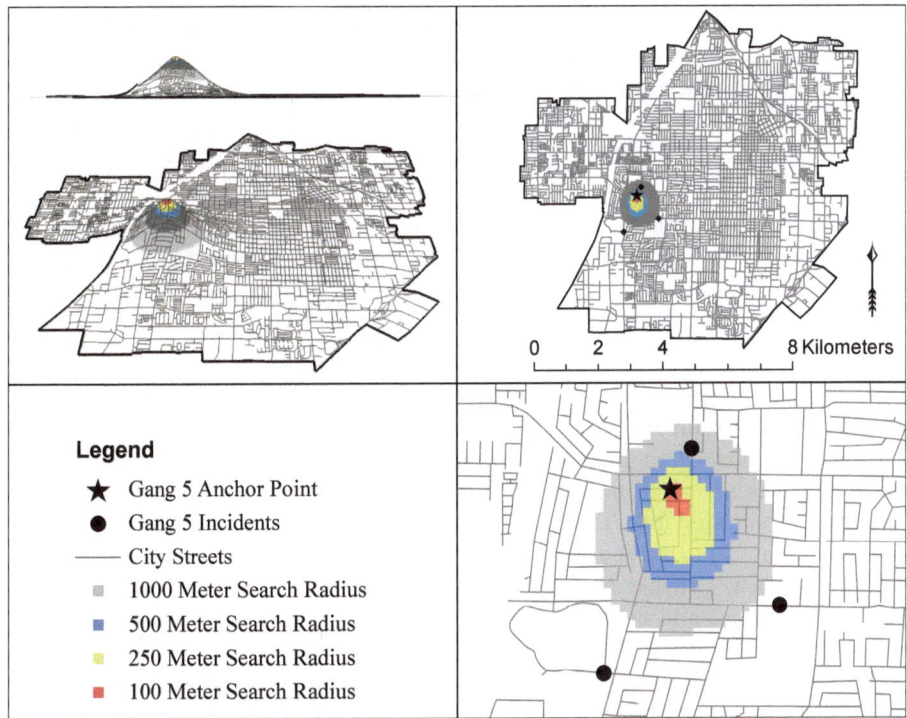

Fig. 1. A geographic profile produced using the CSSB approach for Gang 5's anchor point in 3-D (top left), at the city scale (top right), and at the scale appropriate for tactical level planning (bottom right)

There are several data preparation steps necessary to develop a CSS model of criminal behavior. First, the geographic locations of crimes in the training dataset are converted into a marked spatial point pattern by using a Geographic Information System (GIS) to attach socio-economic features from the US census as well as calculating the Euclidean distance to feature-space predictors [10] such as the responsible party's anchor point, the nearest gang territory, the nearest gang address and other crime generators or attractors [2]. Note that the distance between the crimes and anchor points is only one of many geographic distances considered in the model structure. The response variable y for all of the actual crime locations is recorded as $y_i = 1$. A *null grid* is laid over the study domain at 50-meter intervals and marked with predictive features in the same way. The null grid provides observations for locations where criminals chose not to commit crimes (i.e. the response variable at these locations is recorded as $y_i = 0$). This grid is also used to map probabilistic estimates for gang anchor points by plotting the model predictions over the grid as a raster image, as seen in Figure 1. The training data set therefore contains a response vector Y indexed by i and a predictor matrix X, with each row of X corresponding to a location i and each column of X corresponding to a predictive feature. For notational convenience, the distance between group j's known anchor point z_j and location i is held out of the predictor matrix X and referenced in the notation below as $||i - z_j||$.

3.2   Criminal Site Selection (CSS) Modeling of Group Behavior

Criminal Site Selection (CSS) models calculate the probability that a crime by group j occurs at location i using a logistic regression equation.

$$P_j(y_i = 1|X_i, z_j) = \frac{exp\left(AX_i + B||i - z_j||\right)}{1 + exp\left(AX_i + B||i - z_j||\right)} \tag{1}$$

In the logistic regression above, the A coefficient vector specifies the relationship to the environmental factors associated with a location, which are recorded in vector X_i. The B coefficient captures the distance-decay relationship for the criminal's journey to crime (once the effects of the other environmental factors have been accounted for). This model structure improves current geographic profiling techniques in that it models the effect of the journey to crime relationship after considering other environmental effects such as socio-economic conditions, crime generators, and crime attractors that might affect a criminal's decision-making process.

We fit the logistic regression models in this paper using the stats package in R software, using stepwise regression to automate feature selection. Predictors consistently selected using stepwise regression for the CSS models included distance to the gang's anchor point, distance to other gang addresses (known crime generators and attractors), median home values, the percentage of homes that were owner occupied, the percentage of the population 18-30 years old, the percentage of the population on public assistance, and racial demographics.

3.3   Developing a Geographic Profile from a New Crime Series

The CSS model is a descriptive model of what is generally true about how criminals in a specific geographic area respond to environmental factors, crime attractors, crime generators, and their own geographic anchor points. Once this model is built for a specific geographic region, it can be used to find the geographic anchor points for a newly observed crime series. Three methods for generating a predictive geographic profile are outlined below. The first approach (CSSB) uses the CSS model within a Bayesian framework. The second method (CSSM) uses the CSS model as a mathematical scoring function. The last approach uses the CMD algorithm as a mathematical scoring function. The CMD method is included to enable performance comparison of the CSS modeling approach for geographic profiling to what is widely considered to be the current best method [7, 9, 14, 19].

The input data used to generate the geographic profile for a new group is a crime series by that group. This crime series defines the set of crime locations $S_j = \{s_{j1}, s_{j2}, ...s_{jN_j}\}$. In order to employ the CSS model, we must develop the predictive matrix X used in the CSS model for each of the incidents in S_j, as discussed in Section 3.1 above. This forms the predictive matrix X_j, which contains N_j rows, with each row indexing crime n. Finally, the Euclidean distance $\|s_{jn} - i\|$ is calculated for all i and n.

A Bayesian (CSSB) Method. The first CSS modeling approach places the CSS model within a Bayesian framework to calculate the joint probability of an observed crime series. This approach uses the CSS model and Bayes' Theorem to calculate the posterior probability of an anchor point given a crime:

$$P\left(z_j = i | X_{jn}, s_{jn}\right) = \frac{P\left(s_{jn}|X_{jn}, z_j = i\right) P\left(z_j = i | X_{jn}\right)}{P\left(s_{jn}|X_{jn}\right)} \tag{2}$$

The posterior probability of interest is the probability that the gang's anchor point z_j is located at location i given criminal event s_{jn} and the predictive features at the location of the criminal event described by the matrix X_{jn}. For notational convenience, we define the conditional probability of known crime series event s_{jn}, which occurs at location i, as $P(s_{jn}|X_{jn}, z_j)$ and define that this notation is equivalent to $P_j(y_i = 1|X_i, z_j)$ since when $y_i = 1$ it joins the set S_j. The CSS model shown in Equation 1 provides $P\left(s_{jn}|X_{jn}, z_j = i\right)$. $P\left(z_j = i | X_{jn}\right)$ is defined uniformly to be $1/I$ for all points z_j (i.e. we use a *non-informative prior distribution* for the location of the gang's anchor point).

Estimating the data probability $P\left(s_{jn}|X_{jn}\right)$ has presented some significant challenges in other profiling approaches employing the Bayesian paradigm. As [8] notes, "there is no simple way of estimating the probability of obtaining the information [data] under all possible scenarios." Note that in the way we have developed the problem, the data probability depends only on the feature set of that location. Since this probability is by definition independent of the location of the anchor point, we can assert that for all possible anchor points we evaluate with the posterior probability function, this probability will be the same (unknown) constant, D. Assuming the crime series is a set of independent observations yields:

$$P\left(z_j = i | s_{j1}, s_{j2}, .., s_{jN_j}\right) = \prod_{n=1}^{N_j} \left[\frac{1}{ID_{jn}} P\left(s_{jn}|X_{jn}, z_j = i\right)\right] \tag{3}$$

The product term $\prod_{n=1}^{N_j} \left[1/\left(ID_{jn}\right)\right]$ represents some unknown constant. We can estimate a probability density for the anchor point for unique location, $f(y_i)$ by dropping this constant term. The resulting density estimate at location y_i is the product of the conditionally independent probabilities for the entire crime series and is proportional to the joint posterior probability in Equation 3:

$$f(y_i) = \prod_{n=1}^{N_j} \left[P\left(s_{jn}|X_{jn}, z_j = i\right)\right] \propto P\left(z_j = i | s_{j1}, s_{j2}, .., s_{jN_j}\right) \tag{4}$$

Figure 1 illustrates the resulting mapped probability density surface for one of the criminal gangs. This density surface is sufficient for planning cordon and search operations in an effort to locate the anchor point for a criminal group. To obtain an estimate of the posterior probabilities at the various locations, one can normalize the probability density surface $f(Y)$ to sum to one, producing a probability surface. The resulting mapped joint posterior probability surface is indistinguishable from the mapped density surface shown in Figure 1.

A Mathematical Scoring (CSSM) Method. The second CSS modeling approach replaces the distance-based mathematical scoring functions used in other geographic profiling methods [7,17] with the CSS model. The CSS mathematical scoring method (CSSM) calculates the score L for location i as:

$$L(i) = \sum_{n=1}^{N_j} P\left(s_{jn}|X_{jn}, z_j = i\right) = \sum_{n=1}^{N_j} \left[\frac{exp\left(AX_{jn} + B\|s_{jn} - i\|\right)}{1 + exp\left(AX_{jn} + B\|s_{jn} - i\|\right)} \right] \quad (5)$$

The significant difference between this approach and previous mathematical modeling approaches is the ability of the CSS model to incorporate additional information about environmental factors such as socio-economic conditions, crime generators, and crime attractors. When the scores for all locations are mapped, they produce a geographic profile similar to that illustrated in Figure 1.

The Center of Minimum Distance (CMD) Method. The center of minimum distance (CMD) is calculated as:

$$CMD = \operatorname*{argmin}_{i \in \Re^2} \sum_{n=1}^{N_j} \|s_{jn} - i\| \quad (6)$$

Many researchers have noted that this statistic provides the most accurate point estimate for the location of a serial offender's anchor point [9, 14, 15]. The significant drawback to the use of this statistic has been that it provides only a point estimate for the anchor point's location. However, with some adjustment, the algorithm used to calculate CMD can also be leveraged as a simple heuristic approach for developing a geographic profile:

$$L(i) = \frac{1}{\sum_{n=1}^{N_j} \|s_{jn} - i\|} \quad (7)$$

The simple heuristic in Equation 7 uses the inverse function to reverse the minimization function and maps this calculation for all i. This produces a geographic profile similar to that shown in Figure 1.

3.4   Results

To conduct a performance comparison of the three modeling approaches, we used cross-validation to develop these results by iteratively using the crime series and anchor points for 16 of the gangs to develop a CSS model and then applied that CSS model to predict the "unknown" anchor point of the gang held out of the data set. We assess geographic profiling model performance using three metrics commonly used for geographical profiling models: error distance, search cost, and profile accuracy [16]. Error distance is the Euclidean distance between the point location predicted for the anchor point (the i with the highest geographic profile score) and the actual address for the criminal gang. Search cost is the number of 50 x 50 m grid squares that would have to be searched in order to find the gang anchor point. Table 3.4 provides error distance and search cost performance for each of the gangs.

Table 1. Crime counts, error distance (in meters), and search cost (in count of 50 meter x 50 meter grid cells) by gang for the three geographic profiling methods

Gang	Crime Count	Error Distance			Search Cost		
		CMD	CSSB	CSSM	CMD	CSSB	CSSM
1	8	110	710	184	24	444	83
2	9	572	535	742	435	334	607
3	10	100	100	100	14	13	14
4	8	1535	1535	1564	2418	2761	1794
5	4	69	69	69	10	7	9
6	18	118	273	100	14	90	18
7	22	1010	1112	1067	1307	1653	1047
8	10	2189	2334	1462	6479	6920	4404
9	15	50	20	50	2	1	2
10	5	414	534	387	199	339	195
11	14	1944	1957	1955	4099	4421	2640
12	11	1681	1807	1343	5083	5564	2985
13	14	1517	1632	1490	4012	4219	3394
14	7	459	462	588	289	327	376
15	32	31	73	31	1	3	1
16	15	287	519	183	101	330	23
17	18	216	174	216	46	44	37
Average	13	724	814	678	1443	1616	1037

The CSSM method provides the best overall performance on these metrics. In pairwise comparison, the performance improvements the CSSM method provides over the CMD method are not statistically significant for the the error distance metric but are significant for search cost performance ($p = 0.484$ and 0.032 by Wilcoxon Signed Rank Test). The CSSM approach provides statistically significant performance improvement over the CSSB approach for both error distance

Table 2. Profile accuracy comparison for various search profiles

Search Diameter	Search Blocks	CMD	CSSB	CSSM
100 M	4	12%	12%	12%
250 M	25	35%	24%	35%
500 M	100	41%	35%	47%
1000 M	400	59%	59%	59%

and search cost ($p = 0.032$ and 0.005). The CMD method likewise provides statistically significant performance improvement in both measures over the CSSB method ($p = 0.006$ and 0.004).

Table 2 summarizes the profile accuracy performance for the three geographic profiling methods for profile areas that can be used to define the cordon limits for increasing echelons of military units conducting cordon and search operations. Profile accuracy measures the percentage of criminal gang anchor points that would be found by conducting a cordon and search for a specifically defined region. For example, the geographic search profile for the smallest echelon of military unit that could conduct a cordon and search operation is a diameter of about 100 meters laid over the target location, or the cordon of a neighborhood region containing four 50 x 50 meter search blocks. The CSSM approach again provides the best performance overall performance, although for three of the four search profile zones, the CMD method provides equivalent performance.

Figure 2 visually summarizes the information in Tables 1 and 2. It provides a plot of search cost efficiency: the percentage of gang anchor points in the dataset that are identified when a cordon and search operation of a defined search cost is conducted. As can be seen in the left panel of Figure 2, the CSSM provides better overall performance in search cost efficiency. The right panel illustrates the search cost efficiency over the region applicable to military cordon and search operations. As can be seen in this graphic, the performance of the CSSM approach and CMD approach are very similar in this trade-off space.

3.5 Discussion and Conclusions

Overall, the CSSM model provided the best performance. However, it requires significantly more data than the CMD approach. The CMD method requires only a crime series in excess of three crimes while both of the CSS methods require mapped information about environmental influences (known in military parlance as the *human geography*), a training data set containing crime series linked to their known anchor points, and the ability to fit the CSS statistical model. Therefore, the CMD method provides a simple method that provides good performance for the small search profiles applicable to military cordon and search operations.

A significant shortcoming of all of the modeling approaches demonstrated here (and geographic profiling models in general) is that they cannot accurately identify an anchor point that is not encircled by a crime series (Gangs 8, 10, 12,

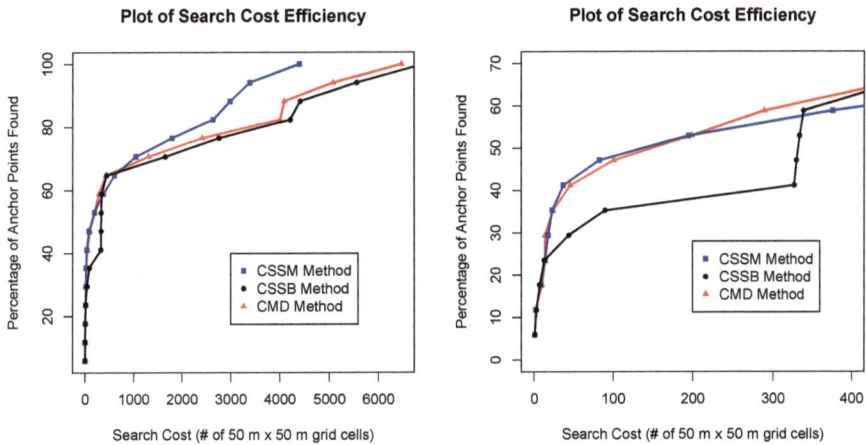

Fig. 2. A graphical search cost efficiency comparison throughout the trade-off space (left panel) and scaled to highlight the region applicable to military and cordon and search operations (right panel)

and 13) and tend to perform poorly when the anchor point is very close to the edge of the crime series (Gangs 7 and 11). Thus, these models are inappropriate for *commuter offenders*, who travel to target areas away from their anchor points to commit their crimes, but are applicable to *marauding offenders* who fan out from a central anchor point in search of criminal opportunities.

While the CSSB method provided the worst performance in this case, a data source important to the CSSB approach was unavailable. One of the strengths of the CSSB approach is the ability to leverage additional information such as an informative distribution for the prior probability for the criminal anchor point for a criminal group. These informative distributions for the prior probability for anchor points could be developed by incorporating data received from additional intelligence sources such as human and signals intelligence (HUMINT/SIGINT). However, the data available for this study did not contain information that could be leveraged in this way.

Both CSS modeling approaches contribute to the geographic profiling model literature by providing a method for modeling the effect of the journey to crime relationship after considering other environmental effects such as socio-economic conditions, crime generators, and crime attractors that might affect a criminal's decision-making process. Future model adaptions include development of informative distributions for the prior probability for anchor points for use in the CSSB method by incorporating data received from additional intelligence sources such as human and signals intelligence (HUMINT/SIGINT) and geographic profiling of individual serial criminals instead of group behavior.

References

1. Arnold, L.H., O'Gwin, C.W., Vickers, J.S.: Small town insurgency: The struggle for information dominance to reduce gang violence. Tech. rep., Monterey, CA (2010)
2. Brantingham, P.J., Brantingham, P.L.: Environmental Criminology. Sage Publications, Thousand Oaks (1981)
3. Brown, D.E., Dalton, J., Hoyle, H.: Spatial forecast methods for terrorist events in urban environments. In: Second Symposium on Intelligence and Security Informatics (June 2004)
4. Freeman, M., Rothstein, H.: Gangs and guerillas. Tech. rep., Monterey, CA (2011)
5. Huddleston, S.H., Brown, D.E.: A statistical threat assessment. IEEE Transactions on Systems, Man, and Cybernetics - Part A: Systems and Humans 39(6) (2009)
6. Huddleston, S.H., Fox, J., Brown, D.E.: Mapping gang spheres of influence. Crime Mapping: A Journal of Research and Practice 4(2) (2012)
7. Levine, N.: Crimestat: A spatial statistics program for the analysis of crime incident locations (versions 1.1 and 3.2). Tech. rep. Ned Levine and Associates (2009)
8. Levine, N.: Introduction to the special issue on Bayesian journey-to-crime modelling. Journal of Investigative Psychology and Offender Profiling 6 (2009)
9. Levine, N., Block, R.: Bayesian journey-to-crime estimation: An improvement in geographic profiling methodology. The Professional Geographer 63(2) (2011)
10. Liu, H., Brown, D.E.: A new point process transition density model for space-time event prediction. IEEE Transactions on Systems, Man, and Cybernetics, Part C: Applications and Reviews 34, 310–324 (2004)
11. Manwaring, M.G.: Street gangs: The new urban insurgency. Tech. rep., Carlisle, PA (2005)
12. Meeker, J.W., Vila, B., Parsons, K.J.: Developing a GIS-based regional gang incident tracking system. In: Responding to Gangs: Evaluation and Research. National Institute of Justice, Washington (2002)
13. Mohler, G., Short, M.B.: Geographic profiling from kinetic models of criminal behavior. SIAM Journal of Applied Mathematics (2012)
14. Paulsen, D.: Connecting the dots: Assessing the accuracy of geographic profiling software. Policing: An International Journal of Police Strategies & Management 29(2) (2006)
15. Paulsen, D.J., Bair, S., Helms, D.: Tactical Crime Analysis: Research and Investigation. CRC Press, Boca Raton (2010)
16. Rich, T., Shively, M.: A methodology for evaluating geographic profiling software. Tech. rep., Cambridge, MA (2004)
17. Rossmo, D.K., Rombouts, S.: Geographic profiling. In: Environmental Criminology and Crime Analysis. Willan Publishing, New York (2008)
18. Smith, M.A., Brown, D.E.: Application of discrete choice analysis to attack point patterns. Information Systems and e-Business Management 5(3) (June 2007)
19. Snook, B., Zito, M., Bennell, C., Taylor, P.J.: On the Complexity and Accuracy of Geographic Profiling Strategies. Journal of Quantitative Criminology 21, 1–26 (2005)
20. U.S. Army: Field Manual 3-24.2: Tactics in Counterinsurgency. Department of the Army, Washington, D.C. (2009)
21. Xue, Y., Brown, D.E.: Spatial analysis with preference specification of latent decision makers for criminal event prediction. Decision Support Systems 41(3) (2006)

Exploiting User Model Diversity in Forecast Aggregation

H. Van Dyke Parunak, Sven A. Brueckner, and Elizabeth Downs

Soar Technology
3600 Green Court, Suite 600
Ann Arbor, MI 48105 USA
van.parunak@soartech.com

Abstract. In many contexts, people generate forecasts about events of interest, and decision-makers wish to aggregate these forecasts to improve their accuracy. These forecasts differ from signals in the physical sciences. In particular, sensor signals are noisy samples from a common underlying distribution, while human-generated forecasts are based on cognitive models that vary from one informant to another. As a result, human forecasts, unlike physical signals, are *not* guaranteed to be statistically independent conditioned on the true outcome. These differences both provide new opportunities for aggregation, and impose restrictions that do not apply to physical signals. This paper describes the difference between forecasts and physical signals, outlines a strategy for exploiting these differences in aggregation, and demonstrates modest but statistically significant gains in the accuracy of aggregated forecasts using data from a large ongoing experiment in forecasting world events.

Keywords: Mental models, generated and interpreted signals, forecast aggregation.

1   Introduction

In many contexts, people forecast events, and decision-makers wish to aggregate these forecasts in the hopes of improving their accuracy. An example question is, "Will Bashar al-Assad be removed from power before 31 December 2013?" Forecaster i's answer is a binomial distribution θ_i over outcomes {Yes, No}. Forecasts to questions with more than two outcomes are multinomial distributions.

The intuitive approach is to average the forecasts. With constant weighting, such an approach is called an Unweighted Linear Opinion Pool (ULinOP). One might vary the weights by a range of factors, including the confidence of individual forecasters [3] or the expected information gain that they have to offer [11].

This intuition, which assumes independence among individual forecasts, is misleading. Information from cognitive processes is likely to be very different from that from physical sensors, both qualitatively and statistically. In particular, it depends on idiosyncratic mental models through which people view the world. Appropriate aggregation across such data should take into account the degree to which informants are working with the same or different mental models.

A.M. Greenberg, W.G. Kennedy, and N.D. Bos (Eds.): SBP 2013, LNCS 7812, pp. 513–522, 2013.

Previous work shows that distinctive characteristics of forecasts derived from cognitive processes (Section 2) urge aggregation methods other than averaging (e.g., voting), and motivate estimations of differences between informants' mental models (Section 3). This paper demonstrates modest but statistically significant gains in forecast aggregation using such estimates in a voting framework (Section 4). Section 5 discusses main lessons.

2   Generated and Interpreted Signals

In this section, we distinguish two main categories of data sources [5] and summarize how differences in their statistical properties affect aggregation [8].

Most analysis methods assume information sources that perform like a physical sensor: given a characteristic of the environment (e.g., temperature t), the sensor "generates" a value $x = t + \varepsilon$, where the error ε is drawn from some distribution. A conventional decision-making process with such "generated" signals compares the signal with a threshold T, and answers "yes" or "no" depending on whether the signal exceeds the threshold. If queried repeatedly, such a source yields answers that are independent of one another, conditioned on the true value of the condition $t > T$.

Some information elicited from people may satisfy this model. For example, in magnitude estimation in psychophysics [4], people confronted with external stimuli (such as sounds of different loudness, or light of different brightness) function as a transducer, converting a simple stimulus into a correlated number.

This model is less satisfying in explaining how people assign probabilities to complex world events. Whether a dictator will leave office depends on a complex web of events, including the country's economy, the level of civil unrest, dissension within the government, and relations with other nations. An intelligence report for decision-makers consists not just of a probability estimate associated with the final event, but also a discussion of the various factors that influence this outcome. Trained analysts describe their work as weighing these factors and the relations among them. In other words, they claim to be *interpreting* the world through an internal cognitive model. The same kind of processing is performed by many AI ("artificial intelligence") systems, which manipulate symbolic information that is mapped onto statements about the world. We call such a signal, an "interpreted signal."

Each source of an interpreted signal may have a different cognitive model, attending to different features of the world. Typically, no single model includes all relevant features, and some features may be invisible to all informants. The responses generated by such a system are not independent, but are guaranteed to be negatively correlated, conditioned on at least one outcome [5].

Interpreted data differ from generated data not only in their correlations, but also in the relation of the optimal aggregation to individual forecasts. Each forecast is a vector of probabilities across the possible outcomes. The space of all such probabilities for a given problem is a "simplex," which is the line segment [0, 1] for a binary problem, an equilateral triangle for a problem with three outcomes, and so forth. For example, Fig. 1 shows three forecasts (*a, b, c*) against a three-outcome question.

The dashed line is the simplex, the space of all triples (p_1, p_2, p_3) such that $p_i \in [0, 1]$ and $\sum p_i = 1$. The corners of the simplex correspond to outcomes. *a* assigns 100% to the first outcome and nothing to the other two. *b* assigns 50% each to the first and third outcomes, and nothing to the second. *c* is the uniform forecast, assigning 33.3% to each outcome.

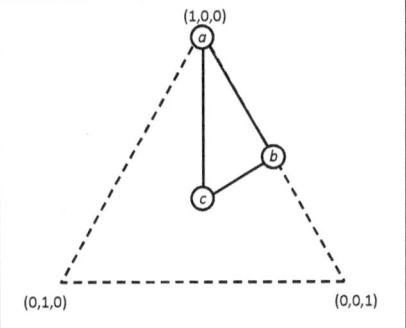

Fig. 1. Three forecasts on the trinomial simplex

Any set of forecasts on the simplex has a convex hull (subject to certain information-geometric refinements [8]). In Fig. 1 the convex hull of (*a*, *b*, *c*) is a solid line.

Under benign constraints, the best aggregation of a set of generated estimates lies within their convex hull, while the best aggregation of a set of interpreted estimates can lie outside the convex hull [8]. Common aggregation methods (such as taking a weighted average with non-negative weights) are constrained to the convex hull, and cannot adequately process interpreted estimates. We explore an alternative approach, voting, which can leave the hull.

These differences between interpreted and generated signals highlight the importance of characterizing forecasters in terms of their individual cognitive models.

A caveat is in order. People can produce generated signals from low-level stimuli, and a response to a forecasting question may include both generated and interpreted components. Such a mixture might result, for example, from emotional stress in the respondent, which in turn might come from a question's subject matter, the complexity of the elicitation environment, or exogenous factors in the respondent's personal life. From our perspective, such a generated component is noise in the signal.

3   Estimating Model Differences

A cognitive model behind an interpreted signal describe the world in terms of *state variables*, mutually exclusive and collectively exhaustive *states*, or non-independent *statements* about the world [7]. Whatever the form, the underlying reasoning deals with a discrete set of features, and in realistic settings, different informants may not attend to the same subset of features.

Estimating a forecaster's actual model would be valuable. For example, a decision-maker who relies on an aggregated forecast is likely to be very interested in understanding the various mental models that support and oppose the conclusion that is presented. However, in this paper we focus simply on the *difference* between pairs of models, to guide aggregation of the resulting forecasts.

We distinguish three broad approaches to estimating model differences [9].

Some approaches to estimating model differences make assumptions about the internal structure of the models. For example, given a graph-structured model with statements as nodes and either conditional probabilities (a Bayesian belief network

[10]) or transition weights (a Narrative Space Model [7]) on the edges, we might ask forecasters what news items they found helpful in responding to a question, and use these to infer what statements a forecaster's model might contain.

At the other extreme, if forecasters are allowed to choose the questions that they address, it is reasonable to assume that forecasters will prefer questions on which their models can shed light. In our experiments, some forecasters are determined to address every question, so their selection does not give a distinct signature. However, for selective forecasters, we can use the overlap in questions between two forecasters to estimate the similarity of their models.

In this paper, we use an intermediate class of measure that requires only the actual forecasts θ_i, θ_j from two forecasters i, j. Each time we aggregate two forecasts from these forecasters, we compute a notional "distance" between their forecasts. One would like such a measure to be invariant with respect to invertible differentiable transformations of the set over which probabilities are defined. This condition allows only a 1-parameter family of geometries [2], characterized by the formula [13]:

$$D_\delta(P, Q) = \frac{1 - \sum_i p_i^\delta q_i^{1-\delta}}{\delta(1-\delta)} \tag{1}$$

where δ is the parameter identifying the geometry. Well-known examples are the Hellinger divergence (for $\delta = 0.5$) and the Kullback-Liebler divergence (or relative entropy, for $\delta = 1$). We use the $\delta = 0.95$ divergence, which we symmetrize (by averaging with the $\delta = 0.5$ divergence) and normalize to yield a true distance in [0, 1].

Let us denote this difference for aggregation episode k as $w_{ij|k}$, or (where there is no risk of confusion) as w_{ij}. w_{ij} is the *within*-event distance between i and j at a single aggregation event. If forecasters i and j participate in multiple such events, we can compute a *cross*-event divergence,

$$c_{ij} = \frac{\sum_k w_{ij|k}}{N} \tag{2}$$

where N is the number of shared aggregation events. c_{ij} estimates the overall tendency of i and j to produce different forecasts on different questions, or on different responses to the same questions (which might reflect differential attention to news events relevant to their respective models). For two forecasters with identical models, $c_{ij} \approx 0$, recognizing the potential noise discussed at the end of Section 2.

Both w_{ij} and c_{ij} are in [0, 1]. How might they compare with each other?

- If both are high, we have different forecasts (high w_{ij}) from forecasters with different models (high c_{ij}), an unremarkable situation.
- We are also not surprised if both are low (similar forecasts from similar models).
- If w_{ij} is high and c_{ij} is low, two forecasters give different results even though their underlying models are similar. This circumstance raises questions about whether their models have purchase on this particular question (or perhaps whether one or the other of them has a high noise component in the forecast in question). In either case, we should discount their divergent contributions to the aggregate.
- If w_{ij} is low and c_{ij} is high, different models lead to a similar answer. Such a circumstance encourages us to pay more attention to their joint opinion.

4   Exploiting Model Diversity in Aggregation

This section outlines how to apply the insights from the previous sections, using diversity measures to modulate voting. We outline our methodology, then define the two dimensions of experimental space: how to select the outcome for which a vote is cast, and how to compute the size of the vote.

4.1   Evaluation Methodology

We analyze responses to 98 forecasting questions from various subsets of 169 forecasters, some responding multiple times to a single question, over time periods ranging from 2 to 245 days. Responses are collected through an on-line interface at https://ace-informed.net. Each question is aggregated once a day while it is active, using the most recent forecast from each forecaster.

We measure forecast accuracy with the Brier score [1], a quadratic error score in [0, 2]. We average the scores from daily aggregation events on a question to give an overall "mean daily error" (MDE) for that question, then average the MDE across all questions to give an overall score, the "mean mean daily error"(MMDE). We measure the performance of an algorithm as the percentage improvement of its overall score (the MMDE) compared to that for ULinOP.

Small improvements may result from a method's exceptional success on a few questions. To estimate the significance of improvements between two methods, we compare the question-by-question MDEs for each method to see whether the method with better MMDE also improves more individual questions. We use the nonparametric Wilcoxon signed rank test [12]. Each question contributes two scores, one from each method being compared. We rank the pairs by the magnitude of their difference. The test statistic W is the difference between the sum of the ranks for which the first method is better and the sum of the ranks for which the second is better. If the two methods are comparable, W should approach 0. The distribution of W approaches normality for $N > 10$, a condition clearly satisfied by our 98 questions, allowing us to estimate the probability of the null hypothesis (that the median of the differences between each pair of results is 0). Small values of $p(H_0)$ increase our confidence that one method is superior to the other.

4.2   Averaging vs. Voting

The ideal aggregate for interpreted signals can leave the convex hull of the individual forecasts [8]. Weighted averages with non-zero weights are confined to the hull, but voting methods can leave the hull. Can voting do better than weighted averages?

Let F_i be the set of forecasters who assign their largest probability to outcome j,

$$F_j = \left\{ i: \overset{argmax}{\underset{k}{}} \left(\theta_i(k) \right) = j \right\}$$ (3)

Assign vote v_i to forecaster i. Then the voting aggregate is

$$\hat{f}_j = \frac{\Sigma_{i \in F_j} v_i}{\Sigma_i v_i} \qquad (4)$$

With unit votes, $\forall i: v_i = 1$, voting outperforms ULinOP by 2.8%, with $p(H_0) = 0.006$. Substantially greater gains (~30%) are achievable by modulating the value of the v_i by various factors, such as a forecaster's historical accuracy, but we have found no benefit to using any diversity measure to change a forecaster-based vote. In the nature of the case, such a vote must be an average across the forecaster's diversity with respect to all other forecasters. We hypothesize that taking such an average discards important information, and we explore here *pairwise* voting methods in which each *pair* of forecasters casts a vote for an outcome derived from their individual forecasts. We explore three ways of selecting the outcome for a given pair of forecasters.

Average Pairwise Voting (APV) votes for the outcome favored by the average of the forecasts. If the forecasts differ, this approach selects the outcome favored by the more extreme forecaster. With unit votes, this approach is 7.4% better than ULinOP with $p(H_0) = 0.002$. To understand this gain, note that the more extreme forecast (the forecast with lower entropy) determines the outcome that receives the vote. Other experiments show that in general, more extreme forecasts (reflecting higher forecaster confidence [3]) are more likely to be correct. To remove this entropy effect, we define two alternative pairwise voting methods.

Split Pairwise Voting (SPV) assigns a vote of 0.5 to the outcome favored by each forecaster. When they agree on the outcome, the entire vote goes to their consensus, but when they do not, each outcome gets half of the credit.

Consensus Pairwise Voting (CPV) gives a vote to a pair only when they agree on the outcome, and registers no vote for pairs who disagree.

4.3   Using Estimates of Model Diversity

We experiment with four vote values. Three use a gain γ reflecting diversity.

The **unit vote** gives each pair has one vote. Combined with SPV, this approach yields the proportion of the forecasters who favor each outcome.

The **ratio γ vote** uses the ratio of c_{ij} and w_{ij}. To avoid singularities, we add a constant ζ (e.g., 0.1) to the numerator and denominator:

$$\gamma_{a|ij} = \frac{c_{ij} + \zeta}{w_{ij} + \zeta} \qquad (5)$$

High values of γ_a indicate that the average forecast of i and j should be given higher credence, while low values indicate that the (divergent) forecasts merit less attention. The unit vote represents the limit of ratio γ when $\zeta \rightarrow \infty$.

The **difference γ vote** transforms (5) so it is bounded to [0, 2] and $c = w \rightarrow \gamma = 1$:

$$\gamma_{ij} = (c_{ij} - w_{ij} + 1) \qquad (6)$$

To derive (6), since c and w are in $[0, 1]$, γ is bounded by

$$\gamma_{max} = \frac{1+\zeta}{\zeta}, \gamma_{min} = \frac{\zeta}{1+\zeta} \tag{7}$$

First, we scale γ to $[0, 1]$:

$$\gamma' = \frac{\gamma - \gamma_{min}}{\gamma_{max} - \gamma_{min}} = \frac{\zeta}{1+2\zeta}\left(\frac{c+c\zeta+\zeta-w\zeta}{w+\zeta}\right) \tag{8}$$

Next, when $c = w$, the parenthesized term in the last element of Equation 4 becomes unity, so we reach $\gamma' = 0.5$ at the midpoint by allowing $\zeta \to \infty$. Taking this limit in (8) and multiplying by 2 so that $\gamma = 1$ at the midpoint, we obtain (6). Scaling before taking the limit retains the influence of c and w, avoiding a unit vote.

The **velocity γ vote** looks at how two forecasters are moving relative to one another on the simplex in their successive forecasts against a single IFP. This vote requires multiple forecasts from at least one forecaster. Consider three conditions:

1. If two forecasters move in synchrony with one another, they probably have very similar models. We will give them a unit vote (comparable to ratio γ when w and c are the same size).
2. If they move apart over time, they are responding differently to events they observe in the world, and we should downweight their vote (a value between 0 and 1).
3. If they move together over time, they started in different positions on the simplex (reflecting different models), but now are coming closer together, suggesting that we should give them a greater vote (say, between 1 and 2).

Let w_{ijlk} be the distance between forecasters i and j at aggregation event k, where k is augmented each time one or the other of them updates a forecast to the IFP in question. When we are considering a single pair, we write simply w_k for their separation at event k. (9) has the required properties:

$$\gamma = 1 + \frac{w_{k-1} - w_k}{w_{k-1} + w_k} \tag{9}$$

When two forecasters start with the same forecast ($w_{k-1} = 0$) and move apart, (9) takes the value 0. When they start with different forecasts and move together ($w_k = 0$), it has the value 2. When they move in synchrony ($w_{k-1} = w_k$), it has the value 1.

The value varies reasonably with distance moved in all cases except when w_{k-1} or w_k is 0. In this case, regardless of the value of the other (non-zero) w, γ will be 0 or 2, depending on the direction of the move. To avoid this, we constrain w to be no less than a minimum value w_0 (0.01 in the experiments reported here). In addition, at the first aggregation event involving a pair of forecasters, we use $\gamma = 1$.

These diversity measures can be applied both to averaging and to voting aggregation. We have explored two averaging approaches. Naively, one might form all possible pairs of forecasts, and weight the average within each pair by γ:

$$\bar{\theta} = \frac{\sum_{ij}\gamma_{ij}(\theta_i + \theta_j)}{2\sum_{ij}\gamma_{ij}} \tag{10}$$

Since γ_{ij} is symmetrical, we can define $\gamma_i \equiv \sum_j \gamma_{ij}$, and (10) becomes

$$\bar{\theta} = \frac{\sum_i \gamma_i \theta_i}{\sum_i \gamma_i} \tag{11}$$

This approach gives no benefit over ULinOP. Assigning a single aggregate γ to each forecaster throws away information about the value of specific pairs of forecasts.

An alternative approach, Cluster-Weighted Aggregation (CWA) [6], agglomeratively clusters individual forecasts using a weighted sum of c_{ij} and w_{ij} as the distance measure. It then aggregates forecasts following the structure of the dendrogram, starting at the leaves, and weighting each cluster by γ before aggregating it into higher-level clusters. This approach offers on the order of 5% improvement over ULinOP with either ratio or velocity γ. Difference γ offers no improvement.

5   Experimental Results

We have ten conditions to compare: ULinOP and 3x3 experimental conditions. Table 1 reports two numbers for each pair of conditions: the lift of the column method over the row method, and the statistical significance ($p(H_0)$ from the signed rank test over individual IFPs). The table is symmetrical, so we report only the upper half.

The various voting schemes with APV dominate everything else, because of the entropy factor, reflecting forecaster confidence. Even with this factor removed, unit voting dominates averaging, decisively for SPV ($p(H_0) = 0.018$), and suggestively for CPV ($p(H_0) = 0.053$). All three forms of γ give positive lift over ULinOP, significantly for difference γ in SPV, for ratio γ in CPV, and for velocity γ in both. Restricting our attention to SPV and CPV, velocity γ dominates the other two forms, and also

Table 1. Lift and Significance of Column over Row (only upper half-table reported). Green bold values have $p(H_0) \leq 0.05$, red values have $p(H_0) > 0.1$.

	APV, Unit	APV, Ratio	APV, Diff.	APV, Velocity	SPV, Unit	SPV, Ratio	SPV, Diff.	SPV, Velocity	CPV, Unit	CPV, Ratio	CPV, Diff.	CPV, Velocity	
ULinOP	.074 / .002	.056 / .002	.072 / .005	.015 / .048	.030 / .018	.013 / .118	.032 / .018	.072 / .003	.015 / .053	.026 / .035	.011 / .061	.034 / .013	ULinOP
APV, Unit		-.018 / .285	-.002 / .463	-.059 / .294	-.044 / .002	-.062 / .001	-.043 / .001	-.002 / .778	-.059 / .353	-.049 / .105	-.063 / .411	-.040 / .003	APV, Unit
APV, Ratio			.015 / .300	-.041 / .257	-.026 / .048	-.044 / .007	-.025 / .086	.015 / .261	-.042 / .269	-.031 / .193	-.046 / .317	-.022 / .064	APV, Ratio
APV, Diff.				-.057 / .269	-.041 / .002	-.059 / .002	-.040 / .002	.000 / .526	-.057 / .319	-.046 / .096	-.061 / .410	-.038 / .004	APV, Diff.
APV, Velocity					.015 / .071	-.003 / .061	.016 / .071	.057 / .378	.000 / .851	.010 / .023	-.005 / .155	.019 / .085	APV, Velocity
SPV, Unit						-.012 / .008	.001 / .025	.041 / .003	-.016 / .089	-.005 / .065	-.020 / .112	.004 / .0001	SPV, Unit
SPV, Ratio							.019 / .010	.059 / .002	.002 / .074	.013 / .049	-.002 / .074	.022 / .003	SPV, Ratio
SPV, Diff.								.041 / .003	-.017 / .092	-.006 / .087	-.021 / .120	.003 / .564	SPV, Diff.
SPV, Velocity									-.057 / .355	-.046 / .117	-.061 / .442	-.038 / .004	SPV, Velocity
CPV, Unit										.011 / .028	-.004 / .036	.019 / .109	CPV, Unit
CPV, Ratio											-.015 / .015	.008 / .089	CPV, Ratio
CPV, Diff.												-.029 / .038	CPV, Diff.
	APV, Unit	APV, Ratio	APV, Diff.	APV, Velocity	SPV, Unit	SPV, Ratio	SPV, Diff.	SPV, Velocity	CPV, Unit	CPV, Ratio	CPV, Diff.	CPV, Velocity	

gives the highest lift against ULinOP (statistically indistinguishable from the impact of entropy in APV with unit voting).

These results invite three observations.

First, the signal in a forecast has an interpreted component that can be exploited in aggregation. Both voting (which can leave the convex hull of individual forecasts) and measures of diversity between forecasters' mental models give statistically significant improvements in forecast aggregation.

Second, our gains over ULinOP are modest. Our limited gains may reflect a generated-signal component in forecasts (e.g., an emotional response to questions). Also, we are hopeful that refinements can increase the contribution of our methods. For example, estimating c_{ij} from multiple instances of w_{ij} is sensitive to small-number effects when forecasters share only a few questions in common. We are exploring techniques for testing whether c_{ij} is converged, and using other estimates of cross-question diversity that are not subject to the convergence challenge.

Third, benefits gained from diversity effects are not always orthogonal to other benefits. There is a suggestion (with very low significance) that APV with any γ vote is worse than with unit voting. All forms of γ dominate ULinOP. Why don't they help with APV? The answer appears to be that the two effects fight against each other. Entropy (certainty) is only exploited when the members of a pair prefer opposite outcomes, while γ is high only when the members of a pair are close together, which usually means they favor the same outcome. Thus the two effects boost different pairs of forecasts, cancelling out the discriminatory information that each of them has to offer. We have observed such interference with other forecast features as well. An important focus of ongoing research is to understand these interaction effects and how to exploit combinations of features for greater performance improvement.

Acknowledgments. This research is supported by the Intelligence Advanced Research Projects Activity (IARPA) via Department of Interior National Business Center contract number D11PC20060. The U.S. Government is authorized to reproduce and distribute reprints for Governmental purposes notwithstanding any copyright annotation thereon. Disclaimer: The views and conclusions contained herein are those of the authors and should not be interpreted as necessarily representing the official policies or endorsements, either expressed or implied, of IARPA, DoI/NBC, or the U.S. Government.

References

1. Brier, G.W.: Verification of Forecasts Expressed in Terms of Probability. Monthly Weather Review 78(2) (1950)
2. Cencov, N.N.: Statistical Decision Rules and Optimal Inference. American Mathematical Society, Rhode Island (1982)
3. Forlines, C., Miller, S., Prakash, S., Irvine, J.: Heuristics for Improving Forecast Aggregation. In: Sun, W. (ed.) AAAI Fall Symposium 2012: Machine Aggregation of Human Judgment, Arlington, VA (2012)
4. Gescheider, G.A.: Psychophysics: The Fundamentals, 3rd edn. Psychology Press (1997)

5. Hong, L., Page, S.E.: Interpreted and Generated Signals. Journal of Economic Theory 144, 2174–2196 (2009)
6. Parunak, H.V.D.: Cluster-Weighted Aggregation. In: Sun, W. (ed.) AAAI Fall Symposium 2012: Machine Aggregation of Human Judgment, Arlington, Virginia (2012)
7. Parunak, H.V.D., Brueckner, S., Downs, L., Sappelsa, L.: Swarming Estimation of Realistic Mental Models. In: Electronic Proceedings, Thirteenth Workshop on Multi-Agent Based Simulation (MABS 2012 at AAMAS) (2012)
8. Parunak, H.V.D., Brueckner, S., Hong, L., Page, S.E., Rohwer, R.: Characterizing and Aggregating Agent Estimates. In: Ito, T., Jonker, C., Gini, M., Shehory, O. (eds.) Twelfth International Conference on Autonomous Agents and Multi-Agent Systems (AAMAS 2013). IFAAMAS, Minneapolis (2013)
9. Parunak, H.V.D., Downs, E.: Estimating Diversity among Forecaster Models. In: Sun, W. (ed.) AAAI Fall Symposium 2012: Machine Aggregation of Human Judgment, Arlington, Virginia (2012)
10. Pearl, J.: Probabilistic Reasoning in Intelligent Systems. Morgan-Kaufmann, San Francisco (1988)
11. Waterhouse, T.: Pay by the Bit: An Information-Theoretic Metric for Collective Human Judgment. In: Sun, W. (ed.) AAAI Fall Symposium 2012: Machine Aggregation of Human Judgment, Arlington, Virginia (2012)
12. Wilcoxon, F.: Individual comparisons by ranking methods. Biometrics Bulletin 1(6), 80–83 (1945)
13. Zhu, H., Rohwer, R.: Measurements of Generalisation based on Information Geometry. In: Ellacott, S.W., Mason, J.C., Anderson, I.J. (eds.) Mathematics of Neural Networks: Models, Algorithms and Applications. Kluwer (1997)

Risk-Based Models of Attacker Behavior in Cybersecurity

Si Li, Ryan Rickert, and Amy Sliva

College of Computer and Information Science, Northeastern University
{xiaolang,rrickert,asliva}@ccs.neu.edu

Abstract. Even as reliance on information and communication technology networks continues to grow, and their potential security vulnerabilities become a greater threat, very little is known about the humans who perpetrate cyber attacks—what are their strategies, resources, and motivations? We present a new framework for modeling such cyber attackers. Utilizing observable information (i.e., network alerts, security implementations, systems logs), we can characterize attackers based on the risk they are willing to incur and delineate them based on skill level. These classifications can facilitate decision-making and resource allocation to counteract cybersecurity incidents. We look at two specific models of attacker risk and discuss empirical results from a prototype implementation of this modeling framework using real-world network data.

1 Introduction

Currently, most research in the field of cyber security has focused on systems security and developing technical approaches to either prevent, respond to, or detect security breaches. While these techniques are crucial for securing our increasingly wired world, they do little to address the human component of cyber attacks. However, when modeling attacker behavior, we are constrained by the limited information—systems logs, firewall alerts, etc.—that we can observe. The identity of an attacker can be easily obscured or anonymized, rendering a complete behavioral profile challenging, if not impossible.

We propose a new metric called *attacker risk* that allows attackers to be classified based on the potential cost they are willing to undertake. In risk management [12,22], *risk* is typically defined as the product of vulnerability, threat, and the value of assets, and has been applied in organizations to quantify vulnerabilities (physical and cyber). We expand these concepts to address the risk of undertaking an attack on a particular system. Different types of attackers may exhibit distinctive behavior related to their willingness and ability to absorb the risk associated with an attack. By estimating an attacker's risk tolerance, we can determine the *cost effectiveness* of an attack, that is, what is the relationship between the potential costs and the goals achieved.

Assuming a basic framework of rational decision-making, this understanding of attacker risk and cost effectiveness provides some indication of other characteristics, such as skill level and resources; a set of operations that is very high-risk

A.M. Greenberg, W.G. Kennedy, and N.D. Bos (Eds.): SBP 2013, LNCS 7812, pp. 523–532, 2013.
© Springer-Verlag Berlin Heidelberg 2013

and low-payoff likely indicates an inexperienced or unskilled attacker, while high-risk, high-reward attacks often require substantial resources to ensure success. A profile based on risk can therefore facilitate classification of intent and skill. Such models can support decision-making and help policy makers or systems administrators allocate defensive resources and mobilize their response teams.

Previous work on cyber attacker behavior has focused mostly on attack features—such as using attack trees or correlating intrusion detection system (IDS) alerts to model patterns or extract anomalies—and making security decisions based on common attacks [16,6,15,26,14,3,24,23]. While some efforts, such as [21] and [25] specifically investigate user intentions, most human factors are not considered. We explicitly address the human component of cybersecurity, modeling the *risk* the attacker incurs. Attacker cost has previously been addressed in game theoretic models, such as [4,10,1,5,11,7,8]. However, these studies make rigid utility assumptions or focus on narrow attack scenarios, while our models are more flexible.

This basic framework is given in Section 2. Sections 3 and 4 discuss risk-based models of behavior and how these models can be used to infer cost-effectiveness and other attacker characteristics. A preliminary implementation is presented in Section 5.

2   Framework for Modeling Cybersecurity Behaviors

Data involving cyber incidents, particularly the behavior and identity of the attacker, is inherently uncertain and incomplete. However, there are several classes of *observable features* that can be captured through logs, intrusion prevention/detection systems, firewall alerts, etc., and analyzed in behavioral models. Three types of observable features are available: *operation, system,* and *consequence attributes*. Let $\mathcal{A}$ be the set of possible operations that can be executed (system calls, attacks, etc.), $\mathcal{S}$ the set of system security attributes (firewall, muti-factor authentication, etc.), and $\mathcal{C}$ the set of possible effects of $\mathcal{A}$ (root, copied files, zombie system, etc.). The operations in $\mathcal{A}$ can either be legitimate or illegitimate, but in either case should be monitored as potential security risks.

Definition 1 (Attack pattern). *Given a set $\mathcal{A}$ of operation attributes, an attack pattern is a sequence A s.t. $\forall a_i \in A, a_i \in \mathcal{A}$.*

An attack pattern $A = \{a_1, a_2, \ldots, a_n\}$ contains actions constituting the entire operation sequence on a system. As we do not know a priori if some actions are malicious, A consists of observable operations, rather than specific well-known attacks. For example, in a denial of service attack using IP spoofing, A might contain "packets received" or "packets with wrong sequence number."

A subset of the available system security attributes is a security policy, i.e., the implemented security features that may be able to either detect or defend against the operations present in A.

Definition 2 (Security policy). *Given a set $\mathcal{S}$ of system security attributes, a security policy is a set $S \subseteq \mathcal{S}$.*

Of course, attempts at hardening the security of a system may not always be successful; some security features may actually introduce new vulnerabilities. To account for this possibility, we assume a *vulnerability factor* v, indicating the potential for vulnerabilities in a security policy. This factor can be estimated in several ways utilizing well-known methods in software engineering for approximating the average number of system bugs or defects in security policies [9,2]. For a security policy S, we say that the *total security* is $T(S) = card(S) - v \cdot card(S)$, i.e., the number of features less their potential for vulnerabilities.

Observed consequences C of an attack pattern are a subset of the possible effects of actions on the system, i.e., $C \subseteq \mathcal{C}$. Similar to A, which only contains low-level observable operations, C contains observable system effects, which may be indicative of malicious behavior. A cyber incident provides a concise representation for a complete attack in a particular security context.

Definition 3 (Cyber incident). *Given an attack pattern A, a security policy S, and consequences C, a cyber incident is a triple $I = (A, S, C)$.*

3 Risk Functions for Cyber Incidents

While we cannot directly determine the motivations of an attacker, it is possible to utilize observable aspects of the cyber domain to analyze their behavior based on the risk or cost they are willing to incur.

Definition 4 (Attacker risk function). *Given attack pattern A and security policy S, an attacker risk function is a mapping $R : \mathcal{A} \times \mathcal{S} \to [0, 1]$ s.t.:*

P1. $R(A, S) \in [0, 1]$. *The value of function R is in the interval $[0, 1]$.*
P2. *For any sequences A_1 and A_2 s.t. $card(A_1) \leq card(A_2)$ and a set S, $R(A_1, S) \leq R(A_2, S)$. For a given S, R increases monotonically with the size of A.*
P3. *For any sets S_1 and S_2 s.t. $T(S_1) \leq T(S_2)$ and a sequence A, $R(A, S_1) \leq R(A, S_2)$. For a given A, R increases monotonically with the total security of S.*

A risk function $R(A, S)$ is a measure of the overall risk (to the attacker) of a cyber incident. **P2** asserts that risk increases with the number of *observable* steps in an attack. Recall that A contains systems operations from logs, firewall alerts, etc.; each element is a possible point of detection for the attacker, increasing the assumed risk. **P3** states that risk is proportional to the amount of security. The above properties provide a flexible framework for analyzing underlying behavior of cyber incidents. Depending on data, expertise, and the domain, a risk function can be defined to suit specific needs. In addition to **P1–P3**, a risk function may satisfy other properties, such as *marginal impact growth* and *risk equality*.

Definition 5 (Marginal impact growth). *Suppose an attack pattern A and security policies $S_1, \cdots, S_n$ s.t. $T(S_n) = T(S_{n-1}) + q, ..., T(S_2) = T(S_1) + q$, where q is a constant. If $R(A, S_2) - R(A, S_1) < ... < R(A, S_n) - R(A, S_{n-1})$, then R shows marginal impact growth w.r.t. the total security of S.*

Marginal impact growth indicates an increasing amount of risk for each constant-sized increase in total security. There is an analogous property w.r.t. the size of A.

The notion of risk in Definition 4 differs from traditional risk management by modeling the cost to the *attacker*. However, in some security situations, the total expected risk is known (i.e., from risk management metrics [12,1,5]), but we do not know how it is distributed. Several risk functions can be derived that model this fixed expected risk. Such models share a property called *risk equality*.

Definition 6 (Expected risk). *For an attacker risk function $R(A, S)$, the expected risk is defined as the expectation*

$$E(R) = \int_{A \text{ s.t. } \forall a_i \in A, a_i \in \mathcal{A}, \ S \subseteq \mathcal{S}} (R(A, S)) d(card(A)) d(T(S))$$

Definition 7 (Risk Equality). *Let R_1 and R_2 be a two attacker risk functions. We say that these functions exhibit* risk equality, *denoted $R_1 \sim R_2$ iff*

$$E(R_1(\_, S)) = E(R_2(\_, S)) \quad \text{for any attack pattern } A \text{ when } T(S) = 0 \quad (1)$$

$$E(R_1(A, \_)) = E(R_2(A, \_)) \quad \text{for any security policy } S \text{ when } card(A) = 0 \quad (2)$$

We can leverage this concept of risk to ultimately determine the utility of a cyber incident for an attacker; the relationship between the payoff and the incurred risk can provide insight into the attacker's strategies and characteristics. Attacker risk, then, can be used to compute the cost-effectiveness of an attack.

Definition 8 (Attacker Cost-effectiveness). *Given a cyber incident $I = (A, S, C)$ and a risk function $R(A, S)$, the attacker's cost-effectiveness γ is*

$$\gamma = \frac{\sum_i C_i * w_i}{R(A, S)}$$

where w_i is the value of consequence C_i.

The value of a consequence is judged by the defender to be the severity of the result, the value of data stolen, cost of damage incurred, etc.

3.1   Basic Attacker Risk Models

As mentioned above, the specific risk function used will often depend on the particular system or organization. In this section we describe two families of risk functions—linear and exponential—that may be useful in a variety of situations. Without loss of generality, we assume in the remainder of this paper that the security policy does not introduce new vulnerabilities, i.e., $T(S) = card(S)$.

Linear Risk Function. The risk of using attack pattern A against security policy S can be a simple linear combination of the attack and system components:

$$R(A, S)_{lin} = \alpha_1 a_1 + \cdots + \alpha_n a_n + \beta_1 s_1 + \cdots + \beta_m s_m \tag{3}$$

where each a_i and s_j are the costs required to execute operation $a_i \in A$ or overcome security feature $s_j \in S$, respectively, and $0 < \sum_{i=1}^{n} \alpha_i, \sum_{j=1}^{m} \beta_j \leq 0.5$.

It may be difficult to assign a particular cost to each element in A and S. One simplification is to assume that each item contributes uniformly to risk, i.e., $a_i = \frac{1}{card(A)}$ and $s_j = \frac{1}{card(S)}$. Since longer, more complicated attacks (or more defenses to overcome) assume more risk, the linear model can be simplified as:

$$R(A, S)_{lin} = (0.5 - \alpha) \frac{card(A)}{card(A)} + \beta \frac{card(S)}{card(S)} \tag{4}$$

where $\alpha = 0.5 - \sum_i \alpha_i$, $\beta = \sum_j \beta_j$ and $0 < \alpha \leq 0.5, 0 < \beta \leq 0.5$.

The values of α and β can be estimated in several ways. One approach is to interpret α and β as the skills and facilities of the attacker and defender. The resources required to compromise a system depend on known vulnerabilities and the skill of the attacker [13]. Since $card(A)$ is the complexity of an attack, combined with the skill α, this total value denotes the attacker's *resources*. As more security is implemented (i.e., as $card(S)$ increases), there will be fewer vulnerabilities to exploit, discovering them will be more difficult, and the risk of detection higher [13]. If $card(S)$ is the number of security mechanisms, and β is the skills of the defender, together these values denote the defender's *resources*.

Exponential Risk Function. The above model assumes that each item in A or S has the same effect. In many real-world cybersecurity scenarios [13], each additional security feature (or action) has a progressively larger impact on risk, i.e., there is marginal impact growth. The risk to an attacker is the function:

$$R(A, S)_{exp} = \frac{d * \frac{1}{\alpha}^{card(A)} \beta^{card(S)}}{\frac{1}{\alpha_{min}}^{card(A)} \beta_{max}^{card(S)}} - e \tag{5}$$

where $1 < \alpha, \beta$, α_{min} is the min value of α, β_{max} is the max value of β, and d, e are constants. The parameters α and β represent skills and facilities s.t. complex security policies or attacks cause an exponential increase in potential risk.

4   Attacker Characterization Based on Risk

A defining characteristic of cyber attackers is their available resources in terms of skills and facilities—knowing if an attack is well orchestrated can help determine how resources should be spent. By estimating α and β, which represent skills or resources, an attacker risk model can classify attackers based on skills and indicate points in the function where security expenditures will be most effective.

Similar to [13], we classify attackers and defenders into three groups according to skill level: novice, intermediate, and expert. A highly skilled attacker requires less time to compromise a system and can better hide his tracks [13], i.e., higher skill corresponds to lower risk $R(A, S)$. Risk will also be higher when more security is in place; however, the impact of additional security will depend on the skill level of the attacker. Let $\alpha_{nov} < \alpha_{int} < \alpha_{exp}$ denote the skill levels for attackers, and define the skill of a defender as $\beta_{nov} < \beta_{int} < \beta_{exp}$. For an attack A and security policy S, the risk $R(A, S)$ will increase as either α or β increases. An ordinal scale such as Table 1 can be used to classify attackers based on skill.

Table 1. Attacker and defender skill levels for ranges of α and β

	α, β
novice	0-0.17
intermediate	0.17-0.34
expert	0.34-0.5

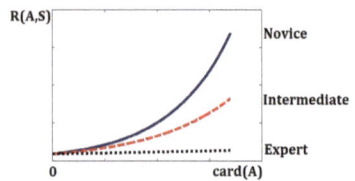

Fig. 1. Attacker risk for varying α skill levels and fixed security resources

From the classic risk management literature [12,22], we can derive our linear model from systems data and the risk metrics many organizations already maintain. However, the exponential model is a more robust model of behavior, cost, and attack complexity. Assuming the linear model represents the total expected risk, according to the risk management metrics, *risk equality* can be exploited to compute α and β for exponential risk. (Details of this procedure are omitted due to space). We can derive an exponential function for each skill level (Figure 1), modeling different *types* of attackers and how risk can be impacted by additional security (i.e., what values of β or $card(S)$ lead to large jumps in attacker risk).

4.1   Inferring Skill from Cost-Effectiveness

In many analyses, we not only want to identify how attackers will respond to security, but need to determine which type of attacker we are facing. The resources dedicated to defensive measures against highly skilled and funded adversaries may be vastly different from those expended on low-skilled, opportunistic attacks. From cost-effectiveness, we can infer the likely skill-level of an attacker.

Since a linear model can be derived from risk management metrics, this function and past data on cyber incidents can produce a range $\Gamma_k = [\gamma_{low}, \gamma_{up}]$ for the cost-effectiveness of each skill level k at a given defensive posture (i.e., value of β). Using Table 1, an attacker can be categorized based on cost-effectiveness:

1. Assume $k = intermediate$ and set $\alpha = min(k)$ and $\alpha = max(k)$
2. Compute cost-effectiveness range $\Gamma'_{int} = [\gamma'_{low}, \gamma'_{up}]$
3. If $\gamma'_{low} < \gamma_{low}$, then $\alpha \in novice$ OR if $\gamma'_{up} > \gamma_{up}$, then $\alpha \in expert$

By assuming an intermediate attacker, we only need to compare the estimated maximum and minimum cost-effectiveness using the α categories from Table 1. If either estimate is outside the range Γ_k, then the attacker is *not* intermediate and must be either an expert or novice. For example, if we compute the actual cost-effectiveness γ'_{up} for $\alpha = 0.34$ from Table 1 s.t. $\gamma'_{up} > \gamma_{up}$, then this higher cost-effectiveness indicates an expert attacker and identifies which risk function to consult when making defensive decisions. We can iteratively subdivide the skill level k into three parts, $[min(k), k_1], [k_1, k_2], [k_2, max(k)]$ and run Steps 1 to 3 again to gradually narrow the estimate of cost effectiveness.

If $\Gamma'_k \subset \Gamma_k$, we cannot assert that the attacker has skill level k, as the ranges Γ_k from data may overlap. Given the distribution of the attacker population over cost-effectiveness, we can compute the probability of an attacker's skill, allowing us allocate resources according to the corresponding function from Figure 1.

Definition 9 (Skill level probability). *Given skill level k with cost-effectiveness $\Gamma_k = [\gamma_{low}, \gamma_{up}]$, and an attacker with estimated cost effectiveness $\gamma'_k \in \Gamma_k$, the probability that the attacker has skill level k is*

$$P(k) = \frac{N_k}{\sum_{i=1}^{N_{k'>max(k)}} \delta(\gamma_{low} \leq \gamma_{k'} \leq \gamma_{up}) + N_k + \sum_{i=1}^{N_{k'<min(k)}} \delta(\gamma_{low} \leq \gamma_{k'} \leq \gamma_{up})}$$

where N_k is the portion of the attacker population with skill level k, $N_{k'>max(k)}$ and $N_{k'<min(k)}$ are the portions with skill levels above and below k, and δ is the distribution over cost-effectiveness.

Intuitively, $\sum_{i=1}^{N_{k'>max(k)}} \delta(\gamma_{low} \leq \gamma_{k'} \leq \gamma_{up})$ is the attacker population whose cost-effectiveness is in the range Γ_k, but whose skill level is above k (e.g., expert).

5   Implementation and Application

A crucial aspect of the proposed framework is its capability of supporting cyber-security decisions utilizing real-world *observable* attributes of the cyber domain. To test the efficacy of the proposed attacker models, we developed a prototype implementation and conducted several empirical analyses utilizing real-world network traffic data (in the form of raw packet captures) from [18] to demonstrate how these models can facilitate policy development or decision-making.

Estimations of attacker risk and cost-effectiveness were implemented using roughly 500 lines of Python code. Snort [20] was used to process the packet capture files and Barnyard2 [17] entered the packet data into a PostgreSQL database. Packets were grouped by time and IP to associate them with an attacker. $card(A)$ was estimated as the number of packets, and $card(S)$ was the number of alert types used in Snort. We trained a linear risk model for various skill levels and derived the exponential model using risk equality.

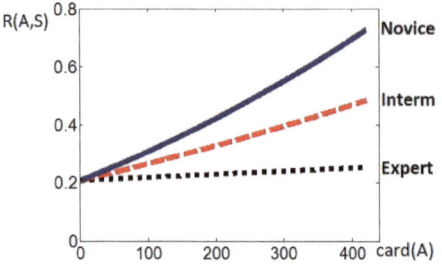
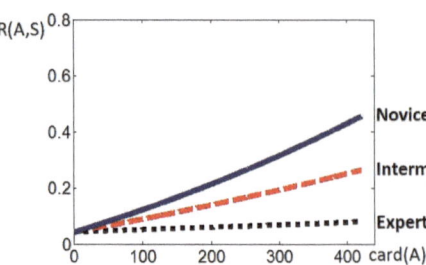

Fig. 2. Exponential risk for all attacker skill levels with an expert defender

Fig. 3. Exponential risk for all attacker skill levels with a novice defender

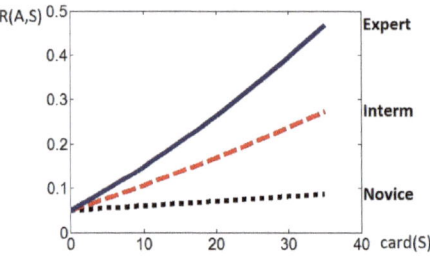
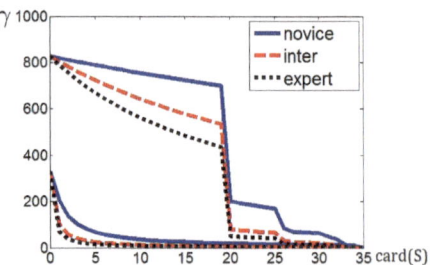

Fig. 4. Exponential risk with varying amounts of security to overcome and defenders of different skill levels

Fig. 5. Bounds of attacker cost-effectiveness as $card(S)$ increases for various defender skill-levels

As described in Section 4, Figures 2 and 3 show the growth in attacker risk as attack complexity ($card(A)$) increases for some fixed amount of security. Using these models of real-world data, we can illustrate how system administrators might utilize the results for decision-making and resource allocation. As noted above, the skill level of the defender has a large impact on the amount of risk; in Figure 2, the overall risk is higher for all possible attackers because the defender is an expert, compared to a novice in Figure 3. Figure 4 shows a similar increase in risk as more security is added. For this data, we begin to see a larger growth in risk at around 20 security features for all attackers, indicating a possible allocation of defensive resources that may be effective on this particular system.

We can also demonstrate cost-effectiveness as a means of inferring attacker skill-level to determine which of the above attacker risk models to consult. Here, we look at possible cost-effectiveness of attacks against defenders of novice ($\beta = 0.085$), intermediate ($\beta = 0.255$), and expert ($\beta = 0.42$) skill levels and use Snort's alert priorities for the value of each consequence. Figure 5 shows the expected reduction in cost-effectiveness as $card(S)$ increases, as well as a downward shift in the cost-effectiveness intervals as the defender's skill increases. We can also observe a narrowing of the upper and lower cost-effectiveness bounds with growing security, with a large effect again at $card(S) = 20$. Once this model is constructed for a system, it can be utilized to support decisions in future attacks. For each security policy, we can use $[\gamma_{low}, \gamma_{up}]$ to define a range of cost-effectiveness for each type of attacker. Then, as new attacks are detected, the the attacker's skill can be estimated with the procedure in Section 4.1. This classification will refer system administrators to the appropriate attacker model, such as those in Figures 2, 3, and 4, to aid in the necessary defensive decisions.

6   Conclusions

We have presented a customizable framework for modeling cyber attacker behavior based on risk, using the cost-effectiveness attackers usually achieve to help delineate other features of their identity. Through empirical analyses of real-world data, we demonstrate how these models can be learned and applied in systems with varying security implementations. Such models can begin to expand our understanding of the human component of cybersecurity and provide system administrators or policy makers with an analytic tool for resource allocation.

There are several possible directions for future work. First, we can expand the cost-effectiveness analysis to find the tightest possible bounds and incorporate more game theoretic aspects, i.e., payoffs and goals, to better classify attacks. The relationship between the linear and exponential models can also be further explored to understand the implications of constant values. Finally, to validate our models, we can collect data from cyber competitions [19] as an approximate "ground truth" for the behavior of attackers (and defenders) of varying skills.

References

1. Al Mannai, W.I., Lewis, T.G.: A general defender-attacker risk model for networks. Journal of Risk Finance (Emerald Group Publishing Limited) 9(3), 244–261 (2008)
2. Alhazmi, O.H., Malaiya, Y.K., Ray, I.: Measuring, analyzing and predicting security vulnerabilities in software systems. Computers and Security 26(3), 219–228 (2007)
3. Bai, H., Wang, K., Hu, C., Zhang, G., Jing, X.: Boosting performance in attack intention recognition by integrating multiple techniques. Frontiers of Computer Science in China, 109–118 (2011)
4. Bao, N., Kreidl, O.P., Musacchio, J.: A network security classification game (2011)
5. Cremonini, M., Nizovtsev, D.: Risks and benefits of signaling information system characteristics to strategic attackers. Journal of Management Information Systems 26(3), 241–274 (2009)
6. Cuppins, F., Miege, A.: Alert correlation in a cooperative intrusion detection framework (2002)
7. Helvik, B.E., Sallhammar, K., Knapskog, S.S.: Incorporating attackers behavior in stochastic models of security. In: International Conference on Security and Management (2005)
8. Wing, J.M., Lye, K.-W.: Game strategies in network security. In: IEEE Computer Security Foundations Workshop (2005)

9. Li, M.N., Malaiya, Y.K., Denton, J.: Estimating the number of defects: a simple and intuitive approach. In: Proc. 7th Intl Symposium on Software Reliability Engineering (ISSRE), pp. 307–315 (1998)

10. Liao, X., Hao, D., Sakurai, K.: Classification on attacks in wireless ad hoc networks: A game theoretic view. In: 2011 7th International Conference on Networked Computing and Advanced Information Management (NCM), pp. 144–149 (2011)

11. Liu, P., Zang, W.: Incentive-based modeling and inference of attacker intent, objectives, and strategies. In: ACM Conference on Computer and Communications Security (CCS) (2003)

12. Major, J.A.: Advanced techniques for modeling terrorism risk. Journal of Risk Finance (Euromoney Institutional Investor PLC) 4(1), 15–24 (2002)

13. McQueen, M.A., Boyer, W.F., Flynn, M.A., Beitel, G.A.: Time-to-compromise model for cyber risk reduction estimation. In: Quality of Protection Workshop. Idaho National Laboratory (INL) (2005)

14. Ning, P., Xu, D.: Learning attack strategies from intrusion alerts. In: Proceedings of the 10th ACM Conference on Computer and Communications Security (2003)

15. Qin, X., Lee, W.: Attack plan recognition and prediction using causal networks. In: Proceedings of the 20th Annual Computer Security Applications Conference (2004)

16. Schneier, B.: Secrets and Lies: Digital Security in a Networked World. John Wiley & Sons (August 2000)

17. Securixlive. Barnyard2©Manual, 2-1.8 edn. (2010)

18. Shiravi, A., Shiravi, H., Tavallaee, M., Ghorbani, A.A.: Toward developing a systematic approach to generate benchmark datasets for intrusion detection. Computers & Security 31(3), 357–374 (2012)

19. Sommestad, T., Hallberg, J.: Cyber security exercises and competitions as a platform for cyber security experiments. In: Jøsang, A., Carlsson, B. (eds.) NordSec 2012. LNCS, vol. 7617, pp. 47–60. Springer, Heidelberg (2012)

20. Sourcefirce, Inc. SNORT®Users Manual, 2.93 edn. (May 2012)

21. Spyrou, T., Darzentas, J.: Intention Modelling: Approximating Computer User Intentions for Detection and Prediction of Intrusions. Chapman & Hall (1996)

22. Woo, G.: Quantifying insurance terrorism risk. In: Lane, M. (ed.) Alternative Risk Strategies, pp. 301–318. Risk Books, London (2002)

23. Wu, Q., Zheng, R., Li, G., Zhang, J.: Intrusion intention identification methods based on dynamic bayesian networks. Procedia Engineering 15 (2011)

24. Yang, S., Stotz, A., Holsopple, J., Sudit, M., Kuhl, M.: High level information fusion for tracking and projection of multistage cyber attacks. Information Fusion 10 (2009)

25. Zhang, Q., Man, D., Yang, W.: Using hmm for intent recognition in cyber security situation awareness. In: Proceedings of the 2009 Second International Symposium on Knowledge Acquisition and Modeling (KAM 2009), vol. 02, pp. 166–169. IEEE Computer Society, Washington, DC (2009)

26. Zhu, B., Ghorbani, A.: Alert correlation for extracting attack strategies. International Journal of Network Security 3(3), 244–258 (2006)

Author Index